Metaheuristics

Patrick Siarry

Editor

Metaheuristics

 Springer

Editor
Patrick Siarry
Laboratory LiSSi (EA 3956)
Université Paris-Est Créteil
Créteil
France

ISBN 978-3-319-83284-5   ISBN 978-3-319-45403-0   (eBook)
DOI 10.1007/978-3-319-45403-0

Printed on acid-free paper

This Springer imprint is published by Springer Nature
The registered company is Springer International Publishing AG
The registered company address is: Gewerbestrasse 11, 6330 Cham, Switzerland

Contents

13 Methodology 357
Eric Taillard

14 Optimization of Logistics Systems Using Metaheuristic-Based Hybridization Techniques.................................... 381
Laurent Deroussi, Nathalie Grangeon and Sylvie Norre

Contributors

Jean-Marc Alliot Institut de Recherche en Informatique de Toulouse, Toulouse, France

Sébastien Aupetit Laboratoire Informatique (EA6300), Université François Rabelais Tours, Tours, France

Sana Ben Hamida Université Paris Ouest, Nanterre, France

Ilhem Boussaïd University of Sciences and Technology Houari Boumediene, Bab-Ezzouar, Algiers, Algeria

Mirsad Buljubašić Centre de Recherche LGI2P, Parc Scientifique Georges Besse, Nîmes Cedex 1, France

Gilles Caporossi GERAD and HEC Montreal, Montreal, Canada

Maurice Clerc Independent Consultant, Groisy, France

Laurent Deroussi Laboratoire LIMOS, IUT d'Allier, Montlucon Cedex, France

Nicolas Durand Laboratoire MAIAA (Ecole Nationale de l'Aviation Civile), Toulouse, France

David Gianazza Laboratoire MAIAA (Ecole Nationale de l'Aviation Civile), Toulouse, France

Jean-Baptiste Gotteland Laboratoire MAIAA (Ecole Nationale de l'Aviation Civile), Toulouse, France

Nathalie Grangeon Laboratoire LIMOS, IUT d'Allier, Montlucon Cedex, France

Pierre Hansen GERAD and HEC Montreal, Montreal, Canada

Nenad Mladenović GERAD and LAMIH, Université de Valenciennes et du Hainaut-Cambrésis, Valenciennes, France

Nicolas Monmarché Laboratoire d'Informatique (EA6300), Université François Rabelais Tours, Tours, France

Sylvie Norre Laboratoire LIMOS, IUT d'Allier, Montlucon Cedex, France

Alain Petrowski Telecom SudParis, Evry, France

Christian Prins ICD-LOSI, UMR CNRS 6281, Université de Technologie de Troyes, Troyes Cedex, France

Caroline Prodhon ICD-LOSI, UMR CNRS 6281, Université de Technologie de Troyes, Troyes Cedex, France

Patrick Siarry Laboratoire Images, Signaux et Systèmes Intelligents (LiSSi, E.A. 3956), Université Paris-Est Créteil Val-de-Marne, Vitry-sur-Seine, France

Mohamed Slimane Laboratoire Informatique (EA6300), Université François Rabelais Tours, Tours, France

Eric Taillard HEIG-VD, Yverdon-les-bains, Switzerland

Charlie Vanaret Laboratoire MAIAA (Ecole Nationale de l'Aviation Civile), Toulouse, France

Michel Vasquez Centre de Recherche LGI2P, Parc Scientifique Georges Besse, Nîmes Cedex 1, France

Chapter 1
Introduction

Patrick Siarry

Every day, engineers and decision-makers are confronted with problems of growing complexity in diverse technical sectors, for example in operations research, the design of mechanical systems, image processing, and, particularly, electronics (CAD of electrical circuits, the placement and routing of components, improvement of the performance or manufacturing yield of circuits, characterization of equivalent schemas, training of fuzzy rule bases or neural networks, …). The problem to be solved can often be expressed as an *optimization problem*. Here one defines an objective function (or several such functions), or cost function, which one seeks to minimize or maximize vis-à-vis all the parameters concerned. The definition of the optimization problem is often supplemented by information in the form of *constraints*. All the parameters of the solutions adopted must satisfy these constraints, otherwise these solutions are not realizable. In this book, our interest is focused on a group of methods, called *metaheuristics* or meta-heuristics, which include in particular the simulated annealing method, evolutionary algorithms, the tabu search method, and ant colony algorithms. These have been available from the 1980s and have a common aim: to solve the problems known as *hard optimization* as well as possible.

We will see that metaheuristics are largely based on a common set of principles which make it possible to design solution algorithms; the various regroupings of these principles thus lead to a large variety of metaheuristics.

P. Siarry (✉)
Laboratoire Images, Signaux et Systèmes Intelligents (LiSSi, E.A. 3956),
Université Paris-Est Créteil Val-de-Marne,
122 rue Paul Armangot, 94400 Vitry-sur-Seine, France
e-mail: siarry@u-pec.fr

© Springer International Publishing Switzerland 2016
P. Siarry (ed.), *Metaheuristics*, DOI 10.1007/978-3-319-45403-0_1

1

1.1 "Hard" Optimization

Two types of optimization problems can be distinguished: "discrete" problems and problems with continuous variables. To be more precise, let us quote two examples. Among the discrete problems, one can discuss the well-known traveling salesman problem: this is a question of minimizing the length of the route of a "traveling salesman," who must visit a certain number of cities before return to the town of departure. A traditional example of a continuous problem is that of a search for the values to be assigned to the parameters of a numerical model of a process, so that the model reproduces the real behavior observed as accurately as possible. In practice, one may also encounter "mixed problems," which comprise simultaneously discrete variables and continuous variables.

This differentiation is necessary to determine the domain of hard optimization. In fact, two kinds of problems are referred to in the literature as hard optimization problems (this name is not strictly defined and is bound up with the state of the art in optimization):

- Certain discrete optimization problems, for which there is no knowledge of an exact *polynomial* algorithm (i.e., one whose computing time is proportional to N^n, where N is the number of unknown parameters of the problem and n is an integer constant). This is the case, in particular, for the problems known as "NP-hard," for which it has been conjectured that there is no constant n for which the solution time is limited by a polynomial of degree n.
- Certain optimization problems with continuous variables, for which there is no knowledge of an algorithm that enables one to definitely locate a global optimum (i.e., the best possible solution) in a finite number of computations.

Many efforts have been made for a long time, separately, to solve these two types of problems. In the field of continuous optimization, there is thus a significant arsenal of traditional methods for *global optimization* [1], but these techniques are often ineffective if the objective function does not possess a particular structural property, such as convexity. In the field of discrete optimization, a great number of *heuristics*, which produce solutions close to the optimum, have been developed; but the majority of them were conceived specifically for a given problem.

The arrival of *metaheuristics* marks a reconciliation of the two domains: indeed, they can be applied to all kinds of discrete problems and they can also be adapted to continuous problems. Moreover, these methods have in common the following characteristics:

- They are, at least to some extent, *stochastic*: this approach makes it possible to counter the *combinatorial explosion* of the possibilities.
- They are generally of discrete origin, and have the advantage, decisive in the continuous case, of being direct, i.e., they do not resort to often problematic calculations of the gradients of the objective function.

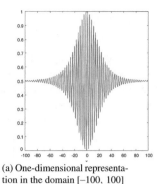
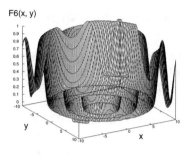

(a) One-dimensional representation in the domain [−100, 100]

(b) Two-dimensional representation in the domain [−10, 10]

Fig. 1.1 Shape of the test function F6

- They are inspired by *analogies*: with physics (simulated annealing, simulated diffusion, …), with biology (evolutionary algorithms, tabu search, …), or with ethology (ant colonies, particle swarms, …).
- They also share the same disadvantages: difficulties of *adjustment* of the parameters of the method, and a *large computation time*.

These methods are not mutually exclusive: indeed, with the current state of research, it is generally impossible to envisage with certainty the effectiveness of a given method when it is applied to a given problem. Moreover, the current tendency is the emergence of *hybrid methods*, which endeavor to benefit from the specific advantages of different approaches by combining them. Finally, another aspect of the richness of metaheuristics is that they lend themselves to all kinds of *extensions*. We can quote, in particular:

- *multiobjective* optimization [6], which is a question of optimizing several contradictory objectives simultaneously;
- *multimodal* optimization, where one endeavors to locate a whole set of global or local optima;
- *dynamic* optimization, which deals with temporal variations of the objective function;
- the use of *parallel implementations*.

These extensions require, for the development of solution methods, various specific properties which are not present in all metaheuristics. We will reconsider this subject, which offers a means for guiding the user in the choice of a metaheuristic, later. The adjustment and comparison of metaheuristics are often carried out empirically, by exploiting analytical sets of test functions whose global and local minima are known. We present in Fig. 1.1 the shape of one of these test functions as an example.

1.2   Source of the Effectiveness of Metaheuristics

To facilitate the discussion, let us consider a simple example of an optimization problem: that of the placement of the components of an electronic circuit. The objective function to be minimized is the length of the connections, and the unknown factors—called "decision variables"—are the sites of the circuit components. The shape of the objective function of this problem can be represented schematically as in Fig. 1.2, according to the "configuration": each configuration is a particular placement, associated with a choice of a value for each decision variable. Throughout the entire book—except where otherwise explicitly mentioned—we will seek in a similar way to minimize an objective. When the space of the possible configurations has such a tortuous structure, it is difficult to locate the global minimum c^*. We explain below the failure of a "classical" iterative algorithm, before commenting on the advantage that we can gain by employing a metaheuristic.

1.2.1   Trapping of a "Classical" Iterative Algorithm in a Local Minimum

The principle of a traditional "iterative improvement" algorithm is the following: one starts from an initial configuration c_0, which can be selected at random, or—for example in the case of the placement of an electronic circuit—can be determined by a designer. An elementary modification is then tested; this is often called a "movement" (for example, two components chosen at random are swapped, or one of them is relocated). The values of the objective function are then compared, before and after this modification. If the change leads to a reduction in the objective function, it is accepted, and the configuration c_1 obtained, which is a "neighbor" of the preceding one, is used as the starting point for a new test. In the opposite case, one returns to the preceding configuration before making another attempt. The process is carried out iteratively until any modification makes the result worse. Figure 1.2 shows that this algorithm of iterative improvement (also known as the *classical method* or *descent method*) does not lead, in general, to the global optimum, but only to one

Fig. 1.2 Shape of the objective function of a hard optimization problem depending on to the "configuration"

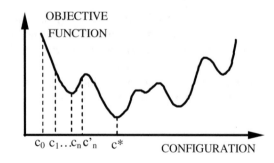

local minimum c_n, which constitutes the best accessible solution taking the initial assumption into account.

To improve the effectiveness of the method, one can of course apply it several times, with arbitrarily selected different initial conditions, and retain as the final solution the best local minimum obtained. However, this procedure appreciably increases the computing time of the algorithm, and may not find the optimal configuration c^*. The repeated application of the descent method does not guarantee its termination and it is particularly ineffective when the number of local minima grows exponentially with the size of the problem.

1.2.2 Capability of Metaheuristics to Extract Themselves from a Local Minimum

Another idea for overcoming the obstacle of local minima has been demonstrated to be very profitable, so much so that it is the basic core of all metaheuristics based on a *neighborhood* (the simulated annealing and tabu methods). This is a question of authorizing, from time to time, movements of *increase*, in other words, accepting a temporary degradation of the situation, during a change in the current configuration. This happens, for example, if one passes from c_n to c'_n in Fig. 1.2. A mechanism for controlling these degradations—specific to each metaheuristic—makes it possible to avoid divergence of the process. It consequently becomes possible to extract the process from a trap which represents a local minimum, to allow it to explore another more promising "valley." The "distributed" metaheuristics (such as evolutionary algorithms) also have mechanisms allowing the departure of a particular solution out of a local "well" of the objective function. These mechanisms (such as *mutation* in evolutionary algorithms) affect the solution in hand; in this case, they help the collective mechanism for fighting against local minima, represented by the parallel control of a "population" of solutions.

1.3 Principles of the Most Widely Used Metaheuristics

1.3.1 Simulated Annealing

Kirkpatrick and his colleagues were specialists in statistical physics, who were interested specifically in the low-energy configurations of disordered magnetic materials, referred to by the term *spin glasses*. The numerical determination of these configurations posed frightening problems of optimization, because the "energy landscape" of a spin glass contains several "valleys" of unequal depth; it is similar to the "landscape" in Fig. 1.2. Kirkpatrick et al. [14] (and, independently, Cerny [2]) proposed to deal with these problems by taking as a starting point the experimental technique

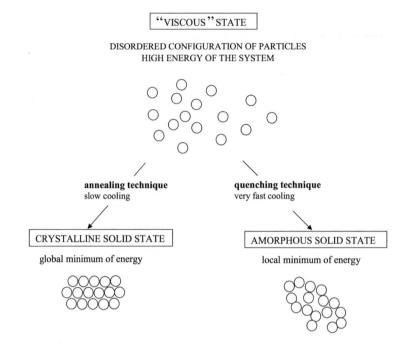

Fig. 1.3 Comparison of the techniques of annealing and quenching

of *annealing* used by metallurgists to obtain a "well-ordered" solid state, of minimum energy (avoiding the "metastable" structures characteristic of local minima of
the energy). This technique consists in heating a material to a high temperature and
then lowering this temperature slowly. To illustrate the phenomenon, we represent
in Fig. 1.3 the effect of the annealing technique and that of the opposite technique of
quenching on a system consisting of a set of particles.

The simulated annealing method transposes the process of annealing to the solution of an optimization problem: the objective function of the problem, similarly to
the energy of a material, is then minimized, with the help of the introduction of a
fictitious *temperature*, which in this case is a simple controllable parameter of the
algorithm.

In practice, the technique exploits the Metropolis algorithm, which enables us
to describe the behavior of a thermodynamic system in "equilibrium" at a certain
temperature. On the basis of a given configuration (for example, an initial placement
of all the components), the system is subjected to an elementary modification (for
example, one may relocate a component or swap two components). If this transformation causes the objective function (or *energy*) of the system to decrease, it is
accepted. On the other hand, if it causes an increase ΔE in the objective function,
it can also be accepted, but with a probability $e^{\frac{-\Delta E}{T}}$. This process is then repeated
in an iterative manner, keeping the temperature constant, until thermodynamic equilibrium is reached, at the end of a "sufficient" number of modifications. Then the

Fig. 1.4 Flowchart of the simulated annealing algorithm

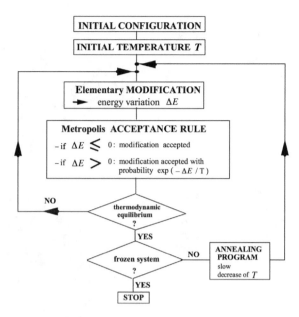

temperature is lowered, before implementing a new series of transformations: the rule by which the temperature is decreased in stages is often empirical, just like the criterion for program termination.

A flowchart of the simulated annealing algorithm is schematically presented in Fig. 1.4. When it is applied to the problem of the placement of components, simulated annealing generates a disorder–order transformation, which is represented in pictorial manner in Fig. 1.5. One can also visualize some stages of this ordering by applying the method of placement of components to the nodes of a grid (see Fig. 1.6).

The disadvantages of simulated annealing lie in the "adjustments," such as the management of the decrease in the temperature; the user must have the know-how about "good" adjustments. In addition, the computing time can become very significant, which has led to parallel implementations of the method. On the other hand, the simulated annealing method has the advantage of being flexible with respect to the evolution of the problem and easy to implement. It has given excellent results for a number of problems, generally of big size.

1.3.2 The Tabu Search Method

The method of searching with tabus, or simply the *tabu search* or *tabu method*, was formalized in 1986 by Glover [10]. Its principal characteristic is based on the use of mechanisms inspired by human memory. The tabu method, from this point of view, takes a path opposite to that of simulated annealing, which does not utilize memory

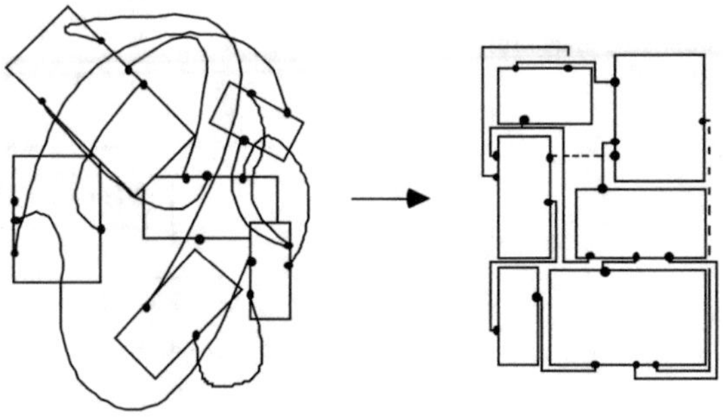

Fig. 1.5 Disorder–order
transformation created by
simulated annealing applied
to the placement of
electronic components

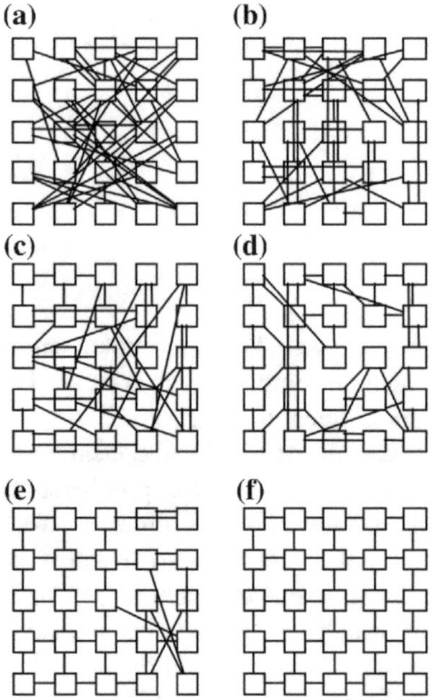

Fig. 1.6 Evolution of a
system at various
temperatures, on the basis of
an arbitrary configuration: L
indicates the overall length
of the connections

(a)   (b)

(c)   (d)

(e)   (f)

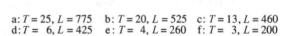

a: $T = 25$, $L = 775$ b: $T = 20$, $L = 525$ c: $T = 13$, $L = 460$
d: $T = 6$, $L = 425$ e: $T = 4$, $L = 260$ f: $T = 3$, $L = 200$

at all, and thus is incapable of learning lessons from the past. On the other hand, the modeling of memory introduces multiple degrees of freedom, which hinders—even in the opinion of the original author [11]—any rigorous mathematical analysis of the tabu method. The guiding principle of the tabu method is simple: like simulated annealing, the tabu method functions at any given time with only one "current configuration" (at the beginning, an unspecified solution), which is updated during successive "iterations." In each iteration, the mechanism of passage of a configuration, called s, to the next one, called t, comprises two stages:

- One builds the set of *neighbors* of s, i.e., the set of the configurations that are accessible in only one elementary movement of s (if this set is too large, one applies a technique for reduction of its size: for example, one may utilize a list of candidates, or extract at random a subset of neighbors of fixed size). Let $V(s)$ be the set (or a subset) of these neighbors.
- One evaluates the objective function f of the problem for each configuration belonging to $V(s)$. The configuration t which succeeds s in the series of solutions built by the tabu method is the configuration of $V(s)$ in which f takes the minimum value. Note that this configuration t is adopted even if it is worse than s, i.e., if $f(t) > f(s)$: this characteristic helps the tabu method to avoid the trapping of f in local minima.

The procedure cannot be used precisely as described above, because there is a significant risk of returning to a configuration already obtained in a preceding iteration, which generates a cycle. To avoid this phenomenon, the procedure requires the updating and exploitation, in each iteration, of a list of prohibited movements, the "tabu list." This list—which gave its name to the method—contains m movements $(t \rightarrow s)$, which are the opposite of the last m movements $(s \rightarrow t)$ carried out. A flowchart of this algorithm, known as the "simple tabu," is represented Fig. 1.7.

The algorithm thus models a rudimentary form of memory, a *short-term memory* of the solutions visited recently. Two additional mechanisms, named *intensification* and *diversification*, are often implemented to equip the algorithm with a *long-term memory* also. This process does not exploit the temporal proximity of particular events more, but rather the frequency of their occurrence over a longer period. Intensification consists in looking further into the exploration of certain areas of the solution space, identified as particularly promising ones. Diversification is, in contrast, the periodic reorientation of the search for an optimum towards areas seldom visited until now.

For certain optimization problems, the tabu method has given excellent results; moreover, in its basic form, the method has fewer adjustable parameters than simulated annealing, which makes it easier to use. However, the various additional mechanisms, such as intensification and diversification, bring a notable amount of complexity with them.

1.3.3   Genetic Algorithms and Evolutionary Algorithms

Evolutionary algorithms (EAs) are search techniques inspired by the biological evolution of species and appeared at the end of the 1950s [9]. Among several approaches [8, 13, 16], genetic algorithms (GAs) constitute certainly the most well-known example, following the publication in 1989 of the well-known book by Goldberg [12]. Evolutionary methods initially aroused limited interest, because of their significant cost of execution. But, in the last ten years, they have experienced considerable development, which can be attributed to the significant increase in the computing power

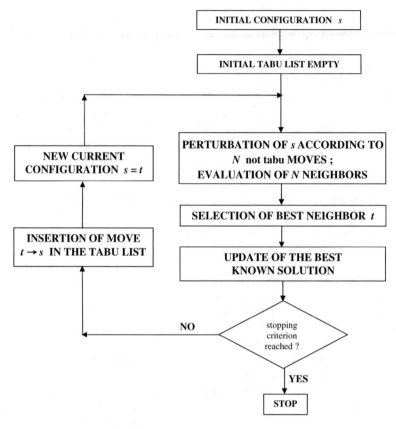

Fig. 1.7 Flowchart of the simple tabu algorithm

of computers, in particular, following the appearance of massively parallel architectures, which exploit "intrinsic parallelism" (see for example [5] for an application to the placement of components). The principle of a simple evolutionary algorithm can be described as follows: a set of N points in a search space, selected a priori at random, constitutes the *initial population*; each individual x of the population has a certain performance, which measures its degree of *adaptation* to the objective aimed at. In the case of the minimization of an objective function f, x becomes more powerful as $f(x)$ becomes smaller. An EA consists in evolving gradually, in successive *generations*, the composition of the population, with its size being kept constant. During generations, the objective is to improve overall the performance of the individuals. One tries to obtain such a result by mimicking the two principal mechanisms which govern the evolution of living beings according to Darwin's theory:

- *selection*, which favors the reproduction and survival of the fittest individuals;
- *reproduction*, which allows mixing, recombination and variation of the hereditary features of the parents, to form descendants with new potentialities.

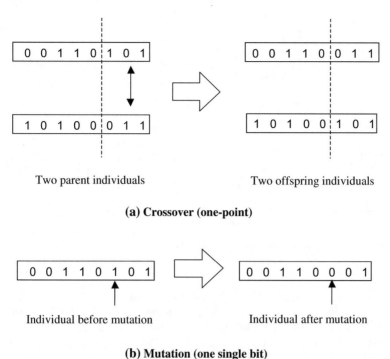

Fig. 1.8 Examples of crossover and mutation operators, in the case of individuals represented by binary strings of eight numbers

In practice, a representation must be selected for the individuals of a population. Classically, an individual can be a list of integers for a combinatorial problem, a vector of real numbers for a numerical problem in a continuous space, or a string of binary numbers for a Boolean problem; one can even combine these representations into complex structures if the need is so felt. The passage from one generation to the next proceeds in four phases: a phase of selection, a phase of reproduction (or variation), a phase of performance evaluation, and a phase of replacement. The selection phase designates the individuals that take part in reproduction. They are chosen, possibly on several occasions, a priori more often the powerful, they are. The selected individuals are then available for the reproduction phase. This phase consists in applying variation operators to copies of the individuals selected to generate new individuals; the operators most often used are *crossover* (or *recombination*), which produces one or two descendants starting from two parents, and *mutation*, which produces a new individual starting from only one individual (see Fig. 1.8 for an example). The structure of the variation operators depends largely on the representation selected for the individuals. The performances of the new individuals are then evaluated during the evaluation phase, starting from the objectives specified. Lastly, the replacement phase consists in choosing the members of the new generation: one can, for example, replace the least powerful individuals of the population

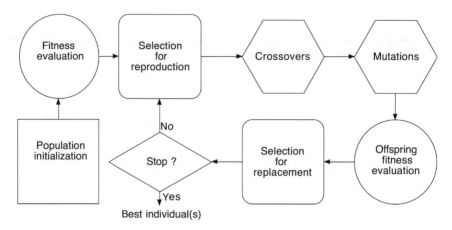

Fig. 1.9 Principle of an evolutionary algorithm

by the best individuals produced, in equal numbers. The algorithm is terminated after a certain number of generations, according to a termination criterion to be specified. Figure 1.9 represents the principle of an evolutionary algorithm.

Because they handle a population of solution instances, evolutionary algorithms are particularly suitable for finding a set of different solutions when an objective function has several global optima. In this case they can provide a sample of compromise solutions when one is solving problems with several objectives, possibly contradictory. These possibilities are discussed more specifically in Chap. 11.

1.3.4 Ant Colony Algorithms

This approach, proposed by Colorni et al. [7], endeavors to simulate the collective capability to solve certain problems observed in colonies of ants, whose members are individually equipped with very limited faculties. Ants came into existence on earth over 100 million years ago and they are one of the most successful species: 10 million billion individuals, living everywhere on the planet. Their total weight is of the same order of magnitude as that of humans! Their success raises many questions. In particular, entomologists have analyzed the collaboration which occurs between ants seeking food outside an anthill. It is remarkable that the ants always follow the same path, and this path is the shortest possible one. This is the result of a mode of indirect communication via the environment called "stigmergy." Each ant deposits along its path a chemical substance, called a pheromone. All members of the colony perceive this substance and direct their walk preferentially towards the more "odorous" areas.

This results particularly in a collective ability to find the shortest path quickly after the original path has been blocked by an obstacle (Fig. 1.10). Although this

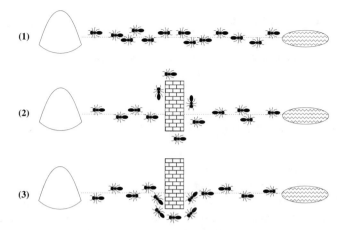

1. Real ants follow a path between the nest and a source of food.
2. An obstacle appears on the path, and the ants choose to turn to the left or right with equal probabilities; the pheromone is deposited more quickly on the shortest path.
3. All the ants choose the shortest path.

Fig. 1.10 Ability of an ant colony to find the shortest path after the path has been blocked by an obstacle

behavior has been taken as a starting point for modeling the algorithm, Colorni et al. [7] proposed a new algorithm for the solution of the traveling salesman problem. Since this research work, the method has been extended to many other optimization problems, some combinatorial and some continuous.

Ant colony algorithms have several interesting characteristics; we shall mention in particular high *intrinsic parallelism*, *flexibility* (a colony of ants is able to adapt to modifications of the environment), *robustness* (a colony is able to maintain its activity even if some individuals fail), *decentralization* (a colony does not obey a centralized authority), and *self-organization* (a colony finds a solution, which is not known in advance, by itself). This method seems particularly useful for problems which are *distributed* in nature, problems of dynamic *evolution*, and problems which require *strong fault tolerance*. At this stage of development of these recently introduced algorithms, however, their application to particular optimization problems is not trivial: it must be the subject of a specific treatment, which can be difficult.

1.3.5   *Other Metaheuristics*

Whether other metaheuristics are variants of the most famous methods or not, they are legion. The interested reader can refer to Chaps. 9 and 10 of this book and three other recent books [15, 17, 19] each one of which is devoted to several metaheuristics.

1.4   Extensions of Metaheuristics

We review here some of the extensions which have been proposed to deal with some special features of optimization problems.

1.4.1   Adaptation for Problems with Continuous Variables

Problems with continuous variables, by far the most which are common ones in engineering, have attracted less interest from specialists in informatics. The majority of metaheuristics, which are of combinatorial origin, can however be adapted to the continuous case; this requires a discretization strategy for the variables. The discretization step must be adapted in the course of optimization to guarantee at the same time the regularity of the progression towards the optimum and the precision of the result. Our proposals relating to simulated annealing, the tabu method, and GAs are described in [3, 4, 21].

1.4.2   Multiobjective Optimization

More and more problems require the simultaneous consideration of several contradictory objectives. There does not exist, in this case, a single optimum; instead, one seeks a range of solutions that are "Pareto optimal," which form the "compromise surface" for the problem considered. These solutions can be subjected to final arbitration by the user. The principal methods of multiobjective optimization (either using a metaheuristic or not) and some applications, in particular in telecommunications, were presented in the book [6].

1.4.3   Hybrid Methods

The rapid success of metaheuristics is due to the difficulties encountered by traditional optimization methods in complex engineering problems. After the initial success of using various metaheuristics, the time came to make a realistic assessment and to accept the complementary nature of these new methods, both with other methods of this type and with other approaches: from this, we saw the current emergence of *hybrid methods* (see for example [18]).

1.4.4 Multimodal Optimization

The purpose of multimodal optimization is to determine a whole set of optimal solutions, instead of a single optimum. Evolutionary algorithms are particularly well adapted to this task, owing to their distributed nature. The variants of the "multi-population" type exploit several populations in parallel, which endeavor to locate different optima.

1.4.5 Parallelization

Multiple modes of parallelization have been proposed for the various metaheuristics. Certain techniques were desired to be general; others, on the other hand, benefit from specific characteristics of the problem. Thus, in problems of placement of components, the tasks can be naturally distributed between several processors: each one of them is responsible for optimizing a given geographical area and information is exchanged periodically between nearby processors (see, for example, [20, 22]).

1.5 Place of Metaheuristics in a Classification of Optimization Methods

In order to recapitulate the preceding considerations, we present in Fig. 1.11 a general classification of mono-objective optimization methods, already published in [6]. In this figure, one can see the principal distinctions made above:

- Initially, combinatorial and continuous optimizations are differentiated.
- For combinatorial optimization, one can approach the problem by several different methods when one is confronted with a hard problem; in this case, a choice is sometimes possible between "specialized" heuristics, entirely dedicated to the problem considered, and a metaheuristic.
- For continuous optimization, we immediately separate the linear case (which is concerned in particular with *linear programming*) from the nonlinear case, where the framework for hard optimization can be found. In this case, a pragmatic solution can be to resort to the repeated application of a local method which may or may not exploit the gradients of the objective function. If the number of local minima is very high, recourse to a global method is essential: those metaheuristics are then found which offer an alternative to the traditional methods of global optimization, those requiring restrictive mathematical properties of the objective function.
- Among the metaheuristics, one can differentiate the metaheuristics of "neighborhood," which make progress by considering only one solution at a time (simulated annealing, tabu search, ...), from the "distributed" metaheuristics, which handle in parallel a complete population of solutions (genetic algorithms, ...).

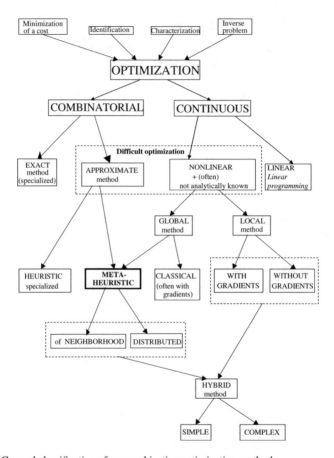

Fig. 1.11 General classification of mono-objective optimization methods

- Finally, hybrid methods often associate a metaheuristic with a local method. This cooperation can take the simple form of relaying between the metaheuristic and the local technique, with the objective of refining the solution. But the two approaches can also be intermingled in a more complex way.

1.6 Applications of Metaheuristics

Metaheuristics are now regularly employed in all sectors of engineering, to such an extent that it is not possible to draw up a complete inventory of the applications here. Several examples will be described in the chapters devoted to various metaheuristics. The last part of this book is devoted to a detailed presentation of three case studies, in the fields of logistics systems, air traffic, and vehicle routing.

1.7   An Open Question: The Choice of a Metaheuristic

This presentation must not ignore the principal difficulty with which an engineer is confronted in the presence of a concrete optimization problem: that of the choice of an "efficient" method, able to produce an "optimal" solution—or one of acceptable quality—in "reasonable" computing time. In relation to this pragmatic concern of the user, the theory is not yet of great help, because convergence theorems are often nonexistent or applicable only under very restrictive assumptions. Moreover, the "optimal" adjustment of the various parameters of a metaheuristic that might be recommended theoretically is often inapplicable in practice, because it induces a prohibitive computing cost. Consequently, the choice of a "good" method, and the adjustment of the parameters of that method, generally calls upon the know-how and "experience" of the user, rather than the faithful application of well-laid-down rules. The research efforts in progress, for example the analysis of the "energy landscape" or the development of a taxonomy of hybrid methods, are aimed at rectifying this situation, which is perilous in the long term for the credibility of metaheuristics. Nevertheless, we will try to outline, in Chap. 13 of this book, a technique that may be of assistance in the selection of a metaheuristic.

1.8   Outline of the Book

This book comprises three parts.

The first part is devoted to a detailed presentation of the more widely known metaheuristics:

- the simulated annealing method (Chap. 2);
- tabu search (Chap. 3);
- variable neighborhood search (Chap. 4);
- the GRASP method (Chap. 5);
- evolutionary algorithms (Chap. 6);
- ant colony algorithms (Chap. 7);
- particle swarm optimization (Chap. 8).

Each one of these metaheuristics is actually a family of methods, the essential elements of which we try to discuss.

In the second part (Chaps. 9–13) we present some other metaheuristics, which are less widespread or still emergent. Then we describe some extensions of metaheuristics (constrained optimization, multiobjective optimization, …) and some ways of searching.

Lastly, we consider the problem of the choice of a metaheuristic, and we describe two unifying methods which may help to reduce the difficulty of this choice.

The last part concentrates on three case studies:

- hybrid metaheuristics designed for optimization of logistics systems (Chap. 14);
- metaheuristics aimed at solving vehicle routing problems (Chap. 15);
- applications in air traffic management (Chap. 16).

References

1. Berthiau, G., Siarry, P.: État de l'art des méthodes d'optimisation globale. RAIRO Operations Research **35**(3), 329–365 (2001)
2. Cerny, V.: Thermodynamical approach to the traveling salesman problem: An efficient simulation algorithm. Journal of Optimization Theory and Applications **45**(1), 41–51 (1985)
3. Chelouah, R., Siarry, P.: A continuous genetic algorithm designed for the global optimization of multimodal functions. Journal of Heuristics **6**, 191–213 (2000)
4. Chelouah, R., Siarry, P.: Tabu Search applied to global optimization. European Journal of Operational Research **123**, 256–270 (2000)
5. Cohoon, J., Hegde, S., Martin, W., Richards, D.: Distributed genetic algorithms for the floorplan design problem. IEEE Transactions on Computer-Aided Design **10**(4), 483–492 (1991)
6. Collette, Y., Siarry, P.: *Multiobjective Optimization*. Springer (2003)
7. Colorni, A., Dorigo, M., Maniezzo, V.: Distributed optimization by ant colonies. In: *Proceedings of the European Conference on Artificial Life, ECAL'91*, pp. 134–142. Elsevier (1991)
8. Fogel, L.J., Owens, A.J., Walsh, M.J.: *Artifical Intelligence Through Simulated Evolution*. Wiley (1966)
9. Fraser, A.S.: Simulation of genetic systems by automatic digital computers. Australian Journal of Biological Sciences **10**, 484–491 (1957)
10. Glover, F.: Future paths for integer programming and links to artificial intelligence. Computers and Operations Research **13**(5), 533–549 (1986)
11. Glover, F., Laguna, M.: *Tabu Search*. Kluwer Academic (1997)
12. Goldberg, D.E.: *Genetic Algorithms in Search, Optimization and Machine Learning*. Addison-Wesley (1989)
13. Holland, J.H.: Outline for logical theory of adaptive systems. Journal of the Association for Computing Machinery **3**, 297–314 (1962)
14. Kirkpatrick, S., Gelatt, C., Vecchi, M.: Optimization by simulated annealing. Science **220**(4598), 671–680 (1983)
15. Pham, D., Karaboga, D.: *Intelligent Optimisation Techniques*. Genetic Algorithms, Tabu Search, Simulated Annealing and Neural Networks. Springer (2000)
16. Rechenberg, I.: *Cybernetic Solution Path of an Experimental Problem*. Royal Aircraft Establishment Library Translation (1965)
17. Reeves, C.: *Modern Heuristic Techniques for Combinatorial Problems*. Advanced Topics in Computer Science Series. McGraw-Hill Ryerson (1995)
18. Renders, J., Flasse, S.: Hybrid methods using genetic algorithms for global optimization. IEEE Transactions on Systems, Man, and Cybernetics, Part B: Cybernetics **26**(2), 243–258 (1996)
19. Saït, S., Youssef, H.: *Iterative Computer Algorithms with Applications in Engineering*. IEEE Computer Society Press (1999)
20. Sechen, C.: *VLSI Placement and Global Routing Using Simulated Annealing*. Kluwer Academic (1988)
21. Siarry, P., Berthiau, G., Durbin, F., Haussy, J.: Enhanced Simulated Annealing for globally minimizing functions of many continuous variables. ACM Transactions on Mathematical Software **238**, 209–228 (1997)
22. Wong, D., Leong, H., Liu, C.: *Simulated Annealing for VLSI Design*. Kluwer Academic (1988)

Chapter 2
Simulated Annealing

Patrick Siarry

2.1 Introduction

The complex structure of the configuration space of a hard optimization problem has inspired people to draw analogies with physical phenomena, which led three researchers at IBM—Kirkpatrick, Gelatt, and Vecchi—to propose in 1982, and to publish in 1983, a new iterative method, the simulated annealing technique [23], which can avoid local minima. A similar method, developed independently at the same time by Cerny [7], was published in 1985.

Since its discovery, the simulated annealing method has proved its effectiveness in various fields, such as the design of electronic circuits, image processing, the collection of household garbage, and the organization of the data-processing network of the French Loto Lottery. On the other hand, it has been found too greedy to solve certain combinatorial optimization problems, which could be solved better by specific heuristics, or completely incapable of solving them.

This chapter starts by initially explaining the principle of the method, with the help of an example of the problem of the layout of an electronic circuit. This is followed by a simplified description of some theoretical approaches to simulated annealing, which underlines its strong points (conditional guaranteed convergence towards a global optimum) and its weak points (tuning of the parameters, which can be delicate in practice). Then various techniques for parallelization of the method are discussed. This is followed by the presentation of some applications. In conclusion, we recapitulate the advantages and the most significant drawbacks of simulated annealing. We put forward some simple practical suggestions, intended for users who are planning to develop their first application based on simulated annealing. In

P. Siarry (✉)
Laboratoire Images, Signaux et Systèmes Intelligents (LiSSi, E.A. 3956),
Université Paris-Est Créteil Val-de-Marne, 122 rue Paul Armangot,
94400 Vitry-sur-Seine, France
e-mail: siarry@u-pec.fr

© Springer International Publishing Switzerland 2016
P. Siarry (ed.), *Metaheuristics*, DOI 10.1007/978-3-319-45403-0_2

Sect. 2.8, we recapitulate the main results of the modeling of simulated annealing based on Markov chains.

This chapter presents, in part, a summary of the review book on the simulated annealing technique [42], which we published at the beginning of 1989; this presentation is augmented by mentioning more recent developments [31, 40]. The references mentioned in the text were selected either because they played a significant role or because they illustrate a specific point in the discussion. A much more exhaustive bibliography—although old—can be found in [37, 42, 47, 50] and in the article [8] on the subject. Interested readers are also recommended to read the detailed presentations of simulated annealing in the article [29] and in Chap. 3 of [31].

2.2 Presentation of the Method

2.2.1 Analogy Between an Optimization Problem and Some Physical Phenomena

The idea of simulated annealing can be illustrated by a picture inspired by the problem of the layout and routing of electronic circuits: let us assume that a relatively inexperienced electronics specialist has randomly spread the components out on a plane, and connections have been made between them without worrying about technological constraints.

It is clear that the solution obtained is an unacceptable one. The purpose of developing a layout-routing program is to transform this disordered situation to an ordered electronic circuit diagram, where all connections are rectilinear, and the components are aligned and placed so as to minimize the length of the connections. In other words, this program must carry out a disorder–order transformation which, starting from a "liquid of components," leads to an ordered "solid."

However, such a transformation occurs spontaneously in nature if the temperature of a system is gradually lowered; there are computer-based digital simulation techniques available which show the behavior of sets of particles interacting in a way that depends on the temperature. In order to apply these techniques to optimization problems, an analogy can be established which is presented in Table 2.1.

Table 2.1 Analogy between an optimization problem and a physical system

Optimization problem	Physical system
Objective function	Free energy
Parameters of the problem	"Coordinates" of the particles
Find a "good" configuration (or even an optimal configuration)	Find the low-energy states

To lead a physical system to a low-energy state, physicists generally use an annealing technique: we will examine how this method of treatment of materials (real annealing) is helpful in dealing with an optimization problem (simulated annealing).

2.2.2   Real Annealing and Simulated Annealing

To modify the state of a material, physicists have an adjustable parameter: the temperature. To be specific, annealing is a strategy where an optimum state can be approached by controlling the temperature. To gain a deeper understanding, let us consider the example of the growth of a monocrystal. The annealing technique consists in heating the material beforehand to impart high energy to it. Then the material is cooled slowly, in a series of stages at particular temperatures, each of sufficient duration; if the decrease in temperature is too fast, it may cause defects which can be eliminated by local reheating. This strategy of a controlled decrease in the temperature leads to a crystallized solid state, which is a stable state, corresponding to an absolute minimum of energy. The opposite technique is that of quenching, which consists in lowering the temperature of the material very quickly: this can lead to an amorphous structure, a metastable state that corresponds to a local minimum of energy. In the annealing technique, the cooling of the material causes a disorder–order transformation, while the quenching technique results in solidifying a disordered state.

The idea of using an annealing technique in order to deal with optimization problems gave rise to the simulated annealing technique. This consists in introducing a control parameter in to the optimization process, which plays the role of the temperature. The "temperature" of the system to be optimized must have the same effect as the temperature of a physical system: it must condition the number of accessible states and lead towards the optimal state if the temperature is lowered gradually in a slow and well-controlled manner (as in the annealing technique), and towards a local minimum if the temperature is lowered abruptly (as in the quenching technique).

To conclude, we have to describe an algorithm in such a way that will enable us to implement annealing on a computer.

2.2.3   Simulated Annealing Algorithm

The algorithm is based on two results from statistical physics.

On one hand, when thermodynamic equilibrium is reached at a given temperature, the probability that a physical system will have a given energy E is proportional to the Boltzmann factor: $e^{\frac{-E}{k_B T}}$, where k_B denotes the Boltzmann constant. Then, the distribution of the energy states is the Boltzmann distribution at the temperature considered.

On the other hand, to simulate the evolution of a physical system towards its thermodynamic equilibrium at a given temperature, the Metropolis algorithm [25]

can be utilized: starting from a given configuration (in our case, an initial layout for all
the components), the system is subjected to an elementary modification (for example,
a component is relocated or two components are exchanged); if this transformation
causes a decrease in the objective function (or "energy") of the system, it is accepted;
in contrast, if it causes an increase ΔE in the objective function, it may also be
accepted, but only with a probability $e^{-\Delta E/T}$. (In practice, this condition is realized
in the following manner: a real number is drawn at random, ranging between 0 and 1,
and a configuration causing a degradation by ΔE in the objective function is accepted
if the random number drawn is less than or equal to $e^{-\Delta E/T}$.) By repeatedly following
this Metropolis rule of acceptance, a sequence of configurations is generated, which
constitutes a Markov chain (in the sense that each configuration depends on only that
one which immediately precedes it). With this formalism in place, it is possible to
show that, when the chain is of infinite length (in practice, of "sufficient" length),
the system can reach (in practice, can approach) thermodynamic equilibrium at the
temperature considered: in other words, this leads us to a Boltzmann distribution of
the energy states at this temperature.

Hence the role given to the temperature by the Metropolis rule is well understood.
At high temperature, $e^{-\Delta E/T}$ is close to 1, and therefore the majority of the moves
are accepted and the algorithm becomes equivalent to a simple random walk in the
configuration space. At low temperature, $e^{-\Delta E/T}$ is close to 0, and therefore the
majority of the moves that increase the energy are rejected. Hence the algorithm
reminds us of a classical iterative improvement. At an intermediate temperature, the
algorithm intermittently allows transformations that degrade the objective function:
hence it leaves a chance for the system to be pulled out of a local minimum.

Once thermodynamic equilibrium is reached at a given temperature, the temperature is lowered "slightly," and a new Markov chain is implemented in this new temperature stage (if the temperature is lowered too quickly, the evolution towards a new
thermodynamic equilibrium is slowed down: the theory of the method establishes a
narrow correlation between the rate of decrease in the temperature and the minimum
duration of the temperature stage). By comparing the successive Boltzmann distributions obtained at the end of the various temperature stages, a gradual increase in the
weight of the low-energy configurations can be noted: when the temperature tends
towards zero, the algorithm converges towards the absolute minimum of energy. In
practice, the process is terminated when the system is "solidified" (which means that
either the temperature has reached zero or no more moves causing an increase in
energy have been accepted during the stage).

2.3   Theoretical Approaches

The simulated annealing algorithm was implemented in many theoretical studies for
the following two reasons: on one hand, it was a new algorithm, for which it was
necessary to establish the conditions for convergence; and on the other hand, the
method contains many parameters and has many variants, whose effect or influence

on the mechanism needed to be properly understood if one wished to implement the method to maximum effect.

These approaches, especially those which appeared during the initial years of its formulation, are presented in detail in the book [42]. Here, we focus on emphasizing on the principal aspects treated in the literature. The theoretical convergence of simulated annealing is analyzed first. Then those factors which are influential in the operation of the algorithm are analyzed in detail: the structure of the configuration space, the acceptance rules, and the annealing program.

2.3.1 Theoretical Convergence of Simulated Annealing

Many mathematicians have invested effort in research into the convergence of the simulated annealing (see in particular [1, 16, 17]) and some of them have even endeavored to develop a general model for the analysis of stochastic methods of global optimization (notably [32, 33]). The main outcome of these theoretical studies is that under certain conditions (discussed later), simulated annealing probably converges towards a global optimum, in the sense that it is possible to obtain a solution arbitrarily close to this optimum with a probability arbitrarily close to unity. This result is, in itself, significant because it distinguishes simulated annealing from other metaheuristic competitors, whose convergence is not guaranteed.

However, the establishment of the "conditions of convergence" is not unanimously accepted. Some of these conditions, such as those proposed by Aarts and Van Laarhoven [1], are based on the assumption of decreasing the temperature in stages. This property enables one to represent the optimization process in the form of completely connected homogeneous Markov chains, whose asymptotic behavior can be described simply. It has also been shown that convergence is guaranteed provided that, on one hand, reversibility is respected (the opposite of any allowed change must also be allowed) and, on the other hand, connectivity of the configuration space is also maintained (any state of the system can be reached starting from any other state with the help of a finite number of elementary changes). This formalization has two advantages:

- it enables us to legitimize the lowering of the temperature in stages, which improves the convergence speed of the algorithm;
- it enables us to establish that a "good"-quality solution (located significantly close to the global optimum) can be obtained by simulated annealing in a polynomial time for certain NP-hard problems [1].

Some other authors, in particular Hajek et al. [16, 17], were interested in the convergence of simulated annealing within the more general framework of the theory of inhomogeneous Markov chains. In this case, the asymptotic behavior was the more sensitive aspect of the study. The main result of this work was the following: the algorithm converges towards a global optimum with a probability of unity if, as the time t tends towards infinity, the temperature $T(t)$ does not decrease more quickly than the expression $C/\log(t)$, where C is a constant related to the depth of the "energy wells" of the problem. It should be stressed that the results of this theoretical work, at present, are not sufficiently general and unambiguous to be used as a guide to an experimental approach when one is confronted with a new problem. For example, the logarithmic law of decrease of the temperature recommended by Hajek is not used in practice for two major reasons: on one hand, it is generally impossible to evaluate the depth of the energy wells of the problem, and, on the other hand, this law leads to an unfavorable increase in computing time.

We now continue this analysis with careful, individual examination of the various components of the algorithm.

2.3.2 Configuration Space

The configuration space plays a fundamental role in the effectiveness of the simulated annealing technique in solving complex optimization problems. It is equipped with a "topology," originating from the concept of proximity between two configurations: the "distance" between two configurations represents the minimum number of elementary changes required to pass from one configuration to the other. Moreover, there is an energy associated with each configuration, so that the configuration space is characterized by an "energy landscape." All of the difficulties of the optimization problem lie in the fact that the energy landscape comprises of a large number of valleys of varying depth, possibly relatively close to each other, which correspond to local minima of energy.

It is clear that the shape of this landscape is not specific to the problem under study, but depends to a large extent on the choice of the cost function and the choice of the elementary changes. However, the required final solution, i.e., the global minimum (or one of the global minima of comparable energy), must depend primarily on the nature of the problem considered, and not (or very little) on these choices. We have shown, with the help of an example problem of placement of building blocks, considered specifically for this purpose, that an apparently sensitive problem can be greatly simplified either by widening the allowable configuration space or by choosing a better adapted topology [42].

Several authors have endeavored to establish general analytical relations between certain properties of the configuration space and the convergence of simulated annealing. In particular, some of their work was directed towards an analysis of the energy landscapes, and they sought to develop a link between "ultrametricity" and simulated annealing [22, 30, 44]: the simulated annealing method would be more effective for those optimization problems whose low local minima (i.e., the required solutions) formed an ultrametric set. Thereafter, Sorkin [45] showed that certain fractal properties of the energy landscape induce polynomial convergence of simulated annealing; Sorkin explained this on the basis of the effectiveness of the method in the field of electronic circuit layouts. In addition, Azencott [3] utilized the "theory of cycles" (originally developed in the context of dynamic systems) to establish general explicit relations between the geometry of the energy landscape and the expected performance of simulated annealing. This work led to the proposal of the "method of distortions" for the objective function, which significantly improved the quality of the solutions for certain difficult problems [11]. However, all these approaches to simulated annealing are still in a nascent stage, and their results have not yet been generalized.

Lastly, another aspect of immediate practical interest relates to the adaptation of simulated annealing to the solution of continuous optimization problems [9, 39]. Here, we stress only the transformations necessary to make the step from "combinatorial simulated annealing" to "continuous simulated annealing." In fact, the method was originally developed for application in the domain of combinatorial optimization problems, where the free parameters can take discrete values only. In the majority of these types of problems encountered in practice, the topology is almost always considered as data for the problem: for example, in the traveling salesman problem, the permutation of two cities has a natural tendency to generate round-trip routes close to a given round-trip route. The same thing occurs in the problem of placement of components when the exchange of two blocks is considered. On the other hand, when the objective is to optimize a function of continuous variables, the topology has to be updated. This gives rise to the concept of "adaptive topology": here, the length of the elementary steps is not imposed by the problem anymore. This choice must instead be dictated by a compromise between two extreme situations: if the step is too small, the program explores only a limited region of the configuration space; the cost function is then improved very often, but by a negligible amount. In contrast, if the step is too large, the test results are accepted only seldom, and they are almost independent of each other. From the point of mathematical interest, it is necessary to mention the work of Miclo [26], which was directed towards the convergence of simulated annealing in the continuous case.

2.3.3   Rules of Acceptance

The principle of simulated annealing requires that one accepts, occasionally and under the control of the "temperature," an increase in the energy of the current state, which enables it to be pulled out of a local minimum. The rule of acceptance generally used is the Metropolis rule described in Sect. 2.2.3. This possesses the advantage that it originates directly from statistical physics. There are, however, several variations of this rule [42], which can be more effective from the point of view of computing time.

Another aspect arises from examination of the following problem: at low temperature, the rate of acceptance of the algorithm becomes very small, and hence the method is ineffective. This is a well-known problem encountered in simulated annealing, which can be solved by substituting the traditional Metropolis rule with an accelerated alternative, called the "thermostat" [42], as soon as the rate of acceptance falls too low. In practice, this methodology is rarely employed.

2.3.4 *Program of Annealing*

The convergence speed of the simulated annealing methodology depends primarily on two factors: the configuration space and the program of annealing. With regard to the configuration space, readers have already been exposed to the effects of topology on convergence and the shape of the energy landscape. Let us discuss the influence of the "program of annealing": this addresses the problem of controlling the "temperature" as well as the possibility of a system reaching a solution as quickly as possible. The program of annealing must specify the following values of the control parameters for the temperature:

- the initial temperature;
- the length of the homogeneous Markov chains, i.e., the criterion for changing to the next temperature stage;
- the law of decrease of the temperature;
- the criterion for program termination.

In the absence of general theoretical results which can be readily exploited, the user has to resort to empirical adjustment of these parameters. For certain problems, the task is complicated even further by the great sensitivity of the result (and the computing time) to this adjustment. This aspect—which unites simulated annealing with other metaheuristics—is an indisputable disadvantage of this method.

To elaborate on the subject a little more, we shall look at the characteristic of the program of annealing that has drawn most attention: the law of decrease of the temperature. The geometrical law of decrease, $T_{k+1} = \alpha \cdot T_k$, $\alpha = $ constant, is a widely accepted one, because of its simplicity. An alternative solution, potentially more effective, is an adaptive law of the form $T_{k+1} = \alpha(T_k) \cdot T_k$, but it is then necessary to exercise a choice from among several laws suggested in the literature. One can show, however, that several traditional adaptive laws, which have quite different origins and mathematical expressions, are in practice equivalent (see Fig. 2.1), and can be expressed in the following generic form:

$$T_{k+1} = \left(1 - T_k \cdot \frac{\Delta(T_k)}{\sigma^2(T_k)}\right) \cdot T_k$$

where

$$\sigma^2(T_k) = \langle f_{T_k}^2 \rangle - \langle f_{T_k} \rangle^2,$$

f denotes the objective function, and $\Delta(T_k)$ depends on the adaptive law selected. The simplest adjustment, $\Delta(T_k) = $ constant, can then be made effective, although it does not correspond to any of the traditional laws.

Owing to our inability to synthesize the results (both theoretical and experimental) presented in the literature, which show some disparities, the reader is referred to Sect. 2.7, where we propose a suitable tuning algorithm for the four parameters of the program of annealing, which can often be useful at least to start with.

Fig. 2.1 Lowering of the temperature according to the number of stages for the geometrical law and several traditional laws

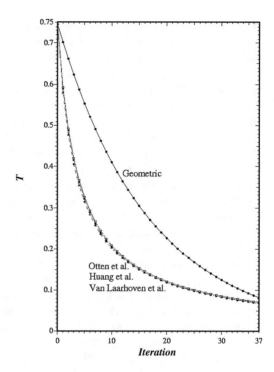

Those readers who are interested in the mathematical modeling of simulated annealing are advised to refer to Sect. 2.8: the principal results produced by the Markov formalism are described there.

2.4 Parallelization of the Simulated Annealing Algorithm

Often, the computing time becomes a critical factor in the economic evaluation of the utility of a simulated annealing technique for applications to real industrial problems. A promising research direction to reduce this time is the parallelization of the algorithm, which consists in simultaneously carrying out several of the calculations necessary for its realization. This step can be considered in the context of the significant activity that has been developing around the algorithms and architectures of parallel computation for quite some time now. This may appear paradoxical, because of the sequential structure of the algorithm. Nevertheless, several types of parallelization have been considered to date. A book [3] completely devoted to this topic has been published; it describes simultaneously the rigorous mathematical results available and the results, of simulations executed on parallel or sequential computers. To provide a concrete idea, we shall describe the idea behind two principal modes of parallelization, which are independent of the problem being dealt with and were suggested very soon after the invention of simulated annealing. The distinction between these two modes remains relevant today, as has been shown in the recent status of the state of the art described by Delamarre and Virot [11].

The first type of parallelization [2] consists in implementing several Markov chain computations in parallel, by using K elementary processors. To implement this, the algorithm is decomposed into K elementary processes, constituting K Markov chains. Let L be the length of these Markov chains, assumed constant, each chain is divided into K subchains of length L/K. The first processor executes the first chain at the initial temperature, and implements the first L/K elements of this chain (i.e., the first subchain); then it calculates the temperature of the following Markov chain, starting from the states already obtained. The second elementary processor then begins executing the second Markov chain at this temperature, starting from the final configuration of the first subchain of the first chain. During this time, the first processor begins the second subchain of the first chain. This process continues for the K elementary processors. It has been shown that this mode of parallelization—described in more detail in [42]—allows one to divide the total computing time by a factor K, if K is small compared with the total number of Markov chains carried out. However, the procedure has a major disadvantage: its convergence towards an optimum is not guaranteed. The formalism of Markov chains enables one to establish that the convergence of simulated annealing is assured provided that the distribution of the states, at the end of each Markov chain is close to the stationary distribution. In the case of the algorithm described here, however, this closeness is not established at the end of each subchain, and the larger the number K of processors in parallel, the larger is the deviation from closeness.

The second type of parallelization [24, 35] consists in carrying out the computation in parallel for several states of the same Markov chain while keeping in mind the following condition: at low temperature, the number of elementary transformations rejected becomes very important; it is thus possible to assume that these moves are produced by independent elementary processes, which may likely be implemented in parallel. Then the computing time can be divided by approximately the number of processes. One strategy consists in subdividing the algorithm into K elementary processes, each of which is responsible for calculating the energy variations corresponding to one or more elementary moves, and for carrying out the corresponding Metropolis tests. Two operating modes are considered:

- At "high temperature," a process corresponds to only one elementary move. Each time K elementary processes are implemented in parallel, one can randomly choose a transition from among those which have been accepted, and the memory, containing the best solution known, is updated with the new configuration.
- At "low temperature," the accepted moves become very rare: less than one transition is accepted for K moves carried out. Each process then consists in calculating the energy variations corresponding to a succession of disturbances until one of them is accepted. As soon as any of the processes succeeds, the memory is updated.

These two operating modes can ensure behavior, and in particular convergence, which is strictly identical to that of sequential algorithms. This type of parallelization has been tested by experimenting on the optimization problem of the placement of connected blocks [35]. We estimated the amount of computing time saved in two cases: the placement of presumed point blocks in predetermined sites and the placement of real blocks on a plane. With five elementary processes in parallel, the saving in computing time was between 60 and 80 %, depending on the program of annealing used. This work was then continued, in the thesis work of Roussel-Ragot [34] by considering a theoretical model, which was validated by programming the simulated annealing using a network of "transputers."

In addition to these two principal types of parallelization of simulated annealing, which should be applicable to any optimization problem, other methodologies have been proposed to deal with specific problems. Some of these problems are problems of placement of electronic components, problems in image processing and problems of meshing of areas (for the finite element method). In each of these three cases, information is distributed in a plane or in space, and each processor can be entrusted with the task of optimizing the data pertaining to a geographical area by simulated annealing; here information is exchanged periodically between neighboring processors.

Another step to reduce the cost of synchronization between processors has been planned: the algorithms known as "asynchronous algorithms" are designed to calculate the energy variations starting from partially out-of-date data. However, it seems very complex and sensitive to control the admissible error, except for certain particular problems [12].

As an example, let us describe the asynchronous parallelization technique suggested by Casotto et al. [6] to deal with the problem of the placement of electronic components. The method consists in distributing the components to be placed into K independent groups, assigned to K respective processors. Each processor applies the simulated annealing technique to seek the optimal site for the components that belong to its group. The processors function in parallel, and in an asynchronous manner with respect to each other. All of them have access to a common memory, which contains the current state of the circuit plan. When a processor plans to exchange the position of a component in its group with that of a component in another group belonging to another processor, it temporarily blocks the activity of that processor. It is clear that the asynchronous working of the processors involves errors, in particular in the calculation of the overlap between the blocks, and thus in the evaluation of the cost function. In fact, when a given processor needs to evaluate the cost of a move (translation or permutation), it will search in the memory for the current positions of all the components in the circuit. However, the information collected is partly erroneous, since certain components are in the course of displacement because of the activities of other processors. In order to limit these errors, the method is supplemented by the following two processes. On one hand, the distribution of the components between the processors is in itself an object of optimization by the simulated annealing technique, which is performed simultaneously with the optimization process already described: in this manner, membership of the components geographically close to the same group can be favored. In addition, the maximum amplitude of the moves carried out by the components is reduced as the temperature decreases. Consequently, when the temperature decreases, the moves relate mainly to nearby components, which thus generally belong to the same group. In this process, the interactions between the groups can be reduced, thus reducing the frequency of the errors mentioned above. This technique of parallelization of simulated annealing was validated using several examples of real circuits: the algorithm functioned approximately six times faster with eight processors than with only one, the results being of comparable quality to those of the sequential algorithm.

2.5   Some Applications

The majority of the preceding theoretical approaches are based on asymptotic behaviors which impose several restrictive assumptions, very often causing excessive increases in computing times. This is why, to solve real industrial problems under reasonable conditions, it is often essential to adopt an experimental approach, which may frequently result in crossing the barriers recommended by the theory. The simulated annealing method has proved to be interesting for solving many optimization problems, both NP-hard and not. Some examples of these problems are presented here.

2.5.1   Benchmark Problems of Combinatorial Optimization

The effectiveness of the method was initially tested on some "benchmark problems" of combinatorial optimization. In this type of problem, the practical purpose is secondary: the initial objective is to develop the optimization method and to compare its performance with that of other methods. We will detail only one example: that of the traveling salesman problem.

The reason for the choice of this problem is that it is very simple to formulate and, at the same time, very difficult to solve: the largest problems for which the optimum has been found and proved comprise a few thousands of cities. To illustrate the disorder–order transformation that occurs in the simulated annealing technique as the temperature goes down, we present in Fig. 2.2 four intermediate configurations obtained by Eric Taillard, in the case of 13 206 towns and villages in Switzerland.

Bonomi and Lutton also considered very high-dimensional examples, with between 1000 and 10 000 cities [4]. They showed that, to avoid a prohibitive computing time, the domain containing the cities can be deconstructed into areas, and the moves for the route of the traveler can be forced so that they are limited to being between cities located in contiguous areas. Figure 2.3 shows the effectiveness of this algorithm for a problem comprising 10 000 cities: the length of this route does not exceed that of the optimal route by more than 2 % (the length of the shortest route can be estimated a priori when the number of cities is large). Bonomi and Lutton compared simulated annealing with traditional techniques of optimization

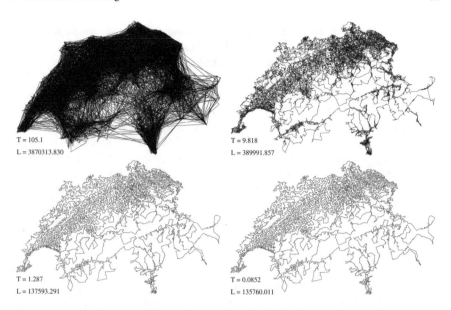

Fig. 2.2 The traveling salesman problem (13 206 cities): the better known configurations (length L) at the end of four temperature stages (T)

Fig. 2.3 The traveling salesman problem: solution, by simulated annealing for a case of 10 000 cities

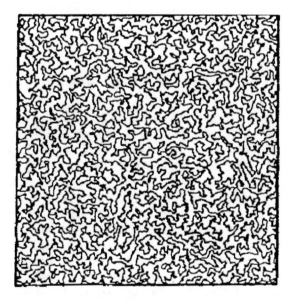

for the traveling salesman problem: simulated annealing was slower for small-dimensional problems (N lower than 100) but, on the other hand, it was far more powerful for higher-dimensional problems (N higher than 800). The traveling salesman problem has been extensively studied to illustrate and establish several experimental and theoretical developments in the simulated annealing method [42].

Many other benchmark problems of combinatorial optimization have also been solved using simulated annealing [29, 42]: in particular, the problems of the "partitioning of a graph," the "minimal coupling of points," and"quadratic assignment." Comparison with the best known algorithms leads to different results, varying according to the problems and the authors. Thus the studies by Johnson et al. [19–21], which were devoted to a systematic comparison of several benchmark problems, conclude that the only benchmark problem for which the results favor simulated annealing is that of the partitioning of a graph. For some problems, promising results were only obtained with the simulated annealing method for high-dimensional examples (a few hundreds of variables), and at the cost of a high computing time. Therefore, if simulated annealing has the merit to be adapted simply to a great diversity of problems, it cannot claim very much to supplement the specific algorithms that already exist for these problems.

We now present the applications of simulated annealing to practical problems. The first significant application of industrial interest was developed in the field of electronic circuit design; this industrial sector still remains the domain in which the greatest number of publications describing applications of simulated annealing have been produced. Two applications in the area of electronics are discussed in detail in the following two subsections. This is followed by discussions of other applications in some other fields.

2.5.2   Layout of Electronic Circuits

The first application of the simulated annealing method to practical problems was developed in the field of the layout and routing of electronic circuits [23, 41, 49]. Numerous studies have now been reported on this subject in several publications and, in particular, two books have been completely devoted to this problem [37, 50]. Detailed bibliographies, concerning the work carried out in the initial period from 1982 to 1988 can be found in the books [37, 42, 47, 50].

The search for an optimal layout is generally carried out in two stages. The first consists in calculating an initial placement quickly, by a constructive method: the components are placed one after another, in order of decreasing connectivity. Then an algorithm for iterative improvement is employed that gradually transforms, by elementary moves (e.g., exchange of components, and operations of rotation or symmetry), the initial layout configuration. The algorithms for iterative improvement of the layout differ according to the rule adopted for the succession of elementary moves. Simulated annealing can be used in this second stage.

Our interest was in a unit of 25 identical blocks to be placed on predetermined sites, which were the nodes of a planar square network. The list of connections was such that, in the optimal configurations, each block would be connected only to its closer neighbors (see Fig. 2.4a): an a priori knowledge of the global minima of the

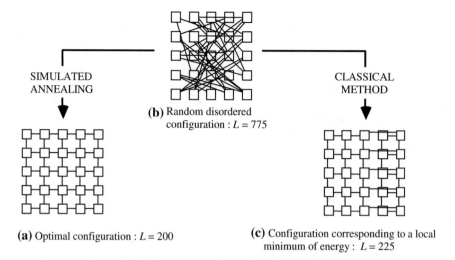

(b) Random disordered configuration : $L = 775$

SIMULATED
ANNEALING

CLASSICAL
METHOD

(a) Optimal configuration : $L = 200$

(c) Configuration corresponding to a local minimum of energy : $L = 225$

Fig. 2.4 The traditional method getting trapped in a local minimum of energy

problem then made it easier to study the influence of the principal parameters of the method on its convergence speed. The cost function was the overall Manhattan length (i.e., the length of L-type) of the connections. The only allowed elementary move was the permutation of two blocks. A detailed explanation for this benchmark problem on layout design—which is a form of "quadratic assignment" problem—can be found in [38, 43]. Here, the discussion will be limited to the presentation of two examples of applications. First of all, to appreciate the effectiveness of the method, we started with a completely disordered initial configuration (Fig. 2.4b), and an initial "elevated" temperature (in the sense that at this temperature 90 % of the moves are accepted). In this example, the temperature profile was that of a geometrical decrease, of ratio 0.9. A global optimum of the problem could be obtained after 12 000 moves, whereas the total number of possible configurations is about 10^{25}.

To illustrate the advantages of the simulated annealing technique, we applied the traditional method of iterative improvement (simulated annealing at zero temperature), with the same initial configuration (see Fig. 2.4b), and allowed the same number of permutations as during the preceding test. It was observed that the traditional method got trapped in a local minimum (Fig. 2.4c); it is clear that shifting from this configuration to the optimal configuration as shown in Fig. 2.4a would require several stages (at least five), the majority of which correspond to an increase in energy, which is inadmissible in the traditional method. This particular problem of placement made it possible to develop empirically a program of "adaptive" annealing, which could achieve a gain in computing time by a factor of 2; the lowering of the temperature was carried out according to the law $T_{k+1} = D_k \cdot T_k$, where:

$$D_k = \min\left(D_0, \frac{E_k}{\langle E_k \rangle}\right)$$

Here, $D_0 = 0.5$ to 0.9, E_k is the minimum energy of the configurations accepted during stage k, and $\langle E_k \rangle$ is the average energy of the configurations accepted during stage k. (At high temperature, $D_k = E_k/\langle E_k \rangle$ is small, and hence the temperature is lowered quickly; at low temperature, $D_k = D_0$, which corresponds to slow cooling).

Then we considered a more complex problem consisting of positioning components of different sizes, with the objective of simultaneous minimization of the length of the necessary connections and of the surface area of the circuit used. In this case, the translation of a block is a new means of iterative transformation of the layout. Here we can observe that the blocks can overlap with each other, which is allowed temporarily, but must generally be excluded from the final layout. This new constraint can be accommodated within the cost function of the problem by introducing a new factor called the "overlapping surface" between the blocks. Calculating this surface area can become very cumbersome when the circuit comprises many blocks. For this reason the circuit was divided into several planar areas, whose size was such that a block could overlap only with those blocks located in the same area or in a very close area. The lists of the blocks belonging to each area were updated after each move, using a chaining method. Moreover, to avoid leading to a circuit obstruction such as an impossible routing, a fictitious increase in the dimensions of each block was introduced. The calculation of the length of the connections consisted in determining, for each equipotential, the barycenter of the terminations, and then adding the distances of L-type of the barycenter with each termination. Lastly, the topology of the problem was adaptive, which can be described in the following manner: when the temperature decreases, the maximum amplitude of the translations decreases, and exchanges are considered more between neighboring blocks only.

With the simulated annealing algorithm, it was possible to optimize industrial circuits, in particular some used in hybrid technology, in collaboration with the Thomson D.C.H. (Department of Hybrid Circuits) company. As an example, we present in Fig. 2.5, the result of the optimization of a circuit layout comprising 41 components and 27 equipotentials: the automated layout design procedure leads to a gain of 18 % in the connection lengths compared with the initial manual layout.

This study showed that the flexibility of the method enables it to take into account not only the rules of drawing, which translate the standards of technology, but also the rules of know-how, which are intended to facilitate routing. In particular, the rules of drawing impose a minimum distance between two components, whereas the rules of know-how recommend a larger distance, allowing the passage of connections. To balance these two types of constraints, the calculation of the area of overlap between the blocks, on a two-to-two basis, was undertaken according to the formula

$$S = S_r + a \cdot S_v,$$

where S_r is the "real" overlapping area, S_v is the "virtual" overlapping area, and a is a weight factor (typically: 0.1).

These areas S_r and S_v were calculated by increasing the dimensions of the components fictitiously, with a larger increase in S_v. This induces some kind of an "intelligent" behavior, similar to that of an expert system. We notice from Fig. 2.5

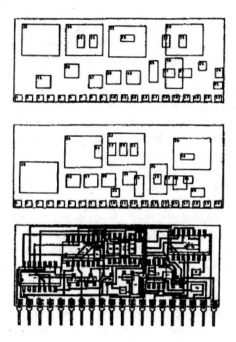

- *Top: initial manual layout; length of connections: 9532,*
- *Middle: final layout, optimized by annealing; length of connections 7861,*
- *Bottom: manual routing using the optimized layout.*

Fig. 2.5 Optimization by simulated annealing of the design of an electronic circuit layout comprising 41 components

a characteristic of hybrid technology which was easily incorporated into the program: the resistances, shown by a conducting link, can be placed under the diodes or integrated circuits.

The observations noted by the majority of authors concerning the application of the simulated annealing technique to the layout design problem agree with our observations: the method is very simple to implement, it can be adapted easily to various evolving technological standards, and the final result is of good quality, but it is sometimes obtained at the cost of a significant computing time.

2.5.3 Search for an Equivalent Schema in Electronics

We now present an application which mixes the combinatorial and the continuous aspects: automatic identification of the "optimal" structure of a linear circuit pattern. The objective was to automatically determine a model which includes the least possible number of elementary components, while ensuring a "faithful" reproduction of experimental data. This activity, in collaboration with the Institute of Fundamental Electronics (IEF, CNRS URA 22, Orsay), began with the integration, in a single software package, of a simulation program for linear circuits (implemented at the IEF) and a simulated annealing-based optimization program developed by us. We initially validated this tool by characterizing models of real components, with a complex structure, described using their distribution parameters S. Comparison with a commercial software package (developed using the gradient method) in use at the time of the IEF showed that simulated annealing was particularly useful if the orders of magnitude of the parameters of the model were completely unknown: obviously the models under consideration were of this nature, since even their structure was to be determined. We developed an alternative simulated annealing method, called logarithmic simulated annealing [9], which allows an effective exploration of the space of variations of the parameters when this space is very wide (more than 10 decades per parameter). Then the problem of structure optimization was approached by the examination—in the case of a passive circuit—of the progressive simplification of a general "exhaustive" model: we proposed a method which could be successfully employed to automate all the simplification stages [10]. This technique is based on the progressive elimination of the parameters according to their statistical behavior during the process of optimization by simulated annealing.

We present here, with the help of illustrations, an example of a search for an equivalent schema for a monolithic microwave integrated circuit (MMIC) inductance, in the frequency range from 100 MHz to 20 GHz. On the basis of an initial "exhaustive" model with 12 parameters, as shown in Fig. 2.6, and allowing each parameter to move over 16 decades, we obtained the equivalent schema shown in Figure 2.7 (the final values of the six remaining parameters are beyond the scope of our present interest: they are specified in [10]). The layouts in the Nyquist plane of the four S parameters of the quadrupole shown in Fig. 2.7 coincided nearly perfectly with the experimental results for the MMIC inductance, and this was true over the entire frequency range [10].

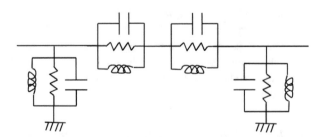

Fig. 2.6 Initial structure with 12 elements

Fig. 2.7 Optimal structure
with six elements

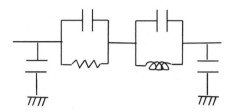

2.5.4 Practical Applications in Various Fields

An important field of application of simulated annealing is image processing: here
the main problem is to restore images, mainly in three-dimensional forms, using a
computer, starting from incomplete or irregular data. There are numerous practical
applications in several domains, such as robotics, medicine (e.g., tomography), and
geology (e.g., prospecting). The restoration of an image using an iterative method
involves, under normal circumstances, the treatment of a large number of variables.
Hence it calls for the development of a suitable method which can limit the computing time of the operation. Based on the local features of the information contained in
an image, several authors have proposed numerous structures and algorithms specifically to address the problem of carrying out calculations in parallel. Empirically, it
appears that the simulated annealing method should be particularly well suited for
this task. A rigorous theoretical justification of this property can be obtained starting
from the concepts of Markovian fields [14], which provide a convenient and coher-
ent model of the local structure of the information in an image. This concept has
been explained in detail in [42]. The "Bayesian approach" to the problem of optimal
restoration of an image, starting from a scrambled image, consists in determining
the image which has "the maximum a posteriori probability." It has been shown that
this problem can ultimately be expressed as a well-known minimization problem of
an objective function, comprising a very large number of parameters, for example
the light intensities of all the "pixels" of an image in case of an image in black and
white. Consequently, the problem can be considered as a typical problem for simulated annealing. The iterative application of this technique consists in updating the
image by modifying the intensities of all of the pixels in turn, in a prespecified order.
This procedure leads to a significant consumption of computing time: indeed, the
number of complete sweeps of the image necessary to obtain a good restoration is
typically about 300 to 1000. But, as the calculation of the energy variation is purely
local in nature, several methods have been proposed to update the image by simultaneously treating a large number of pixels, using specialized elementary processors.
The formalism of Markovian fields has made it possible to treat by simulated annealing several crucial tasks in the automated analysis of images: restoration of scrambled
images, image segmentation, image identification. Apart from this formalism, other
problems in the image-processing domain have also been solved by annealing: for
example, the method has been utilized to determine the geological structure of the
basement, starting from the results of seismic experiments.

To finish, we will mention some specific problems, in very diverse fields, where simulated annealing has been employed successfully: organization of the data-processing network for Loto (this required ten thousand playing machines to be connected to host computers), optimization of the collection of household garbage in Grenoble, timetabling problems (one problem was, for example, to perform the optimal planning of rest days in a hospital), and optimization of structures (in a project on constructing a 17-floor building for an insurance company, it was necessary to distribute the activities among the various parts so that the work output from 2000 employees could be maximized). Several applications of simulated annealing to scheduling problems can be found, in particular, in [5, 18, 27, 48]. The adequacy of the method for this type of problem has also been discussed. For example, Van Laarhoven et al. [48] showed that the computing time involved was unsatisfactory. Moreover, in [13], Fleury underlined several characteristics of scheduling problems which make them unsuitable for simulated annealing and he recommended a different stochastic method for this problem inspired by simulated annealing and tabu search: the "kangaroo method."

2.6   Advantages and Disadvantages of the Method

From the preceding discussion, the principal characteristics of the method can be established. Firstly, the *advantages*: it is observed that the simulated annealing technique generally achieves a good-quality solution (i.e., an absolute minimum or good relative minimum for the objective function). Moreover, it is a general method: it is applicable, to all problems which can potentially employ iterative optimization techniques, and it is easy to implement, under the condition that after each transformation the corresponding change in the objective function can be evaluated directly and quickly (often the computing time becomes excessive if complete re–computation of the objective function cannot be avoided after each transformation). Lastly, it offers great flexibility, as one easily can build new constraints into the program afterwards.

Now, let us discuss the disadvantages. Users are sometimes repelled by the involvement of a great many parameters (initial temperature, rate of decrease of the temperature, length of the temperature stages, termination criterion for the program). Although the standard values published for these parameters generally allow effective operation of the method, the essentially empirical nature of them can never guarantee suitability for a large variety of problems. The second defect of the method—which depends on the preceding one—is the computing time involved, which is excessive in certain applications.

In order to reduce this computing time, we still require an extensive research effort to determine the best values of the parameters of the method beyond the generalized results published so far [39], particularly the law of decrease of the temperature. Any progress in the effectiveness of the technique and in the computing time involved is likely to be obtained by continuing the analysis of the method in three specific directions: the utilization of interactive parameter setting, parallelization of the algorithm, and the incorporation of statistical physics-based approaches to the analysis and study of disordered media.

2.7   Simple Practical Suggestions for Beginners

- *Definition of the objective function*: some constraints can be integrated into the objective function, whereas others constitute a limitation on the form of the disturbances for the problem.
- *Choice of disturbance mechanisms* for the " current configuration": the calculation of the corresponding variation ΔE of the objective function must be *direct* and rapid.
- *Initial temperature T_0*: this may be calculated in a preliminary step using the following algorithm:

  - initiate 100 disturbances at random; evaluate the average $\langle \Delta E \rangle$ of the corresponding variations ΔE;
  - choose an initial rate of acceptance τ_0 of the "degrading perturbations" according to the assumed "quality" of the initial configuration; for example:
    - · "poor" quality: $\tau_0 = 50\%$ (starting at high temperature),
    - · "good" quality: $\tau_0 = 20\%$ (starting at low temperature);
  - deduce T_0 from the relation: $e^{-\langle \Delta E \rangle / T_0} = \tau_0$.

- *Metropolis acceptance rule*: this can be utilized practically in the following manner: if $\Delta E > 0$, a number r in $[0, 1]$ is drawn randomly, and the disturbance is accepted if $r < e^{-\Delta E / T}$, where T indicates the current temperature.
- *Change to next temperature stage*: this can take place as soon as one of the following two conditions is satisfied during a temperature stage:

  - $12 \cdot N$ perturbations accepted;
  - $100 \cdot N$ perturbations attempted, N indicates the number of degrees of freedom (or parameters) of the problem.

- *Decrease of the temperature*: this can be carried out according to the geometrical law $T_{k+1} = 0.9 \cdot T_k$.
- *Program termination*: this can be activated after three successive temperature stages without any acceptances.
- *Essential verifications during the first executions of the algorithm*:

  - the generation of the real random numbers (in $[0, 1]$) must be very *uniform*;
  - the "quality" of the result should not vary significantly when the algorithm is implemented *several times*:
    - · with different "seeds" for the generation of the random numbers,
    - · with different initial configurations;

– for each initial configuration used, the result of simulated annealing should compare favorably, theoretically, with that of the *quenching* ("disconnected" Metropolis rule).

• *An alternative version of the algorithm in order to achieve less computation time*: simulated annealing is greedy and not very effective at low temperature; hence one might be interested in utilizing the simulated annealing technique, prematurely terminated, in cascade with an algorithm of local type for specific optimization of the problem, whose role is to "refine" the optimum.

2.8 Modeling of Simulated Annealing Through the Markov Chain Formalism

Let R be the complete space of all possible configurations of the system, and let $r \in R$ be a "state vector," whose components entirely define a specified configuration (or "state"). Let the set I_R consist of the numbers assigned to each configuration of R:

$$I_R = (1, 2, \ldots, |R|)$$

where $|R|$ is the cardinality of R. Finally, let us denote by $C(r_i)$ the value of the cost function (or "energy") in the state i, where r_i is the state vector for the state, and let $M_{ij}(T)$ be the probability of a transition from the state i to the state j at a "temperature" T. In the case of the simulated annealing algorithm, the succession of states forms a Markov chain, in the sense that the probability of transition from the state i to the state j depends only on these two states, and not on the states previous to i. In other words, all the past information about the system is summarized in the current state. When the temperature T is maintained constant, the probability of a transition $M_{ij}(T)$ is constant, and the corresponding Markov chain is known as *homogeneous*. The probability of a transition $M_{ij}(T)$ from the state i to the state j can be expressed in the following form:

$$M_{ij}(T) = \begin{cases} P_{ij} \cdot A_{ij}(T) & \text{if } i \neq j \\ 1 - \Sigma_{k \neq i} P_{ik} \cdot A_{ik}(T) & \text{if } i = j \end{cases}$$

where P_{ij} is the probability of perturbation, i.e., the probability of generating the state j when one is in the state i, and $A_{ij}(T)$ is the probability of acceptance, i.e., the probability of accepting the state j when one is in the state i at a temperature T.

The first factor, P_{ij}, can be calculated easily. In fact, the system is generally perturbed by randomly choosing a movement from the allowed elementary movements.

The results of this is that

$$P_{ij} = \begin{cases} |R_i|^{-1} & \text{if } j \in I_{R_i} \\ 0 & \text{if } j \notin I_{R_i} \end{cases}$$

where R_i denotes the subset of R comprising all the configurations which can be obtained in only one movement starting from the state i, and I_{R_i} denotes the set of the numbers of these configurations. As for the second factor, $A_{ij}(T)$, this is often defined by the Metropolis rule. Aarts and Van Laarhoven [1] noted that, more generally, the simulated annealing method makes it possible to impose the following five conditions:

1. The configuration space is *connected*, i.e. two unspecified states i and j correspond to a finite number of elementary movements.
2. $\forall i, j \in I_R : P_{ij} = P_{ji}$ (reversibility).
3. $A_{ij}(T) = 1$, if $\Delta C_{ij} = C(r_j) - C(r_i) \leq 0$ (the movements which result in a reduction in energy are systematically accepted).
4. If $\Delta C_{ij} > 0$ $\begin{cases} \lim\limits_{T \to \infty} A_{ij}(T) = 1 \\ \lim\limits_{T \to 0} A_{ij}(T) = 0 \end{cases}$

   (movements which result in an increase in energy are all accepted at infinite temperature, and all refused at zero temperature).
5. $\forall i, j, k \in I_r \mid C(r_k) \geq C(r_j) \geq C(r_i) : A_{ik}(T) = A_{ij}(T) \cdot A_{jk}(T)$.

2.8.1   Asymptotic Behavior of Homogeneous Markov Chains

By using the results obtained for homogeneous Markov chains, one can establish the following properties.

2.8.1.1   Property 1

Consider a Markov process generated by a mechanism of transition which observes the five conditions stated above. This mechanism is applied n times, at constant temperature, starting from a specified initial configuration, arbitrarily chosen. When n tends towards infinity, the Markov chain obtained has one and only one equilibrium vector, called $q(T)$, which is independent of the initial configuration. This vector, which consists of $|R|$ components, is called *distribution of static probability* of the Markov chain. Its ith component, i.e., $q_i(T)$, represents the probability that the system is in the configuration i when, after an infinity of transitions, the steady state is reached.

2.8.1.2   Property 2

$q_i(T)$ is expressed by the following relation:

$$q_i(T) = \frac{A_{i_0 i}(T)}{\sum_{i=1}^{|R|} A_{i_0 i}(T)},$$

where i_0 denotes the number of an optimal configuration.

2.8.1.3   Property 3

When the temperature tends towards infinity or zero, the limiting values of $q_i(T)$ are given by $\lim_{T \to \infty} q_i(T) = |R|^{-1}$ and

$$\lim_{T \to 0} q_i(T) = \begin{cases} |R_0|^{-1} & \text{if } i \in I_{R_0} \\ 0 & \text{if } i \notin I_{R_0} \end{cases}$$

where R_0 denotes the set of the optimal configurations, i.e.,

$$R_0 = \left\{ r_i \in R \mid C(r_i) = C(r_{i_0}) \right\}$$

Property 3 results immediately from property 2 when condition 4 is used. Its interpretation is the following: for larger values of the temperature, all configurations can be obtained with the same probability. On the other hand, when the temperature tends towards zero, the system reaches an optimal configuration with a probability equal to unity. In both cases, the result is obtained at the end of a Markov chain of infinite length.

Remark If one chooses the probability of acceptance $A_{ij}(T)$ recommended by Metropolis (see [1] for a justification for this choice regardless of any analogy with physics),

$$A_{ij}(T) = \begin{cases} e^{-\Delta C_{ij}/T} & \text{if } \Delta C_{ij} > 0 \\ 1 & \text{if } \Delta C_{ij} \leq 0 \end{cases}$$

one finds in property 2 the expression for the Boltzmann distribution.

2.8.2   Choice of Annealing Parameters

We saw in the preceding subsection that the convergence of the simulated annealing algorithm is assured when the temperature tends towards zero. A Markov chain of infinite length undoubtedly ends in the optimal result if it is built at a sufficiently low (though nonzero) temperature. But this result is not of any practical utility because, in this case, the equilibrium is approached very slowly. The Markov chain formalism makes it possible to examine theoretically the convergence speed of the algorithm. One can show that this speed is improved when one starts from a high temperature and this temperature is then decreased in stages. This procedure requires the use of an annealing program, which defines the optimal values of the parameters of the descent in temperature. We will examine four principal parameters of the annealing program:

- the initial temperature;
- the length of the homogeneous Markov chains, i.e., the criterion for changing between temperature stages;
- the law of decrease of the temperature;
- the criterion for program termination.

For each of them, we will indicate first the corresponding result of the theory, which leads to an optimal result but often at the cost of a prohibitive computing time. Then we mention some values obtained by experiment.

2.8.2.1 Initial Temperature

There exists a necessary but not sufficient condition so that the optimization process does not get trapped in a local minimum. The initial temperature T_0 must be sufficiently high that, at the end of the first stage, all configurations can be obtained with the same probability. A suitable expression for T_0 which ensures a rate of acceptance close to 1 is the following:

$$T_0 = r \cdot \max_{ij} \Delta C_{ij}$$

with $r \gg 1$ (typically $r = 10$). In practice, in many combinatorial optimization problems, this rule is difficult to employ, because it is difficult to evaluate $\max_{ij} \Delta C_{ij}$ a priori. The choice of T_0 in this case has to be obtained from an experimental procedure, carried out before the process of optimization itself. During such a procedure, one calculates the evolution of the system during a limited time; one acquires some knowledge about the configuration space, from which one can determine T_0. This preliminary experiment can consist simply in calculating the average value of the variation in energy ΔC_{ij}, with the temperature maintained at zero. Aarts and Van Laarhoven [1] proposed a more sophisticated preliminary procedure: they established an iterative formula which makes it possible to adjust the value of T_0 after each perturbation so that the rate of acceptance is maintained constant. These authors indicated that this algorithm led to good results if the values of the cost function for the various system configurations were distributed in a sufficiently uniform way.

2.8.2.2   Length of the Markov Chains (or Length of the Temperature Stages); Law of Decrease of Temperature

The length of the Markov chain, which determines the length of the temperature stages, and the law of decrease of the temperature, which affects the number of stages, are two parameters of the annealing program that are very closely dependent on each other and which are most critical from the point of view of the computing time involved. An initial approach to the problem is to seek the optimal solution by fixing the length M of the Markov chains so as to reach quasi-equilibrium, i.e. to approach equilibrium to within a short distance ϵ that is fixed a priori and is characterized by the vector of the static probability distribution $q(T)$. One obtains the following condition:

$$M > K \left(|R|^2 - 3|R| + 3 \right)$$

where K is a constant which depends on ϵ. In the majority of combinatorial optimization problems, the total number of configurations $|R|$ is an exponential function of the number N of variables of the system. Consequently, the above inequality leads to an exponential computing time, which has been confirmed by experimental observations in the case of a particular form of the traveling salesman problem (the cities considered occupy all the nodes of a plane square network, which makes it possible to easily calculate the exact value of the global optimum of the cost function: this a priori knowledge of the solution is very useful for analyzing the convergence of the algorithm). These experimental results also show that a considerable gain in CPU time is obtained if one is willing to deviate a little from the optimum. A deviation in the final result of only 2 % compared with the optimum makes it possible to decrease the exponential computing time to a cubic time in N.

This gave rise to the idea of performing the theoretical investigations again, seeking parameters of the annealing program that ensure a deviation from the true optimum, independently of the dimension of the problem considered. The starting postulate of the reasoning is as follows: for each homogeneous Markov chain generated during the process of optimization, the distribution of the states must be close to the static distribution (i.e., the Boltzmann distribution, if one adopts the Metropolis rule of acceptance). This situation can be implemented on the basis of a high temperature (for which one quickly reaches quasi-equilibrium, as indicated by property 3). Then it is necessary to choose the rate of decrease of the temperature such that the static distributions corresponding to two successive values of T are close together. In this way, after each change between temperature stages, the distribution of the states approaches the new static distribution quickly, so that the length of the successive chains can be kept small. Here one can see the strong interaction that exists between the length of the Markov chains and the rate of decrease of the temperature. Let T and T' be the temperatures of two unspecified successive stages and let α be the rate of decrease of the temperature $\left(T' = \alpha T < T \right)$. The condition to be satisfied can be written as

$$\left\| q(T) - q(T') \right\| < \epsilon$$

(ϵ is a positive small number).

This condition is equivalent to the following, which is easier to use:

$$\forall i \in I_R : \frac{1}{1+\delta} < \frac{q_i(T)}{q_i(T')} < 1+\delta$$

(δ is also a positive and small number, called the distance parameter). It can then be shown, with the help of some approximations, that the rate of decrease of the temperature can be written as

$$\alpha = \frac{1}{(1 + T \cdot \ln(1+\delta)/3 \cdot \sigma(T))} \tag{2.1}$$

where $\sigma(T)$ is the standard deviation of the values of the cost function for the states of the Markov chain at a temperature T.

Aarts and van Laarhoven recommend the following choice for the length of the Markov chains:

$$M = \max_{i \in I_R} |R_i| \tag{2.2}$$

where R_i is the subset of R comprising all the configurations that can be obtained in only one movement starting from the state i. The Markov chain formalism thus leads to an annealing program characterized by a constant length of the Markov chain and a variable rate of decrease of the temperature. This result, which is based on theory, differs from the usual empirical approach: in the latter case, one adopts a variable length of the temperature stages and a constant rate α of decrease of the temperature, typically ranging between 0.90 and 0.99. It is observed, however, that the parameter α is not very critical to achieving convergence of the algorithm, provided the temperature stages last long enough.

2.8.2.3   Program Termination Criterion

Quantitative information on the progress of the optimization process can be obtained from the *entropy*, which is a natural measurement of the order of the system. This is defined by the following expression:

$$S(T) = -\sum_{i=1}^{|R|} q_i(T) \cdot \ln(q_i(T))$$

It can be shown that $S(T)$ can be written in the following form:

$$S(T) = S(T_1) - \int_T^{T_1} \frac{\sigma^2(T')}{T'^3} dT'$$

and $\sigma^2(T)$ can easily be estimated numerically using the values of the cost function for the configurations obtained at the temperature T. A termination criterion can then be formulated starting from the following ratio, which measures the difference between the current configuration and the optimal configuration:

$$\frac{S(T) - S_0}{S_\infty - S_0}$$

where S_∞ and S_0 are defined by the relations

$$S_\infty = \lim_{T \to \infty} S(T) = \ln |R|$$
$$S_0 = \lim_{T \to 0} S(T) = \ln |R_0|$$

One can also detect a disorder–order transition (and consequently decide to slow down the cooling) by observing any steep increase in the following parameter, which is similar to the *specific heat*: $\sigma^2(T)/T^2$. If one wishes to perform precise numerical calculations, these criteria are applicable in practice only when the Markov chains are of sufficient length. If this is not the case, another termination criterion can be obtained starting from extrapolation to zero temperature of the smoothed average $C_l(T)$ of the values of the cost function obtained during the process of optimization:

$$\left| \frac{dC_l(T)}{dT} \cdot \frac{T}{C(T_0)} \right| < \epsilon_s \tag{2.3}$$

where ϵ_s is a positive small number, and $C(T_0)$ is the average value of the cost function at the initial temperature T_0.

Remark If one adopts the rate of decrease of the temperature and the termination criterion defined by the relations (2.1) and (2.3), respectively, Aarts and Van Laarhoven showed the existence of an upper limit, proportional to $\ln |R|$, for the total number of temperature stages. Moreover, if the length of the Markov chains is fixed in accordance with the relation (2.2), the execution time of the annealing algorithm is proportional to the following expression:

$$\max_{i \in I_R} |R_i| \cdot \ln |R|$$

But the term $\max |R_i|$ is generally a polynomial function of the number of variables of the problem. Consequently, an annealing program defined by the relations (2.1)–(2.3) allows one to solve the majority of the NP-hard problems while obtaining, in polynomial time, a result which varies by only a few percent from the global optimum, and this is true regardless of the dimension of the problem considered. The above theoretical considerations have been confirmed by the application of this annealing program to the traveling salesman and logical partitioning problems.

2.8.3   Modeling of the Simulated Annealing Algorithm by Inhomogeneous Markov Chains

The results which we have presented up to now are based on the assumption of a decrease of the temperature in stages (which ensures fast convergence of the simulated annealing algorithm, as we have already mentioned). This property makes it possible to represent the process of optimization in the form of a complete set of homogeneous Markov chains, whose asymptotic behavior can be described simply. We have seen that this results in a complete theoretical explanation of the operation of the algorithm, and the development of usable annealing program. Some authors have been interested in the convergence of the simulated annealing algorithm within the more general framework of the theory of inhomogeneous Markov chains. In this case, the study of the asymptotic behavior is more delicate: for example, Gidas [15] showed the possibility of the appearance of phenomena similar to phase transitions. We will be satisfied here with highlighting the main result of this work, of primarily theoretical interest: the annealing algorithm converges towards a global optimum with a probability equal to unity if, as the time t tends towards infinity, the temperature $T(t)$ does not decrease more quickly than the expression $C/\ln(t)$, where C denotes a constant that is related to the depth of the "energy well" of the problem.

2.9   Annotated Bibliography

Reference [42]   This book describes the principal theoretical approaches to simulated annealing and the applications of the method in the early years of its development (1982–1988), when the majority of the theoretical basis was established.

Reference [31]   The principal metaheuristics are described in great detail in this book. An elaborate presentation of simulated annealing is given in Chap. 3. Some applications are presented, in particular, the design of electronic circuits and the treatment of scheduling problems.

Reference [36]   In this book several metaheuristics are extensively described, including simulated annealing (in Chap. 3). The theoretical elements relating to the convergence of the method are clearly presented in detail. The book includes also a study of an application in an industrial context (that of the TimberWolf software package, in connection with the layout-routing problem). This is an invaluable resource for those undertaking academic study of the subject. Each chapter is supplemented with suitable exercises.

Reference [28]   The principal metaheuristics are also described in this book. Chapter 5 is completely devoted to simulated annealing and concludes with an application in the field of industrial production.

Reference [46]   This book is a collection of contributions from a dozen authors. Simulated annealing is not treated in detail, however.

References

1. Aarts, E.H.L., Van Laarhoven, P.J.M.: Statistical cooling: a general approach to combinatorial optimisation problems. Philips Journal of Research **40**, 193–226 (1985)
2. Aarts, E.H.L., De Bont, F.M.J., Habers J.H.A., Van Laarhoven, P.J.M.: A parallel statistical cooling algorithm. In: *Proceeding of the 3rd Annual Symposium on Theoretical Aspects of Computer Science*, Lecture Notes in Computer Science, vol. 210, pp. 87–97 (1986)
3. Azencott, R.: *Simulated Annealing: Parallelization Techniques*. Wiley-Interscience Series in Discrete Mathematics. John Wiley and Sons (1992)
4. Bonomi, E., Lutton, J.L.: The N-city travelling salesman problem, Statistical Mechanics and the Metropolis Algorithm. SIAM Review **26**(4), 551–568 (1984)
5. Brandimarte, P.: Neighbourhood search-based optimization algorithms for production scheduling: a survey. Computer-Integrated Manufacturing Systems **5**(2), 167–176 (1992)
6. Casotto, A., Romea, F., Sangiovanni-Vincentelli, A.: A parallel simulated annealing algorithm for the placement of macro-cells. IEEE Transactions on C.A.D. **CAD-6**(5), 838–847 (1987)
7. Cerny, V.: Thermodynamical approach to the traveling salesman problem: an efficient simulation algorithm. Journal of Optimization Theory and Applications **45**(1), 41–51 (1985)
8. Collins, N.E., Eglese, R.W., Golden, B.: Simulated annealing—An annotated bibliography. American Journal of Mathematical and Management Sciences **8**, 209–307 (1988)
9. Courat, J., Raynaud, G., Mrad, I., Siarry, P.: Electronic component model minimisation based on Log Simulated Annealing. IEEE Transactions on Circuits and Systems I **41**(12), 790–795 (1994).
10. Courat, J., Raynaud, G., Siarry, P.: Extraction of the topology of equivalent circuits based on parameter statistical evolution driven by Simulated Annealing. International Journal of Electronics **79**, 47–52 (1995)
11. Delamarre, D., Virot, B.: Simulated annealing algorithm: technical improvements. Operations Research **32**(1), 43–73 (1998)
12. Durand, M., White, S.: Permissible error in parallel simulated annealing. Technical report, Institut de Recherche en Informatique et Systèmes aléatoires, Rennes (1991)
13. Fleury, G.: Application de méthodes stochastiques inspirées du recuit simulé à des problèmes d'ordonnancement. RAIRO A.P.I.I. (Automatique—Productique—Informatique industrielle) **29**(4–5), 445–470 (1995)
14. Geman, S., Geman, D.: Stochastic relaxation, Gibbs distributions and the Bayesian restoration of images. IEEE Transactions on Pattern Analysis and Machine Intelligence **PAMI-6**, 721–741 (1984)
15. Gidas, B.: Nonstationary Markov chains and convergence of the Annealing Algorithm. Journal of Statiscal Physics **39**, 73–131 (1985)
16. Hajek, B.: Cooling schedules for optimal annealing. Mathematics of Operations Research **13**, 311–329 (1988)
17. Hajek, B., Sasaki, G.: Simulated annealing—to cool or not. Systems and Control Letters **12**, 443–447 (1989)
18. Jeffcoat, D., Bulfin, R.: Simulated annealing for resource-constrained scheduling. European Journal of Operational Research **70**(1), 43–51 (1993)
19. Johnson, D., Aragon, C., McGeoch, L., Schevon, C.: Optimization by simulated annealing: an experimental evaluation—Part I (Graph partitioning). Operational Research **37**(6), 865–892 (1989)

20. Johnson, D., Aragon, C., McGeoch, L., Schevon, C.: Optimization by simulated annealing: an experimental evaluation—Part II (Graph coloring and number partitioning). Operational Research **39**(3), 378–406 (1991)
21. Johnson, D., Aragon, C., McGeoch, L., Schevon, C.: Optimization by simulated annealing: an experimental evaluation—Part III (The travelling salesman problem). Operational Research (1992)
22. Kirkpatrick, S., Toulouse, G.: Configuration space analysis of travelling salesman problems. Journal de Physique **46**, 1277–1292 (1985)
23. Kirkpatrick, S., Gelatt, C., Vecchi, M.: Optimization by simulated annealing. Science **220**(4598), 671–680 (1983)
24. Kravitz, S., Rutenbar, R.: Placement by simulated annealing on a multiprocessor. IEEE Transactions on Computer Aided Design **CAD-6**, 534–549 (1987)
25. Metropolis, N., Rosenbluth, A., Rosenbluth, M., Teller, A., Teller, E.: Equation of state calculations by fast computing machines. Journal of Chemistry Physics **21**, 1087–1090 (1953)
26. Miclo, L.: Évolution de l'énergie libre. Applications à l'étude de la convergence des algorithmes du recuit simulé. Ph.D. thesis, Université de Paris 6 (1991)
27. Musser, K., Dhingra, J., Blankenship, G.: Optimization based job shop scheduling. IEEE Transactions on Automatic Control **38**(5), 808–813 (1993)
28. Pham, D., Karaboga, D.: *Intelligent Optimisation Techniques*. Genetic Algorithms, Tabu Search, Simulated Annealing and Neural Networks. Springer (2000)
29. Pirlot, M.: General local search heuristics in Combinatorial Optimization: a tutorial. Belgian Journal of Operations Research and Computer Science **32**(1–2), 7–67 (1992)
30. Rammal, R., Toulouse, G., Virasoro, M.: Ultrametricity for physicists. Reviews of Modern Physics **58**(3), 765–788 (1986)
31. Reeves, C.: *Modern Heuristic Techniques for Combinatorial Problems*. Advanced Topics in Computer Science Series. McGraw-Hill Ryerson (1995)
32. Rinnooy Kan, A., Timmer, G.: Stochastic global optimization methods—Part I: Clustering methods. Mathematical Programming **39**, 27–56 (1987)
33. Rinnooy Kan, A., Timmer, G.: Stochastic global optimization methods—part II: Multi level methods. Mathematical Programming **39**, 57–78 (1987)
34. Roussel-Ragot, P.: La méthode du recuit simulé: accélération et parallélisation. Ph.D. thesis, Université de Paris 6 (1990)
35. Roussel-Ragot, P., Siarry, P., Dreyfus, G.: La méthode du "recuit simulé" en électronique: principe et parallélisation. In: 2e *Colloque National sur la Conception de Circuits à la Demande*, session G, Article G2, pp. 1–10. Grenoble (1986)
36. Saït, S., Youssef, H.: *Iterative computer Algorithms with Applications in Engineering*. IEEE Computer Society Press (1999)
37. Sechen, C.: *VLSI Placement and Global Routing Using Simulated Annealing*. Kluwer Academic ers (1988)
38. Siarry, P.: La méthode du recuit simulé: application à la conception de circuits électroniques. Ph.D. thesis, Université de Paris 6 (1986)
39. Siarry, P.: La méthode du recuit simulé en électronique. Adaptation et accélération. Comparaison avec d'autres méthodes d'optimisation. Application dans d'autres domaines. Habilitation à diriger les recherches en sciences physiques, Université de Paris-Sud (Orsay) (1994)
40. Siarry, P.: La méthode du recuit simulé: théorie et applications. RAIRO A.P.I.I. (Automatique—Productique—Informatique industrielle) **29**(4–5), 535–561 (1995)
41. Siarry, P., Dreyfus, G.: An application of physical methods to the computer aided design of electronic circuits. Journal de Physique Lettres **45**, L39–L48 (1984)
42. Siarry, P., Dreyfus, G.: La méthode du recuit simulé: théorie et applications. IDSET, ESPCI, Paris (1989)
43. Siarry, P., Bergonzi, L., Dreyfus, G.: Thermodynamic optimization of block placement. IEEE Transactions on Computer Aided Design **CAD-6**(2), 211–221 (1987)
44. Solla, S., Sorkin, G., White, S.: Configuration space analysis for optimization problems. In: E. Bienenstock (ed.) *Disordered Systems and Biological Organization*, pp. 283–292. Springer, New York (1986)

45. Sorkin, G.B.: Efficient simulated annealing on fractal energy landscapes. Algorithmica **6**, 367–418 (1991)
46. Teghem, J., Pirlot, M.: *Optimisation approchée en recherche opérationnelle: recherches locales, réseaux neuronaux et satisfaction de contraintes.* IC2: information, commande, communication. Hermès Science (2002)
47. Van Laarhoven, P., Aarts, E.: *Simulated annealing: theory and applications.* Reidel, Dordrecht (1987)
48. Van Laarhoven, P., Aarts, E., Lenstra, J.: Job-shop scheduling by simulated annealing. Operational Research **40**, 113–125 (1992)
49. Vecchi, M., Kirkpatrick, S.: Global wiring by simulated annealing. IEEE Transactions on C.A.D. **CAD-2**(4), 215–222 (1983)
50. Wong, D., Leong, H., Liu, C.: *Simulated Annealing for VLSI Design.* Kluwer Academic (1988)

Chapter 3
Tabu Search

Eric Taillard

3.1 Introduction

Tabu search was first proposed by Fred Glover in an article published in 1986 [3], although it borrowed many ideas suggested before during the 1960s. The two articles entitled simply "Tabu search" [4, 5] proposed most of tabu search principles which are currently used. Some of these principles did not gain prominence among the scientific community for a long time. Indeed, in the first half of the 1990s, the majority of the research publications on tabu search used a very restricted range of the principles of the technique. They were often limited to a *tabu list* and an elementary *aspiration condition*.

The popularity of tabu search is certainly due to the contribution of de Werra's team at the Federal Polytechnic School of Lausanne during the late 1980s. In fact, the articles by Glover, the founder of the method, were not well understood at the time, when there was not yet a "culture" of metaheuristics-based algorithms. Hence the credit for the popularization of the basic technique must go to [8, 10], which surely played a significant role in the dissemination of the algorithm.

At the same time, competition developed between simulated annealing (which then had an established convergence theorem as its theoretical advantage) and tabu search. For many applications, tabu-search-based heuristics definitely showed more effective results [12–15], which increased the interest in the method among some researchers.

At the beginning of the 1990s, the technique was extensively explored in Canada, particularly in the Center for Research on Transportation in Montreal, where several postdoctoral researchers from de Werra's team worked in this field. This created another focus in the field of tabu search. The technique was then quickly disseminated

E. Taillard (✉)
HEIG-VD, 1401 Route de Cheseaux 1, Cp, Yverdon-les-bains, Switzerland
e-mail: eric.taillard@heig-vd.ch

© Springer International Publishing Switzerland 2016
P. Siarry (ed.), *Metaheuristics*, DOI 10.1007/978-3-319-45403-0_3

among several research communities, and this culminated in the publishing of the first book which was solely dedicated to tabu search [7].

In this chapter, we shall not deal with all of the principles of tabu search presented in the book by Fred Glover and Manuel Laguna [6], but instead we shall focus on the most significant and most general principles.

What unquestionably distinguishes it from the local search technique presented in the preceding chapter is that tabu search incorporates intelligence. Indeed, there is a huge temptation to guide an iterative search in a good, promising, direction, so that it is not guided merely by chance and the value of an objective function to be optimized. Implementing a tabu search gives rise to a couple of challenges: first, as in any iterative search, it is necessary that the search engine, i.e., the mechanism for evaluating neighboring solutions, is effective; and second, pieces of knowledge regarding the problem under consideration should be transmitted to the search procedure so that it will not get trapped in bad regions of the solution space. On the contrary, it should be guided intelligently in the solution space, if such a term is permitted to be used.

Glover proposed a number of learning techniques that can be embedded in a local search. One of the guiding principles is to construct a history of the iterative search or, equivalently, to equip the search with memory.

- *Short-term memory.* The name "tabu search" is a reference to the use of a short-term memory that is embedded in a local search. The idea is to memorize in a data structure T the elements that the local search is prohibited from using. This structure is called a *tabu list*. In its simplest form, a tabu search scans the whole set of neighboring solutions in each iteration and chooses the best that is not forbidden, even if it is worse than the current solution. To prevent the search being blocked or being forced to visit only solutions of bad quality owing to tabu conditions, the number of elements in T is limited. Since the number of prohibitions contained in T is limited—this number is frequently called the *tabu list size*—this mechanism implements a short-term memory.
- *Long-term memory.* A tabu list does not necessarily prevent a cycling phenomenon, that is, visiting a subset of solutions cyclically. If the tabu duration is long enough to avoid a cycling phenomenon, the search may be forced to visit only bad solutions. To avoid both of these complementary phenomena, another kind of memory, operating over a longer term, must be used.
- *Diversification.* A technique for avoiding cycling, which is the basis of variable neighborhood search, is to perform jumps in the solution space. But, in contrast to variable neighborhood search, which performs random jumps, tabu search uses a long-term memory for these jumps, for instance by forcing the use of solution modifications that have not been tried for a large number of iterations. Another diversification technique is to change the modeling of the problem, for instance by accepting nonfeasible solutions but assigning them a penalty.
- *Intensification.* When an interesting solution is identified, an idea is to tentatively examine more deeply the solution space in its neighborhood. Many intensification techniques are used. The simplest one is to come back to the best solution found so far and to change the search parameters, for instance by limiting the tabu duration,

by using a larger neighborhood, or by using a more constrainted model of the problem.

- *Strategic oscillations.* For tackling particularly difficult problems, it is convenient to alternate phases of diversification and intensification. So, the search oscillates between phases where the solution structure is greatly modified and phases where better solutions are built again. This strategy is the origin of other metaheuristics that were proposed later, such as variable neighborhood search, large neighborhood search, and iterated local search.

Some of these principles of tabu search will be illustrated with the help of a particular problem, namely the quadratic assignment problem, so that these principles does not stay in the clouds. We chose this problem for several reasons. First of all, it has applications in multiple fields. For example, the problem of placing electronic modules, which we discussed in Chap. 1 devoted to simulated annealing, is a quadratic assignment problem. In this case, its formulation is very simple, because it deals with finding a permutation. Here, it should be noted that many combinatorial optimization problems can be expressed in the form of searching for a permutation.

3.2   The Quadratic Assignment Problem

Given n objects and a set of flows f_{ij} between objects i and j ($i, j = 1, \ldots, n$), and given n locations with distance d_{rs} between the locations r and s ($r, s = 1, \ldots, n$), the problem deals with placing the n objects on the n locations so as to minimize the sum of the products flows $\times$ distance. Mathematically, this is equivalent to finding a permutation $\mathbf{p}$, whose ith component p_i denotes the position of the object i, which minimizes $\sum_{i=1}^{n} \sum_{j=1}^{n} f_{ij} \cdot d_{p_i p_j}$.

This problem has multiple practical applications; among them, the most popular ones are the assignment of offices or services in a building (e.g., a university campus or hospital), the assignment of departure gates to aircraft at an airport, the placement of logical modules in FPGA (field-programmable gate array) circuits, the distribution of files in a database, and the placement of the keys on typewriter keyboards. In these examples, the flow matrix represents the frequency with which people may move from one office to another, the number of people who may transit from one aircraft to another, the number of electrical connections to be made between two modules, the probability of requesting the access to a second file if one is accessing the first one, and the frequency with which two particular characters appear consecutively in a given language, respectively. The distance matrix has an obvious meaning in the first three examples; in the fourth, it represents the transmission time between databases and, in the fifth, it represents the time separating the striking of two keys.

The quadratic assignment problem is NP-hard. One can easily show this by noting that the traveling salesman problem can be formulated as a quadratic assignment problem. Unless P $=$ NP, there is no polynomial approximation scheme for this problem. This can be shown simply by considering two problem instances which

Table 3.1 Number of connections between modules in the SCR12 problem

Module	1	2	3	4	5	6	7	8	9	10	11	12
1	—	180	120	—	—	—	—	—	—	104	112	—
2	180	—	96	2445	78	—	1395	—	120	135	—	—
3	120	96	—	—	—	221	—	—	315	390	—	—
4	—	2445	—	—	108	570	750	—	234	—	—	140
5	—	78	—	108	—	—	225	135	—	156	—	—
6	—	—	221	570	—	—	615	—	—	—	—	45
7	—	1395	—	750	225	615	—	2400	—	187	—	—
8	—	—	—	—	135	—	2400	—	—	—	—	—
9	—	120	315	234	—	—	—	—	—	—	—	—
10	104	135	390	—	156	—	187	—	—	—	36	1200
11	112	—	—	—	—	—	—	—	—	36	—	225
12	—	—	—	140	—	45	—	—	—	1200	225	—

differ only in the flow matrix. If an appropriate constant is removed from all the flow components of the first problem to obtain the second, the last have an optimum solution value of zero. Consequently, all ϵ-approximations to the second problem has an optimum solution, which is possible to implement in polynomial time only if P = NP. However, problems generated at random (with flows and distances drawn uniformly) satisfy the following property: as $n \to \infty$, the value of any solution (even the worst one) tends towards the value of an optimal solution [1].

3.2.1 Example

Let us consider the placement of 12 electronic modules $(1, \ldots, 12)$ on 12 sites $(a, b, \ldots, l)$. The number of wires connecting any pair of modules is known, and is given in Table 3.1. This problem instance is referred to as SCR12 in the literature.

The sites are distributed on a 3×4 rectangle. Connections can be implemented only horizontally or vertically, implying wiring lengths measured with Manhattan distances. In the solution of the problem represented in Fig. 3.1, which is optimal, module 6 was placed on site a, module 4 on site b, etc.

3.3 Basic Tabu Search

From here onwards and without being restrictive, we can make the assumption that the problem to be solved can be formulated in the following manner:

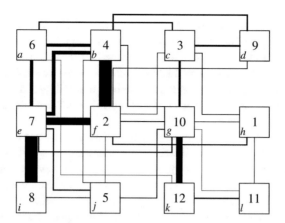

Fig. 3.1 Optimal solution of a problem of connection between electronic modules. The thickness of the lines is proportional to the number of connections

Fig. 3.2 A possibility for creating a neighboring solution in a permutation problem

$$\min_{s \in S} f(s)$$

In this formulation, f denotes the objective function, s a feasible solution of the problem, and S the entire collection of feasible solutions.

3.3.1 Neighborhood

Tabu search is primarily centered on a nontrivial exploration of all the solutions by using the concept of a neighborhood. Formally, one can define, for any solution s of S, a set $N(s) \subset S$ that is a collection of the neighboring solutions of s. For instance, for the quadratic assignment problem, s can be a permutation of n objects and the set $N(s)$ can be the possible solutions obtained by exchanging two objects in a permutation. Figure 3.2 illustrates one of the moves of the set $N(s)$, where objects 3 and 7 are exchanged.

Local search methods are almost as old as the world itself. As a matter of fact, how does a human being behave when they are seeking a solution to a problem for which they cannot find a solution or if they do not have enough patience to find an optimal solution? The person may try to slightly modify the proposed solution and may check whether it is possible to find better solutions by carrying out such local changes. In other words, they will stop as soon as they meet a *local optimum* relative to the modifications to a solution that are allowed. In this process, nothing proves that the solution thus obtained is a *global optimum*—and, in practice, this is seldom the case. In order to find solutions better than the first local optimum met with, one can try to continue the process of local modifications. However, if precautions are not taken, one risks visiting a restricted number of solutions, in a cyclic manner. Simulated annealing and tabu search are two local search techniques which try to eliminate this disadvantage.

Some of these methods, such as, simulated annealing, have been classified as artificial intelligence techniques. However, this classification is certainly incorrect, as they are guided almost exclusively by chance—some authors even compare simulated annealing to the wandering of a person suffering from amnesia moving in fog. Perhaps others describe these methods as intelligent because, often after a large number of iterations during which they have generated several poor-quality solutions, they produce a good-quality solution which would otherwise have required a very large human effort.

In essence, tabu search is not centered on chance, although one can introduce random components for primarily technical reasons. The basic idea of tabu search is to make use of memories during the exploration of some part of the solutions to the problem, which consists in moving repeatedly from one solution to a neighboring solution. It is thus primarily a local search, if we look beyond the limited meaning of this term and dig for a broader meaning. However, some principles that enable one to carry out jumps in the solution space have been proposed; in this respect, tabu search, in contrast to simulated annealing, is not a pure local search.

3.3.2 Moves and Neighborhoods

Local searches are based on the definition of a set $N(s)$ of solutions in the neighborhood of s. But, from a practical point of view, it may be easier to consider the set M of modifications that can be applied to s, rather than the set $N(s)$. A modification made to a solution can be called a *move*. Thus, a modification of a solution of the quadratic assignment problem (see the Fig. 3.2) can be considered as a move characterized by two elements to be transposed in a permutation. Figure 3.3 gives the neighborhood structure for the set of permutations of four elements. This is presented in graphical form, where the nodes represent the solutions and the edges represent the neighbors relative to transpositions.

The set $N(s)$ of solutions in the neighborhood of s can be expressed as the set of feasible solutions that can be obtained by applying a move m to solution s, where m belongs to a set of moves M. The application of m to s can be denoted as $s \oplus m$ and one has the equivalent definition $N(s) = \{s' \mid = s \oplus m, m \in M\}$. When it is possible, expressing the neighborhood in terms of moves facilitates the characterization of the set M. Thus, in the above example of modification of a permutation, M can be characterized by all of the pairs (place 1, place 2) in which the elements are transposed, independently from the current solution. Note that in the case of permutations with

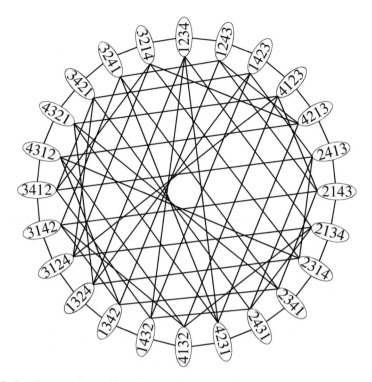

Fig. 3.3 Set of permutations of four elements (represented by nodes) with neighborhood relations relative to transpositions (represented by edges)

a transposition neighborhood, $|S| = n!$ and $|M| = n \cdot (n-1)/2$. Thus, the solution set is much larger than the set of moves, which grows as the square of the number of elements.

However, in some applications this simplification can lead to the definition of moves which would produce unacceptable solutions and, in general, we have $|N(s)| \leq |M|$, without $|M|$ being much larger than $|N(s)|$. For a given problem with few constraints, it is typically the case that $|N(s)| = |M|$.

3.3.2.1 Examples of Neighborhoods for Problems on Permutations

Many combinatorial optimization problems can be formulated naturally as a search for a permutation of n elements. Assignment problems (which include the quadratic assignment problem), and the traveling salesman and scheduling problems are representative examples of such problems. For these problems, several definitions of neighboring solutions are possible; some examples are illustrated in Fig. 3.4. Among the simplest neighborhoods, one can find the inversion of two elements placed

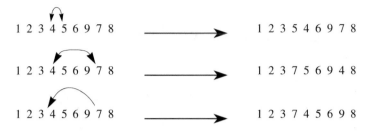

Fig. 3.4 Three possible neighborhoods with respect to permutations (inversion, transposition, displacement)

successively in the permutation, the transposition of two distinct elements, and, finally, the movement of an element to another place in the permutation. Depending on the problem considered, more elaborate neighborhoods that suit the structure of good solutions may be considered. This is typically the case for the traveling salesman problem, where there are innumerable neighborhoods suggested that do not represent simple operations if a solution is regarded as a permutation.

The first type of neighborhood shown in the example in Fig. 3.4 is the most limited one as it is of size $n - 1$. The second type defines a neighborhood with, $n \cdot (n-1)/2$ moves, and the third is of size $n(n - 2) + 1$. The abilities of these various types of neighborhoods to guide a search in a few iterations towards good solutions are very different; generally, the first type shows the worst behavior for many problems since it is a subset of the others. The second type can be better than the third for some problems (such as the quadratic assignment problem), whereas, for scheduling applications, the third type often shows better performance [12].

3.3.3   Neighborhood Evaluation

In order to implement an effective local search engine, it is necessary that the ratio between the quality or suitability of the moves and the computational resources required for their evaluation is as high as possible. If the quality of a type of move can be justified only by intuition and in an empirical manner, the evaluation of the neighborhood can, on the other hand, often be accelerated considerably by algebraic considerations. Let us define $\Delta(s, m) = f(s \oplus m) - f(s)$. In many cases it is possible to simplify the expression $f(s \oplus m) - f(s)$ and thus to evaluate $\Delta(s, m)$ quickly. An analogy can be drawn with continuous optimization: the numerical evaluation of $f(s \oplus m) - f(s)$ would be the equivalent of a numerical evaluation of the gradient, whereas the calculation of the simplified function $\Delta(s, m)$ would be the equivalent of the evaluation of the gradient by means of a function implemented using the algebraic expressions for the partial derivatives.

Moreover, if a move m' was applied to solution s in the previous iteration, it is often possible to evaluate $\Delta(s \oplus m', m)$ for the current iteration as a function of

$\Delta(s, m)$ (which was evaluated in the previous iteration) and to evaluate the entire neighborhood very rapidly, simply by memorizing the values of $\Delta(s, m)$, $\forall m \in M$.

It may appear that $\Delta(s, m)$ would be very difficult and expensive to evaluate. For instance, for vehicle routing problems (see Sect. 13.1), a solution s can consist of partitioning customers demands into subsets whose weights are not more than the capacities of the vehicles. To evaluate $f(s)$, we have to find an optimal order in which one will deliver to the customers for each subset, which is a difficult problem in itself. This is the well-known traveling salesman problem. Therefore, the evaluation of $f(s)$, and consequently that of $\Delta(s, m)$ cannot reasonably be performed for every eligible move (i.e., all moves belonging to M); possibly $\Delta(s, m)$ would need to be calculated for each move selected (and in fact carried out), but, in practice, $f(s)$ is evaluated exactly for a limited number of solutions only. Hence the computational complexity is limited by estimating $\Delta(s, m)$ in an approximate manner.

3.3.3.1 Algebraic Simplification for the Quadratic Assignment Problem

As any permutation is an acceptable solution for the quadratic assignment problem, its modeling is also trivial. For the choices of neighborhood, it should be realized that moving the element into the ith position in the permutation to put it into the jth position implies a very significant modification of the solution. This is because all of the elements between the ith and the jth position are moved. The inversion of the objects in the ith and the $(i + 1)$th position in the permutation generates too limited a neighborhood. In fact, if the objective is to limit ourselves to the neighborhoods in which the sites assigned to two elements only are modified, it is only reasonable to transpose the elements i and j occupying the sites p_i and p_j. Each of these moves can be evaluated in $O(n)$ (where n is the problem size). With a flow matrix $\mathcal{F} = (f_{ij})$ and a distance matrix $\mathcal{D} = (d_{rs})$, the value of move $m = (i, j)$ for a solution $\mathbf{p}$ is given by

$$\Delta(\mathbf{p}, (i, j)) = (f_{ii} - f_{jj})(d_{p_j p_j} - d_{p_i p_i}) + (f_{ij} - f_{ji})(d_{p_j p_i} - d_{p_i p_j}) \\ + \sum_{k \neq i, j} (f_{jk} - f_{ik})(d_{p_i p_k} - d_{p_j p_k}) + (f_{kj} - f_{ki})(d_{p_k p_i} - d_{p_k p_j})$$
$$(3.1)$$

If solution $\mathbf{p}$ was modified into solution $\mathbf{q}$ by exchanging the objects r and s, i.e., $q_k = p_k$ ($k \neq r, k \neq s$), $q_r = p_s$, $q_s = p_r$ in an iteration, it is possible to evaluate $\Delta(\mathbf{q}, (i, j))$ in $O(1)$ in the next iteration by memorizing the value $\Delta(\mathbf{p}, (i, j))$ of the move (i, j) that was discarded:

$$\Delta(\mathbf{q}, (i, j)) = \Delta(\mathbf{p}, (i, j)) \\ + (f_{ri} - f_{rj} + f_{sj} - f_{si})(d_{q_s q_i} - d_{q_s q_j} + d_{q_r q_j} - d_{q_r q_i}) \\ + (f_{ir} - f_{jr} + f_{js} - f_{is})(d_{q_i q_s} - d_{q_j q_s} + d_{q_j q_r} - d_{q_i q_r})$$
$$(3.2)$$

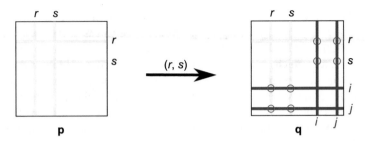

Fig. 3.5 *Left* in *light color*, the elements for which it is necessary to recalculate the scalar product of the matrices to evaluate the move (r, s) applied to **p** (giving the solution **q**). *Right* the *circled* elements are those for which it is necessary to recalculate the product to evaluate the move (i, j) applied to **q**, compared with those which would have been calculated if the move (i, j) had been applied to **p**

Figure 3.5 illustrates the modifications of $\Delta(\mathbf{p}, (i, j))$ that are necessary to obtain $\Delta(\mathbf{q}, (i, j))$ if the move selected for going from **p** to **q** is (r, s). It should be noted here that the quadratic assignment problem can be regarded as the problem of the permutation of the rows and columns of the distance matrix so that the "scalar" product of the two matrices is as small as possible.

Consequently, by memorizing the values of $\Delta(\mathbf{p}, (i, j))$ for all i and j, the complete neighborhood can be evaluated in $O(n^2)$: by using Eq. (3.2), one can evaluate the $O(n^2)$ moves that do not involve the indices r and s, and, by using Eq. 3.1, one can evaluate the $O(n)$ moves which involve precisely these indices.

3.3.4 Neighborhood Limitation: Candidate List

Generally, a local search does not necessarily evaluate all the solutions in $N(s)$ in each iteration, but only a subset. In fact, simulated annealing only evaluate a single neighbor in each iteration. Conversely, tabu search is supposed to make an "intelligent" choice of a solution from $N(s)$. A possible way to acclerate the evaluation of the neighborhood is to reduce its size; this reduction can also have the other goal of guiding the search.

To reduce the number of eligible solutions in $N(s)$, some authors adopt the strategy of randomly selecting from $N(s)$ a number of solutions which is much smaller than $|N(s)|$. If the neighborhood is given by a static collection M of moves, one can also consider partitioning M into subsets; in each iteration, only one of these subsets is examined. In this manner, one can use a partial but cyclic evaluation of the neighborhood, which allows one to choose a move more quickly. This implies a deterioration in quality, since not all moves are not taken into consideration in each iteration. However, at a global level, this limitation might not have too bad an influence on the quality of the solutions produced, because a partial examination can generate a certain diversity in the visited solutions, precisely because the moves which were chosen were not those which would have been chosen, if a complete examination of the neighborhood had been carried out.

Finally, in accordance with Glover's intuition when he proposed the concept of a candidate list, one can make the assumption that a good-quality move for a solution will remain good for solutions that are not too different. Practically, this can be implemented by ordering the entire set of all feasible moves by decreasing quality, in a given iteration. During the later iterations, only those moves that have been classified among the best will be considered. This is implemented in the form of a data structure called a *candidate list*. Naturally, the order of the moves will become degraded during the search, since the solutions become increasingly different from the solution used to build the list, and it is therefore necessary to periodically evaluate the entire neighborhood to preserve a suitable candidate list.

However, for some problems, a static candidate list can be used. For instance, a frequently used technique for speeding up the evaluation of the neighborhood for Euclidean traveling salesman and vehicle routing problems is to consider, for each customer, only the x closest customers. Typically, x is limited to a few dozen. So, the size of the neighborhood grows linearly with the problem size. One form of tabu search exploiting this principle is called *granular tabu search* [17].

3.3.5 Neighborhood Extension: Ejection Chains

An *ejection chain* is a technique for creating potentially interesting neighborhoods by performing a substantial modification of a solution in a compound move. The idea is to remove (eject) an element from a solution and to insert it somewhere else, ejecting another element if necessary. This is repeated until either a suitable solution is found or no suitable ejection can be performed. This process implies a need to manage solutions that are not feasible, called *reference structures* by Glover.

3.3.5.1 Lin and Kernighan Neighborhood for the Traveling Salesman Problem

The best-known ejection chain technique is certainly that of Lin and Kernighan [11] for the traveling salesman problem. The idea is as follows: An edge is removed from a valid tour to obtain a chain (a nonoriented path). One of the extremities of the chain is connected to an internal vertex. The reference structure so obtained is made up of a cycle on a subset of vertices and a chain connected to this cycle. By removing an edge of the cycle adjacent to a node of degree 3 and by connecting both nodes of degree 1, a new feasible tour is obtained. Such a modification corresponds to the traditional 2-opt move.

An interesting exploitation of this reference structure is to transform it into another reference structure. After an edge adjacent to a node of degree 3 has been removed—one gets a chain connecting all vertices—one of the extremities of the chain can be connected to an internal node, creating another reference structure. To guide the ejection chain and determine when to stop, the following rules can be applied:

- The weight of the reference structure must be lower than that of the initial solution.
- Once an edge has been added during an ejection chain, this edge cannot be removed again.
- Once an edge has been removed during an ejection chain, it cannot be added again.
- The ejection chain stops as soon as it is not possible to modify the reference structure while maintaining a weight lower than that of the starting solution or when an improved solution has been found.

This process is illustrated in Fig. 3.6.

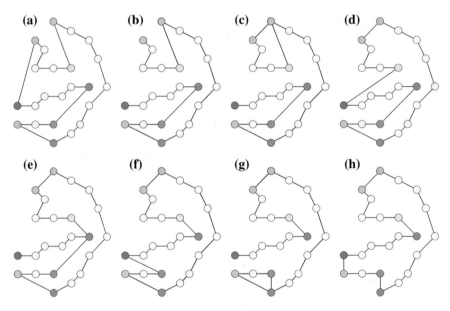

Fig. 3.6 Lin and Kernighan neighborhood for the traveling salesman problem. This neighborhood can be seen as an ejection chain. To start the chain, an edge is removed from the initial solution (**a**) to obtain a chain (**b**). This chain is then transformed into a reference structure (**c**) of weight lower than that of the initial solution by adding an edge. From the reference structure (**c**), it is possible to get either a new tour (**d**) or another reference structure (**e**) by replacing an edge by another one. The process can be propagated to construct solutions that are increasingly different from the initial solution. Solution (**d**) belongs to the 2-opt neighborhood of solution (**a**). Solution (**f**) belongs to the 3-opt neighborhood of (**a**) and solution (**h**) to its 4-opt neighborhood

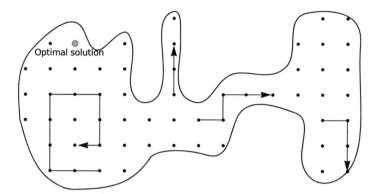

Fig. 3.7 Trajectories that block a search or disconnect it from the optimal solution in a strict tabu search

3.4 Short-Term Memory

When one wishes to make use of memory in an iterative process, the first idea that comes to mind is to check if a solution in the neighborhood has already been visited. However, the practical implementation of this idea can be difficult and, even worse, it may prove not to be very effective. It requires one to memorize every solution visited and, in each iteration, for every eligible solution, to test whether that solution has already been enumerated. This could possibly be done efficiently by using hashing tables, but it is not possible to prevent a significant growth in the memory space requirement, as that increases linearly with the number of iterations. Moreover, the pure and simple prohibition of solutions can lead to absurd situations. Assume that the entire set of feasible solutions can be represented by points whose coordinates are given on a surface in the plane and that one can move from any feasible solution to any other by a number of displacements of unit length. In this case, one can easily find trajectories which disconnect the current solution from an optimal solution or which block an iterative search owing to a lack of feasible neighboring solutions if it is tabu to pass through an already visited solution. This situation is schematically illustrated in Fig. 3.7.

3.4.1 Hash Table

An initial idea for guiding an iterative search, which is very easy to implement, is to prohibit a return to a solution whose value has already been obtained during the last t iterations. Thus one can prevent a cycle of length t or less. This type of prohibition can be implemented in an effective manner: Let L be a integer, relatively large, such that it is possible to store a table of L entries in the main memory of the computer. If $f(s_k)$ is assumed to be the integer value of solution s_k in iteration k (this is not restrictive when one is working with a computer), one can memorize the value $k + t$ in $T[f(s_k)$ modulo $L]$. If a solution s' in the neighborhood of the solution in iteration k' is such that $T[f(s')$ modulo $L)] > k'$, s' is not considered anymore as an eligible solution. This effective method of storing the tabu solutions only approximates the initial intention, which was to prevent a return to a solution of a given value, as not only any solution of the given value is prohibited during t iterations but also all those which have this value modulo L. Nevertheless, only a very moderate modification of the search behavior can be observed in practice if L is selected to be sufficiently large. A benefit of this collateral effect is that it suppresses neutral moves (moves with null cost), which can trap a local search on a large plateau.

This type of tabu condition works only if the objective function has a vast span of values. However, there are many problems where the objective function has a limited span of values. One can circumvent this difficulty by associating with the objective function, and using in its place, another function that has a large span of values. In the case of a problem on permutations, one can associate, for example, the hashing function $\sum_{i=1}^{n} i^2 \cdot p_i$, which takes a number of $O(n^4)$ different values.

More generally, if a solution of a problem can be expressed in the form of a vector $\mathbf{x}$ of binary variables, one can associate the hashing function $\sum_{i=1}^{n} z_i \cdot x_i$ with z_i, a set of n numbers randomly generated at the beginning of the search [18].

When hashing functions are used for implementing tabu conditions, one needs to focus on three points. Firstly, as already mentioned, it is necessary that the function used has a large span of possible values. Secondly, the evaluation of the hashing function for a neighboring solution should not impose a significantly higher computational burden than the evaluation of the objective function. In the case of problems on permutations with a neighborhood based on transpositions, the functions mentioned above can be evaluated in constant time for each neighboring solution if the value of the hashing function for the starting solution is known. Thirdly, it should be noted that even with a very large hashing table, collisions (different solutions with identical hashing values) are frequent. Thus, for a problem on permutations of size $n = 100$, with the transposition neighborhood, approximately five solutions in the neighborhood of the solution in the second iteration will have a collision with the starting solution, if a table of 10^6 elements is used. One technique to reduce the risk of collisions effectively, is to use several hashing functions and several tables simultaneously [16].

3.4.2 Tabu List

As it can be ineffective to restrict the neighborhood $N(s)$ to those solutions which have not yet been visited, tabu conditions are instead based on M, the set of moves applicable to a solution. This set is often of relatively modest size (typically $O(n)$ or $O(n^2)$ if n is the size of the problem) and must have the characteristic of *connectivity*, i.e., an optimal solution can be reached from any feasible solution. Initially, to simplify our analysis, we assume that M also has the property of *reversibility*: for any move m applicable to a solution s, there is a move m^{-1} such that $(s \oplus m) \oplus m^{-1} = s$. As it does not make sense to apply m^{-1} immediately after applying m, it is possible, in all cases, to limit the moves applicable to $s \oplus m$ to those different from m^{-1}. Moreover, one can avoid visiting s and $s \oplus m$ repeatedly in the process if s is a local minimum of the function in the neighborhood selected and if the best neighbor of $s \oplus m$ is precisely s.

By generalizing this technique of limiting the neighborhood, i.e., by prohibiting for several iterations the reverse of a move which has just been made, one can prevent other cycles composed of a number of intermediate solutions. Once it again becomes possible to carry out the reverse of a move, one hopes that the solution has been sufficiently modified that it is improbable—but not impossible—to return to an already visited solution. Nevertheless, if such a situation arises, it is hoped that the tabu list would have changed, and therefore the future trajectory of the search would change. The number of tabu moves must remain sufficiently limited. Let us assume that M does not depend on the current solution. In this situation, it is reasonable to prohibit only a fraction of M. Thus, the tabu list implements a short-term memory, relating typically to a few or a few tens of iterations.

For easier understanding, we have assumed that the reverse moves of those that have been carried out are stored. However, it is not always possible or obvious to define what a reverse move is. Take the example of a problem where the objective is to find an optimal permutation of n elements. A reasonable move could be to transpose the elements i and j of the permutation ($1 \leq i < j \leq n$). In this case, all of the M moves applicable to an unspecified solution are given by the entire set of pairs (i, j). But, thereafter, if the move (i, k) is carried out, the prohibition of (i, j) will prevent the visiting of certain solutions without preventing the cycling phenomenon: indeed, the moves $(i, j)(k, p)(i, p)(k, j)(k, i)(j, p)$ applied successively do not modify the solution. So, the tabu condition must not necessarily prohibit one to perform the reverse of a move too quickly, but it may be defined in such a way to prevent the use of some attribute of the moves or solutions. In the preceding example, if p_i is the position of element i and if the move (i, j) has been performed, it is not the reverse move (i, j) which should be prohibited, but, for example, the simultaneous placing of the element i on position p_i and the element j on position p_j. One can thus at least prevent those cycles which are of length less than or equal to the number of tabu moves, i.e., the length of the tabu list.

3.4.3 Duration of Tabu Conditions

Generally speaking, the short-term memory will prohibit the performance of some moves, either directly by storing tabu moves or tabu solutions, or indirectly by storing attributes of moves or attributes of solutions that are prohibited. If the minimization problem can be represented as a landscape limited to a territory which defines the feasible solutions and where altitude corresponds to the value of the objective function, the effect of this memory is to visit valleys (without always being at the bottom

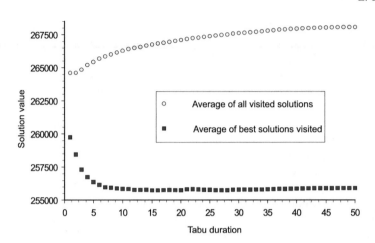

Fig. 3.8 Influence of the number of iterations during which moves are tabu

of the valley, because of tabu moves) and, sometimes, to cross a pass leading to another valley.

The higher the number of tabu moves is, the more likely one is to cross the mountains, but the less thoroughly the valleys will be visited. Conversely, if moves are prohibited for only a few iterations, there will be fewer chances of crossing the passes surrounding the valleys because, almost surely, there will be an allowed move which will lead to a solution close to the bottom of the valley; but, on the other hand, the bottom of the first valley visited will most probably be found.

More formally, for a very small number of tabu moves, the iterative search will tend to visit the same solutions over and over again. If this number is increased, the probability of remaining confined to a very limited number of solutions decreases and, consequently, the probability of visiting several good solutions increases. However, the number of tabu moves must not be very large, because it then becomes less probable that one will find good local optima, for lack of available moves. To some extent, the search is guided by the few allowed moves rather than by the objective function.

Figure 3.8 illustrates these phenomena in the case of the quadratic assignment problem: for each of 3000 instances of size 12, drawn at random, 50 iterations of a tabu search were performed. This figure gives the following two statistics as a function of the number of iterations during which a reverse move is prohibited: firstly, the average value of all the solutions visited during the search and, secondly, the average value of the best solutions found by each search. It should be noted that the first statistic grows with the number of prohibited moves, which means that the average quality of the visited solutions degrades. On the other hand, the quality of the best solutions found improves with an increase in the number of tabu moves, which establishes the fact that the search succeeds in escaping from comparatively poor local optima; then, their quality worsens, but this tendency is very limited here.

Thus, it can be concluded that the size of the tabu list must be chosen carefully, in accordance with the problem under consideration, the size of the neighborhood, the problem instance, the total number of iterations performed, etc. It is relatively easy to determine the order of magnitude that should be assigned to the number of tabu iterations, but the optimal value cannot be obtained without testing all possible values.

3.4.3.1 Random Tabu Duration

To obtain simultaneous benefits from the advantages of a small number of tabu moves—for through exploration of a valley—and a large number—for the ability to escape from the valley—the number of tabu moves can be modified during the search process. Several methodologies can be considered for this choice: for example, this number can be decided at random between a lower and an upper limit, in each iteration or after a certain number of iterations. These limits can often be easily identified; they can also be increased or decreased on the basis of characteristics observed during the search, etc. These were the various strategies employed upto the end of the 1980s [12, 13, 16]. These strategies were shown to be much more efficient than the use of tabu lists of fixed size (often implemented in the form of a circular list, although this may not be the best option, as can be seen in Sect. 3.5.2).

Again for the quadratic assignment problem, Fig. 3.9 gives the average number of iterations necessary for the solution of 500 examples of problems of size 15 generated at random, when the technique was to choose the number of tabu moves at random between a minimum value and that value increased by $Delta$. The size of the dark disks depends on the average number of iterations necessary to obtain optimal solutions for the 500 problems. An empty circle indicates that at least one of the problems was not solved optimally. The size of these circles is proportional to the number of problems for which it was possible to find the optimum. For $Delta = 0$, i.e., when the number of tabu moves is constant, cycles appear. On the other hand, the introduction of a positive $Delta$, even a very small one, can ensure much more protection against cycling. As can be noted in Fig. 3.8, the lower the tabu list size is, the smaller is the average number of iterations required to obtain the optimum. However, below a certain threshold, cycles appear, without passing through the optimum. From the point of view of robustness, one is thus constrained to choose sizes of tabu lists slightly larger than the optimal value (for this size of instances, it seems that this optimal value should be [7, 28] (minimum size $= 7$, $Delta = 21$), but it can be noticed that for [8, 28] a cycle appeared).

This technique of randomly selecting the number of tabu moves can thus guide the search automatically towards good solutions. However, such a mechanism could be described as myopic because it is guided mainly by the value of the objective function. Although it provides very encouraging results considering its simplicity, it

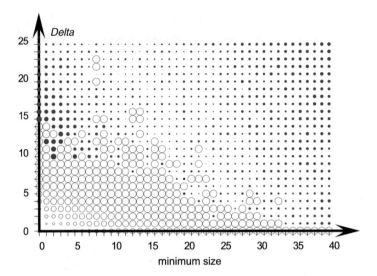

Fig. 3.9 The effect of random selection of the number of iterations during which moves are prohibited, for instances of the quadratic assignment problem of size 15 drawn at random. The number of iterations during which the reverse of a move was prohibited was drawn at random, uniformly between a minimum value and that value increased by *Delta*. The size of the filled disks grows with the average number of iterations necessary for the resolution of the problem until the optimum is found. An *empty circle* indicates that a cycling phenomenon appeared. The size of the circles is proportional to the number of problem instances solved optimally

cannot be considered as an intelligent way of guiding the search, but must rather be viewed as a basic tool for implementing the search process.

3.4.3.2 Type of Tabu List for the Quadratic Assignment Problem

A solution to the quadratic assignment problem can be represented in the form of a permutation **p** of n elements. A type of move very frequently used for this problem is to transpose the positions of two objects i and j. It is possible to evaluate effectively, in $O(n^2)$, the entire set of moves applicable to a solution.

As was discussed earlier, one technique for guiding the search in the short-term is to prohibit, for t iterations, the application of the reverse of the moves which have just been carried out. If a move (i, j) is applied to the permutation **p**, the reverse of the move can be defined as a move which simultaneously places the object i on the site p_i and the object j on the site p_j. There are other possible definitions of the reverse of a move, but this is one of the most effective ones for preventing cycles and appears to be the least sensitive one to the value of the parameter t, the number of iterations during which one avoids applying the reverse of a move. A fixed value of t does not produce a robust search, because the cycles may appear (see Fig. 3.9) even for large values of t. To overcome this problem, it was proposed in [13] that t should be drawn uniformly at random, between $\lfloor 0, 9 \cdot n \rfloor$ and $\lceil 1, 1 \cdot n + 4 \rceil$. In fact, experiments have shown that a tabu duration equal to the size of the problem, or slightly larger for small examples, seems rather effective. This paved the way for the idea of selecting the value of t in a dynamic manner during the search, by choosing a maximum value slightly higher and an average value lower than the value which would have been ideal in the static case.

To implement this tabu mechanism in practice, a matrix $\mathcal{T}$ can be used whose entry t_{ir} gives the iteration number in which the element i was last moved from the site r (to go to the site p_i); this is the number to which one adds the tabu duration t. Thus, the move (i, j) is prohibited if both of the entries t_{ip_j} and t_{jp_i} contain values higher than the current iteration number.

Let us consider the small 5×5 instance of the quadratic assignment problem known in the literature as NUG5, with flow matrix $\mathcal{F}$ and distance matrix $\mathcal{D}$:

$$\mathcal{F} = \begin{pmatrix} 0 & 5 & 2 & 4 & 1 \\ 5 & 0 & 3 & 0 & 2 \\ 2 & 3 & 0 & 0 & 0 \\ 4 & 0 & 0 & 0 & 5 \\ 1 & 2 & 0 & 5 & 0 \end{pmatrix}, \quad \mathcal{D} = \begin{pmatrix} 0 & 1 & 1 & 2 & 3 \\ 1 & 0 & 2 & 1 & 2 \\ 1 & 2 & 0 & 1 & 2 \\ 2 & 1 & 1 & 0 & 1 \\ 3 & 2 & 2 & 1 & 0 \end{pmatrix}$$

With the tabu duration fixed at $t = 5$ iterations, the evaluation of the tabu search is the following.

Iteration 0. On the basis of the initial solution $\mathbf{p} = (5, 4, 3, 2, 1)$, meaning that the first element is placed in position 5, the second in position 4, etc., the value of this solution is 64. The search starts by initializing the matrix $\mathcal{T} = \mathbf{0}$.

Iteration 1. The value of $\Delta(\mathbf{p}, (i, j))$ is then calculated for each transposition m specified by the objects (i, j) exchanged:

m	$(1, 2)$	$(1, 3)$	$(1, 4)$	$(1, 5)$	$(2, 3)$	$(2, 4)$	$(2, 5)$	$(3, 4)$	$(3, 5)$	$(4, 5)$
Cost	-4	-4	16	4	2	14	16	0	14	2

It can be seen that two moves can produce a maximum profit of 4, by exchanging either objects $(1, 2)$ or objects $(1, 3)$. We can assume that it is the first of these moves, $(1, 2)$, which is retained, meaning that object 1 is placed in the position of object 2, i.e., 4, and object 2 is placed in the position of object 1, i.e., 5. It is forbidden for $t = 5$ iterations (i.e., up to iteration 6) to put element 1 in position 5 and element 2 in position 4 simultaneously. The following tabu condition matrix is obtained:

$$T = \begin{pmatrix} 0\,0\,0\,0\,6 \\ 0\,0\,0\,6\,0 \\ 0\,0\,0\,0\,0 \\ 0\,0\,0\,0\,0 \\ 0\,0\,0\,0\,0 \end{pmatrix}$$

Iteration 2. The move chosen in iteration 1 leads to the solution $\mathbf{p} = (4, 5, 3, 2, 1)$, of cost 60. The computation of the value of every move for this new solution gives

m	(1, 2)	(1, 3)	(1, 4)	(1, 5)	(2, 3)	(2, 4)	(2, 5)	(3, 4)	(3, 5)	(4, 5)
Cost	4	10	22	12	−8	12	12	0	14	2
Tabu	Yes									

For this iteration, it should be noted that the reverse of the preceding move is now prohibited. The allowed move (2, 3), giving the minimum cost, is retained, for a profit of 8. The matrix T becomes

$$T = \begin{pmatrix} 0\,0\,0\,0\,6 \\ 0\,0\,0\,6\,7 \\ 0\,0\,7\,0\,0 \\ 0\,0\,0\,0\,0 \\ 0\,0\,0\,0\,0 \end{pmatrix}$$

Iteration 3. The solution $\mathbf{p} = (4, 3, 5, 2, 1)$, of cost 52, is reached, which is a local optimum. Indeed, at the beginning of iteration 3, no move has a negative cost:

m	(1, 2)	(1, 3)	(1, 4)	(1, 5)	(2, 3)	(2, 4)	(2, 5)	(3, 4)	(3, 5)	(4, 5)
Cost	8	14	22	8	8	0	24	20	10	10
Tabu					Yes					

The move (2, 4) selected in this iteration has zero cost. It should be noted here that the move (1, 2), which was prohibited in iteration 2, is again allowed, since the element 5 was never in the third position. The matrix T becomes

$$T = \begin{pmatrix} 0\,0\,0\,0\,6 \\ 0\,0\,8\,6\,7 \\ 0\,0\,7\,0\,0 \\ 0\,8\,0\,0\,0 \\ 0\,0\,0\,0\,0 \end{pmatrix}$$

Iteration 4. The current solution is $\mathbf{p} = (4, 2, 5, 3, 1)$, of cost 52, and the data structure situation is as follows:

m	(1, 2)	(1, 3)	(1, 4)	(1, 5)	(2, 3)	(2, 4)	(2, 5)	(3, 4)	(3, 5)	(4, 5)
Cost	8	14	22	8	8	0	24	20	10	10
Tabu						Yes				

However, it is not possible anymore to choose the move (2, 4) corresponding to the minimum cost, which could bring us back to the preceding solution, because this move is prohibited. Hence we are forced to choose an unfavorable move, (1, 2), that increases the cost of the solution by 8. The matrix T becomes

$$T = \begin{pmatrix} 0 & 0 & 0 & 9 & 6 \\ 0 & 9 & 8 & 6 & 7 \\ 0 & 0 & 7 & 0 & 0 \\ 0 & 8 & 0 & 0 & 0 \\ 0 & 0 & 0 & 0 & 0 \end{pmatrix}$$

Iteration 5. The solution at the beginning of this iteration is $\mathbf{p} = (2, 4, 5, 3, 1)$. The computation of the cost of the moves gives

m	(1, 2)	(1, 3)	(1, 4)	(1, 5)	(2, 3)	(2, 4)	(2, 5)	(3, 4)	(3, 5)	(4, 5)
Cost	−8	4	0	12	10	14	12	20	10	−10
Tabu	Yes									

It can be noticed that the move degrading the quality of the solution in the preceding iteration was beneficial, because it now facilitates arriving at an optimal solution $\mathbf{p} = (2, 4, 5, 1, 3)$, of cost 50, by choosing the move (4, 5).

3.4.4 Aspiration Conditions

Sometimes, some tabu conditions are absurd. For example, a move which leads to a solution better than all those visited by the search in the preceding iterations does not have any reason to be prohibited. In order not to miss this solution, it is necessary to disregard the possible tabu status of such moves. In tabu search terminology, this move is said to be *aspired*. Naturally, it is possible to assume other aspiration criteria, less directly related to the value of the objective to be optimized.

It should be noted here that the first presentations on tabu search insisted heavily on aspiration conditions, but, in practice, these were finally limited to allowing a tabu move which helped to improve the best solution found so far during the search. As this later criterion became implicit, little research was carried out later on defining more elaborate aspiration conditions. On the other hand, aspiration can also sometimes be described as a form of long-term memory, consisting in forcing a move that has never been carried out over several iterations, irrespective of its influence on the objective function.

3.5   Long-Term Memory

In the case of a neighborhood defined by a static set of moves, i.e., when it does not depend on the solution found in the process, the statistics of the moves chosen during the search can be of great utility. If some moves are chosen much more frequently than others, one can suspect that the search is facing difficulties in exploring solutions of varied structure and that it may remain confined in a "valley." In practice, problem instances that include extended valleys are frequently observed. Thus, these can be visited using moves of small amplitude, considering the absolute difference in the objective function. If the only mechanism for guiding the search is to prohibit moves which are the reverse of those recently carried out, then a low number of tabu moves implies that it is almost impossible to escape from some valleys. A high number of tabu moves may force the search procedure to reside often on a hillside, but even if the search can change between valleys, it cannot succeed in finding good solutions in the new valley because numerous moves are prohibited after the visit to the preceding valley. It is thus necessary to introduce other mechanisms to guide a search effectively in the long term.

3.5.1   Frequency-Based Memory

One technique to ensure some diversity throughout the search without prohibiting too many moves, consists in penalizing moves that are frequently used. Several penalization methods can be imagined, for instance a prohibition of moves whose frequency of occurrence during the search exceeds a given threshold, or the addition of a value proportional to their frequency when evaluating moves. Moreover, the addition of a penalty proportional to the frequency has a beneficial effect for problems where the objective function takes only a small number of values, as that situation can generate awkward equivalences from the point of view of guiding the search when several neighboring solutions have same evaluation. In these situations, the search will then tend to choose those moves which are least employed rather than select a move more or less at random.

Figure 3.10 illustrates the effect of a method of penalization of moves which adds a factor proportional to their frequency of usage at the time of their evaluation. For this purpose, the experiment carried out to show the influence of the tabu duration (see Fig. 3.8) was repeated, but this time; the coefficient of penalization was varied;

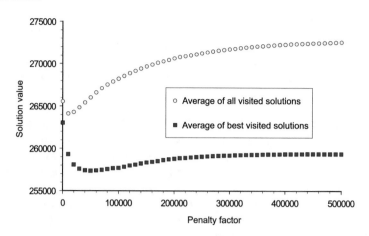

Fig. 3.10 Effect of a coefficient of penalization based on the frequencies of moves

the moves were thus penalized, but never tabu. This experiment relates again to the
3000 quadratic assignment instances of size 12 generated at random. In Fig. 3.10,
the average of the best solutions found after 50 iterations and the average value of all
the solutions visited are given as functions of the coefficient of penalization. It can
be noticed that the behavior of these two statistics is almost the same as that shown
in Fig. 3.8, but overall, the solutions generated are worse than those obtained by the
use of a short-term memory.

Just like what has been done for short-term memory, this method can be generalized to provide a long-term memory of a nonstatic set of moves, i.e., where M
depends on s: in this case the frequency with which certain characteristics of moves
have been employed to is recorded rather than the moves themselves. Here, the similarities in implementing these two forms of memory should be noticed: one method
stores the iteration in which one can use a characteristic of a move again, whereas the
other memorizes the number of times this characteristic has been used in the chosen
moves.

3.5.1.1 Value of the Penalization

It is necessary to tune the value associated with a penalization method based on frequencies. This tuning can be carried out on the basis of the following considerations.
Firstly, if $freq(m)$ denotes the frequency of usage of move m, it seems reasonable
to penalize that move by a factor proportional to $freq(m)$, though other possible
functions can be used, for instance $freq^2(m)$.

Secondly, if the objective function is linear and if a new problem instance is
considered where all the data are multiplied by a constant, it is not desired that

a mechanism of penalization based on frequencies should depend on the value of the constant. In the same way, the mechanism of penalization should not work in a different manner if one adds a constant to the objective function. Consequently, it also seems legitimate to use a penalization which is proportional to the average amplitude of two neighboring solutions.

Thirdly, the larger the neighborhood is, the more the distribution of the frequencies becomes concentrated on small values. The penalty should thus be multiplied by a function that is strictly increasing with the size of the neighborhood, so that the penalty does not become zero when the size of the problem increases. If the identity function proves to be too large in practice (cf. [14, 15]), one can consider, for example, a factor proportional to $\sqrt{|M|}$.

Naturally, the concept of using penalization based on frequencies also requires taking the mechanism of aspiration into consideration. If not, then it is highly likely that we may miss excellent solutions.

3.5.2   Forced Moves

Another long-term mechanism consists in performing a move which has never been used during a large number of iterations, irrespective of its influence on the quality of the solution. Such a mechanism can be useful for destroying the structure of a local optimum, and therefore escaping from the valley in which it was confined. This is also valid for high-dimensional problems, as well as instances that have a more modest size but are very structured (i.e., for which the local optima are separated by very bad solutions).

In the earlier example of the quadratic assignment problem, it is not even necessary to introduce a new data structure to implement this mechanism. In fact, it is enough to implement the tabu list in the form of a matrix with two dimensions (element, position), whose entries indicate in which iteration each element is allowed to occupy a given position, either to decide if a move is prohibited (the entries in the matrix corresponding to the move contain values larger than the number of the current iteration) or, instead, to decide if a given element has not occupied a given position during the last v iterations. If the matrix contains an entry whose value is lower than the number of the current iteration decreased by the parameter v, the corresponding move is chosen, independent of its evaluation. It may happen that several moves could be simultaneously chosen because of this rule. This problem can be solved by imagining that, before the search was started, one had carried out all $|M|$ moves (a static, definite neighborhood of all M moves is assumed) during hypothetical iterations $-|M|, -|M| + 1, \ldots, -1$. Of course, it is necessary that the parameter v be (sufficiently) larger than $|M|$, so that these moves are only imposed after v iterations.

3.6   Convergence of Tabu Search

Formally, one cannot speak about "convergence" for a tabu search, since in each iteration the solution is modified. On the other hand, it is definitely interesting to pass at least once through a global optimum. This was the focus of discussion in [9], on a theoretical level, using an elementary tabu search. It was shown that the search could be blocked if one prohibited passing through the same solution twice. Consequently, it is necessary to allow the same solution to be revisited. By considering a search which memorizes all the solutions visited and which chooses the oldest one that has been visited if all of the neighboring solutions have already been visited, it can be shown that all of the solutions of the problem will be enumerated. This is valid if the set of solutions is finite, and if the neighborhood is either reversible (or symmetric: any neighboring solution to s has s in its neighborhood) or strongly connected (there is a succession of moves that enables one to reach any solution s' starting from any solution s). Here, all of the solutions visited must be memorized (possibly in an implicit form), and it should be understood that this result remains theoretical.

Another theoretical result on the convergence of tabu search was presented in [2]. The authors of that study considered probabilistic tabu conditions. It is then possible to choose probabilities such that the search process is similar to that of simulated annealing. Starting from this observation, it can be expected that the convergence theorems for simulated annealing can easily be adapted to a process called *probabilistic tabu search*. Again, it should be understood that the interest of this result remains of a purely theoretical nature.

3.7   Conclusion

Only some of the basic concepts of tabu search have been presented in this chapter. Other principles may lead to a more effective and intelligent method. When possible, a graphical representation of the solutions visited successively during the search should be used, as it will actively stimulate the spirit of the designer and will suggest, often in an obvious way, how to guide the search more intelligently. The development of a tabu search is an iterative process: it is almost impossible to propose an excellent method at the first attempt. Adaptations, depending on the type of problem as well as on the problem instance considered, will certainly be required. This chapter has described only those principles which should enable a designer to proceed towards an effective algorithm more quickly. Other principles, often presented within the framework of tabu search suggested by F. Glover, such as scatter search, vocabulary building and path relinking, will be presented in Chap. 13, devoted to methodology.

3.8 Annotated Bibliography

Reference [6]   This book is undoubtedly the most important reference on tabu
                     search. It describes the technique extensively, including certain
                     extensions which will be discussed in this book in Chap. 13.

References [4, 5]   These two articles can be considered as the founders of the dis-
                     cipline, even if the name "tabu search" and certain ideas already
                     existed previously. They are not easily accessible; hence certain
                     concepts presented in these articles, such as path relinking and
                     scatter search, were studied by the research community only sev-
                     eral years after their publication.

References

1. Burkard, R.E., Fincke, U.: Probabilistic properties of some combinatorial optimization prob-
    lems. Discrete Applied Mathematics **12**, 21–29 (1985)
2. Faigle, U., Kern, W.: Some convergence results for probabilistic tabu search. ORSA Journal
    on Computing **4**, 32–37 (1992)
3. Glover, F.: Future paths for integer programming and links to artificial intelligence. Computers
    and Operations Research **13**, 533–549 (1986)
4. Glover, F.: Tabu search—Part i. ORSA Journal on Computing **1**, 190–206 (1989)
5. Glover, F.: Tabu search—Part ii. ORSA Journal on Computing **2**, 4–32 (1990)
6. Glover, F., Laguna, M.: *Tabu Search*. Kluwer, Dordrecht (1997)
7. Glover, F., Laguna, M., Taillard, É.D., de Werra, D.: Annals of or 41. In: *Tabu Search*. Baltzer
    (1993)
8. Glover, F., Taillard, É.D., de Werra, D.: A user's guide to tabu search. Annals of Opera-
    tions Research **41**, 1–28 (1993). http://mistic.heig-vd.ch/taillard/articles.dir/GloverTW1993.
    pdf. doi:10.1007/BF02078647
9. Hanafi, S.: On the convergence of tabu search. Journal of Heuristics **7**(1), 47–58 (2001)
10. Hertz, A., de Werra, D.: The tabu search metaheuristic: How we used it. Annals of Mathematics
    Artificial Intelligence **1**, 111–121 (1990)
11. Lin, S., Kernighan, B.W.: An effective heuristic algorithm for the traveling-salesman. Opera-
    tions Research **21**(2), 498–516 (1973)
12. Taillard, E.D.: Some efficient heuristic methods for the flow shop sequencing problem. Euro-
    pean Journal of Operational Research **47**(1), 65–74 (1990)
13. Taillard, E.D.: Robust taboo search for the quadratic assignment problem. Parallel Computing
    17, 443–455 (1991)
14. Taillard, E.D.: Parallel iterative search methods for vehicle routing problems. Networks **23**,
    661–673 (1993)
15. Taillard, E.D.: Parallel taboo search techniques for the job shop scheduling problem. ORSA
    Journal on Computing **6**(2), 108–117 (1994)
16. Taillard, E.D.: Comparison of iterative searches for the quadratic assignment problem. Location
    Science **3**(2), 87–105 (1995)
17. Toth, P., Vigo, D.: The granular tabu search and its application to the vehicle-routing problem.
    INFORMS Journal on Computing **15**, 333–346 (2003)
18. Woodruff, D.L., Zemel, E.: Hashing vectors for tabu search. In: G. Glover, M. Laguna, E.D.
    Taillard, D. de Werra (eds.) *Tabu Search*, no. 41 in Annals of Operations Research, pp. 123–137.
    Baltzer, Basel (1993)

Chapter 4
Variable Neighborhood Search

Gilles Caporossi, Pierre Hansen and Nenad Mladenović

4.1 Introduction

The variable neighborhood search (VNS) metaheuristic was invented by Nenad Mladnović and Pierre Hansen and has been developed at GERAD (Group for Research in Decision Analysis, Montreal) since 1997. From that time on, VNS has various developments and improvements as well as numerous applications. According to Journal Citation Reports, the original publications [8, 13] on VNS were cited more than 600 and 500 times, respectively (and more than 1700 and 1200 times according to Google Scholar), which shows the interest in the method from the development and application points of view.

Applications of VNS may be found in various fields, such as data mining, localization, communications, scheduling, vehicle routine, and graph theory. The reader may refer to [10] for a more exhaustive survey.

VNS has many advantages. The most important, is that it usually provides excellent solutions in a reasonable time, which is also the case for most of the modern metaheuristics, but it is also easy to implement. In fact, VNS is based upon a combination of methods which are quite classical in combinatorial and continuous optimization. It also has very few parameters (and sometimes none) that have to be tuned in order to get good results.

The goal of this chapter is not to provide an exhaustive review of the variants of VNS and their applications, but rather to give the basic rules so as to make its implementation as easy as possible.

The key concepts are presented and illustrated by an example based upon a search for extremal graphs. These illustrations were inspired by an optimization implemented in the first version of the AutoGraphiX (AGX) software package [2]. AGX is dedicated to the search for conjectures in graph theory. We decided to use this example because it involves all the main components of VNS and their use is intuitive.

G. Caporossi (✉) · P. Hansen
GERAD and HEC Montreal, Montreal, Canada
e-mail: gilles.caporossi@gerad.ca

N. Mladenović
GERAD and LAMIH, Université de Valenciennes et du Hainaut-Cambrésis, Valenciennes, France

© Springer International Publishing Switzerland 2016
P. Siarry (ed.), *Metaheuristics*, DOI 10.1007/978-3-319-45403-0_4

4.2   Description of the Algorithm

Like other metaheuristics, VNS is based upon two complementary methods: on one hand, the local search and its extensions aim at improving the current solution, and on the other hand, the perturbations allow the space of explored solutions to be extended. These two principles are usually known as intensification and diversification. In the case of VNS, these principles are combined in an intuitive way that is easy to implement.

4.2.1   Local Search

The local search used in a large number of metaheuristics consists in finding successive improvements of the current solution through an elementary transformation until no improvement is possible. The solution so found is called a local optimum with respect to the transformation used.

Technically, the local search consists in a succession of transformations of the solution in order to improve it. We denote by $\mathcal{N}(S)$ the set of solutions that may be obtained from the solution S by applying the transformations once. We call $\mathcal{N}(S)$ the neighborhood of S. The current solution S is replaced by a better one $S' \in \mathcal{N}(S)$. The process stops when it is no longer possible to find an improving solution in the neighborhood $\mathcal{N}(S)$, as described in Algorithm 4.1.

Input: S
Input: $\mathcal{N}$
Let $imp \leftarrow true$
while $imp = true$ **do**
 $\quad imp \leftarrow false$
 $\quad$ **foreach** $S' \in \mathcal{N}(S)$ **do**
 $\quad\quad$ **if** S' *better than* S **then**
 $\quad\quad\quad S \leftarrow S'$
 $\quad\quad\quad imp = true.$
 $\quad\quad$ **end**
 $\quad$ **end**
end
return S

Algorithm 4.1: Local Search.

After a complete exploration of the neighborhood $\mathcal{N}(S)$, the local search ensures that a local optimum with respect to the transformation considered has been found. Of course, the notion of a local optimum remains relative to the transformation used. A solution may be a local optimum for a given transformation, but not for another one. The choice of the transformations was an important part in the development of the VNS algorithm.

4.2.1.1   Changing the Neighborhoods

A way to improve the quality of the solution is to consider the use of various transformations (and thus various neighborhoods) or various ways to use them.

The performance of the local search, measured according to the computational effort required, the quality of the solution obtained, or the data structure implementation, depends on the transformations and the rules for choosing which one to apply.

It is possible that more than one improving solution may be found in the neighborhood of the current solution. The choice of the best of them seems natural, but this implies the complete exploration of the neighborhood. The computational time required may be too large for the benefit gained. For this reason, it is sometimes better to apply the first improvement found. The reader may refer to [9] for a deep analysis in the case of the traveling salesman problem.

Except when a transformation is integrated into the local search scheme, it is also possible to work on the transformations themselves. Consider a transformation t that may be applied in various ways to the solution S. This transformation allows the construction of a set $\mathcal{N}^t(S)$ of solutions from S.

For a problem with implicit or explicit constraints, given a transformation t, the solutions $S' \in \mathcal{N}^t(S)$ may always, sometimes or never be realizable. According to the nature of the transformation, it may be possible to predict the realizability of a solution in the neighborhood $\mathcal{N}(S)$ of S, and thus to decide whether to use it or not before any attempt.

Apart from the realizability or not of a solution, the transformations have other properties that should be analyzed before implementation.

Each neighborhood corresponds to a set of solutions and the larger this set is, the more we can expect that it contains good solutions. On the other hand, the systematic exploration of this large neighborhood will be time-consuming. The best neighborhood would be one that contains the best solution, and thus improves significantly the current solution while being fast to explore.

Unfortunately, finding such a neighborhood is not always possible. However, machine learning may be used to select the most promising of them during the optimization itself [3].

4.2.1.2   The Variable Neighborhood Descent

To take advantage of the various existing transformations for a given problem and their specificities, it is possible to adapt the local search to use more than just one of them. This is the principle of the variable neighborhood descent (VND).

In the same way as the local search explores the various solutions within a neighborhood, the VND successively explores a series of transformations. Consider the list $\mathcal{N}^t(S)$ for $t = 1, \ldots, T$, where T is the number of transformations considered.

Using successively all the transformations in the list to apply local searches, the VND stops only when no transformation leads to an improved solution. The local optimum obtained after the VND is relative to all the transformations in the list, not just one of them.

The performance of the variable neighborhood descent depends not only on the neighborhoods used, but also on the order in which they are applied. Let us consider two transformations t_1 and t_2 from which two neighborhoods may be built. If $\mathcal{N}^{t_1}(S) \subset \mathcal{N}^{t_2}(S)$, using the neighborhood based upon t_1 after the one based on t_2 will not improve the solution. One might conclude that a local search using the neighborhood $\mathcal{N}^{t_1}(S)$ would be useless, but one may also consider the computational effort that the exploration of these neighborhoods requires.

At the beginning of the search, the solutions are usually of poor quality. If the exploration of $\mathcal{N}^{t_1}(S)$ is faster than that of $\mathcal{N}^{t_2}(S)$, but its improvements are nevertheless good enough, using t_1 as the first step could lead to a faster search than using t_2 solely. If the quality of the solution is the only criterion, $\mathcal{N}^{t_1}(S)$ is not as efficient as $\mathcal{N}^{t_2}(S)$, but using t_1 before t_2 may speed up the process significantly.

If $\mathcal{N}^{t_1}(S)$ contains solutions that are not in $\mathcal{N}^{t_2}(S)$ and vice versa, then the use of both transformations is fully justified.

To improve the global performance of the search, it is often better to use the smallest neighborhoods first, and those which require more computation time after the first has failed.

Adopting this principle, the VND consists in applying such local searches one after another until none succeeds in improving the solution. When a neighborhood fails, the next one is used. If a local search succeeds, the algorithm restarts from the first neighborhood (the fastest to explore). The strongest, but longest to apply, is used when all others fail. The VND is described in Algorithm 4.2 and may be considered as an *mta*-local search.

Input: S
Input: $\mathcal{N}^t$, $t = 1, \ldots, T$
Let $t = 1$
while $t < T$ **do**
  $S' \leftarrow LocalSearch(S, \mathcal{N}^t)$
  if S' *better than* S **then**
    $S \leftarrow S'$
    $t \leftarrow 1$
    $imp \leftarrow true.$
  end
  else
    | $t \leftarrow t + 1$
  end
end
return S

Algorithm 4.2: VND.

4.2.2   Diversification of the Search

Another way to improve the quality of the solution obtained is to change its starting point.

4.2.2.1   The Multistart Search

The first method consists in applying multiple local searches starting from various initial random solutions, and keeping only the best solution obtained, as described in Algorithm 4.3. This is the so-called *multistart* search.

Input: S
Let $S^* \leftarrow S$, the best known solution.
repeat
    Let S be a random solution.
    $S' \leftarrow LocalSearch(S)$
    if S' *better than* S^* **then**
      | $S^* \leftarrow S'$
    end
until *stopping criterion*;
return S^*

Algorithm 4.3: Multistart search.

If the problem has a moderate number of local optima which are far from each other, the multistart algorithm will give good results. Unfortunately, in most cases, owing to the number of local optima and their characteristics, it is unlikely that this approach will give good results.

To illustrate this difficulty, let us consider two optimization problems with only one variable. Figures 4.1 and 4.2 represent the objective function as a function of the decision variable x.

If the initial values of x are chosen randomly in its definition interval, the multistart algorithm will succeed in finding the best solution in the case of problem 1, but it is very unlikely to do so in the case of problem 2. First, the number of local optima

Fig. 4.1 Illustration of
problem 1 with one variable

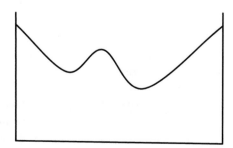

Fig. 4.2 Illustration of
problem 2 with one variable

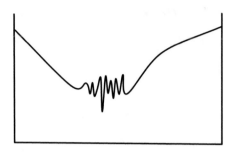

is large, and most initial solutions will lead to the same local optimum (the one on
the right or the one on the left). Because it often provides the same solutions, the
algorithm may also give the impression that it is the best possible one because it is
the easiest to find. In the case of higher dimensions, there is no reason to believe
that the situation will change. On the contrary, it is likely that this phenomenon will
increase, so that generally, the multistart algorithm is a bad choice when local optima
are close to one another.

The first question that arises, then, is whether the problem under study is more
similar to problem 1 or problem 2. Even if there is no general rule on that topic, it
seems that in most cases the local optima share a large number of properties, which
tends to indicate that they are similar to problem 2 in their nature.

Let us consider a few examples:

- *The traveling salesman problem*, a problem in which a salesman needs to meet a
  list of clients and return to his office, minimizing the total length of the journey
  required to visit them. It is likely that in a good solution, the two clients that are
  the furthest away from each other will not appear one after the other in a tour that
  would be found after a local search. In the same way, if two clients are very close
  to each other, it is likely that they will be visited one after the other. Globally, the
  local optima will share some common characteristics.
- *Clustering*, a problem in which one wants to group objects that are alike (homo-
  geneity criterion) and have different objects in different clusters (separation criterion). There exists a wide variety of criteria to evaluate the quality of a partition,
  but in all cases, one can expect that very similar objects will be in the same cluster
  while very different objects will not. Again, whatever the criterion and the local
  search (as long as it is reasonably good) used, all the local optima will have some
  common characteristics. The difference between local optima will likely concern
  only a small portion of the objects.

In these two examples, but many others as well, it is likely that the problem is
more similar to problem 2 than to problem 1 (one must, however, remain cautious).

Not only will a multistart search lead to the solutions that are easiest to find (not
necessarily the best), but also a lot of computational effort will be spent on separating
objects that are very different and on grouping some that are very similar.

4.2.2.2 Perturbations

In order to reduce the computational effort, instead of starting from a random solution in each local search, another approach consists in modifying the best known solution slightly, and starting the next local search from that perturbed solution.

An algorithm that uses multiple local searches starting from solutions that are relatively close to the best known solution will benefit from the characteristics found during the previous local searches. It will therefore use the information gained so far, which is completely ignored in the multistart algorithm. For this reason, VNS does not proceed by successive local searches from random solutions, but from solutions close to the best known one.

The choice of the magnitude of the perturbation is important. If it is too small, only a small portion of the solution space will be explored (we may possibly get back to the previous local optimum after the local search). If, instead, it is too large, the characteristics of the best known solution will be ignored and the perturbation will not be better than a random solution. For that reason, a parameter k is used to indicate the magnitude of the perturbation. The higher k is, the further the perturbed solution will be from the previous one. The neighborhoods used for the perturbations must then have a magnitude related to k. Nested neighborhoods or neighborhoods built by a succession of random transformations as described in Algorithm 4.4 are generally appropriate. A simple but efficient method consists in applying the transformations used in the local search.

Input: S
Input: k
Input: $\mathcal{N}$
repeat k **times**
    Choose randomly $S' \in \mathcal{N}(S)$,
    let $S \leftarrow S'$.
return S

Algorithm 4.4: PERTURB.

If the problem has constraints (either implicit or explicit), it is better to use transformations that preserve the realizability of the solution. For instance, in the the traveling salesman problem, a transformation that creates subtours (a solution that consists of disjoint tours) should be avoided. Note that this description is only an indication. It is likely that other schemes may be better, depending on the problem.

4.2.3 The Variable Neighborhood Search

The variable neighborhood search proceeds by a succession of local searches and perturbations. After each unsuccessful local search, the magnitude of the perturbation

k is increased to allow a wider search. Beyond a maximum value k_{max} which is fixed as a parameter, k is reset to its minimum value to avoid inefficient perturbations that are too large (and would behave like random solutions).

Depending on the application, it may be better to enlarge or reduce the local search, and this provides various formulations of the variable neighborhood search. Although the local search allows an improvement of the current solution, it is nevertheless expensive in term of computation, and there is therefore a trade-off between the quality of the solution and the time required to obtain it. The choice of the transformations to use during the local search is thus important.

Algorithm 4.5 describes the basic variable neighborhood search.

Input: S
Denote by $S^* = S$ the best known solution.
Let $k = 1$
Define k_{max}
repeat
    $S \leftarrow PERTURB(S^*, k)$,
    $S' \leftarrow LocalSearch(S')$.
    if S' *better than* S^* **then**
        $S^* \leftarrow S'$,
        $k \leftarrow 1$.
    end
    else
        $k \leftarrow k + 1$.
        if $k > k_{max}$ **then**
            let $k \leftarrow 1$.
        end
    end
until *stopping criterion*;
return S^*.

Algorithm 4.5: Basic variable neighborhood search.

Based on the structure of the variable neighborhood search, two extensions may be considered. The general variable neighborhood search concentrates on the quality of the solution at the expense of the computational effort. On the other hand, the reduced variable neighborhood search aims at reducing the computational effort at the expense of the quality of the solution.

4.2.3.1 The General Variable Neighborhood Search

In the general variable neighborhood search, the local search is replaced by a variable neighborhood search descent, which may be considered as a meta-local search. Algorithm 4.6 describes the general variable neighborhood search. One of the neighborhoods used for the variable neighborhood descent is usually used for the perturbation.

```
Input: S
Let S* ← S the best known solution.
Let k ← 1
Define k_max
repeat
    S ← PERTURB(S*, k),
    S' ← VND(S').
    if S' better than S* then
        S* ← S',
        k = 1.
    end
    else
        k ← k + 1.
        if k > k_max then
            let k ← 1.
        end
    end
until stopping condition;
return S*.
```

Algorithm 4.6: General variable neighborhood search.

This kind of variable neighborhood search is well suited to situations where the computational effort is not crucial and emphasis is put on the quality of the solutions.

4.2.3.2   The Reduced Variable Neighborhood Search

The variable neighborhood descent gives better results than the local search at the cost of intensive computation. In contrast, for some problems, the reverse is required. For these, a variant of VNS without local search is better, the succession of perturbations playing simultaneously the roles of diversification and stochastic search. This is the reduced variable neighborhood search. Algorithm 4.7 describes the reduced variable neighborhood search.

4.3   Illustration and Extensions

To illustrate how the variable neighborhood search works, let us explain it with some examples. The first example is one of finding extremal graphs. This example is directly inspired by the variable neighborhood search used in the first version of the AutoGraphiX software package [2].

The second example is based upon a possible extension of the k-means clustering algorithm. With its wide use, this algorithm is considered as a reference for clustering. It turns out that this algorithm finds a local optimum which depends strongly on the initial solution. We propose here a way to use k-means within a variable local search

Input: S
Let $S^* \leftarrow S$ the best known solution.
Let $k \leftarrow 1$
Define $k_{\max}$
repeat
    $S \leftarrow PERTURB(S^*, k)$,
    if S' *better than* S^* **then**
        $S^* \leftarrow S'$,
        $k \leftarrow 1$.
    end
    else
        $k \leftarrow k + 1$.
        if $k > k_{\max}$ **then**
          | let $k \leftarrow 1$.
        end
    end
until *stopping criterion*;
return S^*.

Algorithm 4.7: Reduced variable neighborhood search.

algorithm. This adaptation is simple, but allows significant improvement on the performance of k-means.

As a third example, we will then briefly explain how to adapt the variable neighborhood search to continuous optimization problems.

4.3.1 Finding Extremal Graphs with VNS

Let $G = (V, E)$ be a graph with $n = |V|$ vertices and $m = |E|$ edges. An example of a graph with $n = 6$ and $m = 7$ is drawn in Fig. 4.3. Even though we are not working here with labeled vertices (for which each vertex is characterized), the vertices are labeled in order to simplify the descriptions.

Fig. 4.3 A graph G with $n = 6$ vertices and $m = 7$ edges

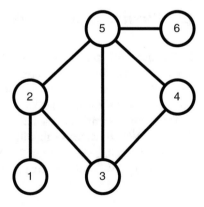

Fig. 4.4 Adjacency matrix
of the graph G

	1	2	3	4	5	6
1	0	1	0	0	0	0
2	1	0	1	0	1	0
3	0	1	0	1	1	0
4	0	0	1	0	1	0
5	0	1	1	1	0	1
6	0	0	0	0	1	0

Since the representation of the graph has no impact on the calculation, the same graph may also be represented by its adjacency matrix (see Fig. 4.4) $A = \{a_{ij}\}$, where $a_{ij} = 1$ if the vertices i and j are adjacent, and $a_{ij} = 0$ otherwise, or by a list indicating which other vertices each vertex is adjacent to. The choice of the representation method (adjacency matrix, adjacency list, or some other) and the choice of the labeling of the vertices are purely arbitrary and have no impact on the computations, since the object under study is independent of its representation.

We denote by $I(G)$ a function which associates with the graph G a value independent of the way the vertices are labeled. Such a function is called an *invariant*. For instance, the number of vertices n and the number of edges m are invariants.

To give some other examples, we can mention the chromatic number $\chi(G)$, the minimum number of colors that are required for a coloring. A coloring assigns a color to each vertex such that two adjacent vertices do not share the same color. Here, $\chi(G) = 3$, and a coloring could consist of assigning blue to vertices 1, 3, and 6, red to vertices 2 and 4, and another color to vertex 5, for example green. As another example, the energy of a graph, $E = \sum_{i=1}^{n} |\lambda_i|$, is the sum of the absolute values of the eigenvalues of the adjacency matrix of G. The number of graph invariants is too large to enumerate them here, but the reader may refer to [5, 16] for a comprehensive survey.

A search for extremal graphs consists in finding a graph that maximizes or minimizes some invariant (or a function of invariants, which is also an invariant), possibly under some constraints. The solutions to those problems are graphs, each different graph being a possible solution to the problem. The graphs with the best values of the objective function (largest or smallest depending on whether the objective is to be maximized or minimized) form the set of optimal solutions.

4.3.1.1 Which Transformations to Use?

The local search can be defined by the addition or removal of an edge, or some more complex transformation. As the neighborhood of a graph varies according to the transformation considered, it is possible that one transformation may not allow any improvement of the current solution while another would. The transformations used in the first version of AGX are described in [2] and presented in Fig. 4.5. We notice, for example, that some neighborhoods preserve the number of vertices while others

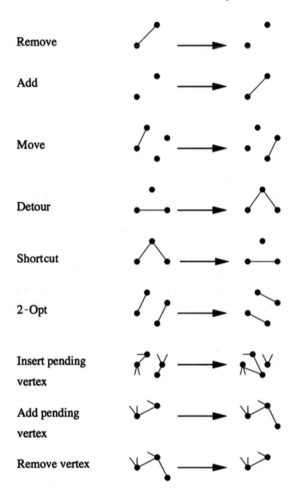

Fig. 4.5 Neighborhoods used in the first version of AGX

do not. As in most cases the number of vertices is fixed, some of these neighborhoods are useless.

We notice also that the transformation 2-opt is the only one that preserves the degree of the vertices; it is thus the only useful transformation in a search for extremal regular graphs when the current graph is also regular. Other transformations may be used to find an initial realizable solution, but would no longer be needed after that step.

In the specific case of regular graphs, it may be interesting to invent other transformations such as the one described in Fig. 4.6. This transformation, based upon a subgraph on five vertices, is computationally demanding, but could be justified in this case.

Fig. 4.6 A transformation
specifically designed for
regular graphs

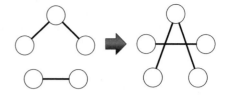

4.3.1.2 In Which Sequence Should the Transformations Be Used?

The choice of the sequence in which the transformations are applied can have an
impact on the performance of the algorithm, since the time required to complete the
search may be significantly affected.

Consider a transformation t_1 which consists of the addition or removal of an edge,
and denote by $\mathcal{N}^{t_1}(G)$ the corresponding neighborhood. Any graph in $\mathcal{N}^{t_1}(G)$ has
one edge more or less than G. Consider now another transformation t_2, defined by a
move of an edge, and denote by $\mathcal{N}^{t_2}(G)$ the corresponding neighborhood. Any graph
in $\mathcal{N}^{t_2}(G)$ has the same number of edges as G. It is clear that these two neighborhoods
are exclusive; no graph belong to both.

Let t_3 be a third transformation which consists of applying t_1 twice, and denote
by $\mathcal{N}^{t_3}(G)$ the corresponding neighborhood.

As an addition followed by a removal could be a transformation from t_3, it turns
out that $\mathcal{N}^{t_2}(G) \subset \mathcal{N}^{t_3}(G)$. Applying t_2 would therefore be useless after t_3 was used.
However, this does not mean that t_2 is useless. In fact, t_3 allows one to find graphs
that cannot be obtained with t_2, while the reverse is not true, but exploring $\mathcal{N}^{t_3}(G)$ is
more time-consuming. Without considering isomorphism (which is difficult to test),
if G has n vertices and m edges, we have:

- $|\mathcal{N}^{t_1}(G)| = n(n-1)/2$,
- $|\mathcal{N}^{t_2}(G)| = m(n(n-1)/2 - m)$, and
- $|\mathcal{N}^{t_3}(G)| = (n(n-1)/2)^2$.

Exploring a neighborhood as large as $\mathcal{N}^{t_3}(G)$ may result in a useless waste of
time. At the beginning of the optimization, when the current solution is not good,
neighborhoods that are easy and fast to explore seem better as they allow a quick
improvement of the solution. At the end of the process, when the exploration has
achieved good solutions, a deeper search is needed. In such a context, the use of
larger neighborhoods may be worthwhile.

4.3.1.3 Using the Basic VNS

To illustrate the basic VNS, consider the graph G shown in Fig. 4.3 as the initial
solution. Suppose that the problem is to a connected graph on six vertices with
minimum energy, and that the local search is based upon the transformation $t\ (=t_1)$
defined as adding or removing an edge.

The value of the objective function is $E(G) = 7.6655$ for this initial solution.

As the problem is restricted to connected graphs, some of the edges cannot be removed (for instance, the edge between vertices 1 and 2, or that between vertices 5 and 6).

The set of graphs in the neighborhood $\mathcal{N}^t(G)$ of G (represented in Fig. 4.3) associated with the transformation t is shown in Fig. 4.7. Owing to isomorphism, it is generally possible that one of these graphs could be obtained in various ways from the same graph, but this is not the case here (see Fig. 4.7), as the graphs all have different values of energy.

At this point, the best known solution is G, so we set $G^* = G$ and $k = 1$.

Comparing the values of the energy for graphs in $\mathcal{N}(G)$ with the value $E(G) = 7.6655$ for the initial graph G, we find that G^1, G^2, G^3, G^4, and G^5 are better solutions than G. The local search could thus proceed from any of these solutions. If the best-first criterion is chosen, one chooses G^3. At the next iteration, we have $G_1 = G^3$ as the current solution for iteration 1, and the value of the objective function is $E(G_1) = E(G^3) = 6.4852$ (Fig. 4.8).

Repeating the process in the next iteration (iteration 2), we get the graph G_2, whose energy is $E(G_2) = 5.81863$. In exploring the neighborhood $\mathcal{N}(G_2)$ of G_2, we notice that no graph is better than G_2, i.e., G_2 is thus a local optimum according to the transformation t_1. As it is the best known solution, we write $G^* = G_2, k = 1$, and the corresponding value of the objective function is $E^* = 5.81862$.

In the following step of the algorithm, we then proceed to a perturbation of the best known solution. Since $k = 1$, this perturbation may consist of adding or removing one edge randomly. After this perturbation, a local search is applied again. If this local search fails, the value k is increased by 1 before the next perturbation is applied (to the best known solution).

During the perturbation, the process of randomly adding or removing an edge is repeated k times, and a local search is applied, etc.

4.3.1.4 The Variable Neighborhood Search Descent

Suppose that a generalized variable neighborhood search had been applied. Instead of entering the perturbation phase after the local search, one would then apply another local search, based upon the next transformation in the list. The perturbation would only be applied when no transformation succeeded.

In the present case, one could try t_2 (moving an edge) and start the search from the graph G_3. Removing the edge between vertices 1 and 2 and inserting it between vertices 1 and 5 would improve the solution again. The solution obtained is then a star on six vertices whose energy is $E = 4.47214$. It is the optimal solution, since it is known that $E \geq 2\sqrt{m}$ [4] and $m \geq n - 1$ for connected graphs.

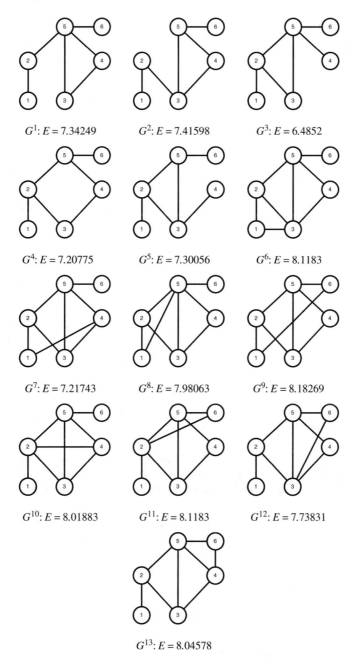

Fig. 4.7 Graphs of the neighborhood $N^t(G)$

Fig. 4.8 Current graph at
each iteration of the local
search, together with its
energy

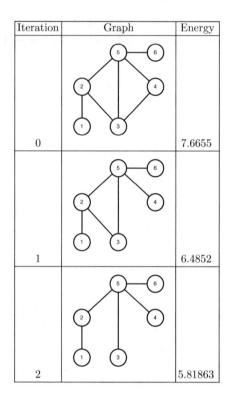

Iteration	Graph	Energy
0		7.6655
1		6.4852
2		5.81863

4.3.1.5 Perturbations

As is often the case with VNS, the perturbations may be built from a transformation
used in the local search. This perturbation is simply applying the transformation
randomly k times. Algorithm 4.8 describes a perturbation that consists of adding or
removing an edge.

Input: G
Input: k
repeat k **times**
  Choose a pair (i, j) of vertices of G.
  if *there is an edge between i and j* **then**
  | Let G' be the graph obtained by removing the edge (i, j).
  end
  else
  | Let G' be the graph obtained by adding the edge (i, j).
  end
return G'

Algorithm 4.8: Algorithme PERTADDREM.

If the number of edges is fixed, it is likely that such a perturbation would not give good results. In such a case, a transformation such as that described by Algorithm 4.9, which preserves the number of edges, would be better.

Input: G
Input: k
repeat k **times**
   | Choose a pair (i, j) of nonadjacent vertices of G.
   | Choose a pair (i', j') of adjacent vertices of G.
   | Let G' be the graph obtained by
   | adding the edge (i, j), and
   | removing the edge (i', j').
return G'

Algorithm 4.9: Algorithm PERTMOVE.

If the degree of every vertex is fixed, the transformation 2-opt would be better, etc.

4.3.2 Improving the k-Means Algorithm

Clustering is one of the important applications of VNS. The most well-known algorithm for clustering is certainly the k-means algorithm [12]. This algorithm is available in most data analysis software and is considered as a reference.

The criterion used by the k-means algorithm is the sum of the squared errors, which means the sum of the squares of the differences between the observations and their barycenter, as described in Algorithm 4.10.

For this problem, a solution is a partition of the observations into P clusters. This partition may be described by the binary variables $z_{ip} = 1$ if and only if the observation i belongs to the cluster p. Each observation is described by a vector of dimension m, denoted by x_{ij}, which gives the value of the variable j for the observation i. We denote by μ_{pj} the average of the variable j for the cluster p (the barycenter μ_p of the cluster p), as defined by Eq. (4.1):

$$\mu_{pj} = \frac{\sum_i z_{ip} x_{ij}}{\sum_i z_{ip}}. \tag{4.1}$$

Algorithm 4.10 is a local search. It turns out that this local search is very fast but that the local optimum it provides depends strongly on the initial solution. Unfortunately, for this kind of problem (and this local search), the number of local optima is large and the solution provided by the k-means algorithm may be very bad. A researcher who believes it produces the best possible solution could make important errors [15].

Input: S
Let $improved \leftarrow true$
while $improved = true$ **do**
   $improved \leftarrow false$
   Step 1:
   Compute μ_{pj} according to Eq. (4.1).
   Step 2:
   foreach $observation$ i **do**
      Let p $such$ $that$ $z_{ip} = 1$
      Let d_{ip} denote the euclidean distance between the observation i and the barycenter
      μ_p.
      if $\exists p'$ $such$ $that$ $d_{ip'} < d_{ip}$ **then**
         $improved \leftarrow true$
         $z_{ip} \leftarrow 0$
         $z_{ip'} \leftarrow 1$.
      end
   end
end
return S

Algorithm 4.10: k-means algorithm.

Instead of trying to find a way to build the initial solution that would lead to a better solution, we propose to use the algorithm as a local search within the VNS scheme. For better results, the reader could consider j-means [6], but the goal of the current section is to show that VNS could easily be implemented starting from an existing local search such as k-means.

Since we have a fast local search, it only remains to build the perturbations. Algorithm 4.11 describes a first perturbation.

Input: k
Input: S
repeat k **times**
   Randomly choose an observation i.
   Randomly choose a cluster p'.
   Let $z_{ip} \leftarrow 0$ $\forall p = 1, \ldots, P$.
   Let $z_{ip'} \leftarrow 1$.
return S

Algorithm 4.11: Algorithm PERTBASE.

The perturbation PERTBASE consists in randomly moving an observation from its cluster to another, k times. Such a perturbation will not be very efficient in practice, as it will not change the structure of the solution very much. The more sophisticated perturbations PERTSPLIT and PERTMERGE are defined in Algorithms 4.12 and 4.13, respectively. Those perturbations modify the solution by taking the nature of the problem into account. They produce a solution of different nature that remains consistent.

Input: k
Input: S
repeat k **times**
| Randomly choose a cluster p_1.
| Move all the observations of p_1 to the closest cluster.
| Randomly choose a cluster $p_2 \neq p_1$.
| Randomly assign the observations of p_2 to p_1 or p_2.
| Apply k-means to observations of clusters p_1 and p_2.
return S

Algorithm 4.12: Algorithm PERTSPLIT.

A perturbation based on PERTSPLIT applies Algorithm 4.12 k times.

One of the reasons these two perturbations are efficient is that k-means usually provides reasonable results if the number of clusters is small. The k-means algorithm is used here to provide some local optimization, restricted to two clusters.

Input: k
Input: S
repeat k **times**
| Randomly choose a cluster p_1.
| Randomly choose an observation i of p_1.
| Denote by p_2 the closest cluster of i (except p_1).
| Randomly assign the observations of p_1 and p_2 to p_1 or p_2 (shuffle observations of p_1 and p_2).
| Apply k-means to observations of clusters p_1 and p_2.
return S

Algorithm 4.13: Algorithm PERTMERGE.

Although the perturbation PERTMERGE seems to provide slightly better results than PERTSPLIT in practice, a combination of the two schemes remains the best option.

4.3.3 Using VNS for Continuous Optimization Problems

Although we have only described the use of VNS in the context of combinatorial optimization, it may also easily be used for continuous optimization [14].

4.3.3.1 Local Search

For continuous optimization problems, the local search cannot be described in the same way as previously. There are two possibilities:

- The objective function is differentiable and descent techniques such as gradient methods or any other descent method (such as the Newton, quasi-Newton, or conjugate gradient method, among others) can be used. The variable neighborhood search can easily be adapted to any kind of local search (as already stressed when we considered the use of the k-means algorithm for a local search). The variable neighborhood descent method may be less justified, even if it could still theoretically be applied. Any continuous optimization technique suitable for the problem may be used to find a local optimum.
- If the objective function is not differentiable, the use of direct search methods is needed. An example of VNS combined with a direct search algorithm is described in [1].

4.3.3.2 Perturbations

In the case of continuous optimization, the perturbations may be as simple as modifying randomly some variables by a magnitude that is proportional to k. However, some more complex strategies may be needed or more efficient.

4.4 Conclusion

In this chapter, the general principles of the variable neighborhood search method were presented, as well as the main variants of the algorithm, namely, the basic VNS, the general VNS, and the reduced VNS.

As demonstrated in numerous papers, VNS usually provides very good results in a reasonable amount of time. There exist a large number of applications for which VNS provides better results than other metaheuristics, and its performance is always respectable.

As shown in Sect. 4.3, dedicated to applications, and mainly the way to improve k-means, the implementation of VNS is relatively easy and comparable to that of multistart. This is certainly one of the main advantages of VNS.

As a final point, we should recall that VNS, like most metaheuristics, proposes a general scheme, a framework that a researcher may use as a starting point. This scheme may and should be adapted for each problem to which it is applied.

4.5 Annotated Bibliography

1. The main reference paper is [8]. This provides a description of the variable neighborhood search, as well as some applications, for instance the traveling salesman problem, the p-median problem, the multisource Weber problem, minimum sum-of-squares clustering (k-means), and bilinear programming under bilinear

constraints. Some extensions are then described, among which are variable neighborhood search with decomposition and nested variable neighborhood search.

2. Further information may be found in [7], which is dedicated to the description of developments related to VNS. The first part deals with the ways in which the method can be implemented, especially when applied to large-size problems. A number of applications are then described, varying from task scheduling on multiprocessor architectures to the problem of the maximum clique and the search for a maximum spanning tree with degree constraints.

3. The use of VNS in a context where the variables are continuous is detailed in [14]. Both the constrained and the unconstrained cases are treated.

4. More recently, two publications [10, 11] have been dedicated to surveys that cover techniques as well as a wide variety of applications. In addition to the extensive text, these two publications provide an important number of references from the technical point of view as well as from the point of view of the applications.

References

1. Audet, C., Béchard, V., LeDigabel, S.: Nonsmooth optimization through mesh adaptive direct search and variable neighborhood search. Journal of Global Optimization **41**(2), 299–318 (2008)
2. Caporossi, G., Hansen, P.: Variable neighborhood search for extremal graphs. 1. The Auto-GraphiX system. Discrete Mathematics **212**(1–2), 29–44 (2000)
3. Caporossi, G., Hansen, P.: A learning optimization algorithm in graph theory—versatile search for extremal graphs using a learning algorithm. In: *LION'12*, pp. 16–30 (2012)
4. Caporossi, G., Cvetković, D., Gutman, I., Hansen, P.: Variable neighborhood search for extremal graphs 2. finding graphs with extremal energy. Journal of Chemical Information and Computer Sciences **39**, 984–996 (1999)
5. Gross, J., Yellen, J., Zhang, P.: *Handbook of Graph Theory*, 2nd edition. Chapman and Hall/CRC (2013)
6. Hansen, P., Mladenović, N.: J-means, a new local search heuristic for minimum sum-of-squares clustering. Pattern Recognition **34**, 405–413 (2001)
7. Hansen, P., Mladenović, N.: *State-of-the-Art Handbook of Metaheuristics*, chapter Developments of Variable Neighborhood Search. Kluwer, Dordrecht (2001)
8. Hansen, P., Mladenović, N.: Variable neighborhood search: Principles and applications. European Journal of Operational Research **130**, 449–467 (2001)
9. Hansen, P., Mladenović, N.: First vs. best improvement: An empirical study. Discrete Applied Mathematics **154**(5), 802–817 (2006)
10. Hansen, P., Mladenović, N., Perez, J.A.M.: Variable neighborhood search: Methods and applications. 4OR **6**(4), 319–360 (2008)
11. Hansen, P., Mladenović, N., Brimberg, J., Perez, J.A.M.: *Handbook of Metaheuristics*, chapter Variable neighborhood search, pp. 61–86. Kluwer (2010)
12. MacQueen, J.: Some methods for classification and analysis of multivariate observations. In: *5th Berkeley Symposium on Mathematics Statistics and Probability*, pp. 281–297 (1967)

13. Mladenović, N., Hansen, P.: Variable neighborhood search. Computers and Operations Research **24**, 1097–1100 (1997)
14. Mladenović, N., Dražić, M., Kovačević-Vujčić, V., Čangalović, M.: General variable neighborhood search for the continuous optimization. European Journal of Operational Research **191**, 753–770 (2008)
15. Steinley, D.: Local optima in k-means clustering: What you don't know may hurt you. Psychological Methods **8**(3), 294–304 (2003)
16. Todeschini, R., Consonni, V.: *Handbook of Molecular Descriptors*. Wiley-VCH (2000)

Chapter 5
A Two-Phase Iterative Search Procedure: The GRASP Method

Michel Vasquez and Mirsad Buljubašić

5.1 Introduction

The GRASP method generates several configurations within the search space of a given problem, based on which it carries out an improvement phase. Being relatively straightforward to implement, this method has been applied to a wide array of hard combinatorial optimization problems, including scheduling [1], quadratic assignment [2], the traveling salesman problem [3], and maintenance workforce scheduling [4]. The interested reader is referred to the annotated bibliography by Festa and Resende [5], who have presented nearly 200 references on the topic. Moreover, the results output by this method are of similar quality to those determined using other heuristic approaches such as simulated annealing, tabu search, and population algorithms.

This chapter will present the principles behind the GRASP method and offer a sample application to the set covering problem.

5.2 General Principle Behind the Method

The GRASP method consists of repeating a constructive phase followed by an improvement phase, provided a stop condition has not yet been met (in most instances, this condition corresponds to a computation time limit expressed, for example, in terms of number of iterations or in seconds). Algorithm 5.1 describes the generic code associated with this procedure.

M. Vasquez (✉) · M. Buljubašić
Centre de Recherche LGI2P, Parc Scientifique Georges Besse, 30035 Nîmes Cedex 1, France
e-mail: michel.vasquez@mines-ales.fr

M. Buljubašić
e-mail: mirsad.buljubasic@mines-ales.fr

© Springer International Publishing Switzerland 2016
P. Siarry (ed.), *Metaheuristics*, DOI 10.1007/978-3-319-45403-0_5

```
input: α, random seed, time limit.
output: best solution found X*
repeat
    X ← Randomized Greedy(α);
    X ← Local Search(X, N);
    if z(X) better than z(X*) then
        X* ← X;
    end
until CPU time > time limit;
```
Algorithm 5.1: GRASP procedure

The constructive phase corresponds to a greedy algorithm, during which the step of assigning the current variable is slightly modified so as to generate several choices rather than just a single one at each iteration. These potential choices constitute a restricted candidate list (or RCL), from which a candidate will be chosen at random. Once the (variable, value) pair has been established, the RCL is updated by taking into account the current partial configuration. This step is then iterated until a complete configuration is obtained. The value associated with the particular (variable, value) pairs (as formalized by the heuristic function $\mathcal{H}$), for the variables still unassigned, reflects the changes introduced by selecting previous elements. Algorithm 5.2 summarizes this configuration construction phase, which will then be improved by a local search (simple descent, tabu search, or any other local-modification-type heuristic). The improvement phase is determined by the neighborhood $\mathcal{N}$ implemented in the attempt to refine the solution generated by the greedy algorithm.

```
input: α, random seed.
output: feasible solution X
X = ∅ ;
repeat
    Assemble the RCL on the basis of heuristic H and α;
    Randomly select an element xₕ from the RCL;
    X = X ∪ {xₕ};
    Update H;
until configuration X has been completed;
```
Algorithm 5.2: Randomized greedy algorithm

The evaluation of the heuristic function $\mathcal{H}$ serves to determine the insertion of (variable, value) pairs into the RCL. The way in which this criterion is taken into account exerts considerable influence on the behavior exhibited during the constructive phase: if only the best (variable, value) pair is selected relative to $\mathcal{H}$, then the same solution will often be obtained, and iterating the procedure will be of rather limited utility. If, on the other hand, all possible candidates were to be selected, the random algorithm derived would be capable of producing quite varied configurations, but of only mediocre quality: the likelihood of the improvement phase being sufficient to yield good solutions would thus be remote. The size of the RCL is therefore a determining parameter of this method. From a pragmatic standpoint, it is simpler to manage a qualitative acceptance threshold (i.e., $\mathcal{H}(x_j)$ better than $\alpha \times \mathcal{H}^*$, where $\mathcal{H}^*$ is the best benefit possible and α is a coefficient lying between 0 and 1) for the random drawing of a new (variable, value) pair to be assigned rather than to implement a list of k potential candidates, which would imply a data sort or the use of more complicated data structures. The terms used in this context are *threshold-based RCL* in the case of an acceptance threshold and *cardinality-based RCL* in all other cases.

The following sections will discuss in greater detail the various components of the GRASP method through an application to the set covering problem.

5.3   Set Covering Problem

Given a matrix (with m rows and n columns) composed solely of 0's and 1's, the objective is to identify the minimum number of columns such that each row contains at least one 1 in the identified columns. One example of minimum set covering problem is shown in Fig. 5.1.

More generally speaking, an n-dimensional vector *cost* has to be considered, containing strictly positive values. The objective then consists of minimizing the total cost of the columns capable of covering all rows: this minimization is known as the set covering problem, see Fig. 5.2 for a linear formulation.

For $1 \le j \le n$, the decision variable x_j equals 1 if column j is selected, and 0 otherwise. In the case of Fig. 5.1, for example, $x = < 101110100 >$ constitutes a solution whose objective value z is equal to 5.

If $cost_j$ equals 1 for each j, then the problem becomes qualified as a *unicost set covering problem*, of the kind stated at the beginning of this section. Both the unicost set covering problem and more general set covering problem are classified as combinatorial NP-hard problems [6]; moreover, once such problems reach a certain size, their solution within a reasonable amount of time becomes impossible by

Fig. 5.1 Incidence matrix for a minimum coverage problem

$$cover = \begin{pmatrix} 0\,1\,1\,1\,0\,0\,0\,0\,0 \\ 1\,0\,1\,0\,1\,0\,0\,0\,0 \\ 1\,1\,0\,0\,0\,1\,0\,0\,0 \\ 0\,0\,0\,0\,1\,1\,1\,0\,0 \\ 0\,0\,0\,1\,0\,1\,0\,1\,0 \\ 0\,0\,0\,1\,1\,0\,0\,0\,1 \\ 1\,0\,0\,0\,0\,0\,0\,1\,1 \\ 0\,1\,0\,0\,0\,0\,1\,0\,1 \\ 0\,0\,1\,0\,0\,0\,1\,1\,0 \\ 1\,0\,0\,1\,0\,0\,1\,0\,0 \\ 0\,1\,0\,0\,1\,0\,0\,1\,0 \\ 0\,0\,1\,0\,0\,1\,0\,0\,1 \end{pmatrix}$$

Fig. 5.2 Mathematical
model for the set covering
problem

$$\min \; z = \sum_{j=1}^{n} cost_j \times x_j,$$

$$\forall i \in [1, m] \quad \sum_{j=1}^{n} cover_{ij} \times x_j \geq 1,$$

$$\forall j \in [1, n] \quad x_j \in \{0, 1\}.$$

means of exact approaches. This observation justifies the implementation of heuristic approaches, such as the GRASP method, to handle these instances of hard problems.

5.4 An Initial Algorithm

This section will revisit the algorithm proposed by Feo and Resende in one of their first references on the topic [7], where the GRASP method was applied to the unicost set covering problem. It will then be shown how to improve the results and extend the study to the more general set covering problem through combining GRASP with the tabu search metaheuristic.

5.4.1 Constructive Phase

Let x be the characteristic vector of all columns X (where $x_j = 1$ if column j belongs to X and $x_j = 0$ otherwise): x is the binary vector in the mathematical model in Fig. 5.2. The objective of the greedy algorithm is to produce a configuration x with n binary components, whose corresponding set X of columns covers all the rows. In each iteration (out of a total of n), the choice of column j to be added to X ($x_j = 1$) will depend on the number of still uncovered rows that this column covers. As an example, the set of columns $X = \{0, 2, 3, 4, 6\}$ corresponds to the vector $x = < 101110100 >$, which is the solution to the small instance shown in Fig. 5.1.

For a given column j, we define the heuristic function $\mathcal{H}(j)$ as follows:

$$\mathcal{H}(j) = \begin{cases} \frac{\mathcal{C}(X \cup \{j\}) - \mathcal{C}(X)}{cost_j} & \text{if } x_j = 0 \\ \frac{\mathcal{C}(X \setminus \{j\}) - \mathcal{C}(X)}{cost_j} & \text{if } x_j = 1 \end{cases}$$

where $\mathcal{C}(X)$ is the number of rows covered by the set of columns X. The list of candidates RCL is managed implicitly: $\mathcal{H}^* = \mathcal{H}(j)$ maximum is first calculated over all columns j such that $x_j = 0$. The next step calls for randomly choosing a column h such that $x_h = 0$ and $\mathcal{H}(h) \geq \alpha \times \mathcal{H}^*$. The pseudocode of the *randomized* greedy algorithm is presented in Algorithm 5.3.

---

input: coefficient $\alpha \in [0, 1]$
output: feasible set X of selected columns
$X = \emptyset$;
repeat
    $j^* \leftarrow$ column, such that $\mathcal{H}(j^*)$ is maximized;
    `threshold` $\leftarrow \alpha \times \mathcal{H}(j^*)$;
    $r \leftarrow$ `rand()` modulo n;
    for $j \in \{r, r+1, \ldots, n-1, 0, 1, \ldots, r-1\}$ **do**
        if $\mathcal{H}(j) \geq$ `threshold` **then**
            `break`;
        end
    end
    $X = X \cup \{j\}$ (add column j to the set $X \Leftrightarrow x_j = 1$);
until *all rows have been covered*;

Algorithm 5.3: `greedy`(α)

Table 5.1 Occurrences of solutions by z value for the instance S45

$\alpha \backslash z$	30	31	32	33	34	35	36	37	38	39	40	41	Total
0.0	0	0	0	0	0	1	9	10	15	17	21	15	88
0.2	0	0	0	1	3	15	34	23	18	5	1	0	100
0.4	0	0	0	5	13	30	35	16	1	0	0	0	100
0.6	0	2	2	45	38	13	0	0	0	0	0	0	100
0.8	0	11	43	46	0	0	0	0	0	0	0	0	100
1.0	0	55	19	26	0	0	0	0	0	0	0	0	100

The heuristic function $\mathcal{H}()$, which determines the insertion of columns into the RCL, is reevaluated at each step so as to take into account only the uncovered rows. This is the property that gives rise to the adaptive nature of the GRASP method.

Let us now consider the instance with $n = 45$ columns and $m = 330$ rows that corresponds to the data file `data.45` (renamed S45) on Beasley's OR-Library site [8], included in the four unicost set covering problems derived from Steiner's triple systems. By choosing the values $0, 0.2, \ldots, 1$ for α and $1, 2, \ldots, 100$ for the seed of the pseudorandom sequence, the results table presented in Table 5.1 was obtained. This table lists the number of solutions whose coverage size z lies between 30 and 41. The quality of these solutions is clearly correlated with the value of the parameter α. For the case $\alpha = 0$ (random assignment), it can be observed that the `greedy()` function produces 12 solutions of a size that strictly exceeds 41. No solution with an optimal coverage size of 30 (known for this instance) is actually produced.

5.4.2 Improvement Phase

The improvement algorithm proposed by Feo and Resende [7] is a simple descent on an elementary neighborhood $\mathcal{N}$. Let x denote the current configuration; then a configuration x' belongs to $N(x)$ if a unique j exists such that $x_j = 1$ and $x'_j = 0$ and, moreover, that $\forall i \in [1, m]$, $\sum_{j=1}^{n} cover_{ij} \times x'_j \geq 1$. Between two neighboring configurations x and x', a redundant column (from the standpoint of row coverage) is deleted.

---

input: characteristic vector x from the set X
output: feasible x without any redundant column
while *redundant columns continue to exist* **do**
    Find redundant $j \in X$ such that $cost_j$ is maximized;
    if j *exists* **then**
       | $X = X \setminus \{j\}$
    end
end

Algorithm 5.4: descent(x)

Algorithm 5.4 describes this descent phase and takes into account the cost of each column, with respect to the column deletion criterion.

The statistical study of the occurrences of the best solutions done with the greedy() procedure on its own (see Table 5.1) was repeated, this time with the addition of the descent() procedure, yielding the results presented in Table 5.2. A leftward shift is observed in the occurrences of the objective value z; such an observation effectively illustrates the benefit of this improvement phase. Before pursuing the various experimental phases, the characteristics of our benchmark will first be presented.

Table 5.2 Occurrences of solutions by z value for the instance S45 with descent ()

$\alpha \backslash z$	30	31	32	33	34	35	36	37	38	39	40	41	Total
0.0	0	0	0	0	1	9	10	15	17	21	15	8	96
0.2	0	0	1	3	15	34	23	18	5	1	0	0	100
0.4	0	0	5	13	30	35	16	1	0	0	0	0	100
0.6	2	2	45	38	13	0	0	0	0	0	0	0	100
0.8	11	43	46	0	0	0	0	0	0	0	0	0	100
1.0	55	19	26	0	0	0	0	0	0	0	0	0	100

Table 5.3 Characteristics of the various instances considered

Instance	n	m	Instance	n	m	Instance	n	m
G1	10,000	1000	H1	10,000	1000	S45	45	330
G2	10,000	1000	H2	10,000	1000	S81	81	1080
G3	10,000	1000	H3	10,000	1000	S135	135	3015
G4	10,000	1000	H4	10,000	1000	S243	243	9801
G5	10,000	1000	H5	10,000	1000			

5.5 Benchmark

The benchmark used for experimentation purposes was composed of 14 instances available on Beasley's OR-Library site [8].

The four instances data.45, data.81, data.135, and data.243 (renamed S45, S81, S135, and S243, respectively) make up the test datasets in the reference article by Feo and Resende [7]: these are all unicost set covering problems. The 10 instances G1, ..., G5 and H1, ..., H5 are considered as set covering problems. Table 5.3 indicates, for each test dataset, the number n of columns and number m of rows.

The GRASP method was run 100 times for each of the three values $0.1, 0.5$, and 0.9 of the coefficient α. The seed g of the srand(g) function took the values $1, 2, \ldots, 100$. For each execution of the method, the CPU time was limited to 10 s. The computer used for this benchmark was equipped with an i7 processor running at 3.4 GHz with 8 GB of hard drive memory. The operating system was Linux, Ubuntu 12.10.

5.6 Experiments with greedy(α)+descent

Algorithm 5.5 shows the pseudocode of the initial version of the GRASP method, GRASP1, which was used for experimentation on the 14 datasets of our benchmark.

The functions srand() and rand() used in the experimental phase were those of *Numerical Recipes* in C [9]. We should point out that the coding of the function $\mathcal{H}$ is critical: the introduction of an incremental computation is essential to obtaining relative short execution times. The values given in Table 5.4 summarize the results output by the GRASP1 procedure. This table of results indicates the following:

- the name of the instance tested;
- the best value z^* known for this particular problem;
- for each value of the coefficient $\alpha = 0.1, 0.5$, and 0.9:

input: α, random seed, time limit.
output: z^{best}
srand(*seed*);
$z^{best} \leftarrow +\infty$;
repeat
　　$x \leftarrow$ greedy(α);
　　$x \leftarrow$ descent(x);
　　if $z(x) < z^{best}$ **then**
　　　　$z^{best} \leftarrow z(x)$;
　　end
until *CPU time > time limit*;

Algorithm 5.5: GRASP1

Table 5.4 Results from greedy(α)+descent

Instance	z^*	$\alpha = 0.1$				$\alpha = 0.5$			$\alpha = 0.9$		
		z	#	$\frac{\sum z_g}{100}$	z	#	$\frac{\sum z_g}{100}$	z	#	$\frac{\sum z_g}{100}$
G1	176	240	1	281.83	181	1	184.16	183	3	185.14
G2	154	208	1	235.34	162	7	164.16	159	1	160.64
G3	166	199	1	222.59	175	2	176.91	176	3	176.98
G4	168	215	1	245.78	175	1	177.90	177	5	178.09
G5	168	229	1	249.40	175	1	178.56	174	6	175.73
H1	63	69	1	72.30	67	29	67.71	67	5	68.19
H2	63	69	2	72.28	66	1	67.71	67	1	68.51
H3	59	64	1	68.80	62	1	64.81	63	34	63.66
H4	58	64	1	67.12	62	18	62.86	63	80	63.20
H5	55	61	1	62.94	59	2	60.51	57	99	57.01
S45	30	30	100	30.00	30	100	30.00	30	100	30.00
S81	61	61	100	61.00	61	100	61.00	61	100	61.00
S135	103	104	2	104.98	104	4	104.96	103	1	104.10
S243	198	201	1	203.65	203	18	203.82	203	6	204.31

- the best value z found using the GRASP method;
- the number of times (#) this value was reached per 100 runs;
- the average of the 100 values produced by this algorithm.

For the four instances S45, S81, S135 and S243 the value displayed in the column z^* is optimal [10]. On the other hand, the optimal value for the other 10 instances (G1, ..., G5 and H1, ..., H5) remains unknown: the z^* values for these instances are the best values published in the article by Azimi et al. [11].

With the exception of instance S243, the best results were obtained using the values 0.5 and 0.9 of the RCL management parameter α. For the four instances derived from the Steiner's triple problem, the values published by Feo and Resende

[7] are corroborated. However, when compared with the results of Azimi et al. [11], performed in 2010, or even those of Caprara et al. [12], dating back to 2000, these results prove to be relatively far from the best published values.

5.7 Local Tabu Search

This section focuses on adding a tabu search phase to the GRASP method in order to generate results that are more competitive with respect to the literature. The algorithm associated with this tabu search is characterized by:

- an infeasible configuration space S, such that $z(x) < z^{min}$;
- a simple move (of the 1-change) type;
- a strict tabu list.

5.7.1 The Search Space

By relying on the configuration x^0 output by the descent phase (corresponding to a set X of columns guaranteeing row coverage), the tabu search explores the space of configurations x with objective value $z(x)$ less than $z^{min} = z(x^{min})$, where x^{min} is the best feasible solution found by the algorithm. The search space S is thus formally defined as follows:

$$S = \{x \in \{0, 1\}^n \, / \, z(x) < z(x^{min})\}$$

5.7.2 Evaluation of a Configuration

It is obvious that the row coverage constraints have been relaxed. The evaluation function $\mathcal{H}$ of a column j now contains two components:

$$\mathcal{H}_1(j) = \begin{cases} \mathcal{C}(X \cup \{j\}) - \mathcal{C}(X) & \text{if } x_j = 0 \\ \mathcal{C}(X \setminus \{j\}) - \mathcal{C}(X) & \text{if } x_j = 1 \end{cases}$$

and

$$\mathcal{H}_2(j) = \begin{cases} cost_j & \text{if } x_j = 0 \\ -cost_j & \text{if } x_j = 1 \end{cases}$$

This step consists of repairing the coverage constraints (i.e., maximizing $\mathcal{H}_1$) at the lowest cost (minimizing $\mathcal{H}_2$).

5.7.3 Managing the Tabu List

This task involves the use of the reverse elimination method proposed by Glover and Laguna [13], which was implemented in order to manage the tabu status of potential moves exactly: a move is forbidden if and only if it leads to a previously encountered configuration. This tabu list is referred to as a strict list.

```
input: j ∈ [0, n − 1]
running list[iteration] = j;
i ← iteration;
iteration ← iteration + 1;
repeat
    j ← running list[i];
    if j ∈ RCS then
        | RCS ← RCS/{j};
    end
    else
        | RCS ← RCS ∪ {j};
    end
    if |RCS| = 1 then
        | j = RCS[0] is tabu;
    end
    i ← i − 1
until i < 0;
```

Algorithm 5.6: updateTabu(j)

The algorithm described in Algorithm 5.6 is identical to one successfully run on another combinatorial problem with binary variables [14]. The *running list* is actually a table in which a recording is made, in each iteration, of the column j targeted by the most recent move: $x_j = 0$ or $x_j = 1$. This column is considered as the move attribute. The RCS (for *residual cancellation sequence*) is another table, in which attributes are either added or deleted. The underlying principle consists of reading past move attributes one by one, from the end of the running list, and adding the RCS should they be absent and removing the RCS if they are already present. The following equivalence is thus derived: $|RCL| = 1 \Leftrightarrow RCL[0]$ prohibited. The interested reader is referred to the academic article by Dammeyer and Voss [15] for further details of this specific method.

5.7.4 Neighborhood

We have made use of an elementary *1-change* move: $x' \in \mathcal{N}(x)$ if $\exists! \, j/x'_j \neq x_j$. The neighbor x' of configuration x differs only by one component yet still satisfies the condition $z(x') < z^{\min}$, where $z^{\min}$ is the value of the best feasible configuration identified. Moreover, the chosen non-tabu column j minimizes the hierarchical criterion $((\mathcal{H}_1(j), \mathcal{H}_2(j)))$. Algorithm 5.7 describes the evaluation function for this neighborhood.

---

input: column interval $[j1, j2]$
output: *best* column identified j^*
$j^* \leftarrow -1;$
$\mathcal{H}_1^* \leftarrow -\infty;$
$\mathcal{H}_2^* \leftarrow +\infty;$
for $j1 \leq j \leq j2$ **do**
 if j *non*-tabu **then**
  if $(x_j = 1) \vee (z + cost_j < z^{\min})$ **then**
   if $(\mathcal{H}_1(j) > \mathcal{H}_1^*) \vee (\mathcal{H}_1(j) = \mathcal{H}_1^* \wedge \mathcal{H}_2(j) < \mathcal{H}_2^*)$ **then**
    $j^* \leftarrow j;$
    $\mathcal{H}_1^* \leftarrow \mathcal{H}_1(j);$
    $\mathcal{H}_2^* \leftarrow \mathcal{H}_2(j);$
   end
  end
 end
end

Algorithm 5.7: $\texttt{eval}\mathcal{H}(j1, j2)$

5.7.5   The Tabu Algorithm

The general $\texttt{Tabu()}$ procedure uses as an argument the solution x produced by the $\texttt{descent()}$ procedure, along with a maximum number of iterations N. Rows 2 through 6 of Algorithm 5.8 correspond to a search diversification mechanism. Each time a feasible configuration is produced, the value $z^{\min}$ is updated and the tabu list is reset to zero.

5.8   Experiments with $\texttt{greedy}(\alpha)\texttt{+descent+Tabu}$

For this second experimental phase, the benchmark was similar to that discussed in Sect. 5.5. The total CPU time remained limited to 10 s, while the maximum number of iterations without improvement for the $\texttt{Tabu()}$ procedure equaled half the number of columns for the instance treated (i.e., $n/2$). The pseudocode of the $\texttt{GRASP2}$ procedure is specified in Algorithm 5.9.

Table 5.5 illustrates the significant contribution of the tabu search to the GRASP method. All values in the z^* column were found using this version of the GRASP method. In comparison with Table 5.4, the parameter α has less influence on the

input: feasible solution x, number of iterations N
output: z^{min}, x^{min}
1 $z^{min} \leftarrow z(x)$;
  $iter \leftarrow 0$;
  repeat
2    $r \leftarrow$ rand() modulo n;
3    $j^* \leftarrow$ eval$\mathcal{H}(r, n-1)$;
4   **if** $j^* < 0$ **then**
5     $|$ $j^* \leftarrow$ eval$\mathcal{H}(0, r-1)$;
   end
   if $x_{j^*} = 0$ **then**
    $|$ add column j^*;
   else
    $|$ remove column j^*;
   end
   if *all the rows are covered* **then**
     $z^{min} \leftarrow z(x)$;
6      $x^{min} \leftarrow x$;
     $iter \leftarrow 0$;
     delete the tabu status;
   end
   updateTabu(j^*);
  until *iter* $\geq N$ *or* $j^* < 0$;

Algorithm 5.8: tabu(x, N)

input: α, random seed *seed*, time limit.
output: z^{best}
$z^{best} \leftarrow +\infty$;
srand(*seed*);
repeat
  $|$ $x \leftarrow$ greedy(α);
  $|$ $x \leftarrow$ descent(x);
  $|$ $z \leftarrow$ Tabu($x, n/2$);
  $|$ **if** $z < z^{best}$ **then**
  $|$ $|$ $z^{best} \leftarrow z$;
  $|$ **end**
until *CPU time* > *time limit*;

Algorithm 5.9: GRASP2

results. It would seem that the multi-start function of the GRASP method is more critical to the tabu phase than control over the RCL. However, as was demonstrated in the following experimental phase, it still appears that rerunning the method, under control of the parameter α, does play a determining role in obtaining the best results (Table 5.6).

Table 5.5 Results from `greedy(α)+descent+Tabu`

Instance	z^*	$\alpha = 0.1$			$\alpha = 0.5$			$\alpha = 0.9$		
		z	#	$\frac{\sum z_g}{100}$	z	#	$\frac{\sum z_g}{100}$	z	#	$\frac{\sum z_g}{100}$
G1	176	176	100	176.00	176	96	176.04	176	96	176.04
G2	154	154	24	154.91	154	32	155.02	154	57	154.63
G3	166	167	4	168.46	167	10	168.48	166	1	168.59
G4	168	168	1	170.34	170	35	170.77	170	29	170.96
G5	168	168	10	169.59	168	7	169.66	168	10	169.34
H1	63	63	11	63.89	63	2	63.98	63	5	63.95
H2	63	63	21	63.79	63	13	63.87	63	5	63.95
H3	59	59	76	59.24	59	82	59.18	59	29	59.73
H4	58	58	99	58.01	58	98	58.02	58	100	58.00
H5	55	55	100	55.00	55	100	55.00	55	100	55.00
S45	30	30	100	30.00	30	100	30.00	30	100	30.00
S81	61	61	100	61.00	61	100	61.00	61	100	61.00
S135	103	103	49	103.51	103	61	103.39	103	52	103.48
S243	198	198	100	198.00	198	100	198.00	198	100	198.00

Table 5.6 Results from `greedy(1)+descent+Tabu`

Instance	z	#	$\frac{\sum z_g}{100}$	Instance	z	#	$\frac{\sum z_g}{100}$	Instance	z	#	$\frac{\sum z_g}{100}$
G1	176	95	176.08	H1	63	2	63.98	S45	30	100	30.00
G2	154	24	155.22	H2	63	4	63.96	S81	61	100	61.00
G3	167	19	168.48	H3	59	36	59.74	S135	103	28	103.74
G4	170	3	171.90	H4	58	91	58.09	S243	198	98	198.10
G5	168	20	169.39	H5	55	97	55.03				

5.9   Experiments with `greedy(1)+Tabu`

To confirm the benefit of this GRASP method, let us now observe the behavior of
Algorithm 5.10, TABU. For each value of the pseudo-random function `rand()` seed
(between 1 and 100), a solution was built using the `greedy(1)` procedure, whereby
redundant x columns were deleted to allow for completion of the `Tabu(x, n)` procedure, provided the CPU time remained less than 10 s.

For this final experimental phase, row 1 in Algorithm 5.8 was replaced by
$z^{\min} \leftarrow +\infty$. Provided the CPU time allocation had not been depleted, the `Tabu()`
procedure was reinitiated starting with the best solution it was able to produce during
the previous iteration. This configuration was saved in row 6. The size of the running
list was twice as long.

```
input: random seed, time limit.
output: z^best
z^best ← +∞;
srand(seed);
x ← greedy(1);
x^min ← descent(x);
repeat
    x ← x^min;
    z, x^min ← Tabu(x, n);
    if z < z^best then
        z^best ← z;
    end
until CPU time > time limit;
```

Algorithm 5.10: TABU

In absolute value terms, these results fall short of those output by Algorithm 5.9, GRASP2. This TABU version produces values of 167 and 170 for instances G3 and G4 versus 166 and 168, respectively, for the GRASP2 version. Moreover, the majority of the average values are of poorer quality than those listed in Table 5.5.

5.10   Conclusion

This chapter has presented the principles behind the GRASP method and has detailed their implementation with the aim of solving large-sized instances associated with a hard combinatorial problem. Section 5.4.1 demonstrated the simplicity of modifying the greedy heuristic proposed by Feo and Resende, namely

$$\mathcal{H}(j) = \begin{cases} \mathcal{C}(X \cup \{j\}) - \mathcal{C}(X) & \text{if } x_j = 0 \\ \mathcal{C}(X \setminus \{j\}) - \mathcal{C}(X) & \text{if } x_j = 1 \end{cases}$$

in order to take the column cost into account and apply the construction phase not only to the minimum coverage problem but also to the set covering problem.

The advantage of enhancing the improvement phase has also been demonstrated by adding, to the general GRASP method loop, a local tabu search on an elementary neighborhood.

A distinct loss of influence of the parameter α was observed when the tabu search was used. This behavior is very likely to be correlated with the instances being treated: Tables 5.1 and 5.2 show that even for the value $\alpha = 1$, the greedy algorithm builds different solutions. The experiments described in Sects. 5.4 and 5.5 nonetheless effectively illustrate the contributions provided by this construction phase. Such contributions encompass both the diversified search space and the capacity of the greedy(α) procedure to produce high-quality initial solutions for the local tabu search, thus yielding a powerful retry mechanism.

In closing, other methods such as Path Relinking have also been proposed for the improvement phase. Moreover, from the implementation perspective, the GRASP method is well suited to parallelism. As regards these last two points, the interested reader will benefit from references [16, 17], to name just two.

5.11   Annotated Bibliography

Reference [1]   In this reference, the GRASP method is introduced to solve a job shop scheduling problem, which entails minimizing the running time of the most heavily used machine in terms of job duration (or makespan). The tasks (jobs) correspond to ordered sequences of operations. During the construction phase, individual operations are scheduled one by one at each iteration on a given machine. The candidate list for this phase is composed of the set of terminal operations sorted in increasing order of their insertion cost (calculated as the makespan value after insertion − the value before insertion). This order, along with the memorization of elite solutions, influences the choice of the next operation to be scheduled. The improvement phase thus comprises a local search on a partial configuration as well as on a complete configuration: this phase involves exchanging tasks within the disjunctive graph, whose objective is to reduce the makespan.

Reference [7]   This is one of the first academic papers on the GRASP method. The principles behind this method are clearly described and illustrated by two distinct implementation cases: one of them inspired the solution of the minimum coverage problem presented in this chapter, and the other was applied to solve the maximum independent set problem an a graph.

Reference [3]   This article presents an application of the GRASP method to the traveling salesman problem. The construction phase relies on a candidate list determined by its size (known as a cardinality-based RCL) and not by any quality threshold. The improvement phase features a local search using a variation of the 2-opt and 3-opt neighborhoods. Rather sophisticated data structures were implemented in this study, and the results obtained on TSPLIB instances are of high quality.

References

1. Binato, S., Hery, W.J., Loewenstern, D.M., Resende, M.G.C.: A greedy randomized adaptive search procedure for job shop scheduling. IEEE Transactions on Power Systems **16**, 247–253 (2001)
2. Pitsoulis, L.S., Pardalos, P.M., Hearn, D.W.: Approximate solutions to the turbine balancing problem. European Journal of Operational Research **130**(1), 147–155 (2001)
3. Marinakis, Y., Migdalas, A., Pardalos, P.M.: Expanding neighborhood GRASP for the traveling salesman problem. Computational Optimization and Applications **32**(3), 231–257 (2005)
4. Hashimoto, H., Boussier, S., Vasquez, M., Wilbaut, C.: A GRASP-based approach for technicians and interventions scheduling for telecommunications. Annals of Operations Research **183**(1), 143–161 (2011)
5. Festa, P., Resende, M.: GRASP: An annotated bibliography. In: *Essays and Surveys on Metaheuristics*, C.C. Ribeiro and P. Hansen (eds.), Kluwer Academic, pp. 325–367 (2002)

6. Garey, M., Johnson, D.: *Computers and Intractability: A Guide to the Theory of NP-Completeness*. Freeman (1979)
7. Feo, T., Resende, M.: Greedy randomized adaptive search procedures. Journal of Global Optimization **6**, 109–134 (1995)
8. Beasley, J.E.: OR-Library: Distributing test problems by electronic mail. Journal of the Operational Research Society **41**(11), 1069–1072 (1990)
9. Press, W.H., Teukolsky, S.A., Vetterling, W.T., Flannery, B.P.: *Numerical Recipes in C*, 2nd edition. Cambridge University Press (1992)
10. Ostrowski, J., Linderoth, J., Rossi, F., Smriglio, S.: Solving large Steiner triple covering problems. Operations Research Letters **39**(2), 127–131 (2011)
11. Azimi, Z.N., Toth, P., Galli, L.: An electromagnetism metaheuristic for the unicost set covering problem. European Journal of Operational Research **205**(2), 290–300 (2010)
12. Caprara, A., Fischetti, M., Toth, P.: Algorithms for the set covering problem. Annals of Operations Research **98**, 2000 (1998)
13. Glover, F., Laguna, M.: *Tabu Search*. Kluwer Academic, vol. 7, pp. 239–240 (1997)
14. Nebel, B. (ed.): *Proceedings of the Seventeenth International Joint Conference on Artificial Intelligence, IJCAI 2001*, Seattle, August 4–10, 2001. Morgan Kaufmann (2001)
15. Dammeyer, F., Voß, S.: Dynamic tabu list management using the reverse elimination method. Annals of Operations Research **41**(2), 29–46 (1993)
16. Aiex, R.M., Binato, S., Resende, M.G.C.: Parallel GRASP with path-relinking for job shop scheduling. Parallel Computing **29**, 393–430 (2002)
17. Crainic, T.G., Mancini, S., Perboli, G., Tadei, R.: GRASP with path relinking for the two-echelon vehicle routing problem. Advances in Metaheuristics, Operations Research/Computer Science Interfaces Series **53**, 113–125 (2013)

Chapter 6
Evolutionary Algorithms

Alain Petrowski and Sana Ben Hamida

6.1 From Genetics to Engineering

Biological evolution has generated extremely complex autonomous living beings which can solve extraordinarily difficult problems, such as continuous adaptation to complex, uncertain environments that are in perpetual transformation. For that purpose, the higher living beings, such as mammals, are equipped with excellent capabilities for pattern recognition, training, and intelligence. The large variety of the situations to which life has adapted shows that the process of evolution is robust and is able to solve many classes of problems. This allows a spectator of the living world to conceive the idea that there are ways other than establishing precise processes, patiently derived from high-quality knowledge of natural laws, to satisfactorily build up complex and efficient systems.

According to Darwin [16], the original mechanisms of evolution of living beings are based on a competition which selects the individuals most well adapted to their environment while ensuring descent, that is, transmission to children of the useful characteristics which allowed the survival of the parents. This inheritance mechanism is based, in particular, on a form of cooperation implemented by sexual reproduction.

The assumption that Darwin's theory, enriched by our current knowledge of genetics, can account for the mechanisms of evolution is still not justified. Nobody can confirm today that these mechanisms are well understood, and that there is no essential phenomenon that remains unexplored.

However, Neo-Darwinism is the only theory of evolution available that has never failed up to now. The development of electronic calculators facilitated the study

A. Petrowski (✉)
Telecom SudParis, 91000 Evry, France
e-mail: Alain.Petrowski@telecom-sudparis.eu

S. Ben Hamida
Université Paris Ouest, 92000 Nanterre, France
e-mail: sbenhami@u-paris10.fr

© Springer International Publishing Switzerland 2016
P. Siarry (ed.), *Metaheuristics*, DOI 10.1007/978-3-319-45403-0_6

of this theory by simulations and some researchers desired to test it on engineering problems, way back in the 1950s. But this work was not convincing, because of insufficient knowledge at that time of natural genetics and also because of the weak performance of the calculators available. In addition, the extreme slowness the evolution crippled the idea that such a process could be usefully exploited.

During the 1960s and 1970s, as soon as calculators of more credible power came into existence, many attempts to model the process of evolution were undertaken. Among those, three approaches emerged independently, mutually unaware of the presence of the others, until the beginning of the 1990s:

- the *evolution strategies* of Schwefel and Rechenberg [9, 45], which were designed in the middle of the 1960s as an optimization method for problems using continuously varying parameters;
- the *evolutionary programming* of Fogel et al. [23], which aimed, during the middle of the 1960s, to make the structure of finite-state automata evolve with iterated selections and mutations; it was intended to provide an alternative to the artificial intelligence of the time;
- the *genetic algorithms*, which were presented in 1975 by Holland [32], with the objective of understanding the underlying mechanisms of self-adaptive systems.

Thereafter, these approaches underwent many modifications according to the variety of the problems faced by their founders and their pupils. Genetic algorithms became extremely popular after the publication of the book *Genetic Algorithms in Search, Optimization and Machine Learning* by Goldberg [26] in 1989. This book, published worldwide, resulted in an exponential growth in interest in this field. Whereas there were about a few hundred publications in this area in the 20 years before this book appeared, there are several hundreds of thousands of references related to evolutionary computation available today, according to the Google Scholar website.[1] Researchers in this field have organized joint international conferences to present and combine their different approaches.

6.1.1 Genetic Algorithms or Evolutionary Algorithms?

The widespread term *evolutionary computation* appeared in 1993 as the title of a new journal published by MIT Press, and it was then widely used to designate all of the techniques based on the metaphor of biological evolution theory. However, some specialists use the term "genetic algorithm" to designate any evolutionary technique even if it has few points in common with the original proposals of Holland and Goldberg.

The various evolutionary approaches are based on a common model presented in Sect. 6.2. Sections 6.3–6.8 describe various alternatives for the selection and variation operators, which are basic building blocks of any evolutionary algorithm. Genetic

---

[1] https://scholar.google.com/scholar?q=genetic+algorithms.

algorithms are the most "popular" evolutionary algorithms. This is why Sect. 6.9 is devoted especially to them. This section shows how it is possible to build a simple genetic algorithm from a suitable combination of specific selection and variation operators.

Finally, Sect. 6.10 presents the *covariance matrix adaptation evolution strategy* (CMA-ES). This powerful method should be considered when one or more optima are sought in $\mathbb{R}^n$. It derives directly from studies aimed at improving the evolution strategies but, strictly speaking, it is not an evolutionary algorithm as defined in Sect. 6.2.

The chapter concludes with a mini-glossary of terminology usually used in the field and a bibliography with accompanying notes.

6.2   The Generic Evolutionary Algorithm

In the world of evolutionary algorithms, the *individuals* subjected to evolution are the solutions, which may be more or less efficient, for a given problem. These solutions belong to the search space of the optimization problem. The set of individuals treated simultaneously by the evolutionary algorithm constitutes a *population*. It evolves during a succession of iterations called *generations* until a termination criterion, which takes into account a priori the quality of the solutions obtained, is satisfied.

During each generation, a succession of operators is applied to the individuals of a population to generate a new population for the next generation. When one or more individuals are used by an operator, they are called the *parents*. The individuals originating from the application of the operator are its *offspring*. Thus, when two operators are applied successively, the offspring generated by one can become parents for the other.

6.2.1   Selection Operators

In each generation, the individuals reproduce, survive, or disappear from the population under the action of two *selection operators*:

- the selection operator for the reproduction, or simply *selection*, which determines how many times an individual will reproduce in a generation;
- the selection operator for replacement, or simply *replacement*, which determines which individuals will have to disappear from the population in each generation so that, from generation to generation, the population size remains constant or, in some cases, is controlled according to a definite policy.

In accordance with the Darwinist creed, the better an individual, the more often it is selected to reproduce or survive. It may be, according to the variant of the algorithm, that one of the two operators does not favor the good individuals compared with the others, but it is necessary that the application of the two operators together during a generation introduces a bias in favor of the best. To make selection possible, a fitness value, which obviously depends on the objective function, must be attached to each individual. This implies that, in each generation, the fitnesses of the offspring are evaluated, which can be computationally intensive. The construction of a good *fitness function* from an objective function is rarely easy.

6.2.2   Variation Operators

In order that the algorithm can find solutions better than those represented in the current population, it is required that they are transformed by the application of *variation operators*, or *search operators*. A large variety of them can be imagined. They are classified into two categories:

- *mutation* operators, which modify an individual to form another;
- *crossover* operators, which generate one or more offspring from combinations of two parents.

  The designations of these operators are based on the real-life concept of the sexual reproduction of living beings, with the difference that evolutionary computation, not knowing biological constraints, can be generalized to implement the combination of more than two parents, and possibly the combination of the entire population.

The way in which an individual is modified depends closely on the structure of the solution that it represents. Thus, if it is desired to solve an optimization problem in a continuous space, for example a domain of $\mathbb{R}^n$, then it will be a priori adequate to choose a vector in $\mathbb{R}^n$ to represent a solution, and the crossover operator must implement a means such that two vectors in $\mathbb{R}^n$ for the parents correspond to one vector (or several) in $\mathbb{R}^n$ for the offspring. On the other hand, if one wishes to use an evolutionary algorithm to solve instances of the traveling salesman problem, it is common that an individual corresponds to a round trip. It is possible to represent this as a vector where each component is a number that designates a city. The variation operators should then generate only legal round trips, i.e., round trips in which each city in the circuit is present only once. These examples show that it is impossible to design universal variation operators, independent of the problem under consideration. They are necessarily related to the *representation* of the solutions in the search space. As a general rule, for any particular representation chosen, it is necessary to define the variation operators to be used, because they depend closely on it.

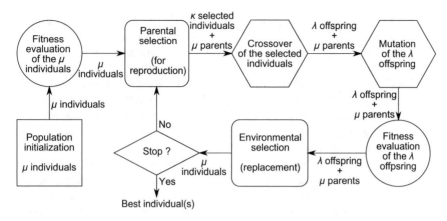

Fig. 6.1 The generic evolutionary algorithm

6.2.3 The Generational Loop

In each generation, an evolutionary algorithm implements a "loop iteration" that incorporates the application of these operators on the population:

1. For reproduction, selection of the parents from among a population of μ individuals to generate λ offspring.
2. Crossover and mutation of the λ selected individuals to generate λ offspring.
3. Fitness evaluation for the offspring.
4. Selection for survival of μ individuals from among the λ offspring and μ parents, or only from among the λ offspring, according to the choice made by the user, in order to build the population for the next generation.

Figure 6.1 represents this loop graphically with the insertion of a stopping test and addition of the phase of initialization of the population. Note that the hexagonal shapes refer to the variation operators, which are dependent on the representation chosen, while the "rounded rectangles" represent the selection operators, which are independent of the solution representation.

6.2.4 Solving a Simple Problem

Following our own way of illustrating the operation of an evolutionary algorithm, let us consider the maximization of the function $C(x) = 400 - x^2$ for x in the interval $[-20, 20]$. There is obviously no practical interest in using this type of algorithm to solve such a simple problem; the objectives here are exclusively didactic. This example will be considered again and commented on throughout the part of this chapter presenting the basics of evolutionary algorithms. Figure 6.2 shows the succession of operations from the initialization phase of the algorithm to the end of the first generation. In this figure, an individual is represented by a rectangle partitioned into two zones. The top zone represents the value of the individual x, ranging between -20 and $+20$. The bottom zone contains the corresponding value of the objective function $C(x)$ after it has been calculated during the evaluation phase. When this value is not known, the zone is shown in gray. As we are confronted with a problem of maximization that is very simple, the objective function is also the fitness function. The 10 individuals in the population are represented in a row, while the vertical axis describes the temporal sequence of the operations.

The reader should not be misled by the choice of using 10 individuals to constitute a population. This choice can be useful in practice when the computation of the objective function takes much time, suggesting that one should reduce the computational burden by choosing a small population size. It is preferred, however, to use populations of the order of at least 100 individuals to increase the chances of discovering an acceptable solution. According to the problem under consideration, the population size can exceed 10 000 individuals, which then requires treatment on a multiprocessor computer (with up to several thousand processing units) so that the execution time is not crippling.

Our evolutionary algorithm works here with an integer representation. This means that an individual is represented by an integer and that the variation operators must generate integers from the parents. To search for the optimum of $C(x) = 400 - x^2$, we have decided that the crossover will generate two offspring from two parents, each offspring being an integer number drawn randomly in the interval defined by the values x of the parents. The mutation is only the random generation of an integer in the interval $[-20, +20]$. The result of this mutation does not depend on the value of the individual before mutation, which could appear destructive. However, one can notice in Fig. 6.2 that mutation is applied seldom in our model of evolution, which makes this policy acceptable.

6.3 Selection Operators

In general, the ability of an individual to be selected for reproduction or replacement depends on its fitness. The selection operator thus determines a number of selections for each individual according to its fitness.

In our "guide" example (see Fig. 6.2), the 10 parents generate eight offspring. This number is a parameter of the algorithm. As shown in the figure, the selection operator thus copies the best parent twice and six other parents once to produce the population of offspring. These are generated from the copies by the variation operators. Then the replacement operator is activated and selects the 10 best individuals from among the parents and the offspring to constitute the population of parents for the next generation. It can be noticed that four parents have survived, while two offspring, which were of very bad quality, have disappeared from the new population.

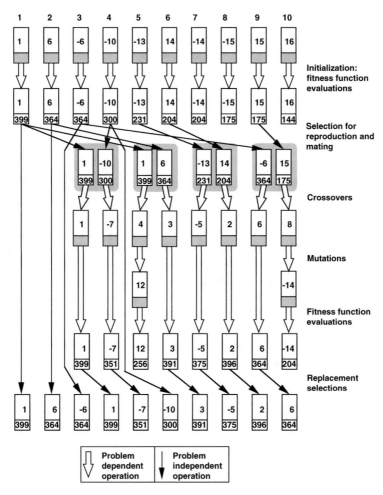

Fig. 6.2 Application of an evolutionary algorithm to a population of $\mu = 10$ parents and $\lambda = 8$ offspring

6.3.1 Selection Pressure

The individuals that have the best fitnesses are reproduced more often than the others and replace the worst ones. If the variation operators were inhibited, the best individual would reproduce more quickly than the others until its copies completely took over the population. This observation led to the first definition of the selection pressure, suggested by Goldberg and Deb in 1991 [27]: the *takeover time* τ^* is defined as the number of generations necessary to fill the population with copies of the best individual under the action of the selection operators only. The selection pressure is higher when τ^* is lower.

The *selection intensity* S provides another method, borrowed from the field of population genetics [31], to define the selection pressure. Let $\bar{f}$ be the average fitness of the μ individuals of the population before a selection. Let $\bar{g}$ be the average fitness of the λ offspring of the population after the selection. Then S measures the increase in the average fitness of the individuals of a population determined before and after selection, with the standard deviation σ_f of the individual fitnesses before selection taken as a unit of measure:

$$S = \frac{\bar{g} - \bar{f}}{\sigma_f}$$

If the selection intensity is computed for the reproduction process, then $\bar{f} = \sum_{i=1}^{\mu} f_i/\mu$, where f_i is the fitness of individual i, and $\bar{g} = \sum_{i=1}^{\lambda} g_i/\lambda$, where g_i is the fitness of individual i.

The definitions presented above are general and are applicable to any selection technique. It is possible also to give other definitions, whose validity may be limited to certain techniques, as we will see later with regard to *proportional selection*.

With a high selection pressure, there is a great risk of *premature convergence*. This situation occurs when the copies of one superindividual, which is nonoptimal but reproduces much more quickly than the others take over the population. Then the exploration of the search space becomes local, since it is limited to a search randomly centered on the superindividual, and there is a large risk that the global optimum will not be approached if local optima exist.

6.3.2 Genetic Drift

Like selection pressure, *genetic drift* is also a concept originating from population genetics [31]. This is concerned with random fluctuations in the frequency of alleles in a population of small size, where an *allele* is a variant of an element of a sequence of DNA having a specific

function. For this reason, hereditary features can disappear or be fixed at random in a small population even without any selection pressure.

This phenomenon also occurs within the framework of evolutionary algorithms. At the limit, even for a population formed from different individuals but of the same fitness, in the absence of variation generated by mutation and crossover operators, the population converges towards a state where all the individuals are identical. This is a consequence of the stochastic nature of the selection operators. Genetic drift can be evaluated from the time required to obtain a homogeneous population using a Markovian analysis. But these results are approximations and are difficult to generalize on the basis of the case studies in the literature. However, it has been verified that the time of convergence towards an absorption state becomes longer as the population size increases.

Another technique for studying genetic drift measures the reduction in the variance of the fitnesses in the population in each generation, under the action of the selection operators only, when each parent has a number of offspring independent of its fitness (neutral selection). This latter condition must be satisfied to ensure that the reduction in variance is not due to the selection pressure. Let r be the ratio of the expectation of the variance of fitness in a given generation to the variance in the previous generation. In this case, Rogers and Prügel–Bennett [47] have shown that r depends only on the variance V_s of the number of offspring of each individual and on the population size, assumed constant:

$$r = \frac{E(V_f(g+1))}{V_f(g)} = 1 - \frac{V_s}{P-1}$$

where $V_f(g)$ is the variance of the fitness distribution of the population in generation g. V_s is a characteristic of the selection operator. It can be seen that increasing the population size or reducing the variance V_s of the selection operator decreases the genetic drift.

The effect of genetic drift is prevalent when the selection pressure is low, and this situation leads to a loss of diversity. This involves a premature convergence, which may a priori be far away from the optimum, since it does not depend on the fitness of the individuals.

In short, in order that an evolutionary algorithm can work adequately, it is necessary that the selection pressure is neither too strong nor too weak for a population of sufficient size, with the choice of a selection operator characterized by a low variance.

6.3.3 Proportional Selection

This type of selection was originally proposed by Holland for genetic algorithms. It is used only for reproduction. The expected number of selections λ_i of an individual i is proportional to its fitness f_i. This implies that the fitness function is positive in the search domain and that it must be maximized, which itself can require some simple transformations of the objective function to satisfy these constraints. Let μ be the population size and let λ be the total number of individuals generated by the selection operator; then λ_i can be expressed as

$$\lambda_i = \frac{\lambda}{\sum_{j=1}^{\mu} f_j} f_i$$

Table 6.1 gives the expected number of selections λ_i of each individual i for a total of $\lambda = 8$ offspring in the population of 10 individuals in our "guide" example.

However, the effective number of offspring can only be an integer. For example, the situation in Fig. 6.2 was obtained with a proportional selection technique. In this figure, individuals 7, 8, and 10, whose respective fitnesses of 204, 175, and 144 are among the worst ones, do not have offspring. Except for the best individual,

Table 6.1 Expected number of offspring in the population of 10 individuals

i	1	2	3	4	5	6	7	8	9	10
f_i	399	364	364	300	231	204	204	175	175	144
λ_i	1.247	1.138	1.138	0.938	0.722	0.638	0.638	0.547	0.547	0.450

Fig. 6.3 RWS method: individual 3 is selected after a random number is drawn

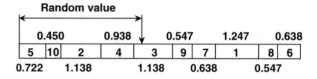

which is selected twice, the others take part only once in the process of crossover. To obtain this type of results, a stochastic sampling procedure constitutes the core of the proportional selection operator. Two techniques are in widespread use and are described below: the *roulette wheel selection* (RWS) method, which is the operator originally proposed for genetic algorithms, but suffers from high variance, and the *stochastic universal sampling* (SUS) method, which guarantees a low variance of the sampling process [7].

6.3.3.1 Proportional Selection Algorithms

The RWS method exploits the metaphor of a biased roulette wheel, which has as many compartments as individuals in the population, and where the size of these compartments is proportional to the fitness of each individual. Once the game has been started, the selection of an individual is indicated by the stopping of the ball in its compartment. If the compartments are unrolled into a straight line segment, the selection of an individual corresponds to choosing, at random, a point on the segment with a uniform probability distribution (Fig. 6.3). The variance of this process is high. It is possible that an individual that has a good fitness value is never selected. In extreme cases, it is also possible, by sheer misfortune, that bad quality individuals are selected as many times as there are offspring. This phenomenon creates a genetic drift that helps some poor individuals to have offspring to the detriment of better individuals. To reduce this risk, the population size must be sufficiently large.

It is the SUS method which was used in our "guide" example. One still considers a straight line segment partitioned into as many zones as there are individuals in the population, each zone having a size proportional to the fitness. But this time the selected individuals are designated by a set of equidistant points, their number being equal to the number of offspring (Fig. 6.4). This method is different from the RWS method because here only one random drawing is required to place the origin of the series of equidistant points and thus to generate all the offspring in the population. In Fig. 6.4, individuals 7, 8, and 10 are not selected, the best individual is selected twice, and the others are selected only once. For an expected number of selections λ_i

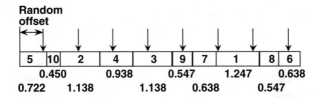

Fig. 6.4 SUS method: the selected individuals are designated by equidistant points

of the individual i, the effective number of selections will be either the integer part of λ_i or the immediately higher integer number. Since the variance of the process is weaker than in the RWS method, the genetic drift appears to be much less and, if $\lambda \geq \mu$, the best individuals are certain to have at least one offspring each.

6.3.3.2 Proportional Selection and Selection Pressure

In the case of proportional selection, the expected number of selections of the best individual, with fitness $\hat{f}$, from among μ selections for a population of μ parents, is appropriate for defining the selection pressure:

$$p_s = \frac{\mu}{\sum_{j=1}^{\mu} f_j} \hat{f} = \frac{\hat{f}}{\bar{f}}$$

where $\bar{f}$ is the average of the fitnesses of the population. If $p_s = 1$, then all the individuals have an equal chance of being selected, indicating an absence of selection pressure.

Let us consider a search for the maximum of a continuous function, for example $f(x) = \exp(-x^2)$. The individuals of the initial population are assumed to be uniformly distributed in the domain $[-2, +2]$. Some of them will have a value close to 0, which is also the position of the optimum, and thus their fitness $\hat{f}$ will be close to 1. The average fitness of the population $\bar{f}$ will be

$$\bar{f} \approx \int_{-\infty}^{+\infty} f(x)p(x)\, dx$$

where $p(x)$ is the probability density of the presence of an individual at x. A uniform density has been chosen in the interval $[-2, +2]$, and thus $p(x)$ is $1/4$ in this interval and 0 elsewhere. Thus

$$\bar{f} \approx \frac{1}{4} \int_{-2}^{+2} e^{-x^2}\, dx$$

that is, $\bar{f} \approx 0.441$, which gives a selection pressure of the order of $p_s = \hat{f}/\bar{f} \approx 2.27$. The best individual will thus have an expected number of offspring close to two (Fig. 6.5a).

Now let us assume that the majority of the individuals of the population are in a much smaller interval around the optimum, for example $[-0.2, +0.2]$. This situation occurs spontaneously after some generations, because of the selection pressure, which favors the reproduction of the best, these being closest to the optimum. In this case, assuming a uniform distribution again, $\bar{f} \approx 0.986$ and $p_s \approx 1.01$ (see Fig. 6.5b). The selection pressure becomes almost nonexistent: the best individual has practically as many expected offspring as any other individual, and it is genetic drift which will prevent the population from converging towards the optimum as precisely as desired.

This undesirable behavior of proportional selection, where the selection pressure decreases strongly when the population approaches the optimum in the case of a continuous function, is overcome by techniques of fitness function scaling.

6.3.3.3 Linear Scaling of the Fitness Function

With a technique of proportional selection, the expected number of selections of an individual is proportional to its fitness. In this case, the effects of a misadjusted selection pressure can be overcome by a linear transformation of the fitness function f. The adjusted fitness value f_i' for an individual i is equal to $f_i - a$, where a is a positive value if it is desired to increase the pressure; otherwise it is negative. a is identical for all individuals. Its value should be chosen so that the selection pressure is maintained at a moderate value, neither too large nor too small, typically about 2. With such a technique, one attention must be paid to the fact that the values of f' are never negative. They can be possibly be bounded from below by 0, or by a small positive value, so that any individual, even of bad quality, has a small chance of being selected. This arrangement contributes to maintenance of the diversity of the

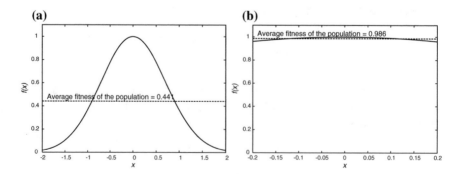

Fig. 6.5 The selection pressure decreases when the population is concentrated in the neighborhood of the optimum

population. Assuming that no individual is given a negative fitness value, the value of a can be calculated in each generation from the value of the desired selection pressure p_s:

$$a = \frac{p_s \bar{f} - \hat{f}}{p_s - 1} \quad \text{with } p_s > 1$$

In the context of the above example, if the individuals are uniformly distributed in the interval $[-0.2, +0.2]$, then $a = 0.972$ for a desired selection pressure $p_s = 2$. Figure 6.6 illustrates the effect of the transformation $f' = f - 0.972$. It can be noticed that there are values of x for which the function f' is negative, whereas this situation is forbidden for proportional selection. To correct this drawback, the fitnesses of the individuals concerned can be kept clamped at zero or at a small constant positive value, which has the side effect of decreasing the selection pressure.

6.3.3.4 Exponential Scaling of the Fitness Function

Rather than performing a linear transformation to adjust the selection pressure, it is also quite common to raise the objective function to a suitable power k to obtain the desired selection pressure:

$$f'_i = f_i^k$$

where the parameter k depends on the problem. Boltzmann selection [19] is another variant, where the scaled fitness is expressed as

$$f'_i = \exp\left(\frac{f_i}{T}\right)$$

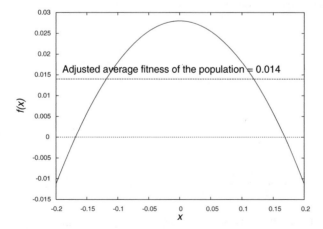

Fig. 6.6 Adjustment of the selection pressure by subtraction of a positive constant from $f(x)$

The value of the parameter T, known as the "temperature," determines the selection pressure. T is usually a decreasing function of the number of generations, thus enabling the selection pressure to grow with time.

6.3.3.5 Rank-based Selection

These techniques for adjusting the selection pressure proceed by ranking the individuals i according to the values of the raw fitnesses f_i. The individuals are ranked from the best (first) to the worst (last). The fitness value f_i' actually assigned to each individual depends only on its rank by decreasing value (see Fig. 6.7) according to, for example, the formula given below, which is usual:

$$f_r' = \left(1 - \frac{r}{\mu}\right)^p$$

Here μ is the number of parents, r is the rank of the individual considered in the population of the parents after ranking, and p is an exponent which depends on the desired selection pressure. After ranking, a proportional selection is applied according to f'. With our definition of the pressure p_s, the relation is $p_s = 1 + p$. Thus, p must be greater than 0. This fitness scaling technique is not affected by a constraint on sign: f_i can be either positive or negative. It is appropriate for a maximization problem as well as for a minimization problem, without the necessity to perform any transformation. However, it does not consider the importance of the differences between the fitnesses of the individuals, so that individuals that are of very bad quality but are not at the last row of the ranking will be able to persist in the population. This is not inevitably a bad situation, because it contributes to better diversity. Moreover, this method does not require an exact knowledge of the objective function, but simply the ability to rank the individuals by comparing each one with

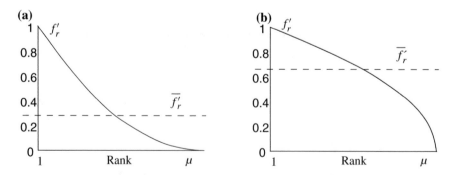

Fig. 6.7 Performance obtained after ranking. **a** $f_r' = (1 - r/\mu)^2$: strong selection pressure; **b** $f_r' = \sqrt{(1 - r/\mu)}$: weak selection pressure

the others. These good properties mean that, overall, it is preferred by the users of evolutionary algorithms over the linear scaling technique.

6.3.4 Tournament Selection

Tournament selection is an alternative to the proportional selection techniques, which, as explained above, present difficulties in controlling the selection pressure during evolution; it is relatively expensive in terms of the computational power involved.

6.3.4.1 Deterministic Tournament

The simplest tournament consists in choosing at random a number k of individuals from the population, and selecting for reproduction the one that has the best fitness. In a selection step, there are as many tournaments as there are individuals, selected. The individuals that take part in a tournament can be replaced in the population or can be withdrawn from it, according the choice made by the user. Drawing without replacement makes it possible to conduct $\lfloor N/k \rfloor$ tournaments with a population of N individuals. A copy of the population is regenerated when it is exhausted, and this is implemented as many times as necessary, until the desired number of selections is reached. The variance of the tournament process is high, which favors genetic drift. It is, however, weaker in the case of drawing without replacement. This method of selection is very much used, because it is much simpler to implement than proportional reproduction with behavior and properties similar to that of ranking selection.

The selection pressure can be adjusted by varying the number of participants k in a tournament. Consider the case where the participants in a tournament are replaced in the population. Then the probability that the best individual in the population is not selected in k drawings is $((N-1)/N)^k$. If we make the assumption that N is very large compared with k, this probability is approximately $1 - k/N$, by a binomial expansion to first order. Thus, the probability that the best individual is drawn at least once in a tournament is close to k/N. If there are M tournaments in a generation, the best individual will have kM/N expected selections, which have a selection pressure of k, according to the definition given earlier for proportional reproduction (with $M = N$). This pressure will necessarily be greater than or equal to 2.

6.3.4.2 Stochastic Tournament

In a stochastic binary tournament, involving two individuals in competition, the best individual wins with a probability p ranging between 0.5 and 1. It is still easy to calculate the selection pressure generated by this process. The best individual takes part in a tournament with a probability of $2/N$ (see Sect. 6.3.4.1). The best individual in the tournament will be selected with a probability p. Since the two events are independent, the probability that the best individual in the population is selected after a tournament is thus $2p/N$. If there are N tournaments, the best will thus have $2p$ expected offspring. The selection pressure will thus range between 1 and 2.

Another alternative, the Boltzmann tournament, ensures that the distribution of the fitness values in a population is close to a Boltzmann distribution. This method makes a link between evolutionary algorithms and simulated annealing.

6.3.5 Truncation Selection

This selection is very simple to implement, as it does nothing but choose the n best individuals from a population, n being a parameter chosen by the user. If the truncation selection operator is used for reproduction to generate λ offspring from n selected parents, each parent will have λ/n offspring. If this operator is used for replacement and thus generates the population of μ individuals for the next generation, then $n = \mu$.

6.3.6 Environmental Selection

Environmental selection, or *replacement selection*, determines which individuals in generation g, from among the offspring and parents, will constitute the population in generation $g + 1$.

6.3.6.1 Generational Replacement

This type of replacement is the simplest, since the population of the parents for the generation $g + 1$ is composed of all the offspring generated in generation g, and only them. Thus, $\mu = \lambda$. The canonical genetic algorithm, as originally proposed, uses generational replacement.

6.3.6.2 Replacement in the Evolution Strategies "(μ, λ)- ES"

Here, a truncation selection of the best μ individuals from among λ offspring forms the population for the next generation. Usually, λ is larger than μ.

6.3.6.3 Steady-State Replacement

Here, in each generation, a small number of offspring (one or two) are generated and they replace a smaller or equal number of parents, to form the population for the next generation. This strategy is useful especially when the representation of a solution is distributed over several individuals, possibly the entire population. In this way, the loss of a small number of individuals (those that are replaced by the offspring) in each generation does not disturb the solutions excessively, and thus they evolve gradually.

The choice of the parents to be replaced obeys various criteria. With uniform replacement, the parents to be replaced are chosen at random. The choice can also depend on the fitness: the worst parent is replaced, or it is selected stochastically according to a probability distribution that depends on the fitness or other criteria.

Steady-state replacement generates a population where the individuals are subject to large variations in their lifespan, measured in number of generations, and thus large variations in number of their offspring. The high variance of these values augments genetic drift, which is especially apparent when the population is small [18].

6.3.6.4 Elitism

An elitist strategy consists in preserving in the population, from one generation to the next, at least the individual that has the best fitness. The example shown in Fig. 6.2 implements an elitist strategy since the best individuals in the population, composed of the parents and the offspring, are selected to form the population of parents for the next generation. The fitness of the best individual in the current population is thus monotonically nondecreasing from one generation to the next. It can be noticed, in this example, that four parents of generation 0 find themselves in generation 1.

There are various elitist strategies. The strategy employed in our "guide" example originates from the class of evolution strategies known as "$(\mu + \lambda)$-ES." In other currently used alternatives, the best parents in generation g are copied systematically into $\mathbf{P}(g + 1)$, the population for generation $g + 1$. Or, if the best individual in $\mathbf{P}(g)$ is better than that in $\mathbf{P}(g + 1)$, because of the action of the variation or selection operators, then the best individual in $\mathbf{P}(g)$ is copied into $\mathbf{P}(g + 1)$, usually by replacing the lowest-fitness individual.

It appears that such strategies improve considerably the performance of evolutionary algorithms for some classes of functions, but prove to be disappointing for other classes, because they can increase the rate of premature convergence. For example, an elitist strategy is harmful for seeking the global maximum of the F5 function of De Jong (Fig. 6.8). In fact, such a strategy increases the exploitation of the best solutions, resulting in an accentuated local search, but to the detriment of the exploration of the search space.

Choosing a nonelitist strategy can be advantageous, but there is then no guarantee that the fitness function of the best individual increases during the evolution. This obviously implies a need to keep a copy of the best solution found by the algorithm

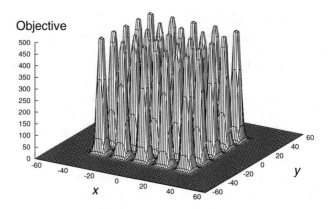

Fig. 6.8 F5 function of De Jong

since the initiation of the evolution, without this copy taking part in the evolutionary process, however.

6.3.7   Fitness Function

The fitness function associates a fitness value with each individual in order to determine the number of times that individual will be selected to be reproduced, or whether it will be replaced or not. In the case of the function $C(x)$ described in Sect. 6.2.4, the fitness function is also the objective function for our maximization problem. This kind of situation is exceptional, however, and it is often necessary to carefully construct the fitness function for a given problem. The quality of this function greatly influences the efficiency of a genetic algorithm.

6.3.7.1   Construction

If a proportional selection method is chosen, it may be necessary to transform the problem under consideration so that it becomes a problem of maximization of a numerical function with positive values in its domain of definition. For example, the solution of a system of equations $\mathbf{S}(\mathbf{x}) = 0$ could be obtained by searching for the maxima of $1/(a + |\mathbf{S}(\mathbf{x})|)$, where the notation $|\mathbf{V}|$ represents the modulus of the vector $\mathbf{V}$, and a is a nonnull positive constant.

The construction of a good fitness function should consider the representation chosen and the nature of the variation operators so that it can give nondeceptive indications of the progress towards the optimum. For example, it may be necessary to try to reduce the presence of local optima on top of broad peaks if a priori knowledge available about the problem allows it. This relates to the study of the fitness landscapes that will be introduced in Sect. 6.4.1, referring to the variation operators.

Moreover, a good fitness function must satisfy several criteria which relate to its complexity, to the satisfaction of the constraints of the problem, and to the adjustment of the selection pressure during the evolution. When the fitness function is excessively complex, consuming considerable computing power, a search for an approximation is desirable, and sometimes indispensable.

6.3.7.2 Reduction of the Computing Power Required

In general, in the case of real-world problems, the evaluation of the fitness function consumes by far the greatest amount of the computing power during an evolutionary optimization. Let us assume that the calculation of a fitness value takes 30 s, that there are 100 individuals in the population, and that acceptable solutions are discovered after a thousand generations, each one implying each time the evaluation of all the individuals; the process will then require 35 days of computation. In the case of real-world problems, the fitness evaluations usually involve computationally intensive numerical methods, for example finite element methods. Various strategies must be used to reduce the computation times. Parallel computing can be considered; this kind of approach is efficient but expensive in terms of hardware. One can also consider approximate calculations of the fitness function, which are then refined gradually as the generations pass. Thus, when a finite element method is being used, for example, it is natural to start by using a coarse mesh at the beginning of the evolution. The difficulty is then to determine when the fitness function should be refined so that the optimizer does not converge prematurely to false solutions generated by the approximations. Another way to simplify the calculation is to make use of a tournament selection or a ranking selection (Sect. 6.3.3.5). In these cases, it is not necessary to know the precise values of the objective function, because only the ranking of the individuals is significant.

6.4 Variation Operators and Representation

6.4.1 Generalities About the Variation Operators

The variation operators belong to two categories:

- crossover operators, which use several parents (often two) to create one or more offspring;
- mutation operators, which transform one individual.

These operators make it possible to create diversity in a population by building "offspring" individuals, which partly inherit the features of "parent" individuals. They must be able to serve two mandatory functions during the search for an optimum:

- exploration of the search space, in order to discover the interesting areas, those which are most likely to contain the global optima;
- exploitation of these interesting areas, in order to concentrate the search there and to discover the optima with the required precision, for those areas which contain them.

For example, a purely random variation operator, where solutions are drawn at random independently of each other, will have excellent qualities of exploration, but will not be able to discover an optimum in a reasonable time. A local search operator that performs "hill climbing" will be able to discover an optimum in an area of the space effectively, but there will be a great risk that it will be a local solution, and the global solution will not be obtained. A good algorithm for searching for the optimum will thus have to find a suitable balance between exploration capabilities and exploitation of the variation operators that it uses. It is not easy to think of how to do this, and the answer depends strongly on the properties of the problem under consideration.

A study of the *fitness landscape* helps us to understand why one variation operator may be more effective than an other operator for a given problem and choice of representation [53]. This notion was introduced in framework of theoretical genetics in the 1930s by Wright [56]. A fitness landscape is defined by:

- a search space Ω, whose elements are called "configurations";
- a fitness function $f: \Omega \rightarrow \mathbb{R}$;
- a relation of neighborhood or accessibility, χ.

It can be noticed that the relation of accessibility is not a part of the optimization problem. This relation depends instead on the characteristics of the variation operators chosen. Starting from a particular configuration in the search space, the application of these stochastic operators potentially gives access to a set of accessible configurations with various probabilities. The relation of accessibility can be formalized in the framework of a discrete space Ω by a directed hypergraph [24], whose hyperarcs have values given by the transition probabilities to an "offspring" configuration from a set of "parent" configurations.

For the mutation operator, the hypergraph of the accessibility relation becomes a directed graph which, starting from an individual or configuration X, represented by a node of the graph, gives a new configuration X', with a probability given by the value of the arc (X, X'). For a crossover operation between two individuals X and Y that produces an offspring Z, the probability of generating Z knowing that X and Y have been crossed is given by the value of the hyperarc $(\{X, Y\}, \{Z\})$.

The definition of the fitness landscape given above shows that it depends simultaneously on the optimization problem under consideration, on the chosen representation, defined by the space Ω, and on the relation of accessibility defined by the variation operators. What is obviously expected is that the application of the latter will offer a sufficiently high probability of improving the fitness of the individuals from one generation to another. This point of view is a useful one to adopt when designing relevant variation operators for a given representation and problem, where one needs to make use of all of the knowledge, formalized or not, that is available for that problem.

After some general considerations regarding crossover and mutation operators, the following subsections present examples of the traditional operators applicable to various popularly used search spaces:

- the space of binary strings;
- real representation in domains of $\mathbb{R}^n$;
- representations of permutations, which can be used for various combinatorial problems such as the traveling salesman problem and problems of scheduling;
- representations of parse trees, for the solution of problems by automatic programming.

6.4.2 Crossover

The crossover operator often uses two parents to generate one or two offspring. The operator is generally stochastic, and hence the repeated crossover of the same pair of distinct parents gives different offspring. As the crossovers in evolutionary algorithms are not subject to biological constraints, more than two parents, and in the extreme case the complete population, can participate in mating for crossover [21].

The operator generally respects the following properties:

- The crossover of two identical parents produces offspring identical to the parents.
- By extension, on the basis of an index of proximity depending on the chosen representation (defined in the search space), two parents which are close together in the search space will generate offspring close to them.

These properties are satisfied by the "classical" crossover operators, such as most of those described in this chapter. They are not absolute, however, as in the current state of knowledge of evolutionary algorithms, the construction of crossover operators does not follow a precise rule.

The *crossover rate* determines the proportion of the individuals that are crossed among the offspring. For the example in Fig. 6.2, this rate was fixed at 1, i.e., all offspring are obtained by crossover. In the simplest version of an evolutionary algorithm, the individuals are mated at random from among the offspring generated by the selection, without taking account of their characteristics. This strategy can prove to be harmful when the fitness function has several optima. Indeed, it is generally not likely that a crossover of high-quality individuals located on different peaks will give good-quality individuals (see Fig. 6.9). A crossover is known as *lethal* if it generates from good parents one or two offspring with too low a fitness to survive.

A solution to avoiding too large a proportion of lethal crossovers consists in preferentially mating individuals that resemble each other. If a distance is defined in the search space, the simplest way to proceed is to select two individuals according to the probability distribution of the selection operator and then to cross them only if the distance between them is lower than a threshold r_c, called the *restriction radius*. If the latter is small, however, this will lower the rate of effective crossover significantly, which can be prejudicial. It is then preferable to select the first parent with the selection operator and then, if there are individuals in its neighborhood, one of them is selected to become the second parent. In all situations, if r_c is selected to be too small, it significantly reduces the exploration of the search space by accentuating the local search, and this can lead to premature convergence. This effect is especially sensitive to the initialization of the evolution, when the crossover of two individuals distant from each other makes it possible to explore new areas of the search space that potentially contain peaks of the fitness function. Thus, to make the technique efficient, the major problem consists in choosing a good value for r_c; however, it depends largely on the fitness landscape, which is in general not known. It is also possible to consider a radius r_c that decreases during the evolution.

6.4.3 Mutation

Classically, the mutation operator modifies an individual at random to generate an offspring that will replace it. The proportion of mutated individuals in the offspring population is equal to the *mutation rate*. Its order of magnitude can vary substantially according to the model of evolution chosen. In the example in Fig. 6.2, two individuals are mutated from among the eight offspring obtained from the selection process. In

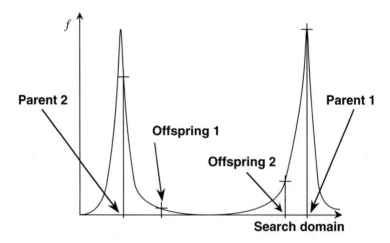

Fig. 6.9 Crossover of two individuals placed on different peaks of a fitness function f

genetic algorithms, mutation is considered as a minor operator, aimed at maintaining a minimum diversity in the population, which crossover cannot ensure. With this model, the mutation rate is typically low, about 0.01–0.1, whereas the crossover rate is high. In contrast, mutation was essential in the original model of the evolution strategies since there was no crossover. The mutation rate in that case was 100 %.

Most of the mutation strategies modify an individual in such a way that the result of the transformation is close to it. In this way, the operator performs a random local search around each individual to be mutated. Mutation can considerably improve the quality of the solutions discovered compared with crossover, which loses its importance when most of the population is located in the neighborhood of the maxima of the fitness function. In fact, the individuals located on the same peak are often identical because of the process of selection for reproduction and do not undergo any modification by the crossover operator. If they belong to different peaks, the offspring generally have low fitness. On the other hand, the local random search due to the mutations gives a chance for each individual to approach the exact position of the maximum, to the extent that the characteristics of the chosen operator allow it.

Mutation with a sufficiently high rate plays an important part in the preservation of diversity, which is useful for efficient exploration of the search space. This operator can fight the negative effects of a strong selection pressure or a strong genetic drift, phenomena which tend to reduce the variance of the distribution of the individuals in the search space.

If the mutation rate is high and, moreover, the mutation is so strong that the individual produced is almost independent of the one which generated it, the evolution of the individuals in the population is equivalent to a random walk in the search space, and the evolutionary algorithm will require an excessive time to converge.

The utilization of mutation as a local search operator suggests combining it with other, more effective, local techniques, although these will be more problem-dependent, such as a gradient technique, for example. This kind of approach has led to the design of *hybrid* evolutionary algorithms.

6.5   Binary Representation

The idea of evolving a population in a space of binary vectors originated mainly from genetic algorithms, which are inspired by the transcription from *genotype* to *phenotype* that occurs in the living world. In the framework of genetic algorithms, the genotype consists of a string of binary symbols or, more generally, a string of symbols belonging to a low-cardinality alphabet. The phenotype is a solution of the problem in a "natural" representation. The genotype undergoes the action of the genetic operators, i.e., selections and variations, while the phenotype is used only for fitness evaluation.

For example, if a solution can be expressed naturally as a vector of real numbers, the phenotype will be that vector. The genotype will thus be a binary string which codes this vector. To code the set of the real variables of a numerical problem as a binary string, the simplest way is to convert each variable into binary format, and then to concatenate these binary numbers to produce the genotype. The most obvious technique to code a real number in binary format is to represent it in fixed-point format with a number of bits corresponding to the desired precision.

6.5.1 Crossover

For a binary representation, there exists three classical variants of crossovers:

- "single-point" crossover;
- "two-point" crossover;
- uniform crossover.

After a pair of individuals has been chosen randomly among the selected individuals, the "single-point" crossover [32] is applied in two stages:

1. Random choice of an identical cut point on the two bit strings (Fig. 6.10a).
2. Cutting of the two strings (Fig. 6.10b) and exchange of the two fragments located to the right of the cut (Fig. 6.10c).

This process produces two offspring from two parents. If only one offspring is used by the evolutionary algorithm employed, this offspring is chosen at random from the pair and the other one is discarded. The "single-point" crossover is the simplest type of crossover and is, traditionally, the one most often used with codings using an alphabet with low cardinality, such as binary coding. An immediate generalization of this operator consists in multiplying the cut points on each string. The "single-point" and "two-point" crossovers are usually employed in practice for their simplicity and their good effectiveness.

The uniform crossover [3] can be viewed as a multipoint crossover where the number of cuts is unspecified a priori. Practically, one uses a "template string," which is a binary string of the same length as the individuals. A "0" at the nth position of the template leaves the symbols in the nth position of the two strings unchanged and a "1" activates an exchange of the corresponding symbols (in Fig. 6.11). The template is generated at random for each pair of individuals. The values "0" and "1" of the elements of the template are often drawn with a probability of 0.5.

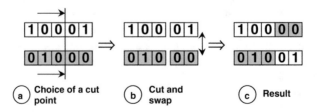

Fig. 6.10 "Single-point" crossover of two genotypes of five bits

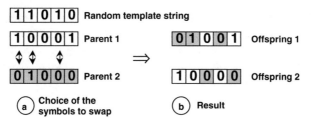

Fig. 6.11 Uniform crossover

6.5.2 Mutation

Classically, the mutation operator on bit strings modifies at random the symbols of a genotype, with a low probability in the framework of genetic algorithms, typically from 0.01 to 0.1 per individual. This probability is equal to the mutation rate. The most common variants are *deterministic mutation* and *bit-flip mutation*. With "deterministic" mutation, a fixed number of bits chosen at random are reversed for each mutated individual, i.e., a "1" becomes "0" and vice versa, while with "bit-flip" mutation, each bit can be reversed independently of the others with a low probability. If the mutation rate is too high, with a large number of mutated bits per individual, the evolution of the individuals in the population is equivalent to a random walk in the search space, and the genetic algorithm loses its effectiveness.

When a bit string represents a vector of integer or real numbers, the positive effects of mutation are opposed by the difficulty of crossing the *Hamming cliffs*, which appear because of the conversion of the bit strings to real-number vectors. For example, let us consider the function $D(x)$, where

$$D(x) = \begin{cases} 256 - x^2 & \text{if } x \leq 0 \\ 0 & \text{otherwise} \end{cases}$$

Let us use a string $b(x) = \{b_1(x), \ldots, b_5(x)\}$ of five bits to represent an individual x that ranges from -16 to $+15$, and thus has 32 possible different values. $b(x)$ can be defined simply as the number $x + 16$ expressed in base 2. The optimum of $D(x)$ is obtained for $x = 0$, which thus corresponds to $b(0) = \{1, 0, 0, 0, 0\}$. The value $x = -1$, obtained from the string $\{0, 1, 1, 1, 1\}$, gives the highest fitness apart from the maximum: this value will thus be favored by the selection operators. However, it can be noticed that there is no common bit between $\{1, 0, 0, 0, 0\}$ and $\{0, 1, 1, 1, 1\}$. This means that there is no other individual with which $\{0, 1, 1, 1, 1\}$ can be mated to give $\{1, 0, 0, 0, 0\}$. The mutation operator will have to change the five bits of the genotype $\{0, 1, 1, 1, 1\}$ simultaneously to give the optimum, because the Hamming distance[2] between the optimum and the individual which has the nearest fitness is equal to the size of the strings. Hence, we encounter a *Hamming cliff* here. It is not

---

[2]The Hamming distance is the number of different bits between two bit strings of the same length.

very likely that we will cross it with a "bit-flip" mutation, and this is impossible with a "deterministic" mutation unless that mutation flips all the bits of a bit string, a form which is never used. But the mutation will be able to easily produce the optimum if there are individuals in the population that differ by only one bit from the optimal string; here, these individuals are:

String $b(x)$	x	$D(x)$
$\langle 0, 0, 0, 0, 0 \rangle$	-16	0
$\langle 1, 1, 0, 0, 0 \rangle$	8	0
$\langle 1, 0, 1, 0, 0 \rangle$	4	0
$\langle 1, 0, 0, 1, 0 \rangle$	2	0
$\langle 1, 0, 0, 0, 1 \rangle$	1	0

Unfortunately, all these individuals have null fitness and thus they have very few chance of "surviving" from one generation to the next.

This tedious phenomenon, which hinders progress towards the optimum, can be eliminated by choosing a *Gray code*, which ensures that two successive integers have binary representations that differ only in one bit. Starting from strings $b(x)$ that represent integer numbers in base 2, it is easy to obtain a Gray code $g(x) = \{g_1(x), \ldots, g_l(x)\}$ by performing, for each bit i, the operation

$$g_i(x) = b_i(x) \oplus b_{i-1}(x)$$

where the operator $\oplus$ implements the "exclusive or" operation and $b_0(x) = 0$. Conversely, the string of l bits $b(x) = \{b_1(x), \ldots, b_l(x)\}$ can be obtained from the string $g(x) = \{g_1(x), \ldots, g_l(x)\}$ by the operation

$$b_i(x) = \bigoplus_{j=1}^{i} g_j(x)$$

The Gray codes of $\{0, 1, 1, 1, 1\}$ and $\{1, 0, 0, 0, 0\}$ are $\{0, 1, 0, 0, 0\}$ and $\{1, 1, 0, 0, 0\}$, respectively. A mutation of the bit g_1 is then enough to reach the optimum. A Gray code is thus desirable from this point of view. Moreover, it modifies the landscape of the fitness function by reducing the number of local optima created by transcribing a real or integer vector into a binary string. It should be noted, however, that Hamming cliffs are generally not responsible for dramatic falls in the performance of the algorithm.

6.6 Real Representation

The real representation allows an evolutionary algorithm to operate on a population of vectors in a bounded search domain Ω included in $\mathbb{R}^n$. Let us assume that any solution $\mathbf{x}$ in a given population is drawn from the search domain according to a probability distribution characterized by a density $p(\mathbf{x})$, where $\mathbf{x}$ is a point in Ω. Assume also that this distribution has an expectation

$$E = \int_\Omega \mathbf{x} p(\mathbf{x}) \, d\mathbf{x}$$

and a total variance

$$V = \int_\Omega \mathbf{x}^2 p(\mathbf{x}) \, d\mathbf{x} - E^2$$

V is also the trace of the covariance matrix of the components of the vectors $\mathbf{x}$. If λ, the size of the population of the offspring, is large enough, these values are approached by the empirical expectation

$$\hat{E} = \frac{\sum_{i=1}^\lambda \mathbf{x}_i}{\lambda}$$

and the empirical total variance

$$\hat{V} = \frac{\sum_{i=1}^\lambda \mathbf{x}_i^2}{\lambda} - \hat{E}^2$$

The empirical variance can be regarded as a measurement of the diversity in the population. If it is zero, then all the individuals are at the same point in Ω. If we adopt a mechanical analogy, $\hat{E}$ is the centroid of the population, where we allot a unit mass to each individual. It is interesting to evaluate these values after application of the variation operators.

6.6.1 Crossover

Let us consider two points $\mathbf{x}$ and $\mathbf{y}$ in the space $\mathbb{R}^n$ corresponding to two individuals selected to generate offspring. After application of the crossover operator, one or two offspring $\mathbf{x}'$ and $\mathbf{y}'$ are drawn randomly, according to a probability distribution which depends on $\mathbf{x}$ and $\mathbf{y}$.

6.6.1.1 Crossover by Exchange of Components

This is a direct generalization of the binary crossovers and consists in exchanging some real components of two parents. One can thus obtain all variants of binary crossover, in particular the "single-point," "two-point," and "uniform" crossovers (see Fig. 6.12). The last variant is also called "discrete recombination" in the terminology of evolution strategies. This type of crossover modifies neither E nor V.

6.6.1.2 BLX-α Crossover

The *BLX-α crossover* was proposed in [22], α being a parameter of the evolutionary algorithm. Two variants are widely mentioned in publications related to evolutionary algorithms. According to the original description of its authors, the first variant randomly generates offspring on a line segment in the search space $\mathbb{R}^n$ passing through the two parents. We refer to this variant as *linear BLX-α crossover*. The second variant randomly generates offspring inside a hyperrectangle defined by the parents. We refer to this as *voluminal BLX-α crossover*.

Voluminal BLX-α crossover. This operator generates offspring chosen uniformly inside a hyperrectangle with sides parallel to the coordinate axes (Fig. 6.13). Let x_i and y_i be the components of the two parents $\mathbf{x}$ and $\mathbf{y}$, respectively, for $1 \leq i \leq n$; the components of an offspring $\mathbf{z}$ are defined as

$$z_i = x_i + (y_i - x_i) \cdot \mathcal{U}(-\alpha, 1 + \alpha)$$

where $\mathcal{U}(-\alpha, 1 + \alpha)$ is a random number drawn uniformly in the interval $[-\alpha, 1 + \alpha]$.

A voluminal BLX-α crossover does not modify E, but changes the value of V. Let V_c be the variance of the distribution of the population after crossover:

$$V_c = \frac{(1 + 2\alpha)^2 + 3}{6} V$$

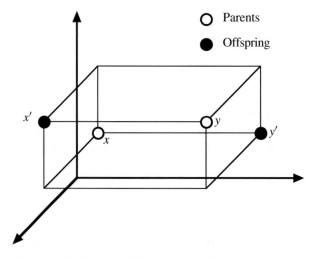

Fig. 6.12 Uniform crossover; an individual x' or y' resulting from the crossover of x and y is located on a vertex of a hyperrectangle with sides parallel to the coordinate axes such that one longest diagonal is the segment (x, y)

Fig. 6.13 BLX-α crossover;
an individual x' or y'
resulting from the crossover
of x and y is located inside a
hyperrectangle with sides
parallel to the coordinate
axes such that one longest
diagonal passes through x
and y

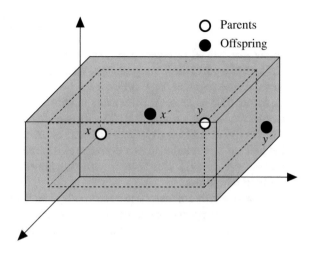

The variance after crossover decreases if

$$\alpha < \frac{\sqrt{3} - 1}{2} \approx 0.366$$

In this case, it is said that the crossover is *contracting*, and the iterative application of the operator alone leads the population to collapse onto its centroid. In particular, if $\alpha = 0$, z is located in a hyperrectangle such that one longest diagonal is the line segment (x, y). In this case, $V_c = \frac{2}{3} V$. After iterative application of this operator alone for g generations, and for an initial population variance V_0, the variance becomes

$$V_{cg} = \left(\frac{2}{3}\right)^g V_0$$

The variance tends quickly towards 0! It can thus be seen that the risk of premature convergence is increased with a BLX-0 operator.

If $\alpha > (\sqrt{3} - 1)/2$, the variance increases if the domain is $\mathbb{R}^n$. In practice, for a bounded search domain Ω, the variance is stabilized at a nonnull value. The "borders" of the search domain can be explored. The possible optima which are there will be found and retained more easily. A commonly used value is $\alpha = 0.5$.

Nomura and Shimohara [41] showed that this operator reduces the possible correlations which may exist between the components of the vectors of the population. Its repeated application makes the coefficients of correlation converge towards zero.

This operator can be seen as a generalization of other crossover operators, such as the *flat crossover* [44], which is equivalent to BLX-0.

Linear BLX-α crossover. If $\mathbf{x}$ and $\mathbf{y}$ are the points corresponding to two individuals in $\mathbb{R}^n$, an individual $\mathbf{z}$ resulting from the linear BLX-α crossover of $\mathbf{x}$ and $\mathbf{y}$ is chosen according to a uniform distribution on a line segment passing through $\mathbf{x}$ and $\mathbf{y}$:

$$\mathbf{z} = \mathbf{x} + (\mathbf{y} - \mathbf{x}) \cdot \mathcal{U}(-\alpha, 1 + \alpha)$$

where $\mathcal{U}(-\alpha, 1 + \alpha)$ is a random number drawn uniformly in the interval $[-\alpha, 1 + \alpha]$. If I is the length of the line segment $[\mathbf{x}, \mathbf{y}]$, $\mathbf{z}$ is on the segment of length $I \cdot (1 + 2\alpha)$ centered on the segment $[\mathbf{x}, \mathbf{y}]$ (Fig. 6.14).

A linear BLX-α crossover does not modify E, but changes the value of V in a way similar to the voluminal BLX-α crossover. On the other hand, it should be noted that the possible correlations that may exist between the components of the individuals of a population do not decrease as a result of the repeated application of the linear operator [41]. This behavior is completely different from that observed for the voluminal operator.

As previously, this operator can be seen as a generalization of some other crossover operators, according to restrictions on the values oα, such as *intermediate recombination* for evolution strategies [8] and *arithmetic crossover* [40], which is equivalent to BLX-0.

6.6.1.3 Intermediate Recombination

This operator is applied to ρ parents and gives one offspring each time it is invoked. ρ is a constant parameter between 2 and the population size. An offspring $\mathbf{z}$ is the centroid of the parents $\mathbf{x}_i$:

$$\mathbf{z} = \frac{1}{\rho} \sum_{i=1}^{\rho} \mathbf{x}_i.$$

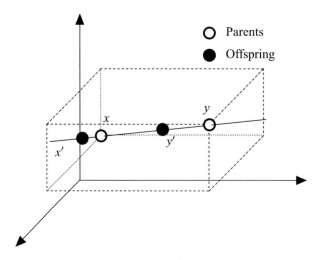

Fig. 6.14 BLX-α crossover; an individual x' or y' resulting from the crossover of x and y is located on the line defined by x and y, possibly outside the segment $[x, y]$

6.6.2 Mutation

Mutation generally consists in the addition of a "small" random value to each component of an individual, according to a zero-average distribution, with a variance possibly decreasing with time. In this way, it is assured that the mutation leaves the centroid of the population unchanged.

6.6.2.1 Uniform Mutation

The simplest mutation technique adds to an individual x, belonging to a domain Ω in $\mathbb{R}^n$, a random variable with a uniform distribution in a hypercube $[-a, +a]^n$. However, such a mutation does not allow an individual trapped in a local optimum located on a peak broader than the hypercube to escape from that local optimum. To avoid this disadvantage, it is preferable to use a distribution with unlimited support.

6.6.2.2 Gaussian Mutation

Gaussian mutation is one of the most widely used types of mutation for the real representation. The simplest form adds a Gaussian random variable $\mathcal{N}(0, \sigma)$, with zero-average and standard deviation σ, to each component of a real-valued vector. The problem is then making a suitable choice of σ. In theory, it is possible to escape from a local optimum irrespective of the width of the peak where it is located, since the support of a Gaussian distribution is unlimited, but if σ is too small that might only happen after far too many attempts. Conversely, if σ is too large, it will be unlikely that an optimum value will be approached accurately within a reasonable time. The value of σ therefore needs to be adapted during the evolution: large at the beginning to quickly explore the search space, and small at the end to accurately approach the optimum. Some adaptation strategies are described in the following.

6.6.2.3 Gaussian Mutation and the 1/5 Rule

Based on a study on two simple, very different test functions with an elitist evolution strategy $(1 + 1)$-ES,[3] Rechenberg [9, 46] calculated optimal standard deviations for each test function that maximized the convergence speed. He observed that for these optimal values, approximately one fifth of the mutations allow one to reduce the distance between the individual and the optimum. He deduced the following rule, termed the "one fifth rule," for adapting σ: *if the rate of successful mutations is larger than 1/5, increase σ, if it is smaller, reduce σ.* The "rate of successful mutations"

---

[3] In $(1 + 1)$-ES: the population is composed of only one parent individual, and this generates only one offspring; the best of both is preserved for the next generation.

Fig. 6.15 Isovalue *ellipse*
$f(x_1, x_2) = 1/2$ when $\mathbf{H}$ is
diagonal with $h_{11} = 1/36$
and $h_{22} = 1$

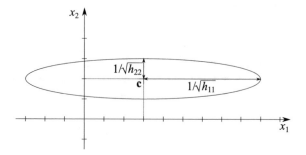

is the proportion of mutations which make it possible to improve the value of the fitness of an individual. Schwefel [52] proposed the following rule in practice:

> Estimate the rates of successful mutations p_s in k mutations
> **if** $p_s < 0.2$ **then**
> $\sigma(g) \leftarrow \sigma(g) \cdot a$
> **else if** $p_s > 0.2$ **then**
> $\sigma(g) \leftarrow \sigma(g)/a$
> **else**
> $\sigma(g) \leftarrow \sigma(g)$

Here, $0.85 \leq a < 1$ according to Schwefel's recommendation; k is equal to the dimension n of the search space when $n > 30$, and g is the index of the current generation. σ should be updated according to the above algorithm every n mutations.

The "one fifth rule" requires that σ should have the same value for all components of a vector x. In this way, the size of the progression step towards the optimum is the same in all directions: the mutation is isotropic. However, isotropy does not make it possible to approach the optimum as quickly as expected when, for example, the isovalues of the fitness function locally take the shape of "flattened" ellipsoids in the neighborhood of the optimum (see Fig. 6.15). If the step is well adapted in a particular direction, it will not be in other directions.

To clarify these considerations with an example, we consider the quadratic performance function defined in $\mathbb{R}^n$ $f(\mathbf{x}) = \frac{1}{2}(\mathbf{x} - \mathbf{c})^T \mathbf{H}(\mathbf{x} - \mathbf{c})$, where $\mathbf{H}$ is a symmetric matrix. This example is interesting because the expression for $f(\mathbf{x})$ is the second-order term of a Taylor expansion near the point $\mathbf{c}$ for any function that is twice continuously differentiable, where $\mathbf{H}$ is the Hessian matrix of this function at $\mathbf{c}$. $f(\mathbf{x})$ is minimal, equal to 0, for $\mathbf{x} = \mathbf{c}$ with $\mathbf{H}$ positive definite. Figure 6.15 represents the isovalue ellipse $f(x_1, x_2) = 1/2$ for a function of two variables obtained when $\mathbf{H}$ is diagonal with $h_{11} = 1/36$ and $h_{22} = 1$.

The condition number $\kappa_{\mathbf{H}}$ is defined as the ratio of the largest eigenvalue of $\mathbf{H}$ to the smallest one: $\kappa_{\mathbf{H}} = \lambda_{\max}/\lambda_{\min}$. In the case shown in Fig. 6.15, the matrix $\mathbf{H}$ is already diagonal, and its eigenvalues are h_{11} and h_{22}. For the $\mathbf{H}$ defined above, the condition number is 36. When the condition number is large compared with 1, the matrix is said to be "ill-conditioned". In real-world applications, the condition number can be larger than 10^{10}, which means that the ratio of the lengths of the major axis and the minor axis of an isovalue hyperellipsoid can be larger than 10^5.

Note that when $\mathbf{H}$ is diagonal, the quadratic function $f(\mathbf{x})$ is an additively separable function: $f(\mathbf{x}) = f(x_1, \ldots, x_i, \ldots, x_n) = \sum_{i=1}^{n} g(x_i)$. Thus, if $\mathbf{H}$ is positive definite, the global minimum of $f(\mathbf{x})$ can be obtained by searching for the minima of n convex functions $f_i(x_i) = f(c_1, \ldots, c_{i-1}, x_i, c_{i+1}, \ldots, c_n)$ with constants $c_1, \ldots, c_{i-1}, c_{i+1}, \ldots, c_n$ arbitrarily chosen. In this case, the optimum of $f(\mathbf{x})$ can be found efficiently in n successive runs of the (1+1)-ES algorithm with the "1/5 rule" to obtain the optimum for each of variables x_i independently of the others. If $\mathbf{H}$ is diagonal, the ratios of the adapted standard deviations σ_i/σ_j in a given generation should ideally be of the order of $\sqrt{h_{jj}}/\sqrt{h_{ii}}$ to reduce the computation time to the greatest possible extent. However, such an approach to solving ill-conditioned problems cannot be applied efficiently when the objective function is not separable.

6.6.2.4 Self-adaptive Gaussian Mutation

Schwefel [52] proposed *self-adaptive Gaussian mutation* to efficiently solve ill-conditioned problems when the objective function is separable or "almost" separable in a neighborhood of the optimum. Self-adaptive mutation should be applicable to a wider range of problems than the "1/5 rule" because that rule was derived from a study of specific objective functions [9].

To implement this adaptation, an individual is represented as a pair of vectors $(\mathbf{x}, \boldsymbol{\sigma})$. Each component σ_i refers to the corresponding component of $\mathbf{x}$. These components σ_i evolve in a similar way to the variables of the problem under the action of the evolutionary algorithm [52]. $\boldsymbol{\sigma}$ is thus likely to undergo mutations. Schwefel proposed that the pair $(x', \boldsymbol{\sigma}')$ obtained after mutation should be such that

$$\sigma_i' = \sigma_i \exp(\tau_0 N + \tau \mathcal{N}(0, 1)) \qquad (6.1)$$

$$\text{with} \qquad \tau_0 \approx \frac{1}{\sqrt{2n}}, \qquad \tau \approx \frac{1}{\sqrt{2\sqrt{n}}}$$

$$x_i' = x_i + \mathcal{N}(0, \sigma_i'^2)$$

where N indicates a Gaussian random variable with average 0 and variance 1, computed for the entire set of n components of $\boldsymbol{\sigma}$, and $\mathcal{N}(0, v)$ represents a Gaussian random variable with average 0 and variance v. σ_i' is thus updated by application of a lognormal perturbation (Eq. (6.1)).

Self-adaptive Gaussian mutation requires a population size μ of the order of the dimension of the search space. Beyer and Schwefel [8] recommended that this mutation should be associated with intermediate recombination (Sect. 6.6.1.3) to prevent excessive fluctuations of parameters that would degrade the performance of the algorithm. As with the "1/5 rule", this operator becomes inefficient when the objective function is ill-conditioned and not separable in the neighborhood of the optimum.

6.6.2.5 Correlated Gaussian Mutation

The self-adaptive mutation described above works best when the matrix $\mathbf{H}$ is diagonal. It is inefficient when there are correlations between variables, as in the case of a fitness function that has the isovalue curve $f(\mathbf{x}) = 1/2$ represented in Fig. 6.16. This case corresponds to a matrix $\mathbf{H} = (\mathbf{DR})^{\mathsf{T}} (\mathbf{DR})$, where $\mathbf{D}$ is the diagonal matrix of the square roots of the eigenvalues of $\mathbf{H}$ and $\mathbf{R}$ is a rotation matrix, with

$$\mathbf{D} = \begin{pmatrix} 1/6 & 0 \\ 0 & 1 \end{pmatrix} \text{ and } \mathbf{R} = \begin{pmatrix} \cos\theta & -\sin\theta \\ \sin\theta & \cos\theta \end{pmatrix} \text{ with } \theta = \frac{\pi}{6} \quad (6.2)$$

The condition number $\kappa_{\mathbf{H}} = (s_{22}/s_{11})^2$ is then equal to 36. This function f, for which there are correlations between variables, is not separable.

Correlated mutation is a generalization of the self-adaptive mutation described above. The mutated vector $\mathbf{x}'$ is obtained from $\mathbf{x}$ by the addition of a Gaussian random vector with zero mean and covariance matrix $\mathbf{C}$:

$$\mathbf{x}' = \mathbf{x} + \mathcal{N}(0, \mathbf{C})$$

The matrix $\mathbf{C}$, which is symmetric positive definite, can always be written as $\mathbf{C} = (\mathbf{SR})^{\mathsf{T}}(\mathbf{SR})$, where $\mathbf{R}$ is a rotation matrix in $\mathbb{R}^n$ and $\mathbf{S}$ is a diagonal matrix with $s_{ii} > 0$ [48].[4] The matrix $\mathbf{R}$ can be computed as the product of $n(n-1)/2$ elementary rotation matrices $\mathbf{R}_{kl}(\alpha_{kl})$:

$$\mathbf{R} = \prod_{k=1}^{n-1} \prod_{l=k+1}^{n} \mathbf{R}_{kl}(\alpha_{kl})$$

Fig. 6.16 An isovalue curve $f(\mathbf{x}) = 1/2$ for $f(\mathbf{x}) = \frac{1}{2}(\mathbf{x} - \mathbf{c})^{\mathsf{T}}\mathbf{H}(\mathbf{x} - \mathbf{c})$, obtained for $\mathbf{H} = (\mathbf{DR})^{\mathsf{T}} (\mathbf{DR})$, where $\mathbf{D}$ and $\mathbf{R}$ are given by Eq. (6.2)

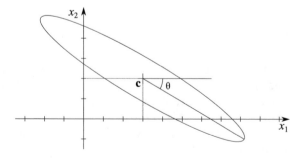

---

[4]Possibly with a column permutation of the matrix $\mathbf{R}$ and the corresponding diagonal coefficients in $\mathbf{S}$.

Here, $\mathbf{R}_{kl}(\alpha_{kl})$ is the matrix for a rotation by an angle α_{kl} in the plane spanned by the vectors k and l. Such a matrix can be written as the identity matrix with the exception of the coefficients $r_{kk} = r_{ll} = \cos(\alpha_{kl})$ and $r_{kl} = -r_{lk} = -\sin(\alpha_{kl})$.

Each individual has its own covariance matrix $\mathbf{C}$ that allows it to mutate. $\mathbf{C}$ is able to adapt itself through mutations of the information that was used to build it. An individual is then defined as a triplet $(\mathbf{x}, \boldsymbol{\sigma}, \boldsymbol{\alpha})$ where $\boldsymbol{\sigma}$ is a vector of n standard deviations, as in self-adaptive mutation, and $\boldsymbol{\alpha}$ is a vector that is a priori composed of $n(n-1)/2$ elementary rotation angles α_{kl} used to build the matrix $\mathbf{R}$. The diagonal coefficients of the matrix $\mathbf{S}$ are $s_{ii} = \sigma_i$.

The vector $\boldsymbol{\sigma}$ evolves under the action of the evolutionary algorithm as described by Eq. (6.1). The Components α_{kl} undergo mutations according to the following formula:

$$\alpha'_{kl} = \alpha_{kl} + \beta \mathcal{N}(0, 1)$$

Schwefel suggests setting β to a value close to 0.087 rad, i.e., $5°$.

In practice, the mutated vector $\mathbf{x}'$ is obtained from $\mathbf{x}$ according to the following expression:

$$\mathbf{x}' = \mathbf{x} + \mathbf{R}'\mathbf{S}'\mathcal{N}(0, \mathbf{I})$$

where $\mathbf{R}'$ and $\mathbf{S}'$ are obtained from $\mathbf{R}$ and $\mathbf{S}$, respectively, after mutation of the angles α_{kl} and standard deviations σ_i. $\mathcal{N}(0, \mathbf{I})$ is a Gaussian random vector with zero mean and variance 1 for each component i.

This technique of mutation, although it seems powerful, is seldom used because of the amount of memory used for an individual, and its algorithmic complexity of the order of n^2 matrix products for a problem of n variables for each generation. Besides, the large number of parameters required for each individual involves a large population size of the order of n^2, where n is the dimension of the search space. The method loses much efficiency when the dimension n increases. It is hardly possible to exceed dimension 10 [30].

The difficulties in the use of the correlated mutation method have prompted a search for new approaches, leading to a major improvement of evolution strategies known as the *covariance matrix adaptation evolution strategy* (CMA-ES), presented in Sect. 6.10.

6.7 Some Discrete Representations for Permutation Problems

There exist many types of combinatorial optimization problems, and it is not possible to describe all of them within a restricted space. We will consider here only the *permutation problem*, which consist in discovering an order of a list of elements that maximizes or minimizes a given criterion. The traveling salesman problem can be considered as an example. Knowing a set of "cities," as well as the distances between these cities, the traveling salesman must discover the shortest possible path passing through each city once and only once. This NP-hard problem is classically used as a benchmark, making it possible to evaluate the effectiveness of an algorithm. Typically, the problems considered comprise several hundreds of cities.

A solution can be represented as a list of integers, each one associated with a city. The list contains as many elements as cities, and each city associated with an element must satisfy the constraint of uniqueness. One has to build individuals that satisfy the structure of the problem, and possibly to specialize the genetic operators.

6.7.1 Ordinal Representation

It is tempting to consider an individual representing an order as an integer vector, and to apply crossovers to the individuals by exchanging components similarly to what is done for binary and real representations (see Sects. 6.5.1 and 6.6.1.1). The ordinal representation makes it possible to satisfy the constraint of uniqueness with the use of these standard crossovers. It is based on a reference order, for example the natural order of the integers. First, a list of the cities O in this reference order is built, for example $O = (123456789)$ for nine cities numbered from 1–9. Then an individual is read from left to right. The nth integer read gives the order number in O of the nth visited city. When a city is visited, it is withdrawn from O. For example, let us consider the individual $\langle 437253311 \rangle$:

- The first integer read from the individual is 4. The first visited city is thus the fourth element of the reference list O, i.e., the city 4. This city is withdrawn from O. One then obtains $O_1 = (12356789)$.
- The second integer read is 3. According to O_1, the second visited city is 3. This city is withdrawn from O_1 to give $O_2 = (1256789)$.
- The third integer read is 7. The third visited city is thus 9, and one obtains $O_3 = (125678)$, which will be used as the reference list for the next step.

We continue in this way until the individual is entirely interpreted. Hence, for this example, the path is $4 \rightarrow 3 \rightarrow 9 \rightarrow 2 \rightarrow 8 \rightarrow 6 \rightarrow 7 \rightarrow 1 \rightarrow 5$.

But, experimentally, this representation associated with the standard variation operators does not give good results. This shows that it is not well adapted to the problem under consideration, and that the simple satisfaction of the uniqueness constraint is not sufficient. Other ways have hence been explored, which enable the offspring to inherit partially the order of the cities or the relations of adjacency which exist in their parents.

6.7.2 Path or Sequence Representation

In this representation, two successive integers in a list correspond to two nodes adjacent to each other in the path represented by an individual. Each number in a list must be present once and only once. Useful information lies in the order of these numbers compared with the others. Many variation operators have been proposed for this representation. A crossover preserving the order and another preserving the adjacencies, chosen from the most common alternatives in the literature, are presented below.

6.7.2.1 Uniform Order-Based Crossover

With uniform order-based crossover, an offspring inherits a combination of the orders existing in two "parent" sequences. This operator has the advantage of simplicity and, according to Davis, one of its proposers [17], it shows good effectiveness. The crossover is done in three stages (Fig. 6.17):

- A binary template is generated at random (Fig. 6.17a).
- Two parents are mated. The "0" and "1" of the binary template define the positions preserved in the sequences of the parents "1" and "2," respectively (Fig. 6.17b).
- To generate the offspring "1" and "2," the non preserved elements of the parents "1" and "2" are permuted in order to satisfy the order they have in the parents "2" and "1" respectively (Fig. 6.17c).

6.7.2.2 Crossover by Edge Recombination

With this class of crossover operators, an offspring inherits a combination of the adjacencies existing in the two parents. This is useful for the nonoriented traveling salesman problem, because the cost does not depend on the direction of the route in a cycle, but depends directly on the weights between the adjacent nodes of a Hamiltonian cycle.

The edge recombination operator was improved by several authors over several years. The "edge-3" version of Mathias and Whitley [39] will now be presented. Let

Fig. 6.17 Uniform order-based crossover

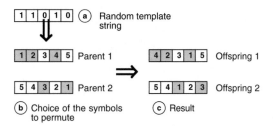

two individuals be selected for mating, for example ⟨ b, g, j, k, i, e, a, c, l, h, f, d⟩ and ⟨ f, c, b, e, k, a, h, i, l, j, g, d⟩. The first action builds an "edge table" of adjacencies (see Table 6.2) such that to each node corresponds a list of adjacent nodes in both parents: the number of such nodes is from two to four. The adjacencies common to both parents are marked by a * in the edge table.

At the time of action 2 of the operator, an initial active node is selected at random and all the references to this node are removed from the table.

Action 3 consists in choosing the edge which leads from the active node to an adjacent node marked by a * or, failing that, has the shortest list of adjacencies. If there are several equivalent options, the choice of the next node is carried out at random. The adjacent node chosen becomes the new active node added in the "offspring" tour. All the references to this node are removed from the adjacency lists in the edge table.

Action 4 builds a string, or possibly a complete tour. It consists of the repetition of action 3 as long as the adjacency list of the active node is nonempty. If the list is empty, then the initial node is reactivated to start again from the beginning of the string, but in the reverse direction, until the adjacency list of the active node is empty again. Then action 4 is concluded. The initial node cannot now be reactivated, because its adjacency list is empty owing to previous removal of the edges.

If a complete tour has not been generated, another active node is chosen at random, from among those which do not belong to any partial tour already built by previous executions of action 4. Then action 4 is initiated again. The application of the operator can thus be summarized as the sequence of actions 1 and 2 and as many actions 4 as necessary.

It is hoped that the operator will create few partial tours, and thus few foreign edges which do not belong to the two parents. The "edge-3" operator is powerful from this point of view.

Nodes	Edge list
a	c, e, h, k
b	d, g, e, c
c	l, a, b, f
d	b, *f, g
e	a, i, k, b
f	*d, h, c
g	*j, b, d
h	f, l, i, a
i	e, k, l, h
j	k, *g, l
k	i, j, a, e
l	h, c, j, i

Table 6.2 Table of adjacencies

Let us assume that node a in the example of Table 6.2 has been selected at random to be the initial active mode. Table 6.3 shows an example of the execution of the algorithm. The progress of the construction of the Hamiltonian cycle is presented in the last row. The active nodes are underlined. When an active node is marked with a superscript (1), this means that the next active node has to be chosen at random because of the existence of several equivalent possibilities. When it is marked with a superscript (2), it is at an end of the string: there is no more possible adjacency, which implies that one needs to move again in the reverse direction by reactivating the initial node a. It was necessary to apply action 4 only once in this case which generated a complete tour $\langle l, i, h, a, c, f, d, g, j, k, e, b \rangle$. Thus, except for the edge (bl), all the edges originate from one of the two parents.

6.7.2.3 Mutations of Adjacencies

The "2-opt" mutation is commonly used with the path representation. It is usually used for the Euclidean traveling salesman problem because of its geometrical properties. It consists in randomly choosing two positions in a sequence and then reversing the subsequence delimited by the two positions. Let the sequence be $\langle 987654321 \rangle$, where the two positions drawn at random are 3 and 8. Then the subsequence located between positions 3 and 8 is reversed, which gives the new sequence $\langle 984567321 \rangle$. Figure 6.18 shows the effect of the operator when applied to this sequence with the path representation. The operator can be generalized by choosing more than two positions for inversion of subsequences.

Table 6.3 Example of algorithm execution

Stage	1	2	3, 4	5, 6	7, 8, 9	10	11
a	c, e, h, k	e, h, k	e, h, k	e, h, k	h		
b	d, g, e, c	d, g, e	g, e	e			
c	l, b, f	l, b, f	l, b	l, b	l	l	l
d	b, *f, g	b, *f, g	b, g	b			
e	i, k, b	i, k, b	i, k, b	i, k, b	i	i	
f	*d, h, c	*d, h	h	h	h		
g	*j, b, d	*j, b, d	*j, b	b			
h	f, l, i	f, l, i	l, i	l, i	l, i	l, i	l
i	e, k, l, h	e, k, l, h	e, k, l, h	e, k, l, h	l, h	l	l
j	k, *g, l	k, *g, l	k, *g, l	k, l	l	l	l
k	i, j, e	i, j, e	i, j, e	i, e	i	i	
l	h, c, j, i	h, j, i	h, j, i	h, i	h, i	i	
Tour:	$\underline{a}^{(1)}$	a, $\underline{c}$	a, c, f, $\underline{d}^{(1)}$	a, c, f, d, g, $\underline{j}^{(1)}$	a, c, f, d, g, j, k, e, $\underline{b}^{(2)}$	$\underline{h}^{(1)}$, a, c, f, d, g, j, k, e, b	$\underline{i}$, h, a, c, f, d, g, j, k, e, b

Fig. 6.18 An example of a
2-opt mutation

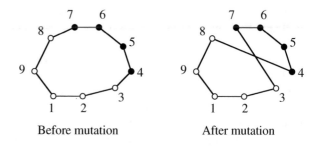

Before mutation   After mutation

6.7.2.4   Mutations of Permutations

If an individual represents a solution for a scheduling problem, the 2-opt operator
modifies the order of a large number of elements, on average $l/2$ if l is the size of
a sequence. However, the route direction of a subsequence, which was irrelevant in
the traveling salesman problem, is essential in this new context. Thus, the modifications which the adjacency mutation applies to a sequence are important. However, a
mutation should be able to apply small perturbations to a solution often, in order to
explore its close neighborhood. This is why other types of mutations have also been
proposed. The simplest one consists in withdrawing an element chosen at random
from a sequence and inserting it into another position. Several operators have been
described in the literature, such as mutation by exchange, where two positions in a
sequence are chosen at random and the elements in these positions are exchanged.
The performance offered by the variants of the mutation operators available depends
very much on the properties of the problem being dealt with.

6.8   Syntax Tree-Based Representation for Genetic
Programming

A dynamic tree-based representation for genetic algorithms was introduced by
Cramer in 1985 [15] in order to evolve sequential subprograms written in a simple programming language. The evolution engine used was the steady-state genetic
algorithm (SSGA) (Sect. 6.3.6.3), whose task was not to find the optimal values for
a problem, but to discover a computer program that could solve the problem.

John Koza adopted the syntax tree representation in 1992 [35] to define genetic
programming as a new evolutionary algorithm. Its main objective was to evolve
subprograms in the LISP language (Fig. 6.19a). He showed empirically that his
approach allows relevant programs to be discovered for a large number of examples
of applications, including the design of complex objects such as electronic circuits,
with an effectiveness significantly higher than what would chance.

Thanks to Koza's book would be expected by [35], the application of genetic programming has expanded to the solution of many types of problems whose solutions

can be represented by syntax tree structures, such as linear functions [42] (Fig. 6.19b), graphs [49, 54], and molecular structures [55].

A syntax tree is composed of a set of leaves, called terminals (T) in genetic programming, and a set of nodes, called nonterminals ($\mathcal{N}$). The two sets T and $\mathcal{N}$ together form the primitive set of a genetic programming system.

Using genetic programming needs the definition of the two sets of nodes $\mathcal{N}$ and leaves T that define the search space. The components of these two collections depend on the problem. For example, for linear functions, a solution is a syntax tree constructed from:

1. A set of nonterminal symbols, which may be arithmetic operators such as $\times, -, \div, +$, or functions with arguments such as sin and cos.
2. A set of terminal symbols, which can be variables, universal constants, or functions without arguments (rnd( ), time( ),...).

For genetic programming to work effectively, the primitive set must respect two important properties: closure and sufficiency [35]. The property of sufficiency requires that the sets of terminal and nonterminal symbols be able to represent any solution of the problem. This means that the set of all possible recursive compositions of the primitives must represent the search space. For example, the set AND, OR, NOT, $X1$, $X2$, ..., XN is a sufficient primitive set for Boolean function induction. The property of closure implies that each node must accept as an argument any type and value that can be produced by a terminal or nonterminal symbol. This means that any leaf or subtree can be used as an argument for every node in the tree.

The shape of the individuals in the genetic programming is very different from those mentioned previously for other representations. The trees must, in particular, have a mechanism for regulating their size. Otherwise, they will have a tendency to grow indefinitely over generations, unnecessarily consuming more memory and

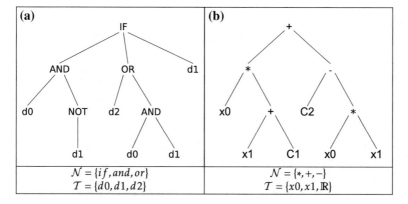

Fig. 6.19 Examples of tree solutions obtained by genetic programming where the search space is the set of LISP subprograms (**a**), and where the space explored is the space of linear functions representing polynomials with two variables (**b**)

computing power. The control mechanism can be implemented simply by giving some additional parameters to the genetic programming system, such as a maximum depth value for the trees or a maximum number of nodes. The genetic operators in the system must be modified to respect these constraints.

6.8.1 Initializing the Population

With the tree-based representation, initializing the population does not follow the same rules as with the binary and real representations. Each tree is generated in two steps: first the nodes, and then the leaves. However, the shape of the tree depends on the initialization approach used. The simplest and earliest three initialization methods are:

- The *Grow* method. The generated trees have irregular shapes; in each step, the selection is done in a uniform manner in the sets of nodes and terminals until the maximum depth is reached, below which only terminals may be chosen (Fig. 6.20a).
- The *Full* method. The trees are balanced and full; for a given node, a terminal is chosen only when the maximum depth is reached (Fig. 6.20b).
- The *Ramped Half and Half* method. Since the two previous methods do not offer a large variety of shapes and sizes of trees, Koza [35] proposed to combine the *Full* and *Grow* methods. In this method, half of the initial population is generated using *Full* and half is generated using *Grow*. The method uses a range of depth limits, which vary between 2 and the maximum depth. Currently, this technique is preferred to the two previous methods.

6.8.2 Crossover

The crossover traditionally used with the syntax tree representation is the *subtree crossover*. This consists of an exchange of two subtrees from the two individuals to be crossed, selected a priori from among the more efficient, and therefore potentially containing interesting subtrees. The crossover point in each parent tree is chosen randomly. An example of subtree crossover is illustrated in Fig. 6.21.

This general principle of crossover, introduced by Cramer in 1985 [15] can be refined with different extensions to constrain the size of the generated offspring. In fact, it is necessary to check the maximum depth for each syntax tree in the new population, so that the size of the individuals does not become unnecessarily large. If the crossover points chosen do not respect the limiting size value, then recombination may not take place. The attitude adopted in this case is a parameter of the crossover. It will be at least one of the two following choices:

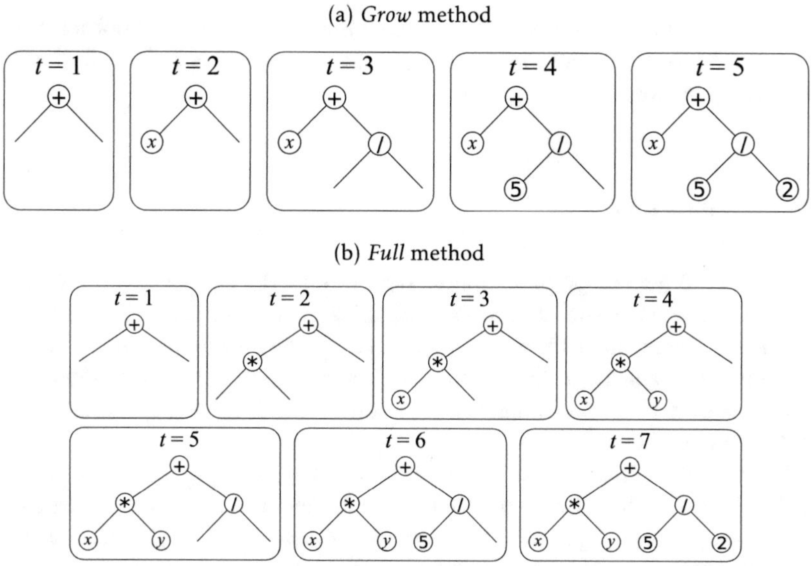

(a) *Grow* method

(b) *Full* method

Fig. 6.20 Construction of a syntax tree with a maximum depth equal to 2, using the *Grow* method **a** and the *Full* method **b**

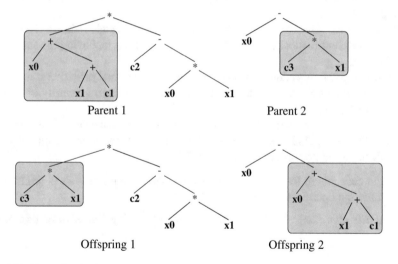

Fig. 6.21 Example of subtree crossover

- Select a new pair of parents and try to reapply the crossover operator until two offspring respecting the size constraint are found.
- Choose new crossover points on the two selected parents until the resulting offspring satisfy the maximum depth constraint.

There are a few other crossover operators that have been proposed for genetic programming systems, such as *context-preserving crossover* and *size-fair crossover* (see [37] for details).

6.8.3 Mutations

The traditional genetic programming system proposed by Koza [35] does not use mutation operators. To ensure access to all primitives of the search language (e.g., LISP) and ensure genetic diversity, it uses a very large population size, to include a large quantity of genetic material. Mutation in genetic programming was introduced for the first time in 1996 by Angeline [4] in order to reduce the population size and thus the computational cost.

Owing to the complexity of the syntax tree in genetic programming, multiple mutation operators have been proposed. Some of them are used for local search, but the majority could be applied for both local and global search. The most commonly used forms of mutation in genetic programming are:

- *Subtree mutation*: the operator randomly selects a mutation point (node) in the tree and substitutes the corresponding subtree with a randomly generated subtree; (Fig. 6.22).
- *Point mutation* (known also as *cycle mutation*): a random node in the tree is replaced with a different random node drawn from the primitive set having the same arity (Fig. 6.23).
- *Grow mutation*: this adds a randomly selected nonterminal primitive at a random position in the tree and adds terminals if necessary to respect the arity of the new node (Fig. 6.24).
- *Shrink mutation*: a randomly chosen subtree is deleted and one of its terminals takes its place. This is a special case of the subtree mutation that is motivated by the desire to reduce program size (Fig. 6.25).

In the case where the leaves of the tree can take numerical values (constants), other mutation operators have been introduced, such as:

- *Gaussian mutation*, which mutates constants by adding Gaussian random noise [4].

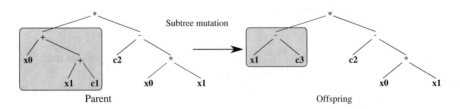

Fig. 6.22 Example of subtree mutation

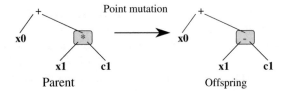

Fig. 6.23 Example of point mutation

- *Optimized constant mutation*, which tunes the solution by trying to find the best values for constants in the tree. This uses a numerical optimization method to reach the nearest local optimum, such as the hill climber method [51] or partial gradient ascent [50].

6.8.4 Application to Symbolic Regression

Given a supervised learning database containing a set of N pairs of vectors $(\mathbf{x}_j, \mathbf{y}_j)$ for $j \in [1, N]$, *symbolic regression* consists in discovering a symbolic expression $\mathcal{S}$ that is able to map the input vector $\mathbf{x}_j$ to the target real value $\mathbf{y}_j$. A priori, there is no constraint on the structure of the expression $\mathcal{S}$ being searched for. For a vector $\mathbf{x}_j$, the expression $\mathcal{S}$ allows one to compute $\hat{\mathbf{y}}_j = \mathcal{S}(\mathbf{x}_j)$, whose gap with respect to $\mathbf{y}_j$ must be minimized for any j by modifying the structure of $\mathcal{S}$.

John Koza [35] has shown that genetic programming can be used advantageously to solve symbolic regression problems. Each tree in the population may represent a

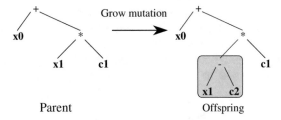

Fig. 6.24 Example of Grow mutation

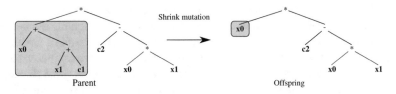

Fig. 6.25 Example of Shrink mutation

mathematical expression. Besides the work of Koza, several studies [11, 28, 34, 38] have shown the benefit of the application of genetic programming in solving symbolic regression problems. Multiple applications in different fields have been presented, for example automatic searching for the structure of filters [43], exploration and prediction of the loads of helicopters [13], forecasting of the production of dairy cows, and determination of the functional associations between groups of proteins [25].

Below we present an example of an application in the field of finance to forecasting market volatility. For financial volatility forecasting, the input set X is a historical financial time series of data ($X = x_1, x_2, \ldots, x_T$), and the output vector $Y = y_1, y_2, \ldots, y_T$ is the implied volatility values computed from the observed data.

6.8.4.1 Implied Volatility Forecasting

One challenge posed by financial markets is to correctly forecast the volatility of financial securities, which is a crucial variable in the trading and risk management of derivative securities. Traditional parametric methods have had limited success in estimating and forecasting volatility as they are dependent on restrictive assumptions and it is difficult to make the necessary estimates. Several machine learning techniques have recently been used to overcome these difficulties [12, 14, 33]. Genetic programming has often been applied to forecasting financial time series and, in some recent work for Abdelmalek and Ben Hamida, it was successfully applied to the prediction of implied volatility [1, 29]. We summarize this work in the following and illustrate the main results.

The data used were daily prices of European S&P500 index call options, from the Chicago Board Options Exchange (CBOE) for a sample period from January 2, 2003 to August 29, 2003. The S&P500 index options are among the most actively traded financial derivatives in the world.

Each formula given by genetic programming was evaluated to test whether it could accurately forecast the output value (*the implied volatility*) for all entries in the training set. To assign a fitness measure to a given solution, we computed the mean squared error (MSE) between the estimated volatility ($\hat{y}_i$) given by the genetic programming solution and the target volatility (y_i) computed from the input data:

$$\text{MSE} = \frac{1}{N} \sum_{N}^{1} (y_i - \hat{y}_i)^2 \tag{6.3}$$

where N is the number of entries in the training data sample.

To generate and evolve the tree-based models, the genetic programming needed a primitive set composed of a terminal set (for the leaves of the tree) and a function set (for the nodes of the tree). The terminal set included the following input variables: the call option price divided by the strike price, C/K; the index price divided by the strike price, S/K; and the time to maturity, τ. The function set included basic mathematical operators and some specific functions that might be useful for implied

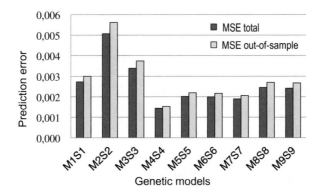

Fig. 6.26 Performance of the volatility models generated by genetic programming according to MSE total and MSE out-of-sample for the nine time series samples

volatility models, such as the components of the Black–Scholes model [10]. The primitive set used in [1] is given in Table 6.4.

The full input sample was sorted by time series and divided chronologically into nine successive subsamples $(S_1, S_2, \ldots, S_9)$ each containing 667 daily observations. These samples were used simultaneously for training and test steps.

Several runs were performed for each training subsample from the time series set $(S_1, S_2, \ldots, S_9)$. Thus, nine best function models were selected for all subsamples, denoted (M1S1 $\cdots$ M9S9). To assess the internal and external accuracy of the functions obtained, two performance measures were used: the "MSE total," computed for the complete sample, and "MSE out-of-sample," computed for samples external to the training sample (the eight samples that were not used for learning). Figure 6.26 describes the evolution pattern of the squared errors for these volatility models.

The following is the function M4S4, which had the lowest MSE total:

$$
\text{M4S4}\left(\frac{C}{K}, \frac{S}{K}, \tau\right) = \exp\left[\left(\ln\left(\Theta\left(\frac{C}{K}\right)\right) \times \sqrt{\tau - 2 \times \frac{C}{K} + \frac{S}{K}}\right) - \cos\left(\frac{C}{K}\right)\right]
$$

All of the models obtained were able to fit well not only the training samples but also the enlarged sample.

Table 6.4 The primitive set used in [1]

Binary functions	Addition, subtraction, multiplication, protected division
Unary functions	Sine, cosine, protected natural log, exponential function, protected square root, normal cumulative distribution Black–Scholes component (*Ncfd*)

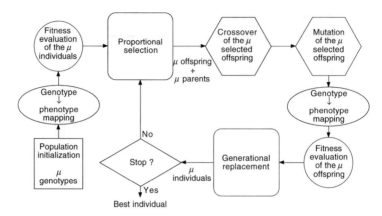

Fig. 6.27 A simple genetic algorithm

The models thus obtained could provide useful information for both speculators and option hedgers that they could provide to investors to help them protect themselves against risk in financial markets. Some simulations of speculation strategies have been carried out to assess the profits that a speculator could achieve based on the estimated volatility generated by these genetic models. The results have shown that genetic models may generate higher yields than conventional models [2].

6.9 The Particular Case of Genetic Algorithms

The simple genetic algorithm follows the outline of an evolutionary algorithm such as the one presented in Fig. 6.1 with a notable original feature: it implements a genotype–phenotype transcription that is inspired by natural genetics. This transcription precedes the phase of evaluation the fitness of the individuals. A *genotype* is often a binary symbol string. This string is decoded to build a solution of a problem represented in its natural formalism: this solution is viewed as the *phenotype* of an individual. This latter one can then be evaluated to give a fitness value that can be exploited by the selection operators.

A flowchart of a simple genetic algorithm is presented in Fig. 6.27. It can be noticed that it implements a proportional selection operator (see Sect. 6.3.3) and a generational replacement, i.e., the population of the offspring replaces that of the parents completely. Another classical version uses a steady-state replacement (Sect. 6.3.6.3). The variation operators work on the genotypes. As these are bit strings, the operators of crossover and mutation presented in Sect. 6.5 related to the binary representation are often used. The crossover operator is regarded as the essential search operator. The mutation operator is usually applied with a small rate, in order to maintain a minimum degree of diversity in the population. Since the representation is based on bit strings, the difficulty is to discover a good coding of the genotype, such that the variation operators in the space of the bit strings produce viable offspring, often satisfying the constraints of the problem. This is generally not a trivial job ...

Holland, Goldberg, and many other authors have worked on a mathematical formalization of genetic algorithms based on a "Schema Theorem" [26], whose utility is controversial. At first glance, it enables us to justify the choice of a binary representation. However, research work using this theorem did not prove in the end to be very useful for modeling an evolution. Many counterexamples showed that conclusions formulated from considerations deduced from this theorem were debatable, in particular even the choice of the binary representation.

Genetic algorithms have been subject to many suggestions for modification in order to improve their performance or to extend their application domain. Thus, bit strings have been replaced by representations closer to the formalism of the problems being dealt with, avoiding the debatable question of the design of an effective coding. For example, research work using "real coded genetic algorithms" uses the real representations discussed in Sect. 6.6. In addition, proportional selection is often replaced by other forms of selection. These modifications are sufficiently significant that the specific features of genetic algorithms compared with the diversity of the other evolutionary approaches disappear.

6.10 The Covariance Matrix Adaptation Evolution Strategy

6.10.1 Presentation of Method

The "covariance matrix adaptation evolution strategy" (CMA-ES) [30] was originally designed to find the global optimum of an objective function in a continuous space such as $\mathbb{R}^n$ with greater efficiency than could be achieved with an evolution strategy using correlated mutation (Sect. 6.6.2.5). In a similar way, the method performs an evolution by building a sample of λ solutions in each generation g. These samples are generated randomly according to the Gaussian distribution $\mathcal{N}(\mathbf{m}(g), \mathbf{C}(g))$, with mean vector $\mathbf{m}(g)$ and covariance matrix $\mathbf{C}(g)$. However, unlike the methods using mutations, the μ best solutions in this sample are then selected to estimate a new Gaussian distribution $\mathcal{N}(\mathbf{m}(g+1), \mathbf{C}(g+1))$, which will be then used in the next generation. There is no more "individual" dependency between "parent" solutions and "offspring" solutions. The distribution $\mathcal{N}(\mathbf{m}(g+1), \mathbf{C}(g+1))$ is constructed to approach (hopefully closely enough) the desired optimum. As in the other evolution strategies (Sect. 6.6.2.4 and following), the CMA-ES algorithms implement a concept of parameter self-adaptation.

In the CMA-ES approach, three parameters $\mathbf{m}$, σ, and $\hat{\mathbf{C}}$ are considered to define the Gaussian distribution $\mathcal{N}(\mathbf{m}, \sigma^2\hat{\mathbf{C}})$, where $\sigma \in \mathbb{R}^+$ is the step size. This decomposition of the covariance matrix $\mathbf{C}$ into two terms makes it possible to separately adjust the parameters σ and $\hat{\mathbf{C}}$ according to different criteria in order to speed up the convergence towards the optimum. The following sections describe the step of selection of the best solutions generated at random and the adaptation mechanisms of $\mathbf{m}$, σ, and $\hat{\mathbf{C}}$.

6.10.1.1   Fitness Function and Selection

At the beginning of generation g, λ solutions $\mathbf{X}_i(g) \in \mathbb{R}^n$ are generated randomly according to the Gaussian distribution $\mathcal{N}(\mathbf{m}(g), \sigma(g)^2\hat{\mathbf{C}}(g))$. A rank i is assigned to solution $\mathbf{X}_i$ such that the objective value $F(\mathbf{X}_i)$ is better than or equal to $F(\mathbf{X}_{i+1})$ for all i. "Better" means "smaller" for a minimization problem and "larger" for a maximization problem. Solution $\mathbf{X}_i$, for $i \in \{1, \ldots, \mu\}$, is associated with fitness values f_i that decrease with index i: $\forall i \in \{1, \ldots, \mu\}, f_i > 0, f_i \geq f_{i+1}$, with $\sum_{i=1}^{\mu} f_i = 1$. The values of f_i depend only on the rank i and are constant throughout evolution. The easiest way is to choose $f_i = 1/\mu$. More sophisticated fitness functions can improve the convergence to the optimum.

The selection is deterministic: it keeps the μ best solutions, which are $\mathbf{X}_1(g)$ to $\mathbf{X}_\mu(g)$.

6.10.1.2   Adaptation of m

The value of $\mathbf{m}(g+1)$ for the next generation is the average weighted by the fitness values f_i of the μ selected solutions $\mathbf{X}_i(g)$. In this way, $\mathbf{m}$ moves from generation to generation according to the optimization path determined by the sequence of sets of the best solutions $\mathbf{X}_i$ which have been selected. We have

$$\mathbf{m}(g+1) = \sum_{i=1}^{\mu} f_i \mathbf{X}_i(g). \tag{6.4}$$

6.10.1.3   Adaptation of σ

The step size $\sigma(g)$ is adapted so that successive vectors

$$\delta(g+1) = \frac{\mathbf{m}(g+1) - \mathbf{m}(g)}{\sigma(g)}$$

according to g are uncorrelated as much as possible. In fact, if the vectors $\delta(g)$ are strongly correlated (with a correlation coefficient close to 1), that means that $\sigma(g)$ is too small because each successive generation leads to progress in the search space in almost the same direction. Thus, $\sigma(g)$ should be increased, thereby reducing the number of evaluations of the objective function for almost the same progression.

However, if successive steps $\delta(g)$ are anticorrelated (with a correlation coefficient close to -1), this leads to variations of $\mathbf{m}(g)$ in almost opposite directions in successive generations, involving too slow a progression in the search space. It can be deduced from this situation that $\sigma(g)$ is too large.

To decide whether the step size $\sigma(g)$ is too small or too large, the designers of the method used the notion of an *evolution path* $\mathbf{p}_\sigma(g)$, which can be calculated as an average of $\delta(g)$ over a few generations. It is then compared with the average progression allowed by independent Gaussian random vectors drawn from distribution $\delta(g)$. As the draws are independent, they are uncorrelated.

If we define $\mu_f = 1/\sum_{i=1}^{\mu} f_i^2$, $\delta(g+1)$ is a random vector drawn from the distribution $\mathcal{N}(0, \hat{\mathbf{C}}/\mu_f)$. In practice, a vector $\delta'(g+1)$ drawn from the distribution $\mathcal{N}(0, \mathbf{I}/\mu_f)$ is calculated as follows:

$$\delta'(g+1) = \hat{\mathbf{C}}(g)^{-1/2}\delta(g+1) = \mathbf{BD}^{-1}\mathbf{B}^{\mathsf{T}}\delta(g+1)$$

where $\mathbf{B}$ and $\mathbf{D}$ are the matrix of eigenvectors and the corresponding diagonal matrix of the square roots of the eigenvalues of $\hat{\mathbf{C}}(g)$, respectively. Thus, $\sqrt{\mu_f}\delta'(g+1)$ is drawn from the distribution $\mathcal{N}(0, \mathbf{I})$. The designers of the method proposed that a weighted average of $\mathbf{p}_\sigma(g)$ and $\sqrt{\mu_f}\delta'(g+1)$ should be calculated recursively to obtain $\mathbf{p}_\sigma(g+1)$:

$$\mathbf{p}_\sigma(g+1) = (1 - c_\sigma)\mathbf{p}_\sigma(g) + \alpha\sqrt{\mu_f}\delta'(g+1)$$

where $c_\sigma \in\ ]0, 1[$ is a parameter of the method. Choosing c_σ close to 0 leads to a smooth but slow adaptation of $\mathbf{p}_\sigma$: the memory effect is important. α is calculated so that when the step size $\sigma(g)$ is well adapted, $\mathbf{p}_\sigma(g)$ and $\mathbf{p}_\sigma(g+1)$ have the same distribution $\mathcal{N}(0, \mathbf{I})$. Now, $\sqrt{\mu_f}\delta'(g+1)$ is also drawn from the distribution $\mathcal{N}(0, \mathbf{I})$. Therefore, $\alpha = \sqrt{1 - (1 - c_\sigma)^2}$, so that the covariance matrix of $\mathbf{p}_\sigma(g+1)$ is $\mathbf{I}$. The following expression for the evolution path $\mathbf{p}_\sigma(g)$ for $g \geq 1$ is thereby obtained:

$$\begin{cases} \mathbf{p}_\sigma(g+1) = (1 - c_\sigma)\mathbf{p}_\sigma(g) + \sqrt{c_\sigma(2 - c_\sigma)\mu_f}\,\mathbf{BD}^{-1}\mathbf{B}^{\mathsf{T}}\dfrac{\mathbf{m}(g+1)-\mathbf{m}(g)}{\sigma(g)} \\ \mathbf{p}_\sigma(1) = 0 \end{cases} \quad (6.5)$$

Then, $\|\mathbf{p}_\sigma(g+1)\|$ is "compared" with $\mathsf{E}\|\mathcal{N}(0, \mathbf{I})\|$, which is the expectation of the norm of the Gaussian random vectors drawn from the distribution $\mathcal{N}(0, \mathbf{I})$, to adapt the value of σ in such a way that it:

- decreases when $\|\mathbf{p}_\sigma(g+1)\|$ is less than $\mathsf{E}\|\mathcal{N}(0, \mathbf{I})\|$,
- increases when $\|\mathbf{p}_\sigma(g+1)\|$ is greater than $\mathsf{E}\|\mathcal{N}(0, \mathbf{I})\|$,
- remains constant when $\mathbf{p}_\sigma(g+1)$ is equal to $\mathsf{E}\|\mathcal{N}(0, \mathbf{I})\|$.

The following expression can perform this adaptation efficiently:

$$\sigma(g+1) = \sigma(g)\exp\left(\frac{c_\sigma}{d_\sigma}\left(\frac{\|\mathbf{p}_\sigma(g+1)\|}{\mathsf{E}\|\mathcal{N}(0, \mathbf{I})\|} - 1\right)\right) \quad (6.6)$$

where d_σ is a damping factor, with a value of around 1. The value of $\sigma(0)$ is problem-dependent; c_σ, d_σ, and $\sigma(0)$ are parameters of the method. A robust strategy for initialization of these parameters is presented in Sect. 6.10.2.

6.10.1.4 Adaptation of $\hat{\mathbf{C}}$

The designers of the method proposed an estimator $\mathbf{C}_\mu(g+1)$ for the covariance matrix $\mathbf{C}(g+1)$ based on the μ best realizations $\mathbf{X}_i(g)$ obtained in generation g:

$$\mathbf{C}_\mu(g+1) = \sum_{i=1}^{\mu} f_i (\mathbf{X}_i - \mathbf{m}(g))(\mathbf{X}_i - \mathbf{m}(g))^{\mathsf{T}}$$

Note that this estimator uses the weighted average $\mathbf{m}(g)$ obtained in the previous generation instead of $\mathbf{m}(g+1)$. Moreover, the contribution of each term $(\mathbf{X}_i - \mathbf{m}(g))$ is weighted by $\sqrt{f_i}$. To see the relevance of this estimator intuitively using an example, we consider the case $\mu = 1$:

$$\mathbf{C}_1(g+1) = f_1 (\mathbf{X}_1 - \mathbf{m}(g))(\mathbf{X}_1 - \mathbf{m}(g))^{\mathsf{T}}$$

The matrix $\mathbf{C}_1(g+1)$ therefore has only one nonzero eigenvalue, for an eigenvector collinear with $(\mathbf{X}_1 - \mathbf{m}(g))$. This means that the Gaussian distribution $\mathcal{N}(\mathbf{m}(g+1), \mathbf{C}_1(g+1))$ will generate realizations $\mathbf{X}_i(g+1)$ only on the line whose direction vector is $(\mathbf{X}_1(g) - \mathbf{m}(g))$, passing through the point $\mathbf{m}(g+1)$. Now, since $\mathbf{X}_1(g)$ is the best solution obtained in the current generation, a heuristic choice of the direction $(\mathbf{X}_1(g) - \mathbf{m}(g))$ to find a better solution $\mathbf{X}_1(g+1)$ is reasonable. However, a priori, this direction is not that of the optimum. To ensure a good exploration of the search space, μ must be large enough, not less than n, so that the covariance matrix $\mathbf{C}_\mu(g+1)$ is positive definite.

Taking into account the step size $\sigma(g)$, with $\mathbf{C}(g) = \sigma(g)^2 \hat{\mathbf{C}}(g)$, the expression for $\hat{\mathbf{C}}_\mu(g+1)$ is

$$\hat{\mathbf{C}}_\mu(g+1) = \sum_{i=1}^{\mu} f_i \frac{\mathbf{X}_i - \mathbf{m}(g)}{\sigma(g)} \left(\frac{\mathbf{X}_i - \mathbf{m}(g)}{\sigma(g)} \right)^{\mathsf{T}}$$

However, giving a large value to μ also increases the number of evaluations of the objective function needed to reach the optimum. To reduce the value of μ while ensuring that the matrix $\hat{\mathbf{C}}(g+1)$ remains positive definite, it is possible to use the matrix $\hat{\mathbf{C}}(g)$ obtained in the previous generation.

Rank-μ update. The designers of the method proposed that $\hat{\mathbf{C}}(g+1)$ should be a weighted average of the matrices $\hat{\mathbf{C}}(g)$ and $\hat{\mathbf{C}}_\mu(g+1)$, with respective weights $1 - c_\mu$ and c_μ, where $c_\mu \in \,]0, 1]$ is a parameter of the method:

The matrix $\hat{\mathbf{C}}(g+1)$ is thereby defined by recurrence for $g \geq 1$.

$$\begin{cases} \hat{\mathbf{C}}(g+1) = (1 - c_\mu)\hat{\mathbf{C}}(g) + c_\mu \hat{\mathbf{C}}_\mu(g+1) \\ \hat{\mathbf{C}}(1) = \mathbf{I} \end{cases} \tag{6.7}$$

The identity matrix is chosen as the initial term because it is symmetric positive definite. By the recurrence relation, $\hat{\mathbf{C}}(g+1)$ is a weighted average of the matrices $\hat{\mathbf{C}}_\mu(i)$ for $i \in \{1, \ldots, g+1\}$.

Thus μ can be much smaller than n while keeping the matrices $\hat{\mathbf{C}}(g+1)$ positive definite. If c_μ is chosen close to 0, the matrix $\hat{\mathbf{C}}(g+1)$ depends strongly on the past and can accept small values of μ. But the evolution will be slow. If c_μ is chosen close to 1, the matrix $\hat{\mathbf{C}}(g+1)$ can evolve quickly, provided μ is large enough to ensure that the matrix $\hat{\mathbf{C}}$ remains positive definite, which ultimately increases the number of evaluations of the objective function that are necessary.

The expression (6.7) is suitable for updating $\hat{\mathbf{C}}(g)$, but at the cost of an excessive number of generations, with a value of μ which needs to be chosen large enough. To reduce the number of evaluations of the objective function needed, an additional adaptation mechanism for $\hat{\mathbf{C}}(g)$ has been used.

Rank-one update. This adaptation mechanism for $\hat{\mathbf{C}}$ consists in generating in every generation a random vector $\mathbf{p}_c(g+1)$ according to the distribution $\mathcal{N}(0, \hat{\mathbf{C}})$. In Sect. 6.10.1.3 we saw that $\delta(g+1) = (\mathbf{m}(g+1) - \mathbf{m}(g))/\sigma(g)$ has the distribution $\mathcal{N}(0, \hat{\mathbf{C}}/\mu_f)$. Similarly to $\mathbf{p}_\sigma(g+1)$, we express $\mathbf{p}_c(g)$ for $g \geq 1$ as an evolution path:

$$\begin{cases} \mathbf{p}_c(g+1) = (1 - c_c)\mathbf{p}_c(g) + \sqrt{c_c(2 - c_c)\mu_f} \, \frac{\mathbf{m}(g+1) - \mathbf{m}(g)}{\sigma(g)} \\ \mathbf{p}_c(1) = 0 \end{cases} \tag{6.8}$$

The expression for $\hat{\mathbf{C}}(g+1)$, which must be of rank n, is expressed as a weighted average of $\mathbf{p}_c(g+1)\mathbf{p}_c(g+1)^\mathsf{T}$, which has rank 1, and of $\hat{\mathbf{C}}(g)$, which is of rank n:

$$\begin{cases} \hat{\mathbf{C}}(g+1) = (1 - c_1)\hat{\mathbf{C}}(g) + c_1 \mathbf{p}_c(g+1)\mathbf{p}_c(g+1)^\mathsf{T} \\ \hat{\mathbf{C}}(1) = \mathbf{I} \end{cases} \tag{6.9}$$

Update of $\hat{\mathbf{C}}$. The combination of the expressions for rank-μ update (Eq. (6.7)) and rank-one update (Eq. (6.9)) gives the complete expression for the updating of $\hat{\mathbf{C}}(g)$, updating for $g \geq 1$:

$$\begin{cases} \hat{\mathbf{C}}(g+1) = (1 - c_1 - c_\mu)\hat{\mathbf{C}}(g) + c_1 \mathbf{p}_c \mathbf{p}_c^\mathsf{T} + c_\mu \sum_{i=1}^{\mu} f_i \mathbf{V}_i \mathbf{V}_i^\mathsf{T} \\ \hat{\mathbf{C}}(1) = \mathbf{I} \end{cases} \tag{6.10}$$

where $\mathbf{V}_i = (\mathbf{X}_i - \mathbf{m}(g))/\sigma(g)$, and c_c, c_1, and c_μ are parameters of the method. A robust strategy for initialization of these parameters is presented in Sect. 6.10.2.

6.10.2   The CMA-ES Algorithm

Algorithm 6.1 implements the CMA-ES method as proposed in [30]. In every generation, λ independent (pseudo-)random solutions $\mathbf{X}_i$ are generated according to the distribution $\mathcal{N}(\mathbf{m}, \sigma^2\hat{\mathbf{C}})$, whose parameters have been determined in the previous generation. The μ best solutions are sorted and returned by the function Selection (Algorithm 6.2) in the form of a matrix $\mathbf{X}$ with n rows and μ columns. Column i of $\mathbf{X}$ gives the solution $\mathbf{X}_i$. Column sorting is done so that if the objective value $F_i = F(\mathbf{X}_i)$ is better than $F_j = F(\mathbf{X}_j)$, then $i < j$.

Inputs: $\mathbf{m}, \sigma, n$ `// n: dimension of search space` Ω

$\lambda, \mu, \mathbf{f}, \mu_f, c_\sigma, d_\sigma, c_c, c_1, c_\mu \leftarrow$ Initialization(n)

$\mathbf{p}_c \leftarrow \mathbf{p}_\sigma \leftarrow 0$

$\hat{\mathbf{C}} \leftarrow \mathbf{B} \leftarrow \mathbf{D} \leftarrow \mathbf{I}$ `// I:` $n \times n$ `identity matrix`

repeat

    $\mathbf{X}, \mathbf{V} \leftarrow$ Selection($\lambda, \mathbf{m}, \sigma, \mathbf{B}, \mathbf{D}$)

    $\mathbf{m}, \delta \leftarrow$ UpdateM($\mathbf{m}, \mu, \mathbf{X}, \mathbf{f}, \sigma$)

    $\sigma, \mathbf{p}_\sigma \leftarrow$ UpdateSigma($\sigma, \mathbf{p}_\sigma, \mathbf{B}, \mathbf{D}, \delta, c_\sigma, d_\sigma, \mu_f$)

    $\hat{\mathbf{C}}, \mathbf{p}_c \leftarrow$ UpdateC($\hat{\mathbf{C}}, \mathbf{p}_c, \mathbf{p}_\sigma, \mathbf{V}, \mathbf{f}, \delta, c_c, c_1, c_\mu, \mu, \mu_f$)

    $\mathbf{B} \leftarrow$ EigenVectors($\hat{\mathbf{C}}$)

    $\mathbf{D} \leftarrow$ EigenValues($\hat{\mathbf{C}}$)$^{1/2}$

    .   `//` $\mathbf{D}$`: diagonal matrix of the eigenvalue root squares`

until *Stopping criterion satisfied*

Algorithm 6.1: The CMA-ES algorithm.

---

for $i = 1 \lambda$ **do**

    $\mathbf{y}_i \leftarrow$ GaussianRandomDraw($0, \mathbf{I}$)   `//` $\mathbf{y}_i$ `has distribution` $\mathcal{N}(0, \mathbf{I})$

    $\mathbf{V}_i \leftarrow \mathbf{B}\mathbf{D}\mathbf{y}_i$ `//` $\mathbf{V}_i$ `has distribution` $\mathcal{N}(0, \hat{\mathbf{C}})$ `with` $\hat{\mathbf{C}} = \mathbf{B}\mathbf{D}^2\mathbf{B}^\mathsf{T}$

    $\mathbf{X}_i \leftarrow \mathbf{m} + \sigma\mathbf{V}_i$ `//` $\mathbf{X}_i$ `has distribution` $\mathcal{N}(\mathbf{m}, \sigma^2\hat{\mathbf{C}})$

    $F_i \leftarrow$ Objective($\mathbf{X}_i$)   `//` F_i `is the objective value associated to` $\mathbf{X}_i$

end

$\mathbf{X}, \mathbf{V} \leftarrow$ Sort($\mathbf{X}, \mathbf{V}, \mathbf{F}$)   `// column sorting of` $\mathbf{X}_i$ `and` $\mathbf{V}_i$ `according to` F_i

return $\mathbf{X}, \mathbf{V}$

Algorithm 6.2: Function Selection($\lambda, \mathbf{m}, \sigma, \mathbf{B}, \mathbf{D}$).

From $\mathbf{X}$, the updates of the parameters $\mathbf{m}$, σ, and $\hat{\mathbf{C}}$ are assigned to the functions UpdateM, UpdateSigma, and UpdateC (Algorithms 6.3, 6.4, and 6.5). These functions do not require special comment: their algorithms are derived directly from analytical expressions given in the previous section.

Let $\mathbf{B}$ be the matrix whose columns i are eigenvectors $\mathbf{b}_i$ of $\hat{\mathbf{C}}$. Let $\mathbf{D}$ be the diagonal matrix such that d_{ii} is the square root of the eigenvalue of $\hat{\mathbf{C}}$ corresponding to eigenvector $\mathbf{b}_i$. The matrices $\mathbf{B}$ and $\mathbf{D}$ are computed as, in particular, they facilitate the independent draws of solutions $\mathbf{X}$ according to the distribution $\mathcal{N}(\mathbf{m}, \sigma^2\hat{\mathbf{C}})$.

$$\mathbf{m}' \leftarrow \mathbf{m}$$
$$\mathbf{m} \leftarrow \sum_{i=1}^{\mu} f_i \mathbf{X}_i$$
$$\delta \leftarrow (\mathbf{m} - \mathbf{m}')/\sigma$$
$$\textbf{return } \mathbf{m}, \delta$$

Algorithm 6.3: Function UpdateM($\mathbf{m}, \mu, \mathbf{X}, \mathbf{f}, \sigma$).

$$\mathbf{p}_\sigma \leftarrow (1 - c_\sigma)\mathbf{p}_\sigma + \sqrt{c_\sigma(2 - c_\sigma)\mu_f}\ \mathbf{B} \cdot \mathbf{D}^{-1} \cdot \mathbf{B}^{\mathsf{T}}\delta$$
$$\sigma \leftarrow \sigma \exp\left(\frac{c_\sigma}{d_\sigma}\left(\frac{\|\mathbf{p}_\sigma\|}{\mathrm{E}\|\mathcal{N}(0,\mathbf{I})\|} - 1\right)\right) \qquad // \ \mathrm{E}\|\mathcal{N}(0,\mathbf{I})\| \approx \sqrt{n}\left(1 - \frac{1}{4n} + \frac{1}{21n^2}\right)$$
$$\textbf{return } \sigma, \mathbf{p}_\sigma$$

Algorithm 6.4: Function UpdateSigma($\sigma, \mathbf{p}_\sigma, \mathbf{B}, \mathbf{D}, \delta, c_\sigma, d_\sigma, \mu_f$).

$$\mathbf{p}_c \leftarrow (1 - c_c)\mathbf{p}_c$$
$$\textbf{if } \|\mathbf{p}_\sigma\| < 1.5\sqrt{n} \textbf{ then}$$
$$\quad \mathbf{p}_c \leftarrow \mathbf{p}_c + \sqrt{c_c(2 - c_c)\mu_f}\ \delta$$
$$\textbf{end}$$
$$\hat{\mathbf{C}} \leftarrow (1 - c_1 - c_\mu)\hat{\mathbf{C}} + c_1\mathbf{p}_c\mathbf{p}_c^{\mathsf{T}} + c_\mu \sum_{i=1}^{\mu} f_i\mathbf{V}_i\mathbf{V}_i^{\mathsf{T}}$$
$$\textbf{return } \hat{\mathbf{C}}, \mathbf{p}_c$$

Algorithm 6.5: Function UpdateC($\hat{\mathbf{C}}, \mathbf{p}_c, \mathbf{p}_\sigma, \mathbf{V}, \mathbf{f}, \delta, c_c, c_1, c_\mu, \mu, \mu_f$).

The setting of the algorithm parameters by the function Initialization depends a priori on the problem to be solved. However, a default initialization, which has proven to be robust and effective and usable for many problems, was proposed in [30]. It is implemented by the function DefaultInitialization (Algorithm 6.6). The values chosen for the parameters λ, μ, $\mathbf{f} = (f_1, \ldots, f_\mu)$, c_σ, d_σ, c_c, c_1, and c_μ can be adapted to the problem to be solve. The proposed values for λ and μ should be considered as minimum values. Larger values improve the robustness of the algorithm, at the cost, however, of a greater number of generations.

The initial values of $\mathbf{m} = (m_1, \ldots, m_n)$ and σ depend on the problem. When the location of the optimum is approximately known, these initial values should be determined so that the optimum lies in the range defined by the intervals $[m_i - 2\sigma, m_i + 2\sigma]$ [30] for each coordinate $i \in \{1, \ldots, n\}$.

$\lambda \leftarrow 4 + \lfloor 3 \ln n \rfloor$ `// ⌊x⌋ is the lower integer part of x`

$\mu \leftarrow \lfloor \lambda/2 \rfloor$

for $i = 1$ μ **do**

   $f_i \leftarrow \frac{\ln(\mu+1)-\ln i}{\sum_{j=1}^{\mu} \ln(\mu+1)-\ln j}$ `// ` $\mathbf{f} = (f_1, \ldots, f_i, \ldots, f_\mu)$

end

$\mu_f \leftarrow 1/\sum_{i=1}^{\mu} f_i^2$

$c_\sigma \leftarrow \frac{\mu_f+2}{n+\mu_f+3}$

$d_\sigma \leftarrow 1 + 2 \max\left(0, \sqrt{\frac{\mu_f-1}{n+1}} - 1\right) + c_\sigma$

$c_c \leftarrow 4/(n+4)$

$c_{\text{cov}} \leftarrow \frac{2}{\mu_f(n+\sqrt{2})^2} + \left(1 - \frac{1}{\mu_f}\right) \min\left(1, \frac{2\mu_f-1}{(n+2)^2+\mu_f}\right)$

$c_1 \leftarrow c_{\text{cov}}/\mu_f$

$c_\mu \leftarrow c_{\text{cov}} - c_1$

return $\lambda, \mu, \mathbf{f}, \mu_f, c_\sigma, d_\sigma, c_c, c_1, c_\mu$

Algorithm 6.6: Function DefaultInitialization(n).

6.10.3 Some Simulation Results

Like all metaheuristics, the CMA-ES method is designed to solve hard optimization problems, at least approximately, in a reasonable time. However, to be convincing, the method must also have acceptable performance on "easier" problems, but of course without using specific properties of them that facilitate the search for an optimum, such as convexity or differentiability. This section aims to give an idea of the ability of CMA-ES to discover the minimum of a set of ill-conditioned, nonseparable quadratic functions of the form $F(\mathbf{X}) = (\mathbf{X} - \mathbf{c})^\mathsf{T} \mathbf{H}(\mathbf{X} - \mathbf{c})$, where $\mathbf{c}$ is the desired optimum and $\mathbf{H}$ is a symmetric positive definite matrix. The isovalue hypersurfaces of $F(\mathbf{X})$ in $\mathbb{R}^n$ are hyperellipsoids (Sect. 6.6.2.5).

6.10.3.1 Parameters of the Quadratic Functions

For each quadratic function $F(\mathbf{X}) = (\mathbf{X} - \mathbf{c})^\mathsf{T} \mathbf{H}(\mathbf{X} - \mathbf{c})$ used in our experiments, each component of the vector $\mathbf{c}$ was a realization of a random variable according to a Gaussian distribution with mean 0 and standard deviation 10. $\mathbf{H}$ was determined by the expression

$$\mathbf{H} = (\mathbf{SR})^\mathsf{T}(\mathbf{SR})$$

where:

- **S** is a diagonal matrix that sets the condition number of **H**. The condition number is the ratio $\kappa_{\mathbf{H}} = \lambda_{\max}/\lambda_{\min}$ of the largest eigenvalue to the smallest eigenvalue of **H**. The diagonal elements s_{ii} of **S** are the square roots of the eigenvalues of **H**. In the experiments, they were expressed as

$$s_{ii} = \kappa_{\mathbf{H}}^{(i-1)/(2(n-1))}$$

Thus, the smallest coefficient s_{ii} was 1, and the highest one was $\sqrt{\kappa_{\mathbf{H}}}$.

- **R** is a rotation matrix, defined as a product of elementary rotation matrices $\mathbf{R}_{kl}$ in the plane defined by the axes k and l, for all $k \in \{1, \ldots, n-1\}$ and $l \in \{k+1, \ldots, n\}$ (Sect. 6.6.2.5). For the experiments using nonseparable objective functions, the angle of each elementary rotation was chosen randomly with a uniform distribution in the interval $[-\pi, \pi]$. When the objective functions were chosen to be separable, the rotation matrix **R** was the identity matrix.

6.10.3.2 Results

An experiment consisted in seeking on optimum with the CMA-ES algorithm for a set of 30 target quadratic functions in $\mathbb{R}^n$, where n was a given parameter. The quadratic functions were obtained by randomly generating the vector **c** and/or the matrix **H** as described in Sect. 6.10.3.1. The result of an experiment was a performance curve expressing the average of the 30 values $F(\mathbf{X}_0)$ of the objective function as a function of the number of objective function evaluations performed. $\mathbf{X}_0$ was the best individual in the population for each of the 30 objective functions. As the optimal value was 0, $F(\mathbf{X}_0)$ was a measure of the error made by the algorithm. The number of evaluations of the objective functions was the product of the generation number and $\lambda = 4 + \lfloor 3 \ln n \rfloor$, as specified in Algorithm 6.6. A series of experiments provided performance curves for dimensions $n = 2, 5, 10, 20, 50,$ and 100.

The CMA-ES algorithm requires one to set the initial values of **m** and σ. For all experiments, $\mathbf{m}(0)$ was the n-dimensional zero vector and $\sigma(0) = 1.0$.

Three series of experiments were performed: the first for nonseparable, ill-conditioned quadratic objective functions with $\kappa_{\mathbf{H}} = 10^6$, the second for separable ill-conditioned functions with $\kappa_{\mathbf{H}} = 10^6$, and the third for well-conditioned quadratic functions with $\kappa_{\mathbf{H}} = 1$ ("sphere functions").

Nonseparable ill-conditioned functions. The results of this first series of experiments are shown in Fig. 6.28. Convergence towards the optimum was obtained in all tests, with an excellent precision of the order of 10^{-20}. This good precision of the results is a feature often observed for the CMA-ES method. It is due to the efficient adaptation of both the step size σ and the covariance matrix $\hat{\mathbf{C}}$. The required computing power remains moderate: the number of generations required to reach a given accuracy as a function of the dimension n of the search space has a complexity a little more than linear.

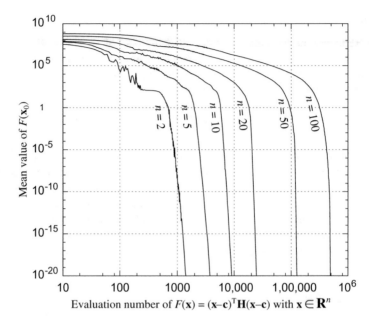

Fig. 6.28 Results of the first series of experiments: average values of $F(\mathbf{X}_0) = (\mathbf{X}_0\text{-}\mathbf{c})^{\mathsf{T}}\mathbf{H}(\mathbf{X}_0\text{-}\mathbf{c})$ as a function of the evaluation number for 30 nonseparable, ill-conditioned objective functions for dimensions 2, 5, 10, 20, 50, and 100

Separable ill-conditioned functions. In this case, the rotation matrix $\mathbf{R}$ was the identity matrix. Thus, $\mathbf{H} = \mathbf{S}^2$. The curves obtained for this series of experiments were indistinguishable from those obtained for the previous series (Fig. 6.28) on nonseparable functions. The adaptation of the matrix $\hat{\mathbf{C}}$ performed in the CMA-ES approach was thus very effective.

Well-conditioned functions. In this series of experiments, the matrix $\mathbf{H}$ was chosen to be proportional to the identity matrix: $\mathbf{H} = 100\,\mathbf{I}$. Thus, $\kappa_{\mathbf{H}} = 1$. The coefficient 100 was chosen so that $F(\mathbf{X}_0)$ in the first generation would be of the same order of magnitude as for in the previous series. Note that when $\mathbf{H} \propto \mathbf{I}$, objective functions are separable. The results of this series of experiments are shown in Fig. 6.29. This time, unlike the "ill-conditioned" case, the number of evaluations of $F(\mathbf{X})$ required to reach a given accuracy as a function of the dimension n of the search space has a complexity a little less than linear.

Compared with the previous two series of experiments, we note that the number of evaluations required to obtain a given accuracy requires less computing power when the quadratic function is well-conditioned. Thus, for dimension 100, 460 000 evaluations of $F(\mathbf{X})$ were needed to reach an accuracy of 10^{-10} in the ill-conditioned case with $\kappa_{\mathbf{H}} = 10^6$, while the well-conditioned "sphere" functions required only 20 000 evaluations to reach the same accuracy, i.e., 23 times less time.

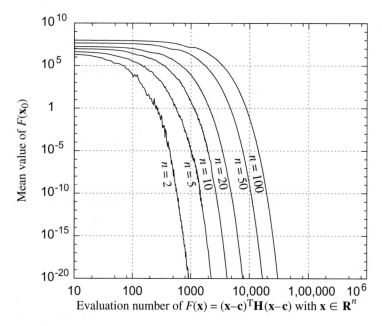

Fig. 6.29 Results of the third series of experiments, in which the objective functions were well-conditioned: $\kappa_{\mathbf{H}} = 1$

6.11 Conclusion

This chapter has presented a set of principles and algorithmic techniques to implement the various operators involved in an evolutionary algorithm. Like building blocks, they can be chosen, configured, and assembled according to the flowchart of the generic evolutionary algorithm (see Fig. 6.1) in order to solve a given problem as efficiently as possible. Obviously, specific choices of operators can reconstitute a genetic algorithm, an evolution strategy, or an evolutionary programming method such as that designed by the pioneers of evolutionary computation in 1960–70. However, the references to these original models, which have merged today into one unifying paradigm, should not disturb the engineer or the researcher when they are making their choices. Engineers and researchers should instead focus on key issues such as the choice of a good representation, a fitness function suitable for the problem to be solved and, finally, the formulation of efficient variation operators for the chosen representation.

The solution of real-world problems, which are typically multicriteria problems, must satisfy constraints and, too often, cannot be completely formalized, requires the implementation of additional mechanisms in evolutionary algorithms. These aspects are treated in Chaps. 11 and 12 of this book.

6.12   Glossary

allele: in the framework of genetic algorithms, a variant of a gene, i.e., the value of a symbol in a specified position of the genotype.

chromosome: in the framework of genetic algorithms, synonymous with "genotype."

crossover: combination of two individuals to form one or two new individuals.

fitness function: a function giving the value of an individual.

generation: iteration of the basic loop of an evolutionary algorithm.

gene: in the framework of genetic algorithms, an element of a genotype, i.e., one of the symbols in a symbol string.

genotype: in the framework of genetic algorithms, a symbol string that generates a phenotype during a decoding phase.

individual: an instance of a solution for a problem being dealt with by an evolutionary algorithm.

locus: in the framework of genetic algorithms, the position of a gene in the genotype, i.e., the position of a symbol in a symbol string.

mutation: random modification of an individual.

phenotype: in the framework of genetic algorithms, an instance of a solution for the problem being dealt with, expressed in its natural representation obtained after decoding the genotype.

population: the set of individuals that evolve simultaneously under the action of an evolutionary algorithm.

recombination: synonymous with "crossover."

replacement operator: this determines which individuals of a population will be replaced by offspring. It thus makes it possible to create a new population for the next generation.

search operator: synonymous with "variation operator."

selection operator: this determines how many times a "parent" individual generates "offspring" individuals.

variation operator: an operator that modifies the structure or parameters of an individual, such as the crossover and mutation operators.

6.13 Annotated Bibliography

References [5, 6]   An "encyclopedia" of evolutionary computation to which, as it should be, the most recognized specialists in this field have contributed. The vision offered by these two volumes is primarily algorithmic.

Reference [20]   A relatively recent reference book (reissued in 2010) dedicated to evolutionary computation. It particularly addresses the important issue of control of the parameter values for the different operators of an evolutionary algorithm. Some theoretical approaches in the field are also discussed.

Reference [26]   The first and the most famous book in the world about genetic algorithms. It was published in 1989, and has not been revised since then. As a result, a large part of the current knowledge about genetic algorithms, a field that is evolving very quickly, is not in this book.

References [35, 36]   Two reference books written by the well-known pioneer of genetic programming. The first volume presents the basic concepts of the genetic programming as viewed by Koza. The second volume introduces the concept of "automatically defined functions." The largest portion of these books, which comprise more than seven hundred pages in each volume, is devoted to the description of examples of applications in a large variety of domains. These are useful for helping the reader to realize the potential of genetic programming. There is also a third volume, published in 1999, which contains a significant part dedicated to the automated synthesis of analog electronic circuits.

References

1. Abdelmalek, W., Ben Hamida, S., Abid, F.: Selecting the best forecasting-implied volatility model using genetic programming. Journal of Applied Mathematics and Decision Sciences **2009** (2009). Article ID 179230
2. Abid, F., Abdelmalek, W., Ben Hamida, S.: Dynamic hedging using generated genetic programming implied volatility models. In: S. Ventura (ed.) *Genetic Programming: New Approaches and Successful Applications*, Chap. 7, pp. 141–172. InTech (2012). doi:10.5772/48148
3. Ackley, D.H.: *A Connectionist Machine for Genetic Hillclimbing*. Kluwer (1987)
4. Angeline, P.J.: Two self-adaptive crossover operators for genetic programming. In: P.J. Angeline, K.E. Kinnear, Jr. (eds.) *Advances in Genetic Programming 2*, Chap. 5, pp. 89–110. MIT Press (1996)
5. Baeck, T., Fogel, D.B., Michalewicz, Z.: *Evolutionary Computation 1: Basic Algorithms and Operators*. Institute of Physics Publishing (2000)
6. Baeck, T., Fogel, D.B., Michalewicz, Z.: *Evolutionary Computation 2: Advanced Algorithms and Operators*. Institute of Physics Publishing (2000)

7. Baker, J.E.: Reducing bias and inefficiency in the selection algorithm. In: J.J. Grefenstette (ed.) *Proceedings of the 2nd International Conference on Genetic Algorithms*, pp. 14–21 (1987)
8. Beyer, H., Schwefel, H.: Evolution strategies—a comprehensive introduction. Natural Computing **1** (1), 3–52 (2002)
9. Beyer, H.G.: *The Theory of Evolution Strategies*, Natural Computing Series. Springer (2001)
10. Black, F., Scholes, M.: The pricing of options and corporate liabilities. Journal of Political Economy **81** (3), 637–654 (1973)
11. Cai, W., Pacheco-Vega, A., Sen, M., Yang, K.T.: Heat transfer correlations by symbolic regression. International Journal of Heat and Mass Transfer **49** (23–24), 4352–4359 (2006). doi:10. 1016/j.ijheatmasstransfer.2006.04.029
12. Chen, Y., Mabu, S., Shimada, K., Hirasawa, K.: A genetic network programming with learning approach for enhanced stock trading model. Expert Systems with Applications **36**(10), 12,537–12,546 (2009). doi:10.1016/j.eswa.2009.05.054
13. Cheung, C., Valdes, J.J., Li, M.: Use of evolutionary computation techniques for exploration and prediction of helicopter loads. In: X. Li (ed.) *Proceedings of the 2012 IEEE Congress on Evolutionary Computation*, pp. 1130–1137. Brisbane, Australia (2012). doi:10.1109/CEC. 2012.6252905
14. Chidambaran, N.K., Triqueros, J., Lee, C.W.J.: Option pricing via genetic programming. In: S.H. Chen (ed.) *Evolutionary Computation in Economics and Finance*. Studies in Fuzziness and Soft Computing, vol. 100, Chap. 20, pp. 383–398. Physica Verlag (2002)
15. Cramer, N.L.: A representation for the adaptive generation of simple sequential programs. In: In J. J. Grefenstette (ed.) *Proceedings of the 1st International Conference on Genetic Algorithms*, pp. 183–187 (1985)
16. Darwin, C.: *On The Origin of Species by Means of Natural Selection or the Preservation of Favored Races in the Struggle for Life*. Murray, London (1859)
17. Davis, L.: *Handbook of Genetic Algorithms*, p. 80. Van Nostrand Reinhold (1991)
18. De Jong, K.A., Sarma, J.: Generation gaps revisited. In: L.D. Whitley (ed.) *Foundations of Genetic Algorithms 2*, pp. 19–28. Morgan Kaufmann (1993)
19. De La Maza, M., Tidor, B.: An analysis of selection procedures with particular attention paid to proportional and Boltzmann selection. In: S. Forrest (ed.) *Proceedings of the 5th International Conference on Genetic Algorithms*, pp. 124–131. Morgan Kaufmann (1993)
20. Eiben, A., Smith, J.: *Introduction to Evolutionary Computing*. Springer (2003).
21. Eiben, A.E., van Kemenade, C.H.M., Kok, J.N.: Orgy in the computer: Multi-parent reproduction in genetic algorithms. In: F. Moran, A. Moreno, J. Merelo, P. Chacon (eds.) *Proceedings of the 3rd European Conference on Artificial Life*. Lecture Notes in Artificial Intelligence, vol. 929, pp. 934–945. Springer (1995)
22. Eshelman, L.J., Schaffer, J.D.: Real-coded genetic algorithms and interval-schemata. In: L.D. Whitley (ed.), pp. 187–202. Morgan Kaufmann (1992)
23. Fogel, L.J., Owens, A.J., Walsh, M.J.: *Artifical Intelligence through Simulated Evolution*. Wiley (1966)
24. Gallo, G., Longo, G., Pallottino, S., Nguyen, S.: Directed hypergraphs and applications. Discrete Applied Mathematics **42**(2–3), 177–201 (1993)
25. Garcia, B., Aler, R., Ledezma, A., Sanchis, A.: Protein–protein functional association prediction using genetic programming. In: M. Keijzer, G. Antoniol, C.B. Congdon, K. Deb, B. Doerr, N. Hansen, J.H. Holmes, G.S. Hornby, D. Howard, J. Kennedy, S. Kumar, F.G. Lobo, J.F. Miller, J. Moore, F. Neumann, M. Pelikan, J. Pollack, K. Sastry, K. Stanley, A. Stoica, E.G. Talbi, I. Wegener (eds.) *GECCO'08: Proceedings of the 10th Annual Conference on Genetic and Evolutionary Computation*, pp. 347–348. ACM, Atlanta, GA (2008). doi:10.1145/1389095. 1389156
26. Goldberg, D.E.: *Genetic Algorithms in Search, Optimization and Machine Learning*. Addison-Wesley (1989)
27. Goldberg, D.E., Deb, K.: A comparison of selection schemes used in genetic algorithms. In: G. Rawlins (ed.) *Foundations of Genetic Algorithms*, pp. 69–93. Morgan Kaufmann (1991)

28. Gustafson, S., Burke, E.K., Krasnogor, N.: On improving genetic programming for symbolic regression. In: D. Corne, Z. Michalewicz, M. Dorigo, G. Eiben, D. Fogel, C. Fonseca, G. Greenwood, T.K. Chen, G. Raidl, A. Zalzala, S. Lucas, B. Paechter, J. Willies, J.J.M. Guervos, E. Eberbach, B. McKay, A. Channon, A. Tiwari, L.G. Volkert, D. Ashlock, M. Schoenauer (eds.) *Proceedings of the 2005 IEEE Congress on Evolutionary Computation*, vol. 1, pp. 912–919. IEEE Press (2005). doi:10.1109/CEC.2005.1554780

29. Hamida, S.B., Abdelmalek, W., Abid, F.: Applying dynamic training-subset selection methods using genetic programming for forecasting implied volatility. Computational Intelligence (2014). doi:10.1111/coin.12057

30. Hansen, N.: The CMA evolution strategy: A comparing review. In: J. Lozano, P. Larraaga, I. Inza, E. Bengoetxea (eds.) *Towards a New Evolutionary Computation*. Studies in Fuzziness and Soft Computing, vol. 192, pp. 75–102. Springer (2006). doi:10.1007/3-540-32494-1_4

31. Hartl, D.L., Clark, A.G.: *Principles of Population Genetics*, 4th edition. Sinauer Associates (2006)

32. Holland, J.H.: *Adaptation in Natural and Artificial Systems*, 2nd edition. MIT Press (1992)

33. Kaboudan, M.A.: Genetic programming prediction of stock prices. Computational Economics **16**(3), 207–236 (2000). doi:10.1023/A:1008768404046

34. Keijzer, M.: Scaled symbolic regression. Genetic Programming and Evolvable Machines **5**(3), 259–269 (2004). doi:10.1023/B:GENP.0000030195.77571.f9

35. Koza, J.R.: *Genetic Programming: On the Programming of Computers by Means of Natural Selection*. MIT Press (1992)

36. Koza, J.R.: *Genetic Programming II: Automatic Discovery of Reusable Programs*. MIT Press (1994)

37. Langdon, W.: Size fair and homologous tree crossovers for tree genetic programming. Genetic Programming and Evolvable Machines **1**(1–2), 95–119 (2000). doi:10.1023/A:1010024515191

38. Lew, T.L., Spencer, A.B., Scarpa, F., Worden, K., Rutherford, A., Hemez, F.: Identification of response surface models using genetic programming. Mechanical Systems and Signal Processing **20**(8), 1819–1831 (2006). doi:10.1016/j.ymssp.2005.12.003

39. Mathias, K., Whitley, D.: Genetic operators, the fitness landscape and the traveling salesman problem. In: R. Manner, B. Manderick (eds.) *Parallel Problem Solving from Nature, 2*, pp. 221–230. Elsevier Science (1992)

40. Michalewicz, Z.: *Genetic Algorithms + Data Structures = Evolution Programs*, revised 3rd edition. Springer (1996)

41. Nomura, T., Shimohara, K.: An analysis of two-parent recombinations for real-valued chromosomes in an infinite population. Evolutionary Computation **9**(3), 283–308 (2001)

42. Nordin, P.: A compiling genetic programming system that directly manipulates the machine code. In: K.E. Kinnear, Jr. (ed.) *Advances in Genetic Programming*, Chap. 14, pp. 311–331. MIT Press (1994)

43. Oakley, H.: Two scientific applications of genetic programming: Stack filters and non-linear equation fitting to chaotic data. In: K.E. Kinnear, Jr. (ed.) *Advances in Genetic Programming*, Chap. 17, pp. 369–389. MIT Press (1994)

44. Radcliffe, N.: Genetic neural networks on MIMD computers. Ph.D. thesis, University of Edinburgh (1990)

45. Rechenberg, I.: *Cybernetic Solution Path of an Experimental Problem*. Royal Aircraft Establishment Library Translation (1965)

46. Rechenberg, I.: *Evolutionsstrategie: Optimierung technischer Systeme nach Prinzipien der biologischen Evolution*. Frommann-Holzboog, Stuttgart (1973)

47. Rogers, A., Prügel-Bennett, A.: Genetic drift in genetic algorithm selection schemes. IEEE Transactions on Evolutionary Computation **3**(4), 298–303 (1999)

48. Rudolph, G.: On correlated mutations in evolution strategies. In: R. Manner, B. Manderick (eds.) *Parallel Problem Solving from Nature*, pp. 105–114. Elsevier (1992)

49. Ryan, C., Collins, J., Collins, J., O'Neill, M.: Grammatical evolution: Evolving programs for an arbitrary language. In: *Proceedings of the First European Workshop on Genetic Programming*. Lecture Notes in Computer Science, vol. 1391, pp. 83–95. Springer-Verlag (1998)

50. Schoenauer, M., Lamy, B., Jouve, F.: Identification of mechanical behaviour by genetic programming. Part II: Energy formulation. Technical report, Ecole Polytechnique, 91128 Palaiseau, France (1995)
51. Schoenauer, M., Sebag, M., Jouve, F., Lamy, B., Maitournam, H.: Evolutionary identification of macro-mechanical models. In: P.J. Angeline, K.E. Kinnear, Jr. (eds.) *Advances in Genetic Programming 2*, Chap. 23, pp. 467–488. MIT Press (1996)
52. Schwefel, H.P.: *Numerical Optimization of Computer Models*. Wiley (1981)
53. Stadler, P.F.: Fitness landscapes. In: M. Lässig, A. Valleriani (eds.) *Biological Evolution and Statistical Physics*, pp. 183–204. Springer (2002)
54. Teller, A., Veloso, M.M.: PADO: Learning tree structured algorithms for orchestration into an object recognition system. Technical Report CMU-CS-95-101, Carnegie-Mellon University, Pittsburgh, PA (1995)
55. Wasiewicz, P., Mulawka, J.J.: Molecular genetic programming. Soft Computing—A Fusion of Foundations, Methodologies and Applications **2**(5), 106–113 (2001)
56. Wright, S.: The roles of mutation, inbreeding, crossbreeeding and selection in evolution. In: D.F. Jones (ed.) *Proceedings of the Sixth International Congress on Genetics*, pp. 356–366 (1932)

Chapter 7
Artificial Ants

Nicolas Monmarché

7.1 Introduction

Ants are social insects with physical and behavioral skills that are still fascinating to human beings (Greek mythology mentioned them!). This fascination is often justified by biological studies and observations: the activity of ants is undoubtedly observable, such as in the huge nests (anthills) that they build, their battles, and their various diets (their "agriculture" when growing fungi, for instance). As was pointed out by Luc Passera [20], our liking for anthropomorphic interpretations leads us to have a globally positive perception of ants, particularly of their activity, which we guess to be ceaseless. But, sometimes, appearances can be misleading: in a colony, in particular in those which are populous, a rather small fraction of the ants actually work. However, thanks to our positive perception of ants and the fact that everyone has been able to recognize an ant since their childhood, this allows us to easily employ the ant metaphor to solve combinatorial problems!

Studies conducted by biologists during the 1980s, more precisely those done by Jean-Louis Deneubourg and colleagues [6, 12], have introduced an "algorithmic" way of thinking about the behavior of ants. This new point of view has led to a new formalism for proposed behavioral models, and this has become accessible to computer simulation. At the same time, computers are starting to be intensively used to explore complex systems and, consequently, it is now possible to study ants *in silico* for their ability to link their nest to various food sources. For instance, in [16], the inherent parallelism of the distributed decisions of ants was studied, but this was not yet a question of optimization.

N. Monmarché (✉)
Laboratoire d'Informatique (EA6300), Université François Rabelais Tours,
64 Avenue Jean Portalis, 37200 Tours, France
e-mail: nicolas.monmarche@univ-tours.fr

© Springer International Publishing Switzerland 2016
P. Siarry (ed.), *Metaheuristics*, DOI 10.1007/978-3-319-45403-0_7

The link between optimization and simulation of the behavior of ants was made at the beginning of the 1990s [5].[1] From this, numerous studies of combinatorial optimization based on the exploitation of a food source by ants followed. The goal of this chapter is to give an outline of these studies and to understand precisely the underlying mechanisms used in this kind of bioinspired metaheuristics.

Before tackling optimization considerations, we start with more details about the behavior and characteristics of ants.

7.2 The Collective Intelligence of Ants

7.2.1 Some Striking Facts

The most ancient known ants are more than 100 million years old, and about 12 000 ant species are known at present [20]. This small number of species is a source of astonishment when compared with the millions of known insect species. However, this apparent evolutionary underperformance has to be set against the huge number of ants that we can find in many ecosystems. The total weight of ants on Earth is probably of the same order of magnitude as the total weight of human beings, and listening biologists (of course they are also myrmecologists) say "the ants represent the greatest ecological success on Earth" [20].

Ants can be found in almost every terrestrial ecosystem and, of course, are subject to the same constraints as other living species: finding food and a place to live, defending themselves, and reproducing. The striking fact with ants is that they respond to all these needs through collective behavior. It is noticeable that all ants live in societies, and this is the main explanation for their ecological success. The collective aspect of their activities can be observed in the division of labor (building the nest, and brood care), information sharing (searching for food, and alerts when attacked) and, what is most fascinating, the fact that the reproductive task is performed by only a few individuals in the colony (most ants in the nest are sterile).

We could spend a lot of time describing the behaviors of ants that can be related to an optimization perspective. For instance, the task regulation performed by ants, i.e., their ability to distribute work without any central supervision, represents an adaptation mechanism to fluctuations in their environment. This can be considered as a distributed problem (several tasks need to be performed at the same time but in different places) which is also dynamic (because needs can evolve with time). However, in this chapter, we will focus on information sharing by ants, i.e., their communication skills, and that already represents a wide topic.

---

[1]This paper is linked to Marco Dorigo's Ph.D. thesis [7].

7.2.2   The Chemical Communication of Ants

The most prominent way that ants have of communicating, without exception, is their ability to employ chemical substances, which are called pheromones. These pheromones are a mix of hydrocarbons secreted by the ants, which are able to lay down these substances on their path and this constitutes an appealing trail for other ants. Pheromones, depending on their composition, have the property of evaporating over time. Thus, a trail which is not reinforced with new deposits disappear.

Pheromones are used on various occasions and by various species. For instance, when an alert message is given, pheromones allow the recruitment of large numbers of ants to defend the nest. Ants use pheromones not only because they can synthesize them but also because they can perceive those substances: their antennae are detectors with a sensitivity beyond the reach of our electronic sensors. Even though ants do not use a nose to sense pheromones, the volatile nature of these chemical substances leads us to say that ants can smell the pheromones they produce. These "odors" are so important that they represent the most prominent way in which ants deal with their identity, i.e., the individual and colonial identity of each ant is linked to its ability to synthesize, share, and smell this chemical mix spread over their cuticle.

The particular example that we are studying in this chapter concerns the form of communication of ants that permits them to set up a mass recruitment. Here, "mass" means that a large number of individuals are involved in exploiting a food source. Basically, exploiting a food source consists, for ants, in getting out of the nest and moving in an environment that is changing, if not dangerous, to reach the location of the food source. Because of the variety of possible ant diets, the description of what can be a food source is beyond the scope of this chapter. The important point to consider is that an ant can capture a small quantity of food. Then the ant brings this food back to the nest in order to feed the (often) large population which does not go outside the nest. A mass recruitment is observed when ants lay pheromones on their way back to the nest. This trail is then perceived by ants which are leaving the nest, and these ants are oriented towards the food source. Then, by a reinforcement mechanism linked to the number of ants looking for food, the greater this number ants looking for food, the more they will be back laying pheromones and the more attractive the trail will be, and so on. We can understand that whenever the colony can send new workers to capture food, the indirect communication of ants can lead to a very efficient method of food gathering.

When the food source disappears (because it has been exhausted or the environment has changed), ants which fail to find food do not reinforce the trail on their way back. After a while, the depleted source and the path to it are abandoned, and another, more attractive source is probably used instead.

The mass recruitment of ants just described can be considered as an interesting model of logistic optimization. However, we can also observe subtle effects in the path built by the ants: we can observe that ants are able to optimize the trajectory between the nest and the food source. This optimization can take place two ways: first, the trajectories that minimize the total distance are most often favored, and second, if an obstacle falls onto the path and modifies it, a shortcut will quickly be found.

Fig. 7.1 Experimental setup with two bridges between the nest and the food source. In this experiment, one of the two paths is clearly more interesting according to the total distance that the ants have to travel

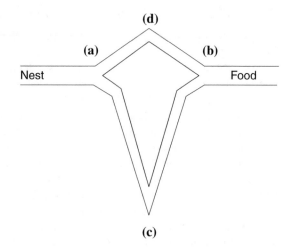

This last capability of ants is related to the ability to solve a dynamic problem. In the remainder of this chapter, however, we will assume that the environmental conditions are not modified (the problem is static).

The conditions that allow ants to find the best path have been studied in the laboratory by Goss et al. [11], by use of an experimental setup with two bridges connecting, without alternative paths, the nest to the food source provided to the ants. Figure 7.1 shows schematically the experiment, in which it has been observed that in the great majority of cases, the ants are able to find the shortest way, that is the path (a)–(d)–(b), and are absent from the path (a)–(c)–(b). This behavior can be explained by the fact that ants which choose the path through (d) reach the food quickly. One can make the hypothesis that the ants all move at the same speed and they always deposit pheromones. At the beginning of the process, ants which are leaving the nest reach point (a) and do not have any information to decide the best direction to choose. So, around one half of the flow of ants chooses (d) and the other half chooses (c). Those which have chosen the shorter path (without knowing it: we recall that ants are considered here as blind) reach the point (b) earlier and, consequently, they get food earlier to bring back to the nest. Then, on their way back, they reach the point (b) and, again, they have to choose between two options. As they lay down pheromones regularly, it is possible that a small difference, in concentration is amplified by the number of ants.

In this experiment, we find all the ingredients of a self-organized system

- a positive reinforcement mechanism: pheromones are attractive to ants which, in turn, lay down pheromones (we speak of an auto-catalytic behavior, i.e., one which reinforces itself);
- a negative feedback mechanism: pheromones evaporate, which limits the phenomenon and allows a loss of memory or even an exit from a stable state;
- random behavior that causes fluctuations in the states of the system;
- a multiplicity of interactions: the ants are numerous.

A collective intelligence is then observed when spatial or temporal structures emerge or appear owing to numerous repeated interactions, direct or indirect, between individuals belonging to the same colony or group. Here, we observe the emergence of a path used by a majority of the ants.

It is obviously the indirect communication mechanism of pheromones that leads to the optimization phenomenon. Ant are then able to find the best path, and this can be translated into combinatorial optimization.

7.3   Modeling the Behavior of Ants

The behavioral analysis work described above can be translated into a behavioral model which does necessarily not mimic the reality of what might be occurring in the heads of the ants. But this model can be used to reconstruct the optimization process with a minimum number of simple rules.

7.3.1   Defining an Artificial Ant

Before modeling the behavior of ants, let us pay attention to the model of one ant, called an "artificial ant" in the following. We use the definition given in the introduction of [17]:

> An artificial ant is an object, virtual or real (for example a software agent or a robot), or symbolic (as a point in a search space) that has a link, a similarity (i.e., a behavior, a common feature) with a real ant.

This definition is sufficiently general to cover various models of ants. The important point is that an artificial ant should not be limited to a system able to mimic food source exploitation behavior.

7.3.2   Ants on a Graph

In order to describe precisely the ant's behavior in the environment we are considering, i.e., the double-bridge experiment, this environment is modeled by a graph (Fig. 7.2).

The behavioral model can be described as follows:

- Ants leave the node labeled "nest" and choose one of the two possible paths.
- The choice of the edge representing the path is influenced by pheromones on the two edges: the ant has a higher probability of choosing the edge with the higher level of pheromones.

Fig. 7.2 A double bridge
modeled by a graph

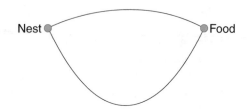

- Pheromones can be modeled by real values, which can be considered as pheromone concentrations on the edges considered.
- The ant goes along the chosen edge while depositing pheromones at each step along its way.
- Once the food has been reached, the ant returns to the nest and chooses its path with the same strategy as it used before.
- The pheromones continuously evaporate: the real value that represents the pheromone concentration decreases.

The bridge example is of course very small (with only two nodes!), but we can imagine the same movement mechanism on a bigger graph (Fig. 7.3). As several ants could run in this graph, one can observe that paths with more pheromones could appear, and so those paths would be used more and more by ants to reach the food (Fig. 7.4).

Fig. 7.3 Modeling of
possible paths for one ant by
a graph

Fig. 7.4 Ants move on the
graph while they deposit
pheromones on the edges
(the *thickness of the lines* is
used to represent the
pheromone concentration on
the edge). The higher the
concentration, the more ants
are attracted by the edge

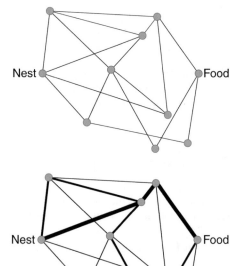

The mechanism that we have just described will now be translated into a meta-heuristic for a combinatorial optimization.

7.4   Combinatorial Optimization with Ants

The graph structure introduced in the previous section will now be developed. More precisely, as was done in the initial research work, we shall use a combinatorial problem as an example which also uses a graph structure, and more precise ant mechanisms will be introduced. This problem is referred to as the traveling salesman problem (TSP), and we start this section with a short description of the problem. Then, the main ant-based algorithms that have been applied to the TSP are detailed.

7.4.1   The Traveling Salesman Problem

The ants' moves between their nest and the food source, and their return moves between the food and the nest are similar to the construction of a cycle in a graph. If we add the constraint that every node must be visited once and only once, then the work of each ant is similar to the construction of a Hamiltonian cycle. If we consider the goal of minimizing the total path length, then the construction of Hamiltonian cycle of minimum length is similar to a very classical combinatorial optimization problem, namely, the TSP. In this problem, the nodes are cities, the edges are roads between the cities, and the goal is to find the path of minimum length for a salesman who needs to visit every city and return home at the end of his journey (i.e., his starting node). Figure 7.5 recalls the formalism of the TSP and how the cost of one solution is calculated.

A TSP instance is determined by:

- n, the number of cities
- The distance matrix d, where $d(i, j)$ is the distance between city i and city j (if $d(i, j) \neq d(j, i)$, the problem is said to be "asymmetric").

A solution of the problem is represented by a permutation σ of the n cities $\sigma = (\sigma_1, \ldots, \sigma_n)$. The objective is to find a solution that minimizes the total length of the complete tour:

$$\min_{\sigma} \left\{ \sum_{i=1}^{n-1} d(\sigma_i, \sigma_{i+1}) + d(\sigma_n, \sigma_1) \right\} \tag{7.1}$$

Fig. 7.5   TSP formalism

Fig. 7.6 Graph in which a
solution, i.e., a permutation
of nodes, or a cycle, is
represented. Here the
solution is
(1, 2, 8, 5, 10, 7, 9, 4, 3, 6)

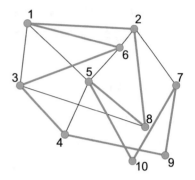

As an example, Fig. 7.6 represents the same graph as seen previously in which a particular solution is highlighted. We see that the ideas of "nest" and "food" can be eliminated because the starting and return nodes are not important.

The interesting property of this problem is that it is easy to explain but it becomes difficult to solve as the number of cities increases: a full enumeration of all possible solutions would require one to generate and evaluate $(n - 1)!/2$ permutations. Figure 7.7 shows an example with 198 cities, where one can observe a nonuniform distribution of cities.

We remark that the problem may be asymmetric: the edges can be oriented and the distances may not necessarily be the same in one direction and the return direction. This new constraint does not change the behavior of the ants, however, since they are forced to move on the edges in the correct direction.

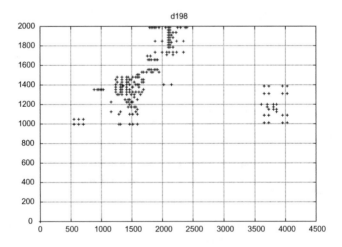

Fig. 7.7 Example of a Euclidean TSP instance with 198 cities (d198)

7.4.2   The ACO Metaheuristic

In this section, we present several algorithms inspired by ant behavior to find a solution to the TSP. The analogy between searching for a cycle of minimum length and ants optimizing their trajectory between the nest and food is immediate. These algorithms share the same inspiration and are gathered together under the acronym ACO, for "ant colony optimization."

7.4.2.1   Ant System Algorithm

We present the ant system (AS) algorithm, which was the first to be proposed to solve a combinatorial optimization problem with artificial ants [5, 9]. This algorithm is not the best one in this category, but it is convenient for introducing the main principles found in several ant-based algorithms which make use of digital pheromones.

In comparison with the ant model described previously, several changes have been made, either for algorithmic reasons or because of the need to build solutions efficiently:

- The nodes that have been passed through are memorized: the ants must memorize the partial path already followed in order to avoid nodes that have already been visited. This memory is not necessary in the original model because when the ants leave the nest, they are looking for the food, and when they have found the food they try to reach the nest.
- Pheromones are deposited after the construction of a complete solution: unlike real ants, which deposit pheromones continuously and independently of the total length of the path, the artificial ants deposit more pheromones on a short path.
- The speed of the artificial ants is not constant: ants move from one node to the next in one unit of time, independently of the length of the edge between the two nodes. This point is easier to implement in the simulation of the ants' moves because, in a synchronous mode where every ant is considered in each iteration, every ant is moved. To compensate for this synchronous simulation, the reinforcement of the path is then proportional to the quality of the solution. The "move duration" is reintroduced into the algorithm in this way.
- The artificial ants are not totally blind: relatively quickly, it became obvious that totally blind ants took a lot of time to find interesting solutions. The notion of visibility was introduced to take account of distance between nodes. Then, in addition to the effect of pheromones, the choices of the ants are also influenced by the distance between two consecutive nodes. This means that artificial ants are not as blind as the initial model suggested.

Building a solution. Each ant builds a solution incrementally, i.e., a permutation of the n nodes (or towns). The starting node is chosen randomly because it does not have a particular role, since a solution is a cycle in the graph. In the example in Fig. 7.8, the ant has already built a partial path and is about to choose between nodes 4, 6, and 10.

Fig. 7.8 Graph on which a partial solution built by one ant has been drawn: (1, 2, 8, 5). The ant is located on node 5 and can choose between nodes 4, 6, and 10. Nodes 1 and 8 cannot be considered because they belong to the partial solution already built

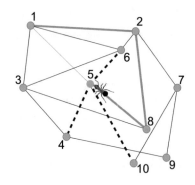

We may notice that the example shown in Fig. 7.8 is a case in which the building of the cycle can lead to a dead end. If the ant chooses node 4 or 6, it will not be able to finish the complete cycle without passing through a node that has already been visited. The problem arises because this example has deliberately been chosen to be simple so that it would be readable and, consequently, it contains a small number of edges. In practice, ant algorithms have been experimented with on graphs which are complete, i.e., in which every node is connected to every other one by one direct edge. This property avoids the problem encountered in our example. If the problem requires that the solution does not contain particular edges, then to avoid incomplete graphs, it is possible to keep these edges but to give them high values of distance. Any ants which chose these penalized edges would build a very bad solution.

From a practical point of view, with each edge (i, j) we associate a quantity of pheromone τ_{ij}, and we define the probability that ant k, located on node/city i, chooses node/city j as

$$p_{ij}^k(t) = \frac{\tau_{ij}(t)^\alpha \times \eta_{ij}^\beta}{\sum_{\ell \in \mathcal{N}_i^k} \tau_{i\ell}(t)^\alpha \times \eta_{i\ell}^\beta} \qquad (7.2)$$

where

- η_{ij} represents the visibility of the ant;
- α and β are two parameters which are used to tune the relative influence of pheromones and visibility;
- $\mathcal{N}_i^k$ is the set of cities which have not yet been visited by ant k (i.e., its memory) when the ant is located on node/city i.

The numerator contains the product of the quantity of pheromones τ_{ij} with the visibility η_{ij}, and hence takes these two kind of information into account in the choice of the ant's moves. In the case of the TSP, the visibility can be estimated using the length of the edge (i, j): $\eta_{ij} = 1/d_{ij}$. The denominator is used to normalize the probabilities so that $\sum_{j \in \mathcal{N}_i^k} p_{ij}^k(t) = 1$.

Updating the pheromones. At the end of the building of one cycle, each ant k deposits a quantity Δ_{ij}^k of pheromone on all the edges (i, j) that belong to its path. This quantity is proportional to the quality of the solution built by the ant (and, consequently, inversely proportional to the length of the complete tour built by the ant)

$$\Delta_{ij}^k = \begin{cases} 1/L^k & \text{if } (i, j) \in T^k \\ 0 & \text{otherwise} \end{cases} \tag{7.3}$$

where T^k is the cycle (also called a tour) built by the ant k, and L^k is its length.

The quantity of pheromone on each edge (i, j) is then updated

$$\tau_{ij}(t+1) \leftarrow (1 - \rho)\tau_{ij}(t) + \sum_{k=1}^{m} \Delta_{ij}^k \tag{7.4}$$

where $\rho \in [0, 1]$ is a parameter for the evaporation rate and m is the number of ants.

Complete algorithm. The framework of the algorithm is given in Algorithm 7.1.

Initialization of pheromones $(\tau_{ij})_{1 \leq i, j \leq n}$
for $t \leftarrow 1$ **to** $t_{\max}$ **do**
    foreach *ant* k **do**
        Build a cycle $T^k(t)$
        Calculate the length $L^k(t)$ of $T^k(t)$
    end
    foreach *edge* (i, j) **do**
        Update pheromones $\tau_{ij}(t)$ with formula (7.4)
    end
end
return *the best found solution*

Algorithm 7.1: Ant system (AS) algorithm.

We have deliberately simplified the presentation of the algorithm to emphasize its structure. Here are, more precisely, the details of its implementation:

- The pheromone values are stored in a matrix because, in the general case, we consider a complete graph. The initialization of the pheromones consists in influencing the behavior of the ants as little as possible, at least in the first few iterations. Then, the pheromones values are used as a collective memory for each ant when its solution, is being built iteratively.
- The number of iterations is fixed by the parameter $t_{\max}$. Of course, as is done in several stochastic population-based metaheuristics for optimization, the stopping criterion for the main generational loop can be improved. For instance, the number of iterations can be linked to a performance measure of the results obtained by the algorithm: the algorithm is stopped when the optimization process does not progress anymore.
- The building of one solution by an ant, i.e., in the case of the TSP, the building of one cycle, is done node by node using formula (7.2). The algorithm presented

Table 7.1 Parameters and corresponding standard values for the ant system algorithm. C represents an estimate of the cost of one solution and n is the number of nodes in the graph (i.e., more or less the size of the problem)

Symbol	Parameter	Values
α	Influence of pheromones	1
β	Influence of visibility	$[2; 5]$
ρ	Evaporation rate	$0, 5$
τ_0	Initial pheromone value	m/C
m	Number of ants	n

here does not explain the details of this building process, but the ant's work can be either synchronous or asynchronous with respect of the activity of other ants. In the synchronous case, each ant takes one step from one node to the next one, ant after ant. In the asynchronous case, each ant builds its cycle independently of the work of other ants.

- The cost of one solution, i.e., the length of the cycle, can be calculated from formula (7.1). Of course, as the goal is to find the shortest tour, the algorithm has to keep in memory the best solution, not only its cost.
- The pheromone update process consists in keeping and sharing information that is useful for optimizing the length of the cycle built by the ants. This collective memory is updated using formulas (7.3) and (7.4).

Choosing the parameters. As with every metaheuristic, the choice of the values of the parameters of the method is crucial. Intervals of possible valid values are often obtained after several experiments. For the AS algorithm applied to the TSP, the standard values are given in Table 7.1.

We can notice that the values linked to the size of the problem n are easier to fix. The initial pheromone value makes use of the value of C, which corresponds to the cost of one solution obtained by a greedy-like heuristic.

The AS algorithm has been a starting point for several improvements. We will now present the main ones.

7.4.2.2   Max–Min Ant System

The max–min ant system ($\mathcal{MMAS}$) [21] has introduced several improvements which have been adopted in other algorithms.

First, pheromone values are bounded so that $\tau_{ij} \in [\tau_{min}, \tau_{max}]$ (which explains the algorithm's name). This allows one to control the difference between preferred edges, i.e., those which belong to the best solutions found, and less-visited ones. Without any limits on the pheromone values, the pheromone value of neglected edges can tend to zero and, consequently, the probability of choosing those edges also tends to zero for every ant. Finally, edges that are neglected at one moment during the search process have the risk of never being visited anymore. This is particularly unwelcome, because this prevents the algorithm from getting out of local minima. The Values of τ_{min} and τ_{max} guarantee that all edges can be reached by the ants. The tuning of τ_{min} and τ_{max} can evolve during the iterations of the algorithm. For instance, τ_{max} can be based on the best solution found so far.

Next, the pheromone update is based on an elitist technique. This mechanism allows the acceleration of the algorithm's convergence. The formula (7.4) is simplified to

$$\tau_{ij}(t+1) \leftarrow (1 - \rho)\tau_{ij}(t) + \Delta_{ij}^+ \tag{7.5}$$

because only the best ant lays down pheromones: $\Delta_{ij}^+ = 1/L^+$ if (i, j) belongs to the path built by the best ant. The best ant since the first iteration of the algorithm or the best ant among all the ants in the current iteration can be used.

Finally, all the pheromone values are initialized to the value τ_{max}, and, in the case of stagnation of the search process, the pheromone values are initialized again to this value to restart the exploration of the whole search space.

Table 7.2 gives the main values used for the parameters.

The $\mathcal{MMAS}$ algorithm has been widely developed, for instance by adding various reinforcement strategies and by adapting it to tackle various problems. Its main point of interest is its precise use of pheromone values and the way in which they evolve and influence future iterations of the algorithm more precisely than in what has been done before.

7.4.2.3 AS$_{\text{rank}}$

The AS$_{\text{rank}}$ algorithm [4] introduced a kind of contribution of the best ants to the pheromones, which is related to the elitist selection by rank found in some other

Table 7.2 Parameters and ranges of value known to be useful for the $\mathcal{MMAS}$ algorithm. C represents an estimate of the cost of one solution and n is the number of vertices in the graph (i.e., the size of the problem). L^{++} is the cost of the best solution found from the beginning of the algorithm, and a is calculated from $\sqrt[n]{0,05}(c-1)/(1 - \sqrt[n]{0,05})$, where c is the mean number of choices encountered by the ant in the current building step

Symbol	Parameter	Values
α	Influence of pheromone	1
β	Influence of visibility	[2; 5]
ρ	Evaporation rate	0, 02
τ_0	Initial pheromone value	$1/\rho C$
m	Number of ants	n
τ_{min}	Lower bound of pheromone values	τ_{max}/a
τ_{max}	Higher bound of pheromones values	$1/\rho L^{++}$

metaheuristics. Thus, the ants are ordered in decreasing order of the lengths L^k of the paths obtained. The pheromone update takes into account the ranks of the σ best solutions:

$$\tau_{ij} \leftarrow (1 - \rho)\tau_{ij} + \frac{\sigma}{L^{++}} + \sum_{k=1}^{\sigma-1} \Delta\tau_{ij}^k \tag{7.6}$$

where L^{++} represents the length of the best path found since the beginning of the algorithm, and the contribution of the $\sigma - 1$ best ants in the current iteration is calculated from

$$\Delta\tau_{ij}^k = \begin{cases} \frac{(\sigma-k)}{L^k} & \text{if } (i, j) \in T^k \\ 0 & \text{otherwise} \end{cases} \tag{7.7}$$

This algorithm has achieved better results than those obtained with the AS algorithm.

7.4.2.4 Ant Colony System

The ant colony system (ACS) was also initially proposed to solve the TSP [8]. It was inspired by the same mechanisms as the AS algorithm, but it follows an opposite direction regarding certain behaviors and also focuses on the goal of combinatorial efficiency. This variant is one of the best-performing ones and, consequently, it is often used to tackle new problems. We shall now describe the ACS algorithm, step by step.

The ants, as in AS, build a cycle in the graph (a Hamiltonian path) iteratively, and they take their decisions according to pheromones and visibility.

Building one solution. The rule for transitions between vertices introduces a bifurcation between two complementary strategies widely used in stochastic optimization methods: at each step in the graph, ants can use either an exploration strategy or an exploitation strategy. Algorithm 7.2 gives details algorithm used to choose the next city according to this exploration/exploitation principle.

We can observe that the parameter $q_0 \in [0, 1]$, which represents the probability of choosing the next vertex with an exploitation strategy, leads to the choice of the vertex which maximizes the quantity $\tau_{i\ell} \times \eta_{i\ell}^\beta$. The notion of visibility is similar to that introduced in the AS algorithm: the distance is used to obtain a value $\eta_{ij} = 1/d(i, j)$. In the case of exploration, the formula (7.9) is very similar to the formula used in the previous algorithms (formula (7.2)); the only difference is that the parameter α has disappeared. But α was always set to 1 in those algorithms, and thus this disappearance is more a simplification.

When the number of vertices is large, particularly at the beginning of the construction of a solution, we expect that the computation time will be costly if all vertices need to be considered (which is the case when the graph is complete). This time can

Let:
- η_{ij} be the visibility of vertex (city) j for the ant located on vertex i;
- β be a parameter used to tune the influence of visibility;
- $\mathcal{N}_i^k$ be the set of the cities which have not yet been visited by the ant k, located on vertex i;
- $q_0 \in [0, 1]$ be a parameter used to tune the exploitation/exploration ratio.

$q \leftarrow$ a real value, uniformly and randomly in the interval $[0, 1]$
if $q \leq q_0$ **then** /* exploitation */
the city j is chosen as follows:

$$j = \arg \max_{\ell \in \mathcal{N}_i^k} \left\{ \tau_{i\ell} \times \eta_{i\ell}^\beta \right\} \tag{7.8}$$

else /* exploration */
city j is chosen according to the probability

$$p_{ij}^k = \frac{\tau_{ij} \times \eta_{ij}^\beta}{\displaystyle\sum_{\ell \in \mathcal{N}_i^k} \tau_{i\ell} \times \eta_{i\ell}^\beta} \tag{7.9}$$

end

Algorithm 7.2: Building of one solution in ACS.

be reduced if candidate lists of cities are used: for each city/vertex, the ant starts by considering a list of d cities, chosen nearby. If this preselection fails, the search is widened to other cities.

Pheromone update. The pheromone update follows the elitism mechanism already described: only the best ant deposits pheromones on the path it has found:

$$\tau_{ij} \leftarrow (1 - \rho)\tau_{ij} + \rho \frac{1}{L^+} \quad \forall (i, j) \in T^+ \tag{7.10}$$

We should notice an important point here: edges (i, j) which do not belong to the path T^+ do not have an evaporation rate given by $\tau_{ij} \leftarrow (1 - \rho)\tau_{ij}$ as in previous algorithms. This represents, in terms of complexity, a particularly interesting point: in each pheromone update step, only n edges are updated, whereas previously n^2 edges were updated.

As a counterpart, evaporation from an edge is applied each time an ant uses the edge. This is quite a misleading idea, since pheromones usually evaporate when ants are not using the trail! This pheromone update, called local pheromone update, is performed in each ant step:

$$\tau_{ij} \leftarrow (1 - \xi)\tau_{ij} + \xi \tau_0 \tag{7.11}$$

In each step, pheromones evaporate and the ant deposits a fixed amount of pheromones $\xi \tau_0$. The formula has the effect of narrowing the pheromone level τ_{ij} towards its initial value τ_0 each time an ant passes along the edge (i, j). As we can see, pheromones are used in the opposite way to before: the more ants use an edge, the close the pheromone value is to τ_0.

Consequently, in ACS the attractive role of pheromones is not the only effect that influences the search: the best solution found increases the pheromone value on edges on its path but if numerous ants use the same edges, the trail will disappear. Of course, this is getting very far from real ants' behavior but, from the point of view, of optimization this permits one to keep diversity in the solutions, that have been built. Without this "strong" evaporation mechanism, when a good solution is found, all the ants are attracted by the edges of this solution, and, after some time, all ants use the same trail, which is quite useless for exploring a large search space.

Local search. The last distinctive point of ACS is that a local heuristic is used to improve the solutions built by the ants. This is a widely recognized principle in the field of metaheuristics: one associates a general search space exploration technique, ensuring broad coverage of the space, with a technique dedicated to the problem under consideration that is capable of exploiting the vicinity of the solutions proposed by the metaheuristic.

In the case of the application of ACS to the TSP, the classical 2-opt and 3-opt heuristics have been used. Without giving too much detail about these simple techniques, we can say in summary that they both consist in trying several permutations of the components of the solutions and keeping those which improve the cost of the best solution. These techniques allow one to reach a local optimum.

Tuning of parameters. As with the previous methods, there are parameter values that have given good results in the case of the TSP. Table 7.3 gives these values. The main difference from the other parameter tuning results is the number of ants used in ACS. This has been is fixed at 10 in ACS, and this is surprising because other ant methods have linked this parameter to the problem size, but no advantage has been demonstrated for this in the ACS case.

Table 7.3 Parameters and range of values known to be useful for the ACS algorithm. C represents an estimate of the cost of one solution and n is the number of vertices in the graph (i.e., the size of the problem)

Symbol	Parameter	Values
β	Influence of visibility	$[2; 5]$
ρ	Evaporation rate	0, 1
τ_0	Initial pheromone value	$1/nC$
m	Number of ants	10
ξ	Local evaporation	0, 1
q_0	Exploitation/exploration ratio	0, 9

Algorithm 7.3 gives the general framework of ACS.

Pheromone initialization: $\tau_{ij} \leftarrow \tau_0 \, \forall i, j = 1, \dots, n$
for $t \leftarrow 1$ **to** $t_{\max}$ **do**
  foreach *ant k* **do**
    Build a cycle $T^k(t)$ using Algorithm 7.2 and updating pheromones in each
      step with formula (7.11)
    Compute the cost $L^k(t)$ of $T^k(t)$
    Perform a local search to possibly improve $T_k(t)$
  end
  Let T^+ be the best solution found since the beginning of the algorithm
  forall *edge* $(i, j) \in T^+$ **do**
    Update the pheromones $\tau_{ij}(t)$ with formula (7.10)
  end
end
return *the best found solution* T^+

Algorithm 7.3: Ant colony system (ACS) algorithm.

Results. Table 7.4 gives the results obtained with ACS [8] on four classical instances of the TSP (we find again the d198 instance shown in Fig. 7.7). The results obtained with ACS are compared with those obtained with the best evolutionary algorithm known to date (STSP-GA). We can notice that ACS, although it is not better than the evolutionary algorithm in terms of quality of solutions found, is comparable to it (only for lin318 does ACS perform as well as STSP-GA). The performance of ACS decreases with increasing problem size but the computation time remains very much less for ACS than for STSP-GA.

We have presented only a very few results here; however, these results illustrate perfectly the reason why several research studies have been conducted with ant algorithms: in a short time (the first publication was in 1991, and ACS was published in 1997), ant-based algorithms for combinatorial optimization became competitive with algorithms based on ideas from the 1960s which had received much more development efforts and been used much more in practice.

Table 7.4 Comparison of results between ACS and an evolutionary algorithm on four instances of the symmetric TSP. The best results are emphasized in bold, the averages were obtained from 10 independent runs, and "duration" represents the mean duration required to obtain the best solution for each run [8]

Problem	ACS+3-opt		STSP-GA	
	Average	Duration (s)	Average	Duration (s)
d198	15781.7	238	**15780**	253
lin318	**42029**	537	**42029**	2054
att532	27718.2	810	**27693.7**	11780
rat783	8837.9	1280	**8807.3**	21210

Starting with these promising results, numerous studies have been done to improve these algorithms but also, mainly, to apply them to numerous combinatorial optimization problems. The interested reader should find references to explore this diversity in the annotated bibliography in this chapter.

7.4.3 Convergence of ACO Algorithm

Theoretical studies that allow one to understand the way ant algorithms work with pheromones are far less numerous than experimental studies tackling various problems. The stochastic component of these algorithms does not facilitate their analysis, but we can give some ideas of the theoretical studies and indicate their direction.

One of the first studies [14, 15] in this direction takes into account a special case of the AS algorithm (called the graph-based ant system), specially modified to obtain convergence results under the following hypotheses:

1. There is only one optimal solution (denoted by $w*$) for the instance of the problem considered.
2. For each edge $(i, j) \in w^*$, we have $\eta_{ij} > 0$ (the visibility is always positive).
3. If $f^* = f^*(m)$ is the best evaluation found during the iterations $1, \ldots, m - 1$, then only paths at least as good as f^* receive reinforcement (we find an elitist strategy here).

Under these conditions, Gutjahr [14, 15] constructed a Markovian process in which each state is characterized by:

- the set of all pheromone values;
- the set of all paths partially built by the ants in the current iteration;
- the best solution found in all of the previous iterations, $f^*(m)$.

This led to a theorem: let P_m be the probability that particular ant follows the optimal path in iteration m, and then the following two assertions are valid:

- for all $\varepsilon > 0$ and with the parameters ρ and β fixed, if we choose a number of ants N large enough, we have $P_m \geq 1 - \varepsilon$ for all $m \geq m_0$ (m_0 is an integer linked to ε);
- for all $\varepsilon > 0$ and with the parameters N and β fixed, if we choose an evaporation factor ρ close enough to 0, we have $P_m \geq 1 - \varepsilon$ for all $m \geq m_0$ (m_0 is again an integer linked to ε).

This theorem means that if we choose correctly, the value of the evaporation parameter or the number of ants, then convergence of the algorithm is guaranteed. However, we do not have an indication of how to choose either one or the other value, nor of the time the algorithm will take: experimental studies remain indispensable.

Similar results were summarized in [10]. That study was based on the properties of the lower bound $\tau_{\min}$ and was then adapted to the $\mathcal{MMAS}$ and ACS algorithms (even though, in the case of ACS, upper and lower bounds $\tau_{\min}$ and $\tau_{\max}$, for the pheromone values are not explicitly given). Convergence in terms of value was proved, but the results do not give information about the time necessary to reach the optimum.

Finally, in [19], upper bounds were given for the time needed to reach the optimum in the case of the ACO algorithm and particular problems (the minimum-weight spanning tree problem, for instance).

7.4.4 Comparison with Evolutionary Algorithms

In conjunction with theoretical studies, such as those presented in the previous section, which allow one to better predict the behavior of stochastic algorithms, it is interesting to study the similarities to and differences from other stochastic methods. If we focus our attention on pheromones, which are central in ACO algorithms, we can notice that the data structure of pheromone values updated according to the activity of ants and used to build new solutions, is very similar to some other structures introduced in other optimization paradigms. For instance, in [10], we find a comparison between ACO metaheuristics and Stochastic Gradient Ascent (SGA) and Cross-Entropy (CE).

We can also make a comparison with some evolutionary algorithms which make use of a probability distribution. Indeed, we can notice that ants, when they build their network of pheromones, build a sort of probability distribution, which is used afterwards to build new solutions. Thus, we can show that the pheromone matrix plays the same role as the probability distribution that was introduced in the early evolutionary algorithms based on such a structure, such as PBIL (population-based incremental learning [2]) and BSC (bit simulated cross-over [23]). Both of these algorithms were proposed for numerical optimization: the goal is to find a minimum value for a function f, defined on a subset of $\mathbb{R}^n$ with its value in $\mathbb{R}$. The coordinates of points in the search space $\mathbb{R}^n$ are translated into binary strings (using discretization on each axis). These algorithms try to find the best distribution for each bit of the binary string (this is not a good idea for performance, but it is useful for study of the algorithms). Instead of maintaining a population of solutions (i.e., of binary strings) as is done in a classical genetic algorithm, a probability distribution is used to represent the genome, and this distribution evolves with the generations. These algorithms are thus called estimation distribution evolutionary algorithms.

If we want to compare the ACO algorithms with PBIL and BSC, we need to define a strategy to manipulate binary strings with pheromones. Figure 7.9 shows a graph framework in which it is possible to generate binary strings.

If we use ACO on this graph, we can define the same algorithmic framework for ACO, PBIL, and BSC [18], and the two main data structures are:

- A real vector $V = (p_1, \ldots, p_m)$ in which each component $p_i \in [0, 1]$ represents the probability of generating a "1". This vector corresponds to the probability distribution, or, in another words, for ACO, to the pheromone matrix.

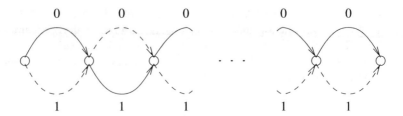

Fig. 7.9 Graph used to generate binary strings. At each vertex, ants have two choices one edge will generate a "0" and the other a "1". In the example shown, the ant has built the solution 01000 in going from *left* to *right* in the graph

- A set of solutions $P = (s_1, \ldots, s_n)$, with $s_i \in \{0, 1\}^m$, which represents a population of n binary strings of length m or, for ACO, the set of solutions built by n ants.

Thus, the three algorithms can be described by the same algorithmic schema, given in Algorithm 7.4. The generation step builds a sample of the population using the distribution V. Then, the evaluation consists in computing the value of the function f to be minimized. Then, the update step (called reinforcement when we are dealing with pheromones) contains the only difference between the three metaheuristics. This step, which is not detailed here, deals with the update of the vector V according to the solutions that have been built. For ACO, we have the same formula as those used for the pheromone updates presented earlier for the AS and ACS algorithms.

Initialization: $V = (p_1, \ldots, p_m) \leftarrow (0.5, \ldots, 0.5)$
while *stop condition is not verified* **do**
    Generate $P = (s_1, \ldots, s_n)$ according to V
    Evaluate $f(s_1), \ldots, f(s_n)$
    Update V according to $(s_1, \ldots, s_n)$ and $f(s_1), \ldots, f(s_n)$
end

Algorithm 7.4: Common algorithm for ACO, PBIL and BSC.

In order to compare the methods, we can study experimentally the role of the update formula for V for ACO, BSC, and PBIL. For instance, we can consider the following function to be minimized:

$$f(\mathbf{x}) = 50 + \sum_{i=1}^{5} \left((x_i - 1)^2 - 10 \cos \left(2\pi (x_i - 1) \right) \right) \text{ with } x_i \in [-5.12, 5.11]$$

$$(7.12)$$

Figure 7.10 gives the results obtained when the optimization problem consists in finding a minimum for the function f with Algorithm 7.4 using the four possible update formulas for V. For each dimension, real values are coded with $m = 10$ bits.

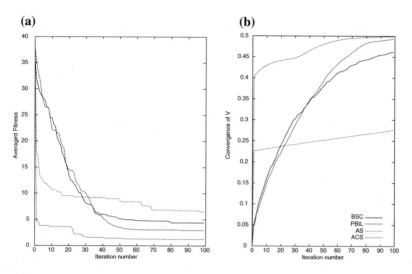

Fig. 7.10 Evolution with time (i.e., the number of generations, here from 0 to 100) of **a** the best solution found and **b** the quantity of information stored in V (also considered as a pheromone matrix for AS and ACS) for each of the methods BSC, PBIL, AS, and ACS

We observe that the ant-based algorithms are distinguishable from the others because their convergence is quicker (Fig. 7.10a), and this is particularly true for ACS. This is probably due to the elitist method used to update the pheromone values in ACS. The BSC and PBIL algorithms behave similarly to each other: their convergence is slow, but this allows these two methods to find a better solution than AS. Figure 7.10b represents the quantity of information in the vector V (i.e., its gap between 1 and 0.5, which means equiprobability). At the beginning, all components of V are set to 0.5, which can be interpreted as no stored information. We observe a big difference in the way pheromone information is acquired, especially for the AS algorithm, for which the process is slower. This can explain its weaker performance. Of course, these results are only related to one function and a small number of iterations: it would be necessary to conduct a wider experimental study to obtain a true comparison of these four methods.

7.5 Conclusion

This chapter was aimed at giving a brief glimpse of metaheuristics inspired by ants for combinatorial optimization. Basic algorithms have been presented in the case of the classical traveling salesman problem, but interested readers will be able to find various algorithms and problems in the annotated bibliography.

In the context of combinatorial problems tackled with artificial ants, we can emphasize the case of network routing problems, which constitutes a very interesting use of ant algorithms. These kinds of problems are intrinsically distributed, and this is also the case for the everyday problems encountered by real ants.

It is interesting to note that one of the first industrial applications was been developed in 1998 in the context of aluminum bar manufacturing in Quebec [13]. The problem addressed in this case was the scheduling of orders for a horizontal casting machine fed by two holding furnaces. Artificial ants were used to find the best schedule for the treatment of orders in order to (1) minimize the unused production capacity due to the different setups, (2) minimize the total lateness of all orders with respect to their due dates, and (3) minimize a penalty function aimed at grouping orders to the transported to a same destination, to maximize the use of truck capacity. The main idea was that each vertex of the graph represented an order and each ant built a sequence of orders. The Pheromones that were laid down depended on the quality of the schedule according to the three objectives.

In the context of combinatorial optimization with artificial ants, the solutions found are not guaranteed to be optimal. Theoretical work and experimental studies have confirmed that ant algorithms converge towards the optimum, but the key point is that the parameter values have to be chosed appropriately in the general case and also adapted to the instance of the problem being considered [22]. Most metaheuristics are coming up against this question.

7.6 Annotated Bibliography

Reference [3]   This book presents several aspects of swarm intelligence and is not limited to optimization. It is a very good starting point for studying and learning how to model collective natural systems.

Reference [10]   This book presents a very good synthesis of the ACO metaheuristic for combinatorial optimization. The basic principles of its application to the TSP are presented, theoretical results are presented, and an overview of problems tackled up to this date, with particular reference to network routing, is given.

Reference [1]   This multiauthor book widens the notion of indirect communication (called "stigmergy") to other collective systems (termites, particle swarms).

Reference [17]   This book presents a recent account of the state of the art of research work on artificial ants. Combinatorial optimization is introduced and several detailed examples are given, both in the field of combinatorial optimization field and various other domains.

References

1. Abraham, A., Grosan, C., Ramos, V. (eds.): *Stigmergic Optimization*. Studies in Computational Intelligence, vol. 31. Springer (2006)
2. Baluja, S., Caruana, R.: Removing the genetics from the standard genetic algorithm. In: A. Prieditis, S. Russell (eds.) *Proceedings of the Twelfth International Conference on Machine Learning (ICML)*, pp. 38–46. Morgan Kaufmann, San Mateo, CA (1995)
3. Bonabeau, E., Dorigo, M., Theraulaz, G.: *Swarm Intelligence: From Natural to Artificial Systems*. Oxford University Press, New York (1999)
4. Bullnheimer, B., Hartl, R., Strauss, C.: A new rank based version of the ant system: A computational study. Central European Journal for Operations Research and Economics **7**(1), 25–38 (1999)
5. Colorni, A., Dorigo, M., Maniezzo, V.: Distributed optimization by ant colonies. In: F. Varela, P. Bourgine (eds.) *Proceedings of the First European Conference on Artificial Life (ECAL)*, pp. 134–142. MIT Press, Cambridge, MA (1991)
6. Deneubourg, J., Goss, S., Pasteels, J., Fresneau, D., Lachaud, J.: Self-organization mechanisms in ant societies (ii): Learning in foraging and division of labor. In: J. Pasteels, J. Deneubourg (eds.) *From Individual to Collective Behavior in Social Insects*. Experientia supplementum, vol. 54, pp. 177–196. Birkhäuser (1987)
7. Dorigo, M.: Optimization, learning and natural algorithms [in Italian]. Ph.D. thesis, Politecnico di Milano, Italy (1992)
8. Dorigo, M., Gambardella, L.: Ant colony sytem: A cooperative learning approach to the travelling salesman problem. IEEE Transactions on Evolutionary Computation **1**(1), 53–66 (1997). ftp://iridia.ulb.ac.be/pub/mdorigo/journals/IJ.16-TEC97.A4.ps.gz
9. Dorigo, M., Maniezzo, V., Colorni, A.: The ant system: Optimization by a colony of cooperating agents. IEEE Transactions on Systems, Man, and Cybernetics, Part B **26**(1), 29–41 (1996)
10. Dorigo, M., Stützle, T.: *Ant Colony Optimization*. MIT Press, Cambridge, MA (2004)
11. Goss, S., Aron, S., Deneubourg, J., Pasteels, J.: Self-organized shortcuts in the Argentine ant. Naturwissenchaften **76**, 579–581 (1989)
12. Goss, S., Fresneau, D., Deneubourg, J., Lachaud, J., Valenzuela-Gonzalez, J.: Individual foraging in the ant *Pachycondyla apicalis*. Œcologia **80**, 65–69 (1989)
13. Gravel, M., Gagné, C.: Ant colony optimization for manufacturing aluminum bars. In: N. Monmarché, F. Guinand, P. Siarry (eds.) *Artificial Ants*. Wiley-Blackwell (2010)
14. Gutjahr, W.: A graph-based ant system and its convergence. Future Generation Computer Systems **16**(8), 873–888 (2000)
15. Gutjahr, W.: ACO algorithms with guaranteed convergence to the optimal solution. Information Processing Letters **82**(3), 145–153 (2002)
16. Manderick, B., Moyson, F.: The collective behavior of ants: An example of self-organization in massive parallelism. In: *Proceedings of the AAAI Spring Symposium on Parallel Models of Intelligence*. American Association of Artificial Intelligence, Stanford, CA (1988)
17. Monmarché, N., Guinand, F., Siarry, P. (eds.): *Artificial Ants: From Collective Intelligence to Real Life Optimization and Beyond*. ISTE-Wiley (2010)
18. Monmarché, N., Ramat, E., Desbarats, L., Venturini, G.: Probabilistic search with genetic algorithms and ant colonies. In: A. Wu (ed.) *Proceedings of the Optimization by Building and Using Probabilistic Models workshop, Genetic and Evolutionary Computation Conference*, Las Vegas, pp. 209–211 (2000)
19. Neumann, F., Witt, C.: *Bioinspired Computation in Combinatorial Optimization, Algorithms and Their Computational Complexity*. Natural Computing Series. Springer (2010)
20. Passera, L.: *Le monde extraordinaire des fourmis*. Fayard (2008)
21. Stützle, T., Hoos, H.: $\mathcal{MAX} - \mathcal{MIN}$ ant system and local search for the traveling salesman problem. In: *Proceedings of the Fourth International Conference on Evolutionary Computation (ICEC)*, pp. 308–313. IEEE Press (1997)

22. Stützle, T., López-Ibáñez, M., Pellegrini, P., Maur, M., Montes de Oca, M., Birattari, M., Dorigo, M.: Parameter adaptation in ant colony optimization. In: Y. Hamadi, E. Monfroy, F. Saubion (eds.) *Autonomous Search*, pp. 191–215. Springer, Berlin, Heidelberg (2012). doi:10.1007/978-3-642-21434-9_8
23. Syswerda, G.: Simulated crossover in genetic algorithms. In: L. Whitley (ed.) *Second Workshop on Foundations of Genetic Algorithms*, pp. 239–255. Morgan Kaufmann, San Mateo, CA (1993)

Chapter 8
Particle Swarms

Maurice Clerc

In this chapter, the reader is assumed to have some basic notions about iterative optimization algorithms, in particular what a definition space and a statistical distribution are. The sections headed "Formalization" can be ignored on first reading.

Preamble

At first, they move at random. Then, each time one of them finds a promising place, she reports it to some other explorers. Not always at the same time, but step by step, all of them will be informed sooner or later, and will be able to take advantage of this information. So, gradually, thanks to this collaboration without exclusion, their quest is usually successful.

8.1 Unity Is Strength

This was the official motto of the future Netherlands as early as 1550, through the Latin expression *Concordia res parvae crescunt,* and was even in fact used by Sallust circa 40 BC [69]. This saying achieved great popularity in politics and sociology, but also—and this is what is interesting for us here—in ethology, more precisely, in the field of the study of animal societies.

In the case of optimization, some methods, particularly genetic algorithms, have been inspired by biological principles such as selection, cross over and mutation. However, more recently, some other methods have tried to take advantage of behaviors that have been proved to be efficient for the survival and development of

M. Clerc (✉)
Independent Consultant, Groisy, France
e-mail: Maurice.Clerc@WriteMe.com

© Springer International Publishing Switzerland 2016
P. Siarry (ed.), *Metaheuristics*, DOI 10.1007/978-3-319-45403-0_8

biological populations. From this point of view, particle swarm optimisation (PSO) is the very first method, dating from 1995, to be based on cooperation without selection.

As we can guess from its name, this method makes use of a population of agents, called "particles," for the underlying metaphor is that during the search process they move like physical particles subject to attractive forces. But, more precisely, this is just a metaphor that guides or intuition (sometimes wrongly), and some other "ingredients" are needed to design an efficient method.

8.2 Ingredients of PSO

As is often the case in scientific research, after the remarkable conceptual break-through of the inventors of PSO, James Kennedy and Russel Eberhart [36], the crucial components of the method were clearly identified only after several years of experiments and theoretical studies. These components can be clustered into three classes: *objects*, *relations* between objects, and *mechanisms* applied to the elements of these two classes. These distinctions may seem a bit arbitrary, but thanks to them a general, modular presentation is possible, so that each component of the system can be modified in order to build variants of the algorithm.

8.2.1 Objects

Recall that the problem to be solved comprises a definition space (a set of elements, which are often points in a multidimensional space of real numbers), and a method by which a numerical value can be assigned to each element of the definition space. Often, this is a calculable function, given by a mathematical formula, but it may also be a mode of an industrial process, or even that process itself.

There are three kinds of objects in the algorithm:

- *Explorers* are particles that fly over the search space. Each explorer has a position, and memorizes the value of this position. Often, it also has a "velocity" (in fact, a movement), which can be seen as an intermediate variable that is manipulated by the algorithm to calculate the successive positions (see Sect. 8.2.3.3). However, this is not always the case, and the position may also be directly calculated [34]. An explorer also has some behavioral features—in practice, numerical coefficients, possibly variables—that are used to compute its movement at each iteration (see Sect. 8.2.3.3).
- *Memorizers* are agents that memorize one or several "good" positions found by the explorers, and also their values. Historically, and still very often, only the last best position found is memorised. But it may be interesting to save more positions to evaluate a tendency and adaptively modify the movement strategy.

- *Random number generators* (RNGs), which are used, in particular, to compute the movements of the explorers.

Note that it is also sometimes useful to define an *evolution space* for the explorers, larger than the definition space. This may seem strange, for the principle of iterative optimization is to find a solution inside the definition space. However, we will see in Sect. 8.2.3.5 that this option may be interesting.

The numbers of explorers and memorizers are not necessarily constant. In the earliest versions of PSO and in the standard versions, this is indeed the case, and they are user-defined parameters. However, in certain adaptive variants there are strategies to increase or decrease them during the optimization process (see Sect. 8.2.3).

A memorizer has a position that may vary in the course of time, so it can be seen as a special kind of particle. A set of particles is called a *swarm*, so one can speak of an *explorer-swarm* and a *memory-swarm* [13, 42]. Classically there is just one swarm of each type, but using more may be interesting.

In the first version of PSO, there was just one RNG, and the numbers were generated according to a uniform distribution in a given interval. Some later variants make use of at least one other RNG to generate nonuniform distributions (Gauss, Lévy, etc.). Also, some studies suggest that it is possible to use generators that are not random, but simply exploit cyclically a short list of predefined numbers. In such cases the generator is deterministic [15].

8.2.2   Relations

We will see that the memorizers receive information in order to guide their moves (see Sect. 8.2.3.3). Moreover, one could envisage that memorizers may exchange information between them. So, it is necessary to define communication links between all the different kinds of particles. In the most general form of the algorithm, these links may be as follows:

- *dynamic*, i.e., they are not necessarily the same in two different time steps;
- *probabilistic*, i.e., they have a certain probability of transmitting a piece of information or not, and this probability may itself be variable.

Nevertheless, in almost all versions of PSO, the number of memorizers is equal to the number of explorers, and each explorer has just one bidirectional information link to "its" memorizer. The description is then simplified by saying that there is just one kind of particle, which could be called "composite," and which combines the two functions of exploration and memorization. In all that follows, except when said otherwise, we will assume that we are considering this particular case. Then, if an explicit information link exists between a composite particle and another one, the beginning of the link is taken to be the memory part of the first particle, and the end of the link the explorer part of the second particle.

To describe the working of the optimizer, it is useful to define is the neighborhood of a particle (in a given time step): this is the set of particles that have an information

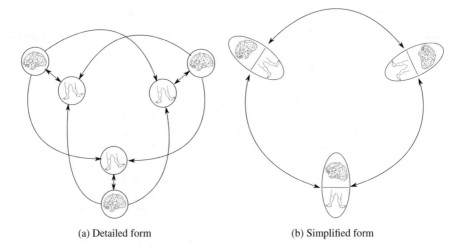

| (a) Detailed form | (b) Simplified form |

Fig. 8.1 Ring topology of explorers and memorizers. In the simplified form, we are assumed to know that each link is in fact from the memory part of one particle to the explorer part of the other particle

link with it. The set of all neighborhoods is the *topology* of the swarm. For example, if "everybody can inform anybody," the topology is said to be global. All other topologies are said to be local. Historically, the first local topology was the *ring*, as in the Fig. 8.1b.

Formalization

In a given time step, the set of relationships can be represented by a valuated graph $G(t) = (S(t), L(t))$, where the nodes S are particles and the edges L are information links. An edge has three components: its source, its sink, and a probability value.

In practice, a topology is often represented by an $n \times n$ square matrix T, where n is the swarm size and $T(i, j)$ the probability that the particle i informs the particle j. In many variants this probability is simply either 0 or 1, and a particle is always assumed to inform itself. The total number of possible topologies is then $2^{n^2 - n}$.

8.2.3 Mechanisms

8.2.3.1 Management of the Particles

At the very least, there must be a mechanism to create the particles that are needed to start the algorithm (initialization phase). There may possibly exist some other mechanisms to create or destroy particles later. Note that in this latter case we are not anymore completely respecting the initial spirit of PSO (cooperation without selection), for usually the particles that are destroyed are the "bad" ones.

For the initial creation, the most classical method is to assign to each particle a random position inside the search space, and, quite often, also a random "velocity."

Several studies [51, 57] have suggested that it would be interesting to initialize the position not according to a uniform distribution, but by using a more "regular" distribution (of low discrepancy, technically speaking). However, some other studies have shown that the influence of the initial distribution decreases very quickly after a few iterations and, in practice, the performance of the algorithm is not significantly improved [54].

8.2.3.2   Management of the Information Links

A cooperation mechanism has to provide three functions for the information links: creation, deletion, and valuation (assigning a probability). Several variants of PSO make use of a swarm of constant size and of fixed topology. In such a case, all information links can be created just once, at the beginning of the process.

However, as can be expected, this method is not very efficient if many different problems have to be solved. An adaptive topology is then used, which may be based on mathematical criteria, such as the pseudo-gradient or others [39, 49], or inspired by social or biological behaviors [5, 8, 16, 32, 33, 68]. In fact, any cooperation model that has been defined in another context can be adapted to PSO. For example, the five models defined in [52] inspired the ones used in [16]. These are based on reciprocity, proximity, relatives, reputation, and complete altruism (which is equivalent to the global topology) (see two examples of topologies on Fig. 8.2)

8.2.3.3   Moves of the Particles

The principle is that each particle is influenced by three tendencies:

- to follow its own velocity;
- to go towards the positions memorized by its neighbors;
- to go towards the best known position.

In practice, there are five steps in the computation and application of particle movements:

- *Select*, in the neighborhood of the particle, the other particles that will be taken into account. Quite often only the best one is used, but more or even all may be taken into account [47].
- For each neighbor taken into account, draw a point "around" its best memorized position. This is usually done at random, and defines a virtual movement to this position. Originally, "around" meant in a domain limited by the current position of the particle and another point a little beyond the selected memorized position (see Fig. 8.3).

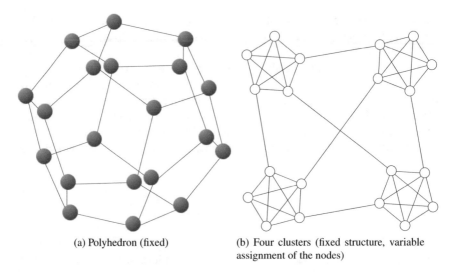

(a) Polyhedron (fixed)   (b) Four clusters (fixed structure, variable
                                          assignment of the nodes)

Fig. 8.2 Two examples of topologies. The circles are particles. All links are bidirectional. In **b**, the structure is fixed, but the particles are assigned to the nodes according to their values. This is therefore a semi-dynamical topology

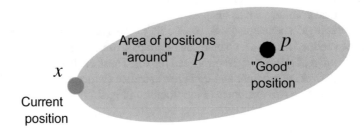

Fig. 8.3 Illustration of "around" in the computation of a movement

- Combine all the virtual movements and, partly, the current velocity. This gives the real movement.
- Apply this movement to the particle.
- If the particle flies outside the search space, a confinement mechanism may also be applied (see Sect. 8.2.3.5).

Figure 8.4 visualizes this process. An RNG is used to define a point that is "near to" a given one. Typically, when the distribution is not uniform, one make use of a distribution whose density decreases with distance.

The combination of the virtual movements and the velocity is usually a linear one. When applied to the current position, this movement gives the new position or, more precisely, one position from amongst the ones that are possible, because of the use of the RNG. This set of positions, more or less probable, positions is called the distribution of the next possible Positions (DNPP).

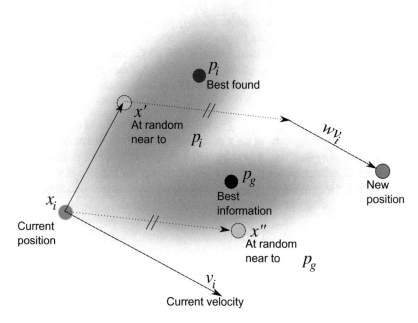

Fig. 8.4 Movement of a particle when just one informer (neighbor) is used, with a linear combination. The three tendencies are represented by three vectors, which are added

8.2.3.4 Management of the Parameters

The computation of the movement usually makes use of two or three numerical parameters (see Sect. 8.3.1). In the simplest case these parameters are constant and user-defined, but many variants have been proposed. In the most rudimentary variants, the parameter values depend only on the number of iterations [1, 31, 59, 75, 77, 79]. More sophisticated variants adapt the values according to the information that is collected during the process. One area of research is how to define an algorithm that is as adaptive as possible, so that the user does not have to tune any mechanism [9, 18, 21, 45, 60, 71].

8.2.3.5 Confinement and Constraints

The kind of optimization that we are studing here is always under constraints, for the solution is in a bounded search space. Usually, for each variable, an interval of values is given, or a finite list of acceptable values, but constraints may be more complicated, and given as relations between variables.

When a particle reaches a position that is not acceptable, there are two options:

- Let it fly and do not evaluate the new position. From the point of view of the particle, it is as if the search space, initially equal to the definition space, has been extended by a plateau. As the particle is constantly attracted by memorized positions that are in the definition space, it will come back sooner or later. This method does not need any parameter, but the convergence may be quite slow.
- Apply a confinement method, where the confinement may be either immediately complete or progressive.

Most of these methods can be used by any iterative algorithm and are therefore not presented here. However, some of them are specific to PSO, particularly because they modify not only the position but also the velocity, an element that does not exist in all algorithms. The simplest methods stop the particle at the frontier of the definition space and either the velocity is set to zero or its direction is inverted, sometimes more or less at random [12, 28].

8.3   Some Versions of PSO

We now have all the elements needed to describe the working of some versions of PSO. It is, of course, not possible to present all of them. For the interested reader, several more or less complete reviews have been published (see, for example, [23, 24, 60]). Here, we will just explain in detail the successive versions that can be called "standard," for they are very near to the historical version [36].

8.3.1   1998. A Basic Version

The features of this version are the following:

- The size of the swarm is constant, and defined by the user.
- The positions and velocities are initialized at random according to uniform distributions.
- Each particle memorizes the best position it has ever found (at the beginning, this is of course the same as the initial position).
- The topology is global, i.e., each particle informs all the others (and therefore is informed by all the others).
- The information that is transmitted is the best position memorized in the neighborhood (which contains the particle itself).
- The movement of a particle is computed independently for each dimension of the search space, by linearly combining three components: the current velocity, the best position memorized, and the best position memorized in the neighborhood (which is the whole swarm here), using *confidence coefficients*. The coefficient for the velocity is often called the *inertia weight*. The other two have equal values, given by the maximum value of a uniform random variable. If needed, the movement is

bounded, so that it does not exceed a predefined maximum value. Indeed, without this, the swarm could easily tend to "explode."

- The stop criterion is either a maximum number of iterations or a minimum value to be reached (in the case of minimization).

Formalization

Let us suppose we are looking for the global minimum of a function f whose definition space is $E = \prod_{d=1}^{D} [x_{min,d}, x_{max,d}]$:

```
Elements
Position of a particle i at time t: x_i (t) = (x_{i,1} (t),...,x_{i,D} (t))
Velocity of a particle i at time t: v_i (t) = (v_{i,1} (t)...,xv_{i,D} (t))
Best position memorized by the particle
k at time t: p_k (t) = (p_{k,1} (t),...,p_{k,D} (t))
Index of the particle that memorizes the best position
over the whole swarm: g(t)
Parameters
Swarm size n
Maximum movement (absolute value) v_max
Inertia weight: 0 < w < 1
Cognitive confidence coefficient: c_1 > 1
Social confidence coefficient: c_2 = c_1
```

The usual values are 0.72 for w, and 1.2 for c_1 and c_2.

```
Initialization
For each particle and each dimension d
x_{i,d} (0) = U (x_{min,d}, x_{max,d})  (U = uniform distribution)
p_{i,d} (0) = x_{i,d} (0)
v_{i,d} (0) = (U (x_{min,d}, x_{max,d}) - x_{i,d} (0)) /2 (for certain variants
we have v_{i,d} (0) = 0)
Index of the best memorized position: g(0)
```

To simplify the formulae, we will now write g instead of $g(t)$.

```
Movement
For each particle i and each dimension d
v_{i,d} (t+1) = wv_{i,d} (t) + c_1 (p_{i,d} (t) - x_{i,d} (t)) + c_2 (p_{g,d} (t) - x_{i,d} (t))
v_{i,d} (t+1) > v_max ⇒ v_{i,d} (t+1) = v_max
v_{i,d} (t+1) < -v_max ⇒ v_{i,d} (t+1) = -v_max
x_{i,d} (t+1) = x_{i,d} (t) + v_{i,d} (t)
Confinement
For each particle i and each dimension d
x_{i,d} (t+1) > x_{max,d} ⇒ x_{i,d} (t+1) = x_{max,d} and v_{i,d} (t+1) = 0
x_{i,d} (t+1) < x_{min,d} ⇒ x_{i,d} (t+1) = x_{min,d} and v_{i,d} (t+1) = 0
Memorization
For each particle i
```

```
if  x_i ∈ E
f (x_i (t + 1)) < f (p_i (t)) ⇒ p_i (t + 1) = x_i (t + 1)  (else  p_i (t + 1) = p_i (t))
f (p_i (t + 1)) < f (p_g (t)) ⇒ g = i  (else  g  does  not  change)
Iteration
As long as no stop criterion is satisfied
Repeat Movement and Memorization
```

8.3.2 Two Improved "Standard" Versions

The use of the basic version has made some defects evident:

1. Because of the global topology, there is often a premature convergence to a point that is not even always a local minimum [40].
2. The maximum movement is arbitrarily defined, and modifying its value may modify the performances.
3. The behavior depends on the system of coordinates.

The last point deserves to be explained, for there are many ways to interpret it [3, 76]. The sensitivity to rotation is due to the fact that the movement is computed independently for each dimension. However, as the definition space is almost never a hypersphere centered on the center of rotation, rotating the coordinate system modifies the landscape of the function in which we are looking for the position of the minimum. In fact, this position may even not be in the new definition space anymore, and in that case the new optimal position is therefore another one. In other words, the rotated problem is not identical to the initial one, and performance comparisons are then not very meaningful.

After a rotation of the problem, the performance of the optimizer may deteriorate, but if may also be improved because, in particular, and for quite subtle mathematical reasons, PSO finds solution points near to an axis or even a diagonal, more easily, and, a fortiori, near to the center of the coordinate system [12, 50, 72].

Point 1 quickly led to the use of fixed local topologies, such as the ring that we have already seen, and a lot of others. A review can be found in [46]. New ones are regularly proposed, but to solve large classes of problems more efficiently, variable topologies have been defined. Concerning point 2, several studies have shown that arbitrarily defining a maximum movement can be avoided by using mathematical relationships between the confidence coefficients (the concept of constriction). The first of these relationships was put on line in 1999, used a little later in published papers [7, 22], and itself published a year afterwards [17]. Some other studies have then simplified it [74], or generalized it [4, 64]. The point 3 was also taken into account quite early (2003), but in a completely adaptive variant (swarm size, topology, numerical parameters) that was significantly different from the basic version.

We present here first a version that modifies the basic version as little as possible to take points 1 and 2 into account, at least partly, and then a version that copes with point 3. For this last point, the formulae have to be modified.

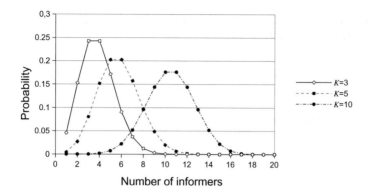

Fig. 8.5 SPSO 2007. Probability distribution of the size of the neighborhood of a particle in a given time step. The parameter K is the number of information links generated by each particle. The swarm size is 20. The bigger K is, the larger the divergence; sometimes this is beneficial

8.3.2.1   SPSO 2007

Here, the topology is not fixed anymore, but modified after every iteration that has not improved the best known position. In that case, each particle generates a given number of links (typically 3, 4, or 5) to particles that are randomly chosen. As a result, the size of the neighborhood of a particle can take any value between 1 and the swarm size, but it does so according to a statistical nonuniform distribution (a bell curve, for which the extreme values are of low probability), as can be seen in Fig. 8.5. This method was defined in [11]. It was shown later that it could be seen as a particular case of the "stochastic star" presented in [48], even though in that method the authors is used, the reverse strategy is used, i.e., the topology is modified only if the iteration has in fact improved the best position.

The confidence coefficients are not chosen arbitrarily. Depending on the variant, they are either bounded by a constriction relation [17] or defined directly by a theoretical stagnation analysis. In other words, there is just one user-defined value, from which the other is computed.

Formalization

For the simple form of constriction, the inertia weight w is user-defined, typically between 0.7 and 0.9, and the other confidence coefficients are derived from:

$$c_1 = c_2 = \frac{(w+1)^2}{2} \tag{8.1}$$

The usual value of w is 0.72, which gives $c_1 = c_2 = 1.48$. Note that the relation (8.1) is often given by the inverse formula, i.e., w as a function of c_1, which is more complicated. In fact, the theoretical constriction criterion is the inequality $c_1 + c_2 \le (w+1)^2$. Given like that, it can easily be extended to variants in which more than just one informer is taken into account [47]. If there are m such informers, the movement is computed as follows:

$$v_{i,d}(t+1) = w v_{i,d}(t) + \sum_{k=1}^{m} c_k \left(p_{\alpha(k),d}(t) - x_{i,d}(t) \right)$$

and the criterion becomes

$$\sum_{k=1}^{m} c_k \leq (w+1)^2.$$

When the coefficients are obtained from a stagnation analysis, SPSO 2007 uses the following values:

$$\begin{cases} w = \dfrac{1}{2\ln(2)} \simeq 0.721 \\ c_1 = c_2 = \dfrac{1}{2} + \ln(2) \simeq 1.193 \end{cases}$$

Note that for the same w value, the coefficients c are smaller than with the constriction approach. The search space is explored more slowly, but the risk of not "seeing" an interesting position is reduced.

8.3.2.2 SPSO 2011

This is similar to SPSO 2007 with respect to the topology and the parameters. The initialization of the velocity is a little different, to ensure that a particle will not leave the definition space in the very first iteration. The main difference is that the movement is not anymore computed dimension by dimension anymore, but directly as a vector, by using hyperspheres (Fig. 8.6). So, the behavior of the algorithm (the sequence of successive positions) no longer dependent on the coordinate system.

The complete computation needs several steps, which are described in the "Formalization" section below. Choosing a point at random in a hypersphere can be done according to several statistical distributions. It seems that a nonuniform one whose density is a decreasing function of the distance from the center, may be more efficient. For more details, see the source code on Particle Swarm Central [66].

Formalization

Here we give only what is different from SPSO 2007:

```
If necessary, the search space is normalized
in order to be a hypercube [x_min, x_max]^D.
Initialization of the velocity: v_{i,d}(0) =
U (x_min - x_{i,d}(0), x_max - x_{i,d}(0))
```

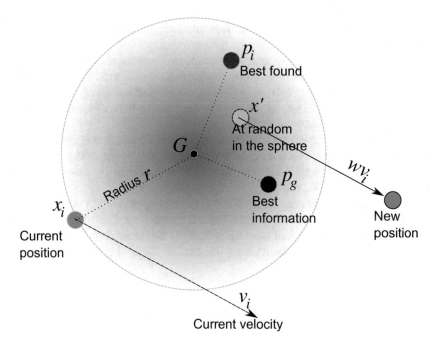

Fig. 8.6 SPSO 2011. Movement computation. G is the center of gravity of the three positions that are used in the basic version of PSO. A point is drawn at random in a hypersphere of center G and of radius r, and then moved in parallel with the current velocity, in order to take inertia into account

```
Movement
For each particle i
Hypersphere center G: if i = g,  G =
```
$\frac{x_i(t)+p_i(t)}{2}$, else $G = \frac{x_i(t)+p_i(t)+p_g(t)}{3}$
```
Radius r = ‖xᵢ (t) − G‖
Select a point x′ at random
in the hypersphere of center G and of radius r
Partly add the velocity xᵢ (t + 1) = x′ + wv (t)
New velocity vᵢ (t + 1) = xᵢ (t + 1) − xᵢ (t)
```

8.4 Applications and Variants

It is not the purpose of this chapter to make an inventory of all possible applications. Lists are regularly published, and these are papers that show how wide and various are the domains in which PSO has been used [2, 37, 38, 63, 80]. This is due to the fact that the prerequisites for the method are very simple: as explained in Sect. 8.2.1, we just need a definition space, and a way to assign a value to any element of this search space. Nevertheless, of course, it is interesting to design some variants for more efficiency.

In the simplest varients only the confidence coefficients are modified, for example by using an inertia weight which is a decreasing function of the number of iterations [70]. More examples are presented in [24]. Some variants are specifically designed for a given type of problem, for example multiobjective [67], dynamic [6, 41], or discrete or even combinatorial problems.

In this last case, some variants define the "movements" (which are permutations or elements) and their combinations differently [10, 20, 62]. Some others, in contrast, are aimed at being usable on a large spectrum of problems by using adaptation mechanisms [19, 27, 30, 78].

For problems that are continuous, discrete or heterogeneous (but not combinatorial), the standard versions that we have seen here have been defined to be reference methods that any other variant must outperform to be interesting. They are in fact easy to improve. As an example, a simple variant of SPSO 2011 is given in Sect. 8.6.2.

8.5 Going Further

In addition to Particle Swarm Central [66], which has already been mentioned, the interested reader could look with profit at some other sources of information that are more complete, including numerous papers and books that can be found on the Internet, particularly those dedicated to specific applications.

The earliest books are still useful for understanding the basic principles:

- *Swarm Intelligence* [35], by the inventors of the method;
- *Particle Swarm Optimization* [13], the first book entirely devoted to PSO.

Some later books are these, enriched by numerous theoretical and experimental studies, are:

- *Particle Swarms: The Second Decade* [65] (book);
- *Particle Swarm Optimization and Intelligence: Advances and Applications* [58] (book);
- Development of efficient particle swarm optimizers and bound handling methods [55] (thesis);
- "Particle swarm optimization in stationary and dynamic environments" [41] (thesis);
- "Particle swarms for constrained optimization" [26] (thesis);
- "Development and testing of a particle swarm optimizer to handle hard unconstrained and constrained problems" [29] (thesis);
- *Particle Swarm Optimization: Theory, Techniques and Applications* [53] (book);
- *Handbook of Swarm Intelligence* [56] (book).

8.6 Appendix

8.6.1 A Simple Example

The Tripod function is defined on $[-100, 100]^2$ by the following formula:

```
if  x₂ < 0 then  f (x₁, x₂) = |x₁| + |x₂ + 50|
else, if  x₁ < 0 then  f (x₁, x₂) = 1 + |x₁ + 50| + |x₂ − 50|
else  f (x₁, x₂) = 2 + |x₁ − 50| + |x₂ − 50|
```

It has a global minimum at $(0, -50)$ (value 0), and two local minima, $(-50, 50)$ of value 1 and $(50, 50)$ of value 2, whose basins of attraction are $[-100, 100] \times [0, -50]$, $[0, 100]^2$, and $[-100, 0] \times [0, 100]$, respectively.

The size of the first basin is twice the size of the other two, so a good algorithm must have a success rate greater than 50 % when the acceptable error is smaller than 0.0001 and the number of points evaluated, greater than 10000. This is not very easy (see for example the results from SPSO 2007 in the Table 8.1).

On this kind of problem, classical PSO has two handicaps. The first one, which is common to all stochastic methods, is the risk of converging to a local minimum. The second one, which is more specific, is due to the fact that, roughly speaking, all velocities tend to zero. If this decrease is too quick, the particles may converge to any point, as shown in the Fig. 8.7b.

8.6.2 SPSO 2011 with Distance–Fitness Correlation

In the versions that we have seen, the probabilities on the information links are simply 0 or 1, and the links are purely "social" ones, not related to any distance. A particle may inform another one in the same way, no matter whether it is very far or or very near. Nevertheless, it may happen that the nearer a particle is near to a good position, the more its own value improves.

In fact, for any classical iterative algorithm, this property has to be true on average for the efficiency to be better than the at of a pure random search [14]. Therefore, it is tempting to take advantage of this kind of information. For example, in [61], some "distance versus value" ratios were computed for each dimension, and a new position was built from those of the particles that had the best ratios.

For SPSO 2011, to keep the property that the algorithm does not depend on the system of coordinates, one can simply evaluate a distance–value correlation around a good position, and modify accordingly the center and the radius of the hypersphere that is the support of the DNPP. The idea is that the higher the correlation, the more interesting it is to focus the search around this good position.

Fig. 8.7 SPSO 2011 on Tripod. Two trajectories. The starting point is ✫ and the final point is ◆. Out of 40 particles, only about ten converge to the solution, as in (**a**), the others coverage to a point that is not even a local minimum, as in (**b**)

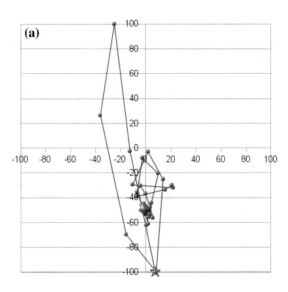

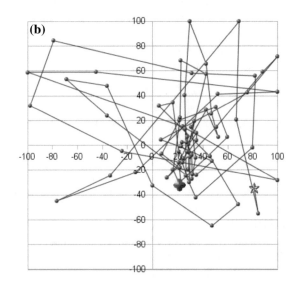

Formalization

Selection of the D particles $\left(x_{\alpha_1}, \ldots, x_{\alpha_D}\right)$
that are nearest to $p_g(t)$.
The values of their positions are $\left(f_{\alpha_1}, \ldots, f_{\alpha_D}\right)$
Compute the Euclidean distances $d_{\alpha_j} = \left\| x_{\alpha_j} - p_g(t) \right\|$
Compute the averaged correlation coefficient

$$\rho = \frac{\sum_{j=1}^{D}\left(f_{\alpha_j} - f\left(p_g(t)\right)\right)d_{\alpha_j}}{\text{var}\left(f_{\alpha_1} - f\left(p_g(t)\right), \ldots, f_{\alpha_D} - f\left(p_g(t)\right)\right)\text{var}\left(d_{\alpha_1}, \ldots, d_{\alpha_D}\right)}$$

```
Center of the hypersphere, depending on ρ:
if  ρ ≥ 0,  G (ρ) = G (0) + ρ (pg (t) − G (0)),  else   G (ρ) =
G (0) − ρ (x (t) − G (0)),
where  G (0) is the center  G  as computed in SPSO 2011
Radius of the hypersphere, depending on ρ:
r (ρ) = rmax − (ρ+1)/2 rmax
where  rmax is the radius  r  as computed in SPSO 2011
```

In this rudimentary linear formalization, the radius is zero when the correlation is perfect ($\rho = 1$). In that case, the particle simply keeps moving in the same direction, but more slowly.

8.6.3   Comparison of Three Simple Variants

Table 8.1 give the success rates (on 13 problems, with 100 runs for each) of the three algorithms that we have seen, namely SPSO 2007, SPSO 2011, and its variant with distance–fitness correlation. The details of these problems are not important (the last six are in [73], except that the last two ones are not rotated). What matters is the variation of the success rate.

Even though, on average, there is an improvement, this is not always the case for each problem, taken one by one. In fact, this test bed was constructed precisely to show that the way the problems are chosen is important (see the discussion in the Sect. 8.6.4.3). In passing, we can see that these versions of PSO are not suitable for binary problems (Network) or for very difficult ones (Shifted Rastrigin). That is why more specific variants are sometimes needed.

8.6.4   About Some Traps

When using PSO, researchers and users can fall into some traps.

In fact, these kinds of pitfalls can occur in many other methods as well. For a researcher, the deceitfully intuitive character of the method can lead them to take some particular behavior for granted, when this is not true in reality. A user, who thinks, on the evidence of a published article, that a particular variant should be effective in the scenario, described there can be disappointed because the benchmark used in the article is not representative enough of real problems.

In both cases, the use of pseudo-randomness can lead to surprises because the various generators are not equivalent. Let us give some examples here.

Table 8.1 Success rates on 13 problems. There is not always an improvement in all cases, but on average each variant is probably better than the previous one

		Space	Number of evaluations accuracy	SPSO 2007	SPSO 2011	SPSO 2011 + correlation
1	Tripod	$[-100, 100]^2$	10 000 0.0001	49	79	62
2	Network	$\{0, 1\}^{38} \times [0, 20]^4$	5 000 0	0	0	0
3	Step	$[-100, 100]^{10}$	2500 0.0	100	99	100
4	Lennard–Jones (5 atoms)	$[-2, 2]^{15}$	635 000 10^{-6}	94	50	100
5	Gear train (complete)	$\{12, \ldots 60\}^4$	20 000 10^{-13}	8	58	30
6	Perm (complete)	$\{-5, \ldots, 5\}^5$	10 000 0	14	36	49
7	Compression spring	$\{1, \ldots, 70\} \times [0, 6, 3] [0, 207, 0, 208, \ldots, 0, 5]$	20 000 10^{-10}	35	81	88
8	Shifted sphere	$[-100, 100]^{30}$	200 000 0.0	100	100	100
9	Shifted Rosenbrock	$[-100, 100]^{10}$	100 000 0.01	71	50	74
10	Shifted Rastrigin	$[-5, 12, 5, 12]^{30}$	300 000 0.01	0	0	0
11	Shifted Schwefel	$[-100, 100]^{10}$	100 000 10^{-5}	100	100	100
12	Shifted Griewank	$[-600, 600]^{10}$	100 000 0.01	7	39	33
13	Shifted Ackley	$[-32, 32]^{10}$	100 000 10^{-4}	99	100	100
	Mean			**52.07**	**60.9**	**63.02**

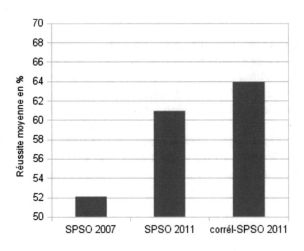

8.6.4.1 Exploitation and Exploration

Two important features of an iterative algorithm are exploration (of the search space) and exploitation (around a promising position), sometimes called diversification and intensification.

We sometimes assert that the balance between the two is crucial for the efficiency of the algorithm, without verifying this statement with the help of a measurable definition of these notions. In the context of PSO, this is nevertheless rather easy, because the promising positions are the best positions stored by the particles.

Thus, it is enough to formalize the expression "around" to define exploitation and, by complementarity, the exploration. When the algorithm works dimension by dimension, as in numerous versions, we may use, for example, a hyperparallelepiped containing the promising position. When the algorithm is independent of the system of coordinates, we use a hypersphere. We can then calculate a ratio of exploitation to exploration, follow its evolution, and look to see if there is a correlation between this ratio and the efficiency.

The important point is that, experimentally, no correlation of this kind seems to exist, for PSO. The ratio can, evolve in a very unbalanced way and the algorithm can nevertheless be very effective, and the opposite can also be true. This is because the optimum can in fact be found in several different ways, such as "in passing" by a single particle of not insignificant speed, or collectively by a set of particles whose speeds tend towards zero. There is a trap here for the researcher who, in finalising a new variant, assumes intuitively, but maybe wrongly, that it is necessary to improve the balance between exploitation and exploration to be more efficient.

Formalization in the Dimension-by-Dimension Case

For every dimension d, we sort the coordinates of the "good" stored positions $p_j (t)$ in increasing order. We then have

$$x_{\min,d} \leq p_{\alpha_d(1),d} (t) \leq \cdots \leq p_{\alpha_d(S),d} \leq x_{\max,d}$$

By convention, we write $p_{\alpha_d(0),d} = x_{\min,d}$ and $p_{\alpha_d(S+1),d} = x_{\max,d}$. We then say that $x_i (t + 1)$ is an exploitation point around $p_j (t)$ if, for all dimensions d, if $\alpha_d (k) = j$, then

$$p_{j,d} (t) - \delta \left(p_{j,d} (t) - p_{\alpha(k-1),d} (t) \right) \leq x_{i,d} (t + 1) \leq p_{j,d} (t) + \delta \left(p_{\alpha(k+1),d} (t) - p_{j,d} (t) \right)$$

where δ is a coefficient smaller than $1/2$ (typically $1/3$).

8.6.4.2 Success Rate

When random numbers are used, we can calculate an estimate of the success rate for a given problem by executing the algorithm several times, having defined what "success" means. Classically, in the case of minimization, this is a question of finding

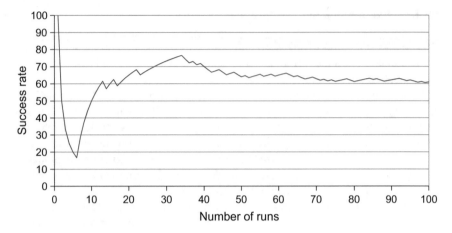

Fig. 8.8 Typical evolution of a success rate with the number of runs. After 30 runs, the estimated rate is 73 %, while the real value is about 61 %

a position in the definition space whose value is lower than a predefined threshold. If we run the algorithm M times and there are m successes, the estimate of the rate is m/M. But how much can we trust it?

If, for a given problem and a given algorithm we draw a curve of success rate versus number of runs, very generally it oscillates before it more or less stabilizes, sometimes after only a lot of runs, as shown in Fig. 8.8. It is thus useful to estimate statistical distribution and, at least, its mean and standard deviation.

To do this, we can launch 100 runs 100 times (of course, without resetting the RNG) for example. The 100 rates obtained allow us to draw a curve such as that in Fig. 8.9. It is not rare that for 30 executions, the standard deviation is at least 10 % and, so that it is lower than 1 %, more than 1000 executions are sometimes necessary.

So, for a given problem, if we run Algorithm A 30 times with a success rate of, let us say, 60 %, and then run Algorithm B 30 times with a 55 % success rate, it is risky to conclude from this that A is better than B.

8.6.4.3   The Benchmark

No benchmark can reflect the diversity of real problems. Nevertheless, we can try to make it representative enough so that if the results of an algorithm are satisfactory with it, there is a good chance that this is also the case for the end user's problems.

In this respect, as already indicated, numerous variants of PSO show the inconvenient feature that they can find the solution more easily if it has a "special" position, for example on an axis or a diagonal of the definition, space or, to an even greater extent, in the center. Such problems are said to be *biased* [72] and thus must be banned thus totally in a benchmark.

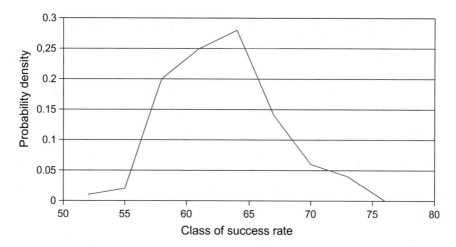

Fig. 8.9 Typical statistical distribution of a success rate. Estimation according to 100 series of 100 runs. Here, the standard deviation remains even greater than 4%

Besides, it is very easy, and even very attractive, to choose a benchmark which overestimates the capacities of an algorithm. For example, if we consider the algorithms used to obtain the results shown in Table 8.1, and if we remove the Problems 4, 9, and 13, we can conclude that SPSO 2011 is always much better than SPSO 2007, which is wrong.

8.6.5   On the Importance of the Generators of Numbers

Almost all versions of PSO are stochastic and presume that their random number generators are perfect. This is obviously wrong when the RNGs are coded, rather than being derived from material systems (quantum, for example). For example, many coded RNGs are cyclic.

Two consequences of this imperfect character must be taken into account. On one hand, the results of statistical tests which assume independence of the successive runs must be considered with caution because, once the number of generated numbers is approximately equal to the half of the length of the cycle, the runs cannot validly be considered as independent anymore. On the other hand, for the same problem and the same algorithm, the use of different generators can give results that are themselves appreciably different, as shown in the Table 8.2 and Fig. 8.10.

As Hellekalek [25] writes:

Do not trust simulation results produced by only one (type of) generator, check the results with widely different generators before taking them. seriously

Table 8.2 Comparison of results given by SPSO 2011 on 13 problems with two RNGs, KISS [43] and Mersenne-Twister [44]. The success rate is given as a percentage. For certain problems (Gear Train, Shifted Griewank), the difference is significant

Problem	KISS		Mersenne Twister	
	Success rate	Mean error	Success rate	Mean error
Tripod	79	0.146	72	0154
Network optimization	0	108.7	0	111.8
Step	99	0.01	99	0.01
Lennard–Jones	50	0.168	48	0.189
Gear train	58	1.9×10^{-11}	46	2.6×10^{-11}
Perm function	36	308.78	29	342.79
Compression spring	81	0.0033	79	0.0035
Shifted sphere	100	0	100	0
Shifted Rosenbrock[a]	50	57.67	46	59.46
Shifted Rastrigin	0	51.2	0	48.7
Schwefel	100	8.63×10^{-6}	100	9.81×10^{-6}
Shifted Griewank	39	0.0216	32	0.0223
Shifted Ackley	100	8.76×10^{-5}	100	8.86×10^{-5}
Mean	**60.9 %**		**57.8 %**	

[a]For this problem the mean values are not significant, for over 100 runs, the variance is extremely large

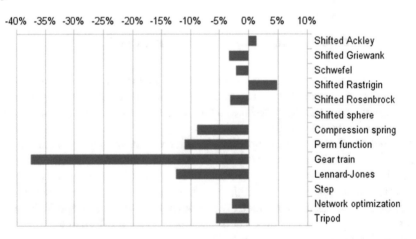

Fig. 8.10 SPSO 2011. Relative differences of means for 13 problems, over 100 runs, with the RNGs KISS and Mersenne-Twister

References

1. Al-Sharhan, S., Omran, M.: A parameter-free barebones particle swarm algorithm for unsupervised pattern classification. International Journal of Hybrid Intelligent Systems **9**, 135–143 (2012)
2. AlRashidi, M.R., El-Hawary, M.E.: A survey of particle swarm optimization applications in electric power systems. IEEE Translation on Evolutionary Computation **13**(4), 913–918 (2009). doi:10.1109/TEVC.2006.880326
3. Auger, A., Hansen, N., Perez Zerpa, J.M., Ros, R., Schoenauer, M.: Empirical comparisons of several derivative free optimization algorithms. In: *Acte du 9me colloque national en calcul des structures*, vol. 1, pp. 481–486. Giens, France (2009). In Practice (2009) Volume: 5526, Publisher: Springer Berlin Heidelberg, Pages: 3–15
4. Van den Bergh, F.: An analysis of particle swarm optimizers. Ph.D. thesis, University of Pretoria, Pretoria, South Africa (2002)
5. Bird, S., Li, X.: Adaptively choosing niching parameters in a PSO. In: *GECCO 2006 - Genetic and Evolutionary Computation Conference*, vol. 1, pp. 3–9 (2006)
6. Blackwell, T.M., Bentley, P.J.: Dynamic search with charged swarms. In: *Genetic and Evolutionary Computation Conference*, pp. 19–26. Morgan Kaufmann, San Francisco (2002)
7. Carlisle, A., Dozier, G.: An off-the-shelf PSO. In: *Workshop on Particle Swarm Optimization*, Purdue School of Engineering and Technology, INPUI, Indianapolis (2001)
8. Carvalho, D.F.de, Bastos-Filho, C.J.A.: Clan particle swarm optimization. International Journal of Intelligent Computing and Cybernetics **2**(2), 197–227 (2009). doi:10.1108/17563780910959875
9. Clerc, M.: TRIBES—Un exemple d'optimisation par essaim particulaire sans paramètres de contrôle. In: *OEP'03 (Optimisation par Essaim Particulaire)*, Paris (2003). http://www.particleswarm.info/oep_2003/
10. Clerc, M.: Discrete particle swarm optimization, illustrated by the traveling salesman problem. In: *New Optimization Techniques in Engineering*, pp. 219–239. Springer, Heidelberg (2004)
11. Clerc, M.: *L'optimisation par essaims particulaires. Versions paramétriques et adaptatives.* Hermés Science (2005)
12. Clerc, M.: *Confinements and Biases in Particle Swarm Optimisation.* Technical report, Open Archive HAL (2006). https://hal.archives-ouvertes.fr/hal-00122799
13. Clerc, M.: Particle swarm optimization. In: *ISTE (International Scientific and Technical Encyclopedia)* (2006)
14. Clerc, M.: When nearer is better (2007). http://hal.archives-ouvertes.fr/hal-00137320.
15. Clerc, M.: *List Based PSO for Real Problems.* Technical report, Open Archive HAL (2012). http://hal.archives-ouvertes.fr/docs/00/76/49/94/PDF/List_Based_PSO.pdf
16. Clerc, M.: *Cooperation Mechanisms in Particle Swarm Optimisation, in Nature Inspired Computing: Theory and Industrial Application.* CNRS, Centre pour la Communication Scientifique Directe (CCSD) (2013). http://hal.archives-ouvertes.fr/hal-00868161
17. Clerc, M., Kennedy, J.: The particle swarm-explosion, stability, and convergence in a multidimensional complex space. IEEE Transactions on Evolutionary Computation **6**(1), 58–73 (2002)
18. Cooren, Y., Clerc, M., Siarry, P.: Initialization and displacement of the particles in TRIBES, a parameter-free particle swarm optimization algorithm. In: C. Cotta, M. Sevaux, K. Srensen (eds.) *Adaptive and Multilevel Metaheuristics*, Studies in Computational Intelligence, vol. 136, pp. 199–219. Springer, Berlin, Heidelberg (2008)
19. Cooren, Y., Clerc, M., Siarry, P.: Performance evaluation of TRIBES, an adaptive particle swarm optimization algorithm. Swarm Intelligence **3**, 149–178 (2009). doi:10.1007/s11721-009-0026-8
20. Deroussi, L., Gourgand, M., Kemmoe S., Quilliot, A.: *Discrete Particle Swarm Optimization for the Permutation Flow Shop Problem.* Technical report, LIMOS CNRS UMR 6158 (2006)
21. Dos Santos Coelho, L., Alotto, P.: TRIBES optimization algorithm applied to the Loney's solenoid. IEEE Transactions on Magnetics **45**(3), 1526–1529 (2009)

22. Eberhart, R.C., Shi, Y.: Comparing inertia weights and constriction factors in particle swarm optimization. In: *International Congress on Evolutionary Computation*, pp. 84–88. IEEE Press, Pistacataway, NJ (2000)
23. El-Abd, M.: Cooperative models of particle swarm optimizers. Ph.D. thesis, University of Waterloo, Ontario, Canada (2008)
24. Eslami, M., Shareef, H., Khajehzadeh, M., Mohamed, A.: A survey of the state of the art in particle swarm optimization. Research Journal of Applied Sciences, Engineering and Technology **4(9)**, 1181–1197 (2012)
25. Hellekalek, P.: Good random number generators are (not so) easy to find. Mathematics and Computers in Simulation, **46**, 485–505 (1998)
26. Helwig, S.: Particle swarms for constrained optimization. Ph.D. thesis, Technischen Fakultät der Universität Erlangen-Nürnberg (2010)
27. Helwig, S., Neumann, F., Wanka, R.: Particle swarm optimization with velocity adaptation. In: *International Conference on Adaptive and Intelligent Systems (ICAIS)*, Klangenfurt, pp. 146–151. IEEE press (2009)
28. Helwig, S., Wanka, R.: Particle swarm optimization in high-dimensional bounded search spaces. In: *Proceedings of the 2007 IEEE Swarm Intelligence Symposium*, Honolulu, pp. 198–205. IEEE Press (2007)
29. Innocente, M.S.: Development and testing of a particle swarm optimizer to handle hard unconstrained and constrained problems. Ph.D. thesis, Swansea University, UK (2010)
30. Ismail, A., Engelbrecht, A.P.: The self-adaptive comprehensive learning particle swarm optimizer. In: M. Dorigo, M. Birattari, C. Blum, A.L. Christensen, A.P. Engelbrecht, R. Gross, T. Stützle (eds.) *Swarm Intelligence*, Lecture Notes in Computer Science, vol. 7461, pp. 156–167. Springer, Berlin, Heidelberg (2012)
31. Iwasaki, N., Yasuda, K., Ueno, G.: Dynamic parameter tuning of particle swarm optimization. IEEJ Transactions on Electrical and Electronic Engineering **1(4)**, 353–363 (2006)
32. Janson, S., Middendorf, M.: A hierarchical particle swarm optimizer and its adaptive variant. IEEE Transactions on Syststems, Man and Cybernetics B: Cybernetics **35**(6), 1272–1282 (2005)
33. Jordan, J., Helwig, S., Wanka, R.: Social interaction in particle swarm optimization, the ranked FIPS, and adaptive multi-swarms. In: *Proceedings of the Genetic and Evolutionary Computation Conference (GECCO08)*, Atlanta, Georgia, pp. 49–56. ACM Press (2008)
34. Kennedy, J.: Bare bones particle swarms. In: *IEEE Swarm Intelligence Symposium*, pp. 80–87 (2003).
35. Kennedy, J., Eberhart, R., Shi, Y.: *Swarm Intelligence*. Morgan Kaufmann (2001)
36. Kennedy, J., Eberhart, R.C.: Particle swarm optimization. In: *IEEE International Conference on Neural Networks*, vol. IV, pp. 1942–1948. IEEE Press, Piscataway, NJ (1995)
37. Kothari, V., Anuradha, J., Shah, S., Mittal, P.: A survey on particle swarm optimization in feature selection. In: P.V. Krishna, M.R. Babu, E. Ariwa (eds.) *Global Trends in Information Systems and Software Applications*. Communications in Computer and Information Science, vol. 270, pp. 192–201. Springer, Berlin, Heidelberg (2012)
38. Kulkarni, R.V., Venayagamoorthy, G.K.: Particle swarm optimization in wireless-sensor networks: A brief survey. IEEE Transaction on Systems Man, Cybernetics Part C **41**(2), 262–267 (2011). doi:10.1109/TSMCC.2010.2054080
39. Lane, J., Andries, E., Gain, J.: Particle swarm optimization with spatially meaningful neighbours. In: *Proceedings of the 2008 IEEE Swarm Intelligence Symposium*, pp. 1–8. IEEE Press, Piscataway, NJ (2008)
40. Langdon, W.B., Poli, R.: Evolving problems to learn about particle swarm optimizers and other search algorithms. IEEE Transactions on Evolutionary Computation **11**(5), 561–578 (2007)
41. Li, C.: Particle swarm optimization in stationary and dynamic environments. Ph.D. thesis, University of Leicester, UK (2010)
42. Li, X.: A multimodal particle swarm optimizer based on fitness Euclidean-distance ratio. In: *Proceedings of the 9th Annual Conference on Genetic and Evolutionary Computation, GECCO '07*, pp. 78–85. ACM, New York (2007). doi:10.1145/1276958.1276970

43. Marsaglia, G., Zaman, A.: *The KISS Generator*. Technical report, Deptartment of Statistics, University of Florida (1993)
44. Matsumoto, M., Nishimura, T.: Mersenne Twister: A 623-dimensionally equidistributed uniform pseudo-random number generator. *ACM Transactions on Modeling and Computer Simulation* **8**(1), 3–30 (1998)
45. Mekni, S., Chaâr, B.F., Ksouri, M.: Flexible job-shop scheduling with TRIBES-PSO approach. Journal of Computing **3**(6), 97–105 (2011)
46. Mendes, R.: Population topologies and their influence in particle swarm performance. Ph.D. thesis, Universidade do Minho (2004)
47. Mendes, R., Kennedy, J., Neves, J.: Fully informed particle swarm: simpler, maybe better. IEEE Transactions on Evolutionary Computation **8**, 204–210 (2004)
48. Miranda, V., Keko, H., Duque, A.J.: Stochastic star communication topology in evolutionary particle swarms (EPSO). International Journal of Computational Intelligence Research, **4**(2), 105–116 (2008)
49. Mohais, A.: Random dynamic neighbourhood structures in particle swarm optimisation. Ph.D. thesis, University of the West Indies (2007)
50. Monson, C.K., Seppi, K.D.: Exposing origin-seeking bias in PSO. In: *GECCO'05*, Washington, DC, pp. 241–248 (2005)
51. Nguyen, X.H., Nguyen, Q.U., McKay, R.I.: PSO with randomized low-discrepancy sequences. In: *Proceedings of the 9th Annual Conference on Genetic and Evolutionary Computation, GECCO '07*, pp. 173–173. ACM, New York (2007). doi:10.1145/1276958.1276987
52. Nowak, M.A.: Five rules for the evolution of cooperation. Science **314(5805)**, 1560–1563 (2006)
53. Olsson, A.: *Particle Swarm Optimization: Theory, Techniques and Applications. Engineering Tools, Techniques and Tables*. Nova Science (2011)
54. Omran, M.G.H., al Sharhan, S., Salman, A., Clerc, M.: Studying the effect of using low-discrepancy sequences to initialize population-based optimization algorithms. Computational Optimization and Applications, **56**(2), 457–480 (2013). doi:10.1007/s10589-013-9559-2
55. Padhye, N.: Development of efficient particle swarm optimizers and bound handling methods. Master's thesis, Indian Institute of Technology, Kanpur 208016, India (2010)
56. Panigrahi, B.K., Shi, Y., Lim, M.H. (eds.): *Handbook of Swarm Intelligence: Concepts, Principles and Applications*. Springer (2011)
57. Pant, M., Thangaraj, R., Grosan, C., Abraham, A.: Improved particle swarm optimization with low-discrepancy sequences. In: *IEEE Congress on Evolutionary Computation, 2008, CEC 2008, (IEEE World Congress on Computational Intelligence)*, pp. 3011–3018 (2008). doi:10.1109/CEC.2008.4631204
58. Parsopoulos, K., Vrahatis, M.: *Particle swarm optimization and intelligence: advances and applications*. IGI Global (2009)
59. Parsopoulos, K.E., Vrahatis, M.N.: Parameter selection and adaptation in unified particle swarm optimization. Mathematical and Computer Modelling **46**, 198–213 (2007)
60. Parsopoulos, K.E., Vrahatis, M.N. (eds.): *Particle swarm optimization and intelligence: advances and applications*. Information Science Reference, Hershey, NY (2010)
61. Peram, T., Veeramachaneni, K., Mohan, C.K.: Fitness-distance-ratio based particle swarm optimization. In: *Proceedings of the 2003 IEEE Swarm Intelligence Symposium (SIS '03)* (2003)
62. Pierobom, J.L., Delgado, M.R., Kaestner, C.A.: Particle swarm optimization applied to task assignment problem. In: *10th Brazilian Congress on Computational Intelligence (CBIC' 2011)*, Fortaleza, Cear. Brazilian Society of Computational Intelligence (SBIC) (2011)
63. Poli, R.: Analysis of publications on particle swarm optimisation applications. Journal of Artificial Evolution and Applications, Article ID 685175 (2008)
64. Poli, R.: Dynamics and stability of the sampling distribution of particle swarm optimisers via moment analysis. Journal of Artificial Evolution and Applications, Article ID 761459 (2008)
65. Poli, R., Kennedy, J., Blackwell, T.: *Particle Swarms: The Second Decade*. Hindawi (2008)
66. Particle Swarm Central: Home Page. http://www.particleswarm.info

67. Reyes-Sierra, M., Coello, C.A.C.: Multi-objective particle swarm optimizers: a survey of the state-of-the-art. International Journal of Computational Intelligence Research **2**(3), 287–308 (2006)
68. Richards, M., Ventura, D.: Dynamic sociometry and population size in particle swarm optimization. pp. 1557–1560. In: *Sixth International Conference on Computational Intelligence and Natural Computing*, pp. 1557–1560 (2003)
69. Sallust: La guerre de Jugurtha. Belles Lettres (2002)
70. Shi, Y.H., Eberhart, R.C.: A Modified Particle swarm optimizer. In: *International Conference on Evolutionary Computation*, pp. 69–73. IEEE Press, Piscataway, NJ (1998)
71. Souad Larabi, M.S., Ruiz-Gazen, A., Berro, A.: TRIBES : une méthode d'optimisation efficace pour révéler des optima locaux d'un indice de projection. In: *ROADEF* (2010)
72. Spears, W.M., Green, D.T., Spears, D.F.: Biases in particle swarm optimization. International Journal of Swarm Intelligence Research **1**(2), 34–57 (2010)
73. Suganthan, P., Hansen, N., Liang, J., Deb, K., Chen, Y., Auger, A., Tiwari, S.: *Problem Definitions and Evaluation Criteria for the CEC 2005 Special Session on Real Parameter Optimization*. Technical report, Nanyang Technological University, Singapore (2005)
74. Trelea, I.C.: The particle swarm optimization algorithm: Convergence analysis and parameter selection (2003)
75. Ueno, G., Yasuda, K., Iwasaki, N.: Robust adaptive particle swarm optimization. In: Systems, Man and Cybernetics, 2005 IEEE International Conference on, vol. 4, pp. 3915–3920 Vol. 4 (2005). doi:10.1109/ICSMC.2005.1571757
76. Wilke, D.N., Kok, S., Groenwold, A.A.: Comparison of linear and classical velocity update rules in particle swarm optimization: notes on scale and frame invariance. International Journal for Numerical Methods in Engineering **70**(8), 985–1008 (2007). doi:10.1002/nme.1914. Linear PSO
77. Xie, X.F., Zhang, W., Yang, Z.L.: Adaptive particle swarm optimization on individual level. In: International Conference on Signal Processing (ICSP 2002). Beijing, China (2002)
78. Yasuda, K., Yazawa, K., Motoki, M.: Particle swarm optimization with parameter self-adjusting mechanism. IEEJ Transactions on Electrical and Electronic Engineering **5**(2), 256–257 (2010). doi:10.1002/tee.20525
79. Zhan, Z.H., Zhang, J., Li, Y., Chung, H.H.: Adaptive particle swarm optimization. IEEE Transactions on Systems, Man, and Cybernetics, Part B **39**(6), 1362–1381 (2009)
80. Zou, Q., Ji, J., Zhang, S., Shi, M., Luo, Y.: Model predictive control based on particle swarm optimization of greenhouse climate for saving energy consumption. In: World Automation Congress (WAC), 2010, pp. 123–128 (2010)

Chapter 9
Some Other Metaheuristics

Ilhem Boussaïd

9.1 Introduction

In the last thirty years, great interest has been shown in metaheuristics. We can try to indicate some of the steps that have marked the history of metaheuristics. One pioneering contribution was the proposal of the simulated annealing method by Kirkpatrick et al. in 1983 [46]. In 1986, the tabu search method was proposed by Glover [28], and the artificial immune system was proposed by Farmer et al. [21]. In 1988, Koza registered his first patent on genetic programming, later published in 1992 [48]. In 1989, Goldberg published a well-known book on genetic algorithms [30]. In 1992, Dorigo completed his Ph.D. thesis, in which he described his innovative work on ant colony optimization [20]. In 1993, the first algorithm based on bee colonies was proposed by Walker et al. [98]. Another significant step was the development of the particle swarm optimization method by Kennedy and Eberhart in 1995 [44]. In the same year, Hansen et al. proposed CMA-ES (Covariance Matrix Adaptation Evolution Strategy) [35] and [22] proposed the GRASP method (Greedy Randomized Adaptive Search Procedure). In 1996, Mühlenbein and Paaß proposed the estimation of distribution algorithm [59]. In 1997, Storn and Price proposed the differential evolution algorithm [89]. In 2001, Geem et al., inspired by the improvisation process of music performers, proposed the harmony search algorithm [27]. In 2002, Passino introduced an optimization algorithm based on the social foraging behavior of *Escherichia coli* bacteria [64]. In 2006, a new population-based optimization algorithm, called the *group search optimizer*, which was based on the *producer–scrounger* model,[1] was proposed by He et al. [40]. In

---

[1]Group Search Optimizer is based on the behavior of animals living in groups, where producers search to find food and scroungers search for joining opportunities.

I. Boussaïd (✉)
University of Sciences and Technology Houari Boumediene, El-Alia BP 32,
16111 Bab-Ezzouar, Algiers, Algeria
e-mail: iboussaid@usthb.dz

© Springer International Publishing Switzerland 2016
P. Siarry (ed.), *Metaheuristics*, DOI 10.1007/978-3-319-45403-0_9

2008, Simon proposed a biogeography-based optimization algorithm [86], which was strongly influenced by the equilibrium theory of island biogeography [55]. In 2009, Xin-She Yang and Suash Deb proposed cuckoo search [104]. This algorithm is based on the breeding behavior of some cuckoo species in combination with the Lévy flight behavior of some birds and fruit flies. In the same year, Rashedi et al. [71] proposed the *gravitational search algorithm* based on a simulation of the behavior of Newton's gravitational force. In 2010, Yang [103] proposed a new metaheuristic method, the *bat-inspired algorithm*, based on the echolocation behavior of bats.

The better-known metaheuristics have been presented in the first part of this book. We present in this chapter a nonexhaustive collection of some other metaheuristics. The omission of some algorithms does not mean they are not popular, but it is not possible to include all algorithms. In what follows we present:

- artificial immune systems;
- the differential evolution algorithm;
- the bacterial foraging optimization algorithm;
- biogeography-based optimization;
- cultural algorithms;
- coevolutionary algorithms.

9.2 Artificial Immune Systems

The immune system is a network of cells, tissues, and organs that work together to protect organisms from *pathogens* (harmful microorganisms such as bacteria and viruses) without prior knowledge of their structure. Recent work in immunology has characterized the immune system as a cognitive network capable of adaptation, recognition, learning, and memory, as pointed out by Anspach and Varela [3]. The organization of the immune system into a communication network gives it three essential properties: (1) a large capacity for *information exchange*; (2) strong *regulation* to maintain a *steady state* equilibrium in the immune system in order to achieve an *adaptive* immune response; and (3) powerful *effector* functions that are essential for maintaining the integrity of the immune system. These properties, along with the highly distributed and self-organizing nature of the immune system offers rich metaphors for its artificial counterpart. Research into artificial immune systems (AIS) emerged in the mid 1980s with articles authored by Farmer, Packard, and Perelson [21]. It attempts to apply principles of the immune system to optimization and machine learning problems.

Below we present the immune system from the biological viewpoint by introducing some of the basic principles.

The immune system is often divided into two distinct but interrelated subsystems, namely the *innate* (or nonspecific) immune system and the *adaptive* (specific or acquired) immune system. The innate immune system constitutes the first line of defense of the host in the early stages of infection. It is so called because the body is born with the ability to recognize certain microbes and immediately destroy them. The innate immune response is mediated primarily by phagocytic cells and antigen-presenting cells (APCs), such as granulocytes, macrophages, and dendritic cells (DCs). In order to detect invading pathogens and initiate the innate immune response, the immune system is equipped with receptors called *pattern recognition receptors*. These receptors are activated by *pathogen-associated molecular patterns* (PAMPs) present in microbial molecules or by *damage-associated molecular patterns* (DAMPs) exposed on the surface of, or released by, damaged cells.

In contrast to innate immunity, specific immunity allows a targeted response against a specific pathogen. The adaptive immune system is organized around two classes of specialized *lymphocytes*, more specifically B lymphocytes (or B-cells) and T lymphocyte (or T-cells). B-cells can mature and differentiate into plasma cells capable of secreting special Y-shaped proteins called *antibodies*. Unlike B-cells, which attack targets indirectly by secreting antibodies, T-cells directly attack the invading organism; however, they are not able to recognize antigens without the help of other cells. These cells, known as APCs, process and display the antigen to which the T-cell is specific. In each case, the B- or T-cell is specific to a particular antigen. The specificity of binding resides in the antigen receptors on B-cells (the B-cell receptor) and T-cells (the T-cell receptor). See Fig. 9.1 for an illustration of the recognition of an antigen by B-cell and T-cell receptors.

Principles drawn from immune systems, including *pattern recognition*, *hypermutation*, *clonal selection*, the *danger theory*, the theory of *immune networks*, and many others, have inspired many researchers in the design of tools to solve complex tasks. The existing algorithms can be divided into four categories:

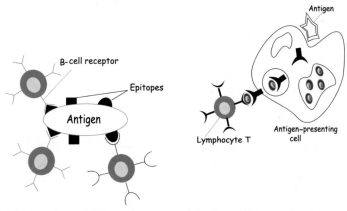

(a) An antigen with its multiple epitopes recognized by different B-cells.

(b) Presentation of antigens by specialized accessory cells, termed antigen-presenting cells (APC) to T-cells.

Fig. 9.1 Antigen recognition by B-cell and T-cell receptors. An epitope is the part of an antigen that is recognized by the immune system. There are recognized by specific T-cells, B-cells, and the antibody produced by B-cells

1. Negative selection-based algorithms [24].
2. Artificial immune networks [42].
3. Clonal selection-based algorithms [10].
4. Danger theory-inspired algorithms [2] and dendritic cell algorithms [32].

Readers who wish to obtain more detailed information about AIS algorithms are recommended to examine [94]. There are several reviews of AIS research [37, 38, 93, 107], and a number of books including [8, 17], covering the field. The most recent and most comprehensive survey of AIS is possibly that of Dasgupta et al. [18].

9.2.1 Negative-Selection-Based Algorithms

The key feature of a healthy immune system is its remarkable ability to distinguish between the body's own cells, recognized as "self," and foreign cells, or "nonself." Negative selection is the main mechanism in the thymus that eliminates self-reactive cells, i.e., T-cells whose receptors recognize and bind with self antigens presented in the thymus. Thus, only T-cells that do not bind to self-proteins are allowed to leave the thymus. These mature T-cells then circulate throughout the body, performing immunological functions and protecting the body against foreign antigens.

The negative selection algorithm is based on the principles of self–nonself discrimination in the immune system and was initially introduced by Forrest et al. in 1994 [24] to detect data manipulation caused by a virus in a computer system. The starting point of this algorithm is the production of a set of self strings, (S), that define the normal state of the system. The task is then to generate a set of detectors, (D), that only bind/recognize the complement of (S). These detectors can then be applied to new data in order to classify them as being self or nonself. This negative selection algorithm is summarized in Algorithm 9.1.

---

input : S_{seen} = set of seen known self elements

output: D = set of generated detectors

repeat

    Randomly generate potential detectors and place them in a set (P)

    Determine the affinity of each member of (P) with each member of the self set (S_{seen})

    if At least one element in (S) recognizes a detector in (P) according to a recognition threshold **then**
    | The detector is rejected

    else
    | The detector is added to the set of available detectors (D)

    end

until stopping criteria have been met

Algorithm 9.1: Generic negative selection algorithm.

A diverse family of negative selection algorithms has been developed and has been used extensively in anomaly detection. A survey of negative selection algorithms was published in [43]. Some other researchers have proposed negative selection algorithms, which can be found in [18, 94].

9.2.2   Clonal Selection-Based Algorithms

The clonal selection theory postulates that a vast repertoire of different B-cells, each encoding antibodies with a predetermined shape and specificity, is generated prior to any exposure to an antigen. Exposure to an antigen then results in the proliferation, or *clonal expansion*, of only those B-cells with antibody receptors capable of reacting with part of the antigen. However, any clone of activated B-cells with antigen receptors specific to molecules of the organism's own body (self-reactive receptors) is eliminated. Here, the affinity maturation of the B-cells takes place. During proliferation, a *hypermutation* mechanism becomes activated, which diversifies the repertoire of the B-cells. The antigen ensures that only those B-cells with high-affinity receptors are selected to differentiate into plasma cells and memory cells. Memory B-cells are developed to generate a more effective immune response to antigens that have previously been encountered. Figure 9.2 illustrates the principle of clonal selection.

Many algorithms have been inspired by the adaptive immune mechanisms of B-cells [18]. The general algorithm, named CLONALG [10], is based on the principles of clonal selection and affinity maturation. One generation of cells in this algorithm includes the initiation of candidate solutions, selection, cloning, mutation, reselection, and population replacement; these processes are somewhat similar to what happens in evolutionary algorithms. When applied to pattern matching, CLONALG produces a set of memory antibodies, (M), that match the members in a set (S) of patterns considered to be antigens. Algorithm 9.2 outlines the working of CLONALG.

Fig. 9.2 Principles of clonal selection and affinity maturation

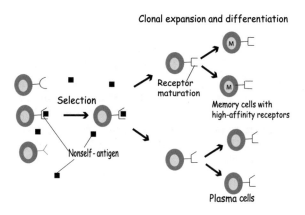

Many other clonal selection-based algorithms have been introduced in the literature and have been applied to a wide range of optimization and clustering problems [38]. A summary of the basic features of these algorithms, their areas of application, and their hybridization was published in [97].

---

input : (S) = set of patterns to be recognized, n number of worst elements to be selected for removal

output: (M) = set of memory detectors capable of classifying unseen patterns

Create an initial random set of antibodies (A)

forall the patterns in (S) **do**

    Determine the affinity with each antibody in (A)

    Generate clones of a subset of the antibodies in (A) with the highest affinity. The number of clones of an antibody is proportional to its affinity

    Mutate attributes of these clones in a manner inversely proportional to its affinity

    Add these clones to the set (A), and place a copy of the highest-affinity antibodies in (A) into the memory set (M)

    Replace the n lowest-affinity antibodies in (A) with new randomly generated antibodies

end

Algorithm 9.2: Generic clonal selection algorithm.

9.2.3 Artificial Immune Networks

The immune network theory (Fig. 9.3), as originally proposed by Jerne [42], states that the immune system is a network in which antibodies, B-cells, and T-cells recognize not only things that are foreign to the body but also each other, creating a structurally and functionally plastic network of cells that adapts dynamically to stimuli over time. It is thus the interactions between cells that give rise to the emergence of complex phenomena such as memory and other functionalities such as tolerance and reactivity [36].

The paper by Farmer et al. [21] is considered to be a pioneering work and has inspired a variety of immune network algorithms. One algorithm that has received much attention is *aiNet*, first developed by de Castro and Von Zuben for the task of data clustering [19] and then specialized into a series of algorithms for optimization and data mining in a variety of domains over the following years [9, 12]. aiNet is a simple extension of CLONALG but exploits interactions between B-cells according to the immune network theory. The aiNet algorithm is illustrated in Algorithm 9.3. A review of several different artificial immune network models is presented in the paper by Galeano et al. [25]. Some other existing immune network models can be found in [18].

Fig. 9.3 The immune
network theory. The
recognition of an antigen by
an antibody or cell receptor
leads to network activation,
whereas the recognition of
an idiotope by the antibody
results in network
suppression. The antibody
Ab_2 is considered as the
internal illustration of the
external antigen Ag, since
Ab_1 is able of recognizing
the antigen and Ab_2 also

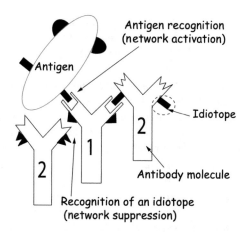

---

inputs : (S) = set of patterns to be recognized, nt = network affinity threshold, ct = clonal
pool threshold, h = number of highest-affinity clones, a = number of new antibodies
to be introduced

output: (N) = set of memory detectors capable of classifying unseen patterns

Create an initial random set of network antibodies (N)

repeat

    forall the patterns in (S) **do**

        Determine the affinity with each antibody in (N)

        Generate clones of a subset of the antibodies in (N) with the highest affinity. The
number of clones of an antibody is proportional to its affinity

        Mutate attributes of these clones in a manner inversely proportional to its affinity, and
place the number h of highest-affinity clones into a clonal memory set (C)

        Eliminate all members of (C) whose affinity with the antigen is less than a predefined
threshold ct

        Determine the affinity amongst all the antibodies in (C) and eliminate those
antibodies whose affinity with each other is less than the prespecified threshold ct

        Incorporate the remaining clones in (C) into (N)

    end

    Determine the affinity between each pair of antibodies in (N) and eliminate all antibodies
whose affinity is less than a prespecified threshold nt

    Introduce a number a of new randomly generated antibodies into (N)

until stopping condition has been met

Algorithm 9.3: Generic immune network algorithm.

9.2.4 Danger-Theory-Inspired Algorithms

The danger theory was first proposed by Polly Matzinger in 1994 [56]. This theory
attempts to explain the nature and workings of an immune response in a way different

from the widely held self–nonself viewpoint. According to the self–nonself theory, an organism does not trigger an immune response against its own constituents (self), whereas it triggers an immune response against all foreign (nonself) elements. The danger theory does not deny the existence of self–nonself discrimination, but rather states that self constituents can trigger an immune response if they are dangerous (e.g., cellular stress and some autografts), and nonself constituents can be tolerated if they are not dangerous (e.g., a fetus or commensal bacteria).

The danger model suggests that the immune response is due to the emission, within the organism, of *danger signals*. Damaged or dying cells release DAMPs, which serve as *endogenous danger signals* that alert the innate immune system to stress, unscheduled cell death, and microbial invasion. In contrast, PAMPs provide *exogenous signals* that alert the immune system to the presence of pathogenic bacteria or viruses.

Danger-theory-inspired algorithms are still in their infancy. The first paper that proposed an application of the danger theory was published in 2002 by Aickelin and Cayzer [2]. In 2003, Aickelin et al. proposed the Danger Project [1], an interdisciplinary project which aims at understanding from an immunological perspective the mechanisms of intrusion detection in the human immune system and applying these findings to AIS with a view to improving applications to computer security (see, for example, [33, 45]). Secker et al. [83] explored the relevance of the danger theory to the application domain of web mining. In [106], a novel immune algorithm inspired by danger theory was proposed for solving two-class classification problems online.

9.2.5 Dendritic Cell Algorithms

Dendritic cells (DCs) are immune cells that form part of the mammalian immune system. Their main function is to process antigen material and present it on their surface to other cells of the immune system, thus functioning as antigen-presenting cells and regulators of the adaptive immune system through the production of immunoregulatory cytokines (immune messenger proteins). DCs are responsible for some of the initial pathogenic recognition processes, by sampling the environment and differentiating depending on the concentration of signals or the perceived misbehavior in host tissue cells.

The maturation of immature DCs is regulated in response to various safety and danger signals. DCs can combine these signals with bacterial signatures (or PAMPs) to generate different output concentrations of costimulatory molecules, semimature cytokines, and mature cytokines.

The dendritic cell algorithm (DCA) is based on an abstraction of the functionality of biological DCs. It was first conceptualized and developed by Greensmith et al. [32] (see Algorithm 9.4), who introduced the notion of danger signals, safety signals, and PAMPs, which all contribute to the context of a data signal at any given time.

As stated in [31], most of the studies in which the DCA has been applied have been related to computer security, but there are also applications to wireless sensor networks, robotics, and scheduling of processes.

---

inputs : S = set of data items to be labelled safe or dangerous

output: L = set of data items labelled safe or dangerous

Create an initial population of dendritic cells (DCs), D

Create a set to contain migrated DCs, M

forall the data items in S **do**
> Create a set of DCs randomly sampled from D, P
>
> **forall the** DCs in P **do**
> > Add data items to collected list of DCs
> >
> > Update danger, PAMP, and safe signal concentrations
> >
> > Update concentrations of output cytokines
> >
> > Migrate dendritic cell from D to M and create a new DC in D if concentration of costimulatory molecules is above a threshold
> >
> **end**
>
end

forall the DCs in M **do**
> Set DC to be semimature if output concentration of semimature cytokines is greater than mature cytokines, otherwise set as mature
>
end

forall the data items in S **do**
> Calculate number of times data item is presented by a mature DC and a semimature DC
>
> Label data item as safe if presented by more semimature DCs than mature DCs, otherwise label it as dangerous
>
> Add data item to labelled set M
>
end

Algorithm 9.4: Generic dendritic cell algorithm.

Over the last few years, important investigations have focused on the proposal of theoretical frameworks for the design of AIS [8]; theoretical investigations into existing AIS can be found in [26, 95]. Other newly developed models have recently been reported in the literature, for example, the humoral immune response and pattern recognition receptor models. The interested reader is referred to [18] for a detailed discussion of these models.

9.3 Differential Evolution

The differential evolution (DE) algorithm is one of the most popular algorithms for continuous global optimization problems. It was proposed by Storn and Price in the 1990s [89] in order to solve the Chebyshev polynomial fitting problem and has proven to be a very reliable optimization strategy for many different tasks.

As in any evolutionary algorithm, a population of candidate solutions for the optimization task to be solved is arbitrarily initialized. DE uses N D-dimensional real-valued vectors $\mathbf{X}_{i,g} = X_{i,1,g}, X_{i,2,g}, \ldots, X_{i,D,g}$, where g denotes the current generation and N the number of individuals in the population. In each generation of the evolution process, new individuals are created by applying reproduction operators (crossover and mutation). The fitness of the resulting solutions is evaluated, and each individual $\mathbf{X}_{i,g}$ (*the target individual*) of the population competes against a new individual $\mathbf{U}_{i,g}$ (the trial individual) to determine which one will be maintained into the next generation $(g + 1)$. The trial individual is created by recombining the target individual with another individual $\mathbf{V}_{i,g}$ created by mutation (called the *mutant individual*). Different variants of DE have been suggested by Price et al. [70] and are conventionally named DE/*x*/*y*/*z*, where DE stands for differential evolution; x represents a string that denotes the base vector, i.e., the vector being perturbed, which may be *rand* (a randomly selected population vector) or *best* (the best vector in the population with respect to fitness value); y is the number of difference vectors considered for perturbation of the base vector x; and z denotes the crossover scheme, which may be *binomial* or *exponential*. A description of the DE algorithm is outlined in Algorithm 9.5.

---

input : N = population size, f = objective function, F = constant of differentiation, CR = crossover control parameter.

output: $\mathbf{X}_{\text{opt}}$, which minimizes f

Initialization: Initialize the whole vector population randomly

Set the generation counter/$g = 0$

Evaluate the fitness of each vector in the population

repeat

   for $i = 1$ to N **do**

      Mutation: Compute a mutant vector $\mathbf{V}_{i,g}$. A target vector $\mathbf{X}_{i,g}$ is mutated using a difference vector (obtained as a weighted difference between the selected individuals)

      $\mathbf{X}_{i,g} \Rightarrow \mathbf{V}_{i,g} = V_{i,1,g}, V_{i,2,g}, \ldots, V_{i,D,g}$

      Crossover: Create a trial vector $\mathbf{U}_{i,g}$ by the crossover of $\mathbf{V}_{i,g}$ and $\mathbf{X}_{i,g}$

   end

   for $i = 1$ to N **do**

      Evaluate the trial vector $\mathbf{U}_{i,g}$

      Selection: Replace the population vector $\mathbf{X}_{i,g}$ by its corresponding trial vector $\mathbf{U}_{i,g}$ if the fitness of the trial vector is better than that of its population vector:

      if $f(\mathbf{U}_{i,g}) < f(\mathbf{X}_{i,g})$ **then**

         | $\mathbf{X}_{i,g+1} \leftarrow \mathbf{U}_{i,g}$

      end

   end

   $g = g + 1$

until the stopping criterion is satisfied

return the best found solution $\mathbf{X}_{\text{opt}}$

Algorithm 9.5: Differential evolution (DE).

9.3.1   Mutation Schemes

For each target individual $\mathbf{X}_{i,g}$ in the current generation, its associated mutant individual $\mathbf{V}_{i,g}$ is obtained through the differential mutation operation. The mutation strategies most often used in the DE algorithm are listed below [70]:

DE/rand/1. This mutation scheme involves three distinct randomly selected individuals in the population. Only one weighted difference vector is used to perturb a randomly selected vector. The scaling factor F controls the amplification of the differential evolution:

$$\mathbf{V}_{i,g} = \mathbf{X}_{r_1,g} + F(\mathbf{X}_{r_2,g} - \mathbf{X}_{r_3,g}) \tag{9.1}$$

DE/rand/2. In this mutation scheme, to create $\mathbf{V}_{i,g}$ for each ith member $\mathbf{X}_{i,g}$, a total of five other distinct vectors (say the $r_1, r_2, r_3, r_4,$ and r_5th vectors) are chosen in a random way from the current population:

$$\mathbf{V}_{i,g} = \mathbf{X}_{r_1,g} + F(\mathbf{X}_{r_2,g} - \mathbf{X}_{r_3,g}) + F(\mathbf{X}_{r_4,g} - \mathbf{X}_{r_5,g}) \tag{9.2}$$

DE/best/1. Here the vector to be perturbed is the best vector $\mathbf{X}_{\text{best},g}$ of the current population, and the perturbation is done by using a single difference vector:

$$\mathbf{V}_{i,g} = \mathbf{X}_{\text{best},g} + F(\mathbf{X}_{r_1,g} - \mathbf{X}_{r_2,g}) \tag{9.3}$$

DE/best/2. In this mutation scheme, the mutant vector is formed by using two difference vectors, chosen at random, as shown below:

$$\mathbf{V}_{i,g} = \mathbf{X}_{\text{best},g} + F(\mathbf{X}_{r_1,g} - \mathbf{X}_{r_2,g}) + F(\mathbf{X}_{r_3,g} - \mathbf{X}_{r_4,g}) \tag{9.4}$$

DE/current-to-best/1. The mutant vector is created using any two randomly selected members of the population as well as the best vector in the current generation:

$$\mathbf{V}_{i,g} = \mathbf{X}_{i,g} + F(\mathbf{X}_{\text{best},g} - \mathbf{X}_{i,g}) + F(\mathbf{X}_{r_1,g} - \mathbf{X}_{r_2,g}) \tag{9.5}$$

DE/rand-to best/2. The mutant vector is created using the best solution in the population and a total of five randomly selected members of the population:

$$\mathbf{V}_{i,g} = \mathbf{X}_{r_1,g} + F(\mathbf{X}_{\text{best},g} - \mathbf{X}_{i,g}) + F(\mathbf{X}_{r_2,g} - \mathbf{X}_{r_3,g}) + F(\mathbf{X}_{r_4,g} - \mathbf{X}_{r_5,g}) \tag{9.6}$$

DE/current-to-rand/1. The mutant vector is determined by the following formula:

$$\mathbf{V}_{i,g} = \mathbf{X}_{i,g} + K(\mathbf{X}_{r_1,g} - \mathbf{X}_{i,g}) + F'(\mathbf{X}_{r_2,g} - \mathbf{X}_{r_3,g}) \tag{9.7}$$

where K is the combination coefficient, chosen with a uniform random distribution from $[0, 1]$, and $F' = KF$. For this special mutation, the mutated solution does not undergo a crossover operation.

DE/rand/1/either-or. This mutation scheme is formulated as

$$
\mathbf{V}_{i,g} = \begin{cases} \mathbf{X}_{r_1,g} + F(\mathbf{X}_{r_2,g} - \mathbf{X}_{r_3,g}) & \text{if } U(0, 1) < P_F \\ \mathbf{X}_{r_3,g} + K(\mathbf{X}_{r_1,g} + \mathbf{X}_{r_2,g} - 2\mathbf{X}_{r_3,g}) & \text{otherwise} \end{cases} \tag{9.8}
$$

For a given value of F, $K = 0.5(F + 1)$ [70]. As in DE/current-to-rand/1, when this mutation scheme is applied, it is not followed by a crossover.

The indices r_1, r_2, r_3, r_4, r_5 are randomly chosen from $[1, N]$ and should all be different from the running index i; $F \in [0, 1]$ is a real constant scaling factor in the range $[0, 2]$, usually less than 1, which controls the amplification of the difference between two individuals so as to avoid stagnation of the search process; and $\mathbf{X}_{\text{best},g}$ is the vector with the best fitness value in the population in generation g.

Figure 9.4 illustrates the distribution of mutant vectors in the search space. The mutation schemes presented above may be classified according to the location of the vectors generated as follows [100]:

- Schemes where the vector which has the best performance ($\mathbf{X}_{\text{best},g}$) is used as a base vector, such as *DE/best/1* and *DE/best/2*. These schemes tend to generate descendants around the best individuals.
- Schemes using a random vector as a base vector, such as DE/rand/1, DE/rand/2, and DE/rand-to-best/2. The mutant vectors can potentially be generated anywhere in the vicinity of the population.
- Schemes using the current solution as a base vector, such as DE/current-to-rand/1 and DE/current-to-best/1, which can be considered as an intermediate between the two categories above, since the mutant vectors are generated in the vicinity of the current solution.
- Schemes involving the best solution without using it as a base vector. These schemes consider the direction of the best individual without restricting the area explored to its immediate vicinity.

9.3.2 Crossover

Based on the mutant vector, a trial vector $\mathbf{U}_{i,g}$ is constructed through a crossover operation which combines components from the ith population vector $\mathbf{X}_{i,g}$ and its corresponding mutant vector $\mathbf{V}_{i,g}$. Two types of crossover operators are widely used in DE, *binomial* (or *uniform*) and *exponential* ones. Comparative studies of the role of crossover in differential evolution have been presented in [49, 105].

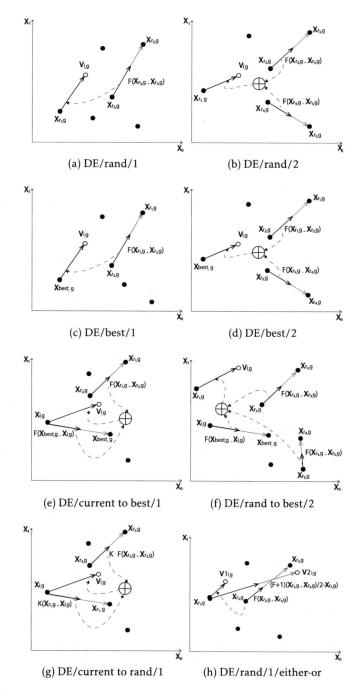

(a) DE/rand/1

(b) DE/rand/2

(c) DE/best/1

(d) DE/best/2

(e) DE/current to best/1

(f) DE/rand to best/2

(g) DE/current to rand/1

(h) DE/rand/1/either-or

Fig. 9.4 Mutation schemes [100]

In the basic version, DE employs a binomial crossover defined using the following rule:

$$U_{i,j,g} = \begin{cases} V_{i,j,g} & \text{if}(rand(0, 1) \le CR) \text{ or } (j = j_{\text{rand}}) \\ X_{i,j,g} & \text{otherwise} \end{cases} \quad (9.9)$$

The crossover factor CR is taken randomly from the interval $[0, 1]$ and represents the probability of creating parameters for trial vectors from a mutant vector; the index j_{rand} is a randomly chosen integer in the range $[1, N]$, and is responsible for ensuring that the trial vector $U_{i,g}$, containing at least one parameter from the mutant vector $V_{i,g}$, does not duplicate $X_{i,g}$; $rand(0, 1)$ is a uniform random number in the range $[0, 1]$; and $j = 1, 2, \ldots, D$.

The exponential crossover is a two-point crossover where the first cut point is selected randomly from $1, \ldots, D$ and copied from the mutant vector to the corresponding trial parameter, so that the trial vector will be different from the target vector $X_{i,g}$ with which it will be compared. The second cut point is determined such that L consecutive components (counted in a circular manner) are taken from the mutant vector $V_{i,g}$. The value of L is determined randomly by comparing CR with a uniformly distributed number j_{rand} between 0 and 1 that is generated anew for each parameter. As long as $j_{\text{rand}} \le CR$, parameters continue to be taken from the mutant vector, but the first time that $j_{\text{rand}} > CR$, the current *and all remaining* parameters are taken from the target vector [70].

In both the binomial crossover and the exponential crossover, the crossover rate parameter CR determines the distance between the generated trial vector $U_{i,g}$ and the reference vector $X_{i,g}$. Small values of CR, close to zero, result in very small exploratory moves, aligned with a small number of axes of the search space, while large values of CR enable a wider range of exploration of the search space [58].

The main advantage of differential evolution consists in its small number of control parameters. It has only three input parameters controlling the search process, namely the population size N; the constant of differentiation F, which controls the amplification of the differential variation; and the crossover control parameter CR. In the original version of DE, the control parameters were kept fixed during the optimization process. It is not obvious how to define a priori which parameter settings should be used, as this task is problem-specific. Therefore, some researchers (see, for example [7, 51, 92]) have developed various strategies to make the setting of the parameters self-adaptive according to a learning experience.

DE is currently one of the most popular heuristics for solving single-objective optimization problems in continuous search spaces. Owing to this success, its use has been extended to other types of problems, such as multiobjective optimization [57]. However, DE has certain flaws, such as slow convergence and stagnation of the population. Several modified versions of DE are available in the literature for improving the performance of the basic DE. One class of such algorithms includes hybridized versions, where DE is combined with some other algorithm to produce a new algorithm. For a more detailed description of many of the existing variants and major application areas of DE, readers should refer to [11, 16, 60].

9.4 Bacterial Foraging Optimization Algorithm

The bacterial foraging optimization algorithm (BFOA), introduced by Passino in 2002 [64], is a relatively new paradigm for solving optimization problems. It is inspired by the social foraging behavior of the *Escherichia coli* (*E. coli*) bacteria present in human intestines.

For many organisms, the survival-critical activity of foraging involves aggregations of organisms into groups and trying to find and consume nutrients in a manner that maximizes the energy obtained from nutrient sources per unit time spent foraging, while at the same time minimizing the exposure to risks from predators [65]. A particularly interesting group of types of social foraging behavior has been demonstrated for several motile species of bacteria, including *E. coli*. During foraging, individual bacteria move by taking small steps to acquire nutrients and avoid danger. Their motion is determined by the rotation mode of the flagellar filaments that help these bacteria to move, so that they move in alternating periods of *runs* (relatively long intervals during which the bacteria swim smoothly in a straight line) and *tumbles* (short intervals during which the bacteria change direction to start another smooth run). This alternation between the two modes is called *chemotaxis*. Figure 9.5 illustrates the principle.

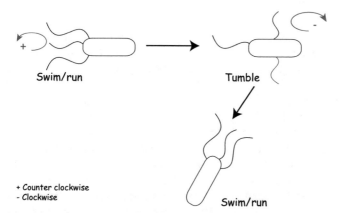

Fig. 9.5 Chemotaxis of *E. coli*. When the flagella rotate counterclockwise, they cause a swimming motion, and when they rotate clockwise, they cause tumbling

Bacteria may respond directly to local physical cues, such as the concentration of nutrients or the distribution of some chemicals (which may be laid down by other individuals). In the absence of a stimulus (i.e., no attractant or repellent is present, or else there is a constant, uniform concentration, so that they move no gradient), an *E. coli* bacterium swims in a random walk by alternating runs and tumbles. In the presence of a concentration gradient of an attractant (food sources such as sugars or amino acids), the bacteria are able to bias their random walk by changing their tumbling frequency. When moving toward an increasing attractant concentration or decreasing repellent concentration, the bacteria tumble less frequently, thereby increasing the lengths of their runs in the direction of increasing attractant concentration. As the concentration of the attractant decreases, the tendency to tumble is enhanced.

To facilitate the migration of bacteria on viscous substrates, such as semisolid agar surfaces, *E. coli* cells arrange themselves in a traveling ring and move over the surface in a coordinated manner called *swarming* motility. This is in contrast to swimming motility, which represents the motility of individual cells in an aqueous environment [6]. After a bacterium has collected a sufficient amount of nutrients, it can reproduce itself and divide into two. The population of bacteria can also suffer a process of *elimination*, through the appearance of a noxious substance, or disperse, through the action of another substance, generating the effects of elimination and dispersion.

Based on these biological concepts, the BFOA is formulated on the basis of the following steps: chemotaxis, swarming, reproduction, and elimination–dispersal. The general procedure of the BFO algorithm is outlined in Algorithm 8.6.

9.4.1 Chemotaxis

Chemotaxis is the process by which bacteria direct their movements according to certain chemicals in their environment. This is important for allowing bacteria to find food by climbing up nutrient hills and at the same time avoid noxious substances. The sensors they use are receptor proteins, which are very sensitive and possess high gain. That is, a small change in the concentration of nutrients can cause a significant change in behavior [52].

Suppose that we want to find the minimum of $J(\theta)$, where $\theta \in \mathbb{R}^D$ is the position of a bacterium in a D-dimensional space and the cost function $J(\theta)$ is an attractant–repellent profile (i.e., it represents where nutrients and noxious substances are located). Then $J(\theta) \leq 0$ represents a nutrient-rich environment, $J(\theta) = 0$ represents a neutral medium, and $J(\theta) > 0$ represents the presence of noxious substances.

Let $\theta^i(j, k, l)$ represent the ith bacterium in the jth chemotactic, kth reproductive, and lth elimination–dispersal step. The position of the bacterium in the $(j+1)$th chemotactic step is calculated in terms of the position in the previous chemotactic step and the step size $C(i)$ (termed the run length unit) applied in a random direction $\phi(i)$:

$$\theta^i(j+1,k,l) = \theta^i(j,k,l) + C(i)\phi(i) \tag{9.10}$$

The function $\phi(i)$ is a unit length random direction to describe tumbling and is given by

$$\phi(i) = \frac{\Delta(i)}{\sqrt{\Delta^T(i)\Delta(i)}} \tag{9.11}$$

where $\Delta(i) \in \mathbb{R}^D$ is a randomly generated vector with elements in the interval $[-1, 1]$. The cost of each position is determined by the following equation:

$$J(i,j,k,l) = J(i,j,k,l) + J_{cc}\left(\theta, \theta^i(j,k,l)\right) \tag{9.12}$$

It can be noticed in Eq. (9.12) that the cost of any particular position $J(i,j,k,l)$ is also affected by attractive and repulsive forces between the bacteria in the population, given by J_{cc} (see Eq. (9.13)). If the cost of the location of the ith bacterium in the $(j+1)$th chemotactic step, denoted by $J(i, j+1, k, l)$, is better (lower) than that for the position $\theta^i(j,k,l)$ in the jth step, then the bacterium will take another chemotactic step of size $C(i)$ in the same direction, up to a maximum number of permissible steps, denoted by N_s.

9.4.2 Swarming

Swarming is a particular type of motility that is promoted by flagella and allows bacteria to move rapidly over and between surfaces and in viscous environments. Under certain conditions, cells of chemotactic strains of *E. coli* excrete an attractant, aggregate in response to gradients of that attractant, and form patterns of varying cell density. Central to this self-organization into swarm rings is chemotaxis. The cell-to-cell signaling in an *E. coli* swarm may be represented by the following function:

$$
\begin{aligned}
J_{cc}(\theta, \theta^i(j,k,l)) = \sum_{i=1}^{s}\left[-d_{\text{attractant}}\exp\left(-w_{\text{attractant}}\sum_{m=1}^{D}\left(\theta_m - \theta_m^i\right)^2\right)\right] \\
+ \sum_{i=1}^{s}\left[h_{\text{repellent}}\exp\left(-w_{\text{repellent}}\sum_{m=1}^{D}\left(\theta_m - \theta_m^i\right)^2\right)\right]
\end{aligned} \tag{9.13}
$$

where $\theta = [\theta_1, \theta_2, \ldots, \theta_D]^T$ is a point in the D-dimensional search space, $J_{cc}(\theta, \theta^i(j, k, l))$ is to be added to the actual objective function, and $d_{attractant}$, $w_{attractant}$, $h_{repellent}$, and $w_{repellent}$ are coefficients which determine the depth and width of the attractant and the height and width of the repellent. These four parameters need to be chosen judiciously for any given problem. θ_m^i is the mth dimension of the position of the ith bacterium θ^i in the population of the S bacteria.

9.4.3 Reproduction

After N_c chemotaxis steps (steps comprising the movement of each bacterium and determination of the cost of each position), the bacteria enter into the reproductive step. Suppose there are N_{re} reproduction steps. For reproduction, the population is sorted in order of ascending accumulated cost J_{health} (higher cost means lower health):

$$J_{health}(i) = \sum_{j=1}^{Nc+1} J(i, j, k, l) \tag{9.14}$$

The least healthy bacteria will die; these are the bacteria that could not gather enough nutrients during the chemotactic steps, and they are replaced by the same number of healthy ones, and thus the population size remains constant. The healthiest bacteria (those having sufficient nutrients and yielding lower values of the fitness function) split asexually into two bacteria and are placed in the same location.

9.4.4 Elimination and Dispersal

Changes in the environment can influence the proliferation and distribution of bacteria. So, when a local environmental change occurs, either gradually (e.g., via consumption of nutrients) or suddenly for some other reason (e.g., a significant local rise in temperature), all the bacteria in a region may die or disperse into some new part of the environment. This dispersal has the effect of destroying all the previous chemotactic processes. However, it may have a good impact too, since dispersal may place bacteria in a nutrient-rich region.

Let N_{ed} be the number of elimination–dispersal events and, for each elimination–dispersal event, let each bacterium in the population be subjected to elimination–dispersal with a probability P_{ed}, in such a way that, at the end of the process, the number of bacteria in the population remains constant (if a bacterium is eliminated, another one is dispersed to a random location).

Initialize parameters: $D, S, N_c, N_s, N_{re}, N_{ed}, P_{ed}, C(i), \theta^i$ $(i = 1, 2, \ldots, S)$

while terminating condition is not reached **do**

    `Elimination-dispersal loop`

    for $l = 1, \ldots, N_{ed}$ **do**

        `Reproduction loop`

        for $k = 1, \ldots, N_{re}$ **do**

            `Chemotaxis loop`

            for $j = 1, \ldots, N_c$ **do**

                foreach bacterium $i = 1, \ldots, S$ **do**

                    Compute fitness function J (i, j, k, l) using Eq. (9.12)

                    $J_{last} = J(i, j, k, l)$

                    Tumble: Generate a random vector $\Delta(i) \in \mathbb{R}^D$

                    Move: Compute the position of the bacterium θ^i $(j + 1, k, l)$ in $(j+1)$th chemotactic step using Eq. (9.10)

                    Compute fitness function J $(i, j + 1, k, l)$ using Eq. (9.12)

                    Swim: $m = 0$ `//m: counter for swim length`

                    while $m < N_s$ **do**

                        $m = m + 1$

                        if J $(i, j + 1, k, l) < J_{last}$ **then**

                            $J_{last} = J$ $(i, j + 1, k, l)$

                            Move: Compute the position of the bacterium θ^i $(j + 1, k, l)$ in $(j+1)$th chemotactic step using Eq. (9.10)

                            Compute fitness function J $(i, j + 1, k, l)$ using Eq. (9.12)

                      else

                        $m = N_s$

                      end

                    end

                end

            end

            for $i = 1, \ldots, S$ **do**

                Reproduction: $J_{health}(i) = \sum_{j=1}^{Nc+1} J(i, j, k, l)$

            end

            Sort bacteria in order of ascending J_{health}. The least healthy bacteria die and the other, healthier bacteria each split into two bacteria, which are placed in the same location

        end

        for $i = 1, \ldots, S$ **do**

            Elimination–dispersal: Eliminate and disperse the ith bacterium, with probability P_{ed}

        end

    end

end

Algorithm 9.6: BFO algorithm.

In [15], Das et al. discussed some variations on the original BFOA algorithm, and hybridizations of BFOA with other optimization techniques. They also provided an account of most of the significant applications of BFOA. However, experimentation with complex optimization problems reveals that the original BFOA algorithm possesses poor convergence behavior compared with other nature-inspired algorithms, such as genetic algorithms and particle swarm optimization, and its performance also decreases heavily with increasing dimensionality of the search space.

9.5   Biogeography-Based Optimization (BBO)

The various species of plants and animals are not uniformly distributed over the globe's surface: each one occupies a region or a habitat of its own. The current geographic distribution of organisms is influenced by internal factors specific to the organisms (such as their capacity for propagation and their ecological amplitude) and by external factors related to their environment (such as predation, disease, and competition for resources such as food and water). Biogeography is the study of this *spatial distribution* of species in geographic space, and its causes. The theory of biogeography grew out of the work of Wallace [99] and Darwin [14] in the past and that of McArthur and Wilson [55] more recently.

Strongly influenced by the equilibrium theory of island[2] biogeography [55], Dan Simon developed the biogeography-based optimization (BBO) algorithm [86]. The basic premise of this theory is that the rate of change of the number of species on an island depends critically on the balance between the immigration of new species onto the island and the extinction of established species. Figure 9.6 shows the basic idea of the equilibrium condition.

The BBO algorithm operates upon a population of individuals called *islands* (or *habitats*). Each habitat represents a possible solution for the problem at hand, and each feature of the habitat is called a *suitability index variable* (SIV). A quantitative performance index, called the *habitat suitability index* (HSI), is used as a measure of how good a solution is; this is analogous to the fitness in other population-based optimization algorithms. The greater the total number of species in the habitat, which corresponds to a high HSI, the better the solution it contains.

According to MacArthur and Wilson's theory of island biogeography, the number of species present on an island is determined by a balance between the rate at which new species arrive and the rate at which old species become extinct on the island. In BBO, each individual has its own immigration rate λ and emigration (or extinction)

---

[2]The term "island" is used descriptively rather than literally here. That is, an island is not just a segment of land surrounded by water, but any habitat that is geographically isolated from other habitats, including lakes and mountaintops. The theory of island biogeography has also been extended to peninsulas, bays, and other only partially isolated areas.

Fig. 9.6 The equilibrium model of the species present on a single island. The equilibrium number of species is reached at the point of intersection between the rate of immigration and the rate of extinction [55]

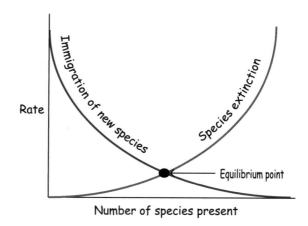

Number of species present

rate μ. These parameters are affected by the number of species S in a habitat and are used to share information probabilistically between habitats. Habitats with smaller populations are more vulnerable to extinction (i.e., the immigration rate is high). But as more species inhabit the habitat, the immigration rate reduces and the emigration rate increases. In BBO, good solutions (i.e., habitats with many species) tend to share their features with poor solutions (i.e., habitats with few species), and poor solutions accept a lot of new features from good solutions. The maximum immigration rate I occurs when the habitat is empty and decreases as more species are added; the maximum emigration rate E occurs when all possible species, with a number S_{max} are present on the island. The immigration and emigration rates when there are S species in the habitat vary linearly with the species number according to the following equation:

$$\lambda_S = I \left(1 - \frac{S}{S_{\max}}\right)$$
$$\mu_S = E \left(\frac{S}{S_{\max}}\right).$$

(9.15)

For the sake of simplicity, the original BBO considered a linear migration model (Fig. 9.7) where the immigration rate λ_S and the emigration rate μ_S are linear functions of the number of species S in the habitat, but different mathematical models of biogeography that include more complex variables are presented in [55]. There are in fact other important factors which influence migration rates between habitats, including the distance to the nearest neighboring habitat, the size of the habitat, climate (temperature and precipitation), plant and animal diversity, and human activity. These factors make immigration and emigration curves complicated, unlike those described in the original BBO paper [86]. To study the influence of different migration models on the performance of BBO, Haiping Ma [54] explored the behavior of

Fig. 9.7 Relationship between species count, immigration rate, and emigration rate

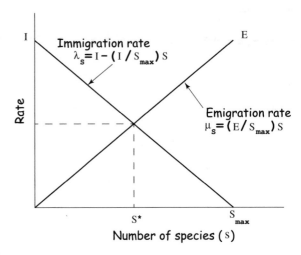

six different migration models and investigates the performance on 23 benchmark functions with a wide range of dimensions and diverse complexities. The experimental results clearly showed that different migration models resulted in significant changes in performance, and BBO migration models which were closer to nature (that is, nonlinear) were significantly better than linear models for most of the benchmarks.

We now consider the probability P_S that the habitat contains exactly S species. The number of species will change from time t to time $(t + \Delta t)$ as follows:

$$P_S(t + \Delta t) = P_S(t)(1 - \lambda_S \Delta t - \mu_S \Delta t) + P_{S-1}\lambda_{S-1}\Delta t + P_{S+1}\mu_{S+1}\Delta t \quad (9.16)$$

which states that the number of species in the habitat in one time step is based on the current total number of species in the habitat, the number of new immigrants, and the number of species that leave during that time period. We assume here that Δt is small enough that the probability of more than one immigration or emigration can be ignored. In order to have S species at time $(t + \Delta t)$, one of the following conditions must hold:

- There were S species at time t, and no immigration or emigration occurred between t and $(t + \Delta t)$.
- One species immigrated into a habitat already occupied by $(S - 1)$ species at time t.
- One species emigrated from a habitat occupied by $(S + 1)$ species at time t.

The limit of Eq. (9.16) as $\Delta t \to 0$ is given by the following equation:

$$\dot{P}_S = \begin{cases} -(\lambda_S + \mu_S)P_S + \mu_{S+1}P_{S+1} & \text{if } S = 0 \\ -(\lambda_S + \mu_S)P_S + \lambda_{S-1}P_{S-1} + \mu_{S+1}P_{S+1} & \text{if } 1 \le S \le S_{\max} - 1 \\ -(\lambda_S + \mu_S)P_S + \lambda_{S-1}P_{S-1} & \text{if } S = S_{\max} \end{cases} \quad (9.17)$$

Fig. 9.8 The migration
process in BBO

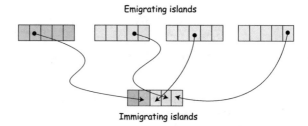

The system of Eq. (9.17) can be written as a matrix equation in the form

$$
\begin{bmatrix} \dot{P_0} \\ \dot{P_1} \\ \vdots \\ \vdots \\ \dot{P_n} \end{bmatrix} = \begin{bmatrix} -(\lambda_0 + \mu_0) & \mu_1 & 0 & \cdots & & 0 \\ \lambda_0 & -(\lambda_1 + \mu_1) & \mu_2 & \cdots & & \vdots \\ \vdots & & \ddots & \ddots & \ddots & \vdots \\ \vdots & & \ddots & \lambda_{n-2} & -(\lambda_{n-1} + \mu_{n-1}) & \mu_n \\ 0 & \cdots & 0 & & \lambda_{n-1} & -(\lambda_n + \mu_n) \end{bmatrix} \begin{bmatrix} P_0 \\ P_1 \\ \vdots \\ \vdots \\ P_n \end{bmatrix}
$$

(9.18)

For brevity of notation, we write simply $n = S_{\max}$.

The BBO algorithm is described overall in Algorithm 8.7. The two basic operators that govern the working of BBO are *migration* and *mutation*. Migration is used to modify existing islands by mixing features within the population, where $rand(0, 1)$ is a uniformly distributed random number in the interval [0, 1] and $X_{i,j}$ is the jth SIV of the solution $\mathbf{X}_i$. The migration strategy in BBO is similar to the global recombination approach in evolutionary strategies [4], in which many parents can contribute to a single offspring. The main difference is that recombination is used to in evolutionary strategies create new solutions, while in BBO migration is used to change existing solutions (Fig. 9.8).

A habitat's HSI can change suddenly owing to apparently random events (unusually large flotsam arriving from a neighboring habitat, disease, natural catastrophes, etc.). BBO models this phenomenon as *SIV mutation*, and uses species count probabilities to determine mutation rates. The species count probability P_S indicates the likelihood that a given solution S is expected a priori to exist as a solution for the given problem. In this context, it should be remarked that very high-HSI solutions and very low-HSI solutions are both equally improbable. Medium-HSI solutions are relatively probable. If a given solution has a low probability, then it is likely to mutate to some other solution. Conversely, a solution with a high probability is less likely to mutate. Mutation is used to enhance the diversity of the population, thereby preventing the search from stagnating. If a habitat S is selected for execution of the mutation operation, then a chosen variable (SIV) is randomly modified based on its associated probability P_S. The mutation rate $m(S)$ is inversely proportional to the probability P_S of the solution:

$$m(S) = m_{\max} \left(1 - \frac{P_S}{P_{\max}} \right) \tag{9.19}$$

where $m_{\max}$ is a user-defined parameter, and $P_{\max} = \max\limits_{S} P_S$, $S = 1, ..., S_{\max}$.

---

input : N = populationsize, f = objective function, I = maximum immigration rate, E = maximum emigration rate.

output: X_{opt} which minimizes f

Initialize a set of solutions (habitats) to a problem

while termination condition not met **do**

    Evaluate the fitness (HSI) for each solution

    Compute the number of species S, λ, and μ for each solution (see Eq. (9.15))

    Migration: modify habitats based on λ and μ

    for $i = 1$ to N **do**

        Use λ_i to decide probabilistically whether to immigrate to X_i

        if $rand(0, 1) < \lambda_i$ **then**

            for $j = 1$ to N **do**

                Select the emigrating island X_j with probability $\propto \mu_j$

                if $rand(0, 1) < \mu_j$ **then**

                    Replace a randomly selected decision variable (SIV) of X_i with its corresponding variable in X_j

                end

            end

        end

    end

    Mutation:

    Probabilistically perform mutation based on the mutation probability given in Eq. (9.19)

    Elitism:

    Implement elitism to retain the best solutions in the population from one generation to the next

end

return the best solution found X_{opt}

Algorithm 9.7: Biogeography-based optimization (BBO) algorithm.

The BBO algorithm has demonstrated good performance on various unconstrained and constrained benchmark functions. It has also been applied to real-world optimization problems, including sensor selection, economic load dispatch problems, satellite image classification, and power system optimization. The website http://embeddedlab.csuohio.edu/BBO/ is dedicated to BBO and related material.

9.6   Cultural Algorithms

The term *culture* was first introduced by the anthropologist Edward B. Taylor in his book *Primitive Culture* [96]. Taylor offered a broad definition, stating that culture is "that complex whole which includes knowledge, belief, art, morals, law, custom, and any other capabilities and habits acquired by man as a member of society." Cultural algorithms (CAs), introduced by Robert G. Reynolds, are a class of computational models derived from observing the cultural evolution process in nature [74].

The term *cultural evolution* has been used more recently to refer to the idea that the processes that produce cultural stability and change are analogous in important respects to those of biological evolution. In this view, just as biological evolution is characterized by changing frequencies of genes in populations over time as a result of such processes as natural selection, so cultural evolution refers to the changing distributions of cultural attributes in populations, which are likewise affected by processes such as natural selection but also by others that have no analogue in genetic evolution. Using this idea, Reynolds developed a computational model in which cultural evolution is seen as an inheritance process that operates at both a *microevolutionary* level, in terms of transmission of genetic material between individuals in a population, and a *macroevolutionary* level, in terms of knowledge acquired based upon individual experiences. A fundamental part of the macroevolutionary level is Renfrew's notion of an individual's mental *mappa*, a cognitive map or worldview, which is based on experience of the external world and shapes the individual's interactions with it [72]. Individual mappas can be merged and modified to form *group mappas* in order to direct the future actions of a group and its individuals.

CAs consist of three components:

1. A *population space*, at the microevolutionary level, that maintains a set of individuals to be evolved and the mechanisms for its evaluation, reproduction, and modification. In the population space, any evolutionary algorithms can be adopted, and evolutionary operators are defined with the aim of obtaining a set of possible solutions to the problem.
2. A *belief space*, at the macroevolutionary level, that represents the knowledge that has been acquired by the population during the evolutionary process. The main principle is to preserve beliefs that are socially accepted and to discard unacceptable beliefs. There are at least five basic categories of cultural knowledge that are important in the belief space of any cultural evolution model: situational, normative, topographic or spatial, historical or temporal, and domain knowledge [76].
3. A *communications protocol*, including an acceptance and influence phase, is used to determine the interaction between the population and the beliefs.

The basic framework of a CA is shown in Algorithm 8.8. In each generation, individuals in the population space are first evaluated using an evaluation or performance function (*Evaluate()*). An *acceptance* function (*Accept()*) is then used to determine which of the individuals in the current population will be able to contribute their knowledge to the belief space. Experiences of those selected individuals are then added to the contents of the belief space via the function *Update()*. The function *Generate()* includes the influence of the knowledge in the belief space, through the *Influence()* function, on the generation of offspring. The *Influence* function acts in such a way that the individuals resulting from the application of the variation operators (i.e., recombination and mutation) tend to approach a desirable behavior while staying away from undesirable behaviors. Such desirable and undesirable behaviors are defined in terms of the information stored in the belief space. The two functions *Accept()* and *Influence()* constitute the communication link between the population space and the belief space. This supports the idea of dual inheritance in that the population and the belief space are updated in each time step based upon feedback from each other. Finally, in the replacement phase, a selection function (*Select()*) is applied to the current and the new populations. The CA repeats this process for each generation until the prespecified termination condition is met.

Set the generation counter $g = 0$

Initialize the population (POP(g))

Initialize belief space (Beliefs(g))

repeat

> Evaluate population: *Evaluate* (POP(g))
>
> Update(Beliefs(g), *Accept*(POP(g)))
>
> Generate(POP(g), *Influence*(Beliefs(g)))
>
> $g = g + 1$
>
> *Select*(POP(g) from POP($g - 1$))

until a termination condition is achieved

Algorithm 9.8: Cultural algorithm.

As such, cultural algorithms are based on hybrid evolutionary systems that integrate evolutionary search and symbolic reasoning [90]. They are particularly useful for problems whose solutions require extensive domain knowledge (e.g., constrained optimization problems [13]) and dynamic environments [82]. The performance CAs has been studied using benchmark optimization problems [75], as well as applied successfully in a number of diverse application areas, such as modeling the evolution of agriculture [73], the job shop scheduling problem [79], reengineering of large-scale semantic networks [81], combinatorial optimization problems [62], multiobjective optimization problems [77], and agent-based modeling systems [78]. Recently, many optimization methods have been combined with CAs, such as evolutionary programming [13], particle swarm optimization [50], differential evolution algorithms [5], genetic algorithms [102], and local search [61]. Adaptations of CAs have also been proposed (see, for example, [34] for multipopulation CAs).

9.7 Coevolutionary Algorithms

When organisms that are ecologically intimate — for example, predators and prey, hosts and parasites, or insects and the flowers that they pollinate — influence each other's evolution, we say that *coevolution* is occurring. The biological coevolution encountered in many natural processes has been an inspiration for coevolutionary algorithms (CoEAs), where two or more populations of individuals, each adapting to changes in the other, constantly interact and coevolve simultaneously, in contrast to traditional single-population evolutionary algorithms.

Significant research into CoEAs began in the early 1990s with the seminal work of Hillis [41] on sorting networks. Unlike conventional evolutionary algorithms, in which individuals are evaluated independently of one another through an absolute fitness measure, the fitness of individuals in CoEAs is *subjective*, in the sense that it is a function of the interactions of the individual with other individuals.

Many variants of CoEAs have been implemented since the beginning of the 1990s. These variants fall into two categories: *competitive coevolution* and *cooperative coevolution*. In the case of the competitive approaches, the different populations compete into solve the global problem and individuals are rewarded at the expense of those with which they interact. In the case of the cooperative approaches, however, the various isolated populations are coevolved to solve the problem cooperatively; therefore, individuals are rewarded when they work well with other individuals and punished when they perform poorly together.

Competitive coevolution is usually used to simulate the behavior of competing forces in nature, such as that of predators and prey, where there is a strong evolutionary pressure for prey to defend themselves better as future generations of predators develop better attacking strategies. Competitive coevolution can lead to an *arms race*, in which the two populations have opposing interests and the success of one population depends on the failure of the other. The idea is that continued minor adaptations in some individuals will force competitive adaptations in others, and these reciprocal forces will drive the algorithm to generate individuals with ever-increasing performance. The fitness of individuals is evaluated through competition with other individuals in the population. In other words, fitness signifies only the relative strengths of solutions; an increased fitness for one solution leads to a decreased fitness for another. This inverse interaction of fitnesses will increase the capabilities of each population until the global optimal solution is attained [88]. Competitive coevolutionary models are especially suitable for problem domains where it is difficult to explicitly formulate an objective fitness function. The classic example of competitive coevolution, given in [41], coevolved a population of sorting networks against a population of test cases. Competitive coevolution has since been successfully applied to game-playing strategies [66, 80], evolving better pattern recognizers [47], coevolving complex behaviors of agents [87], coevolutionary interactions between neural networks and their training data [63], etc.

Cooperative coevolution is inspired by the ecological relationship of symbiosis, where different species live together in a mutually beneficial relationship. A general framework for cooperative coevolutionary algorithms was introduced by Potter and De Jong [69] in 1994 for evolving solutions in the form of coadapted subcomponents. Potter and De Jong's model is usually applied in situations where a complex problem can be decomposed into a collection of easier subproblems.[3] Each subproblem is assigned to a population such that the individuals in a given population represent potential components of a larger solution. Evolution of these populations occurs almost simultaneously, but in isolation form one another, the populations interact only when the fitness is obtained. Such a process can be static, in the sense that the divisions between the separate components are decided a priori and never altered, or dynamic, in the sense that populations of components may be added or removed as a run progresses [101]. This model has been analyzed from the perspective of evolutionary dynamics in [53, 101]. Cooperative CoEAs have had success in adversarial domains (see for example [68] and [91]). The influence of the design decisions on the performance of CoEA has been studied in [67]. Some variants of cooperative CoEAs have been proposed, such as coevolutionary particle swarms [39] and coevolutionary differential evolution [84]. A combination of competitive and cooperative mechanisms has been proposed by Goh and Tan [29] to solve multiobjective optimization problems in a dynamic environment.

Furthermore, both styles of coevolution (i.e., competitive and cooperative) can use multiple, reproductively isolated populations; both can use similar patterns of interpopulation interaction, similar diversity maintenance schemes, and so on. Aside from the novel problem decomposition scheme of cooperative coevolution, the most salient difference between cooperative and competitive coevolution resides primarily in the game-theoretic properties of the domains in which these algorithms are applied [23].

9.8   Conclusion

A wide range of metaheuristic algorithms have emerged over the last thirty years, and many new variants are continually being proposed. We have presented a description of a collection of optimization approaches in this chapter. Some of them are inspired by natural processes such as evolution and others by the behavior of biological systems. There are also other well-established optimization algorithms that we have not addressed in this chapter, including harmony search, the group search optimizer, cuckoo search, the gravitational search algorithm and the bat-inspired algorithm. Readers interested in these modern techniques can refer to more advanced literature.

---

[3]The decomposition of the problem consists in determining an appropriate number of subcomponents and the role each will play. The mechanism for dividing the optimization problem f into n subproblems and treating them almost independently of one another depends strongly on the properties of the function f.

9.9 Annotated Bibliography

Reference [8]   This book provides an introduction to artificial immune systems that is accessible to all. It gives a clear definition of an AIS, sets out the foundations (including the basic algorithms), and analyzes how the immune system relates to other biological systems and processes. No prior knowledge of immunology is required — all essential basic information is covered in the introductory chapters.

Reference [17]   This book provides an overview of AIS and their applications.

Reference [70]   This book deals with the differential evolution method. The authors claim that this book is designed to be easy to understand and simple to use, and they have in fact achieved their goal. The book is enjoyable to read, and is fully illustrated with figures and pseudocode. This book is primarily addressed to engineers. In addition, those interested in evolutionary algorithms should certainly find this book both interesting and useful.

Reference [85]   This book discusses the theory, history, mathematics, and programming of evolutionary optimization algorithms. Featured algorithms include differential evolution, biogeography-based optimization, cultural algorithms, and many others.

Reference [101]   This thesis offers a detailed analysis of cooperative coevolutionary algorithms.

References

1. Aickelin, U., Bentley, P., Cayzer, S., Kim, J., Mcleod, J.: Danger theory: The link between AIS and IDS? In: J. Timmis, P. Bentley, E. Hart (eds.) *Artificial Immune Systems*, Lecture Notes in Computer Science, pp. 147–155. Springer (2003)
2. Aickelin, U., Cayzer, S.: The danger theory and its application to artificial immune systems. In: *Proceedings of the 1st International Conference on Artificial Immune Systems*, pp. 141–148 (2002)
3. Anspach, M., Varela, F.: Le systme immunitaire : un soi cognitif autonome. In: D. Andler (ed.) *Introduction aux sciences cognitives*, p. 514. Gallimard, Paris (1992)
4. Bäck, T.: *Evolutionary Algorithms in Theory and Practice: Evolution Strategies, Evolutionary Programming, Genetic Algorithms*. Oxford University Press (1996)
5. Becerra, R.L., Coello, C.A.C.: A cultural algorithm with differential evolution to solve constrained optimization problems. In: *IBERAMIA*, pp. 881–890 (2004)
6. Brenner, M.P., Levitov, L.S., Budrene, E.O.: Physical mechanisms for chemotactic pattern formation by bacteria. Biophysical Journal **74**(4), 1677–1693 (1998)
7. Brest, J., Maucec, M.: Self-adaptive differential evolution algorithm using population size reduction and three strategies. Soft Computing: A Fusion of Foundations, Methodologies and Applications **15**(11), 2157–2174 (2011)
8. de Castro, L.N.: *Artificial Immune Systems: A New Computational Intelligence Approach*. Springer, London (2002)
9. de Castro, L.N., Von Zuben, F.J.: aiNet: An artificial immune network for data analysis. In: H.A. Abbass, R.A. Sarker, C.S. Newton (eds.) *Data Mining: A Heuristic Approach*, Chap. 12, pp. 231–259. Idea Group (2001)
10. de Castro, L.N., Von Zuben, F.J.: Learning and optimization using the clonal selection principle. IEEE Transactions on Evolutionary Computation **6**(3), 239–251 (2002)

11. Chakraborty, U.: *Advances in Differential Evolution*, 1st edn. Springer (2008)
12. Coelho, G.P., Zuben, F.V.: omni-aiNet: An immune-inspired approach for omni optimization. In: *Proceedings of the 5th International Conference on Artificial Immune Systems*, pp. 294–308. Springer (2006)
13. Coello Coello, C.A., Becerra, R.L.: Adding knowledge and efficient data structures to evolutionary programming: A cultural algorithm for constrained optimization. In: *GECCO*, pp. 201–209 (2002)
14. Darwin, C.: *Origin of Species*. Gramercy, New York (1995)
15. Das, S., Biswas, A., Dasgupta, S., Abraham, A.: Bacterial foraging optimization algorithm: Theoretical foundations, analysis, and applications. In: A. Abraham, A.E. Hassanien, P. Siarry, A. Engelbrecht (eds.) *Foundations of Computational Intelligence*. Studies in Computational Intelligence, vol. 3, pp. 23–55. Springer, Berlin, Heidelberg (2009)
16. Das, S., Suganthan, P.N.: Differential evolution: A survey of the state-of-the-art. IEEE Transactions on Evolutionary Computation **15**(1), 4–31 (2011)
17. Dasgupta, D.: *Artificial Immune Systems and Their Applications*. Springer, New York (1998)
18. Dasgupta, D., Yu, S., Nino, F.: Recent advances in artificial immune systems: Models and applications. Applied Soft Computing **11**(2), 1574–1587 (2011)
19. de Castro, L.N., Zuben, F.J.V.: An evolutionary immune network for data clustering. In: *Proceedings of the 6th Brazilian Symposium on Neural Networks*, pp. 84–89. IEEE Computer Society Press (2000)
20. Dorigo, M.: Optimization, learning and natural Algorithms. Ph.D. thesis, Politecnico di Milano, Italy (1992)
21. Farmer, J.D., Packard, N.H., Perelson, A.S.: The immune system, adaptation, and machine learning. Phys. D **2**(1–3), 187–204 (1986)
22. Feo, T.A., Resende, M.G.C.: A probabilistic heuristic for a computationally difficult set covering problem. Operations Research Letters **8**(2), 67–71 (1989)
23. Ficici, S.G.: Solution concepts in coevolutionary algorithms. Ph.D. thesis, Brandeis University, Waltham, MA (2004). AAI3127125
24. Forrest, S., Perelson, A.S., Allen, L., Cherukuri, R.: Self-nonself discrimination in a computer. In: *Proceedings of the Symposium on Research in Security and Privacy*, pp. 202–212 (1994)
25. Galeano, J.C., Veloza-Suan, A., González, F.A.: A comparative analysis of artificial immune network models. In: *Proceedings of the 2005 Conference on Genetic and Evolutionary Computation, GECCO '05*, pp. 361–368. ACM, New York (2005)
26. Garrett, S.M.: How do we evaluate artificial immune systems? Evolutionary Computation **13**(2), 145–177 (2005)
27. Geem, Z.W., Kim, J.H., Loganathan, G.V.: A new heuristic optimization algorithm: Harmony search. Simulation **76**(2), 60–68 (2001)
28. Glover, F.: Future paths for integer programming and links to artificial intelligence. Computers and Operations Research **13**(5), 533–549 (1986)
29. Goh, C.K., Tan, K.C.: A competitive–cooperative coevolutionary paradigm for dynamic multi-objective optimization. IEEE Transactions on Evolutionary Computation **13**, 103–127 (2009)
30. Goldberg, D.E.: *Genetic Algorithms in Search, Optimization, and Machine Learning*, 1st edn. Studies in Computational Intelligence. Addison-Wesley Longman (1989)
31. Greensmith, J., Aickelin, U.: The deterministic dendritic cell algorithm. In: P.J. Bentley, D. Lee, S. Jung (eds.), *Artificial Immune Systems*. LNCS, vol. 5132, pp. 291–302. Springer (2008)
32. Greensmith, J., Aickelin, U., Cayzer, S.: Introducing dendritic cells as a novel immune-inspired algorithm for anomaly detection. In: C. Jacob, M. Pilat, P. Bentley, J. Timmis (eds.), *Artificial Immune Systems*, LNCS, vol. 3627, pp. 153–167. Springer (2005)
33. Greensmith, J., Aickelin, U., Twycross, J.: Detecting danger: Applying a novel immunological concept to intrusion detection systems. In: *Proceedings of the 6th International Conference on Adaptive Computing in Design and Manufacture (ACDM2004)*, Bristol, UK (2004)
34. Guo, Y., Cheng, J., Cao, Y., Lin, Y.: A novel multi-population cultural algorithm adopting knowledge migration. Soft Computing **15**, 897–905 (2011)

35. Hansen, N., Ostermeier, A., Gawelczyk, A.: On the adaptation of arbitrary normal mutation distributions in evolution strategies: The generating set adaptation. In: *Proceedings of the 6th International Conference on Genetic Algorithms*, pp. 57–64. Morgan Kaufmann, San Francisco (1995)

36. Hart, E., Bersini, H., Santos, F.: Structure versus function: A topological perspective on immune networks. Natural Computing **9**, 603–624 (2010)

37. Hart, E., McEwan, C., Timmis, J., Hone, A.: Advances in artificial immune systems. Evolutionary Intelligence **4**(2), 67–68 (2011)

38. Hart, E., Timmis, J.: Application areas of AIS: The past, the present and the future. Applied Soft Computing **8**(1), 191–201 (2008)

39. He, Q., Wang, L.: An effective co-evolutionary particle swarm optimization for constrained engineering design problems. Engineering Applications of Artificial Intelligence **20**, 89–99 (2007)

40. He, S., Wu, Q., Saunders, J.: A novel group search optimizer inspired by animal behavioural ecology. In: *Proceedings of 2006 IEEE Congress on Evolutionary Computation*, pp. 16–21, Vancouver (2006)

41. Hillis, W.D.: Co-evolving parasites improve simulated evolution as an optimization procedure. Physica D **42**, 228–234 (1990)

42. Jerne, N.K.: Towards a network theory of the immune system. Annals of Immunology **125C**(1–2), 373–389 (1973)

43. Ji, Z., Dasgupta, D.: Revisiting negative selection algorithms. Evolutionary Computation **15**(2), 223–251 (2007)

44. Kennedy, J., Eberhart, R.: Particle swarm optimization. In: *IEEE International Conference on Neural Networks*, vol. 4, pp. 1942–1948 (1995)

45. Kim, J., Greensmith, J., Twycross, J., Aickelin, U.: Malicious code execution detection and response immune system inspired by the danger theory. CoRR **abs/1003.4142** (2010)

46. Kirkpatrick, S., Gelatt, C., Vecchi, M.: Optimization by simulated annealing. Science **220**(4598), 671–680 (1983)

47. Kowaliw, T., Kharma, N.N., Jensen, C., Moghnieh, H., Yao, J.: Using competitive co-evolution to evolve better pattern recognisers. International Journal of Computational Intelligence and Applications **5**(3), 305–320 (2005)

48. Koza, J.R.: *Genetic Programming: On the Programming of Computers by Means of Natural Selection (Complex Adaptive Systems)*, 1st edn. MIT Press (1992)

49. Lin, C., Qing, A., Feng, Q.: A comparative study of crossover in differential evolution. Journal of Heuristics **17**(6), 675–703 (2011). doi:10.1007/s10732-010-9151-1

50. Lin, C.J., Chen, C.H., Lin, C.T.: A hybrid of cooperative particle swarm optimization and cultural algorithm for neural fuzzy networks and its prediction applications. IEEE Transactions on Systems, Man, and Cybernetics, Part C **39**, 55–68 (2009)

51. Liu, J., Lampinen, J.: A fuzzy adaptive differential evolution algorithm. Soft Computing **9**, 448–462 (2005)

52. Liu, Y., Passino, K.: Biomimicry of social foraging bacteria for distributed optimization: Models, principles, and emergent behaviors. Journal of Optimization Theory and Applications **115**, 603–628 (2002)

53. Luke, S., Wiegand, P.R.: When coevolutionary algorithms exhibit evolutionary dynamics. In: A.M. Barry (ed.) *GECCO 2002: Proceedings of the Bird of a Feather Workshops, Genetic and Evolutionary Computation Conference*, pp. 236–241. AAAI, New York (2002)

54. Ma, H.: An analysis of the equilibrium of migration models for biogeography-based optimization. Information Sciences **180**(18), 3444–3464 (2010)

55. MacArthur, R., Wilson, E.: *The Theory of Biogeography*. Princeton University Press, Princeton, NJ (1967)

56. Matzinger, P.: Tolerance, danger, and the extended family. Annual Review of Immunology **12**, 991–1045 (1994)

57. Mezura-Montes, E., Reyes-Sierra, M., Coello Coello, C.: Multi-objective optimization using differential evolution: A survey of the state-of-the-art. In: U. Chakraborty (ed.) *Advances in*

Differential Evolution. Studies in Computational Intelligence, vol. 143, pp. 173–196. Springer, Berlin (2008)

58. Montgomery, J., Chen, S.: An analysis of the operation of differential evolution at high and low crossover rates. In: *2010 IEEE Congress on Evolutionary Computation (CEC)*, pp. 1–8 (2010). doi:10.1109/CEC.2010.5586128

59. Mühlenbein, H., Paaß, G.: From recombination of genes to the estimation of distributions. I. Binary parameters. In: *Proceedings of the 4th International Conference on Parallel Problem Solving from Nature, PPSN IV*, pp. 178–187. Springer, London (1996)

60. Neri, F., Tirronen, V.: Recent advances in differential evolution: A survey and experimental analysis. Artificial Intelligence Review **33**, 61–106 (2010)

61. Nguyen, T., Yao, X.: Hybridizing cultural algorithms and local search. In: E. Corchado, H. Yin, V. Botti, C. Fyfe (eds.) *Intelligent Data Engineering and Automated Learning, IDEAL 2006*. Lecture Notes in Computer Science, vol. 4224, pp. 586–594. Springer, Berlin, (2006)

62. Ochoa, A., Ponce, J., Hernández, A., Li, L.: Resolution of a combinatorial problem using cultural algorithms. JCP **4**(8), 738–741 (2009)

63. Paredis, J.: Steps towards co-evolutionary classification neural networks. In: R.A. Brooks, P. Maes (eds.) *Proceedings of the Fourth International Workshop on the Synthesis and Simulation of Living Systems (Artificial Life IV)*, pp. 102–108. Cambridge, MA (1994). http://www.mpi-sb.mpg.de/services/library/proceedings/contents/alife94.html

64. Passino, K.M.: Biomimicry of bacterial foraging for distributed optimization and control. IEEE Control Systems Magazine **22**(3), 52–67 (2002). doi:10.1109/MCS.2002.1004010

65. Passino, K.M.: Bacterial foraging optimization. International Journal of Swarm Intelligence Research **1**(1), 1–16 (2010)

66. Pollack, J.B., Blair, A.D.: Co-evolution in the successful learning of backgammon strategy. Machine Learning **32**, 225–240 (1998)

67. Popovici, E., De Jong, K.: The effects of interaction frequency on the optimization performance of cooperative coevolution. In: *Proceedings of the 8th Annual Conference on Genetic and Evolutionary Computation, GECCO '06*, pp. 353–360. ACM, New York (2006)

68. Potter, M.A., De Jong, K.A.: Cooperative coevolution: An architecture for evolving coadapted subcomponents. Evolutionary Computation **8**(1), 1–29 (2000)

69. Potter, M.A., Jong, K.A.D.: A cooperative coevolutionary approach to function optimization. In: *Proceedings of the International Conference on Evolutionary Computation, Third Conference on Parallel Problem Solving from Nature, PPSN III*, pp. 249–257. Springer, London, (1994)

70. Price, K.V., Storn, R.M., Lampinen, J.A.: *Differential Evolution: A Practical Approach to Global Optimization*. Natural Computing Series. Springer, Berlin (2005)

71. Rashedi, E., Nezamabadi-pour, H., Saryazdi, S.: GSA: A gravitational search algorithm. Information Sciences **179**(13), 2232–2248 (2009). doi:10.1016/j.ins.2009.03.004

72. Renfrew, A.: Dynamic modeling in archaeology: what, when, and where? In: *Dynamical Modeling and the Study of Change in Archaelogy* (1994)

73. Reynolds, R.G.: An adaptive computer model of plan collection and early agriculture in the eastern valley of Oaxaca. In: G. Naquitz (ed.) *Archaic Foraging and Early Agriculture in Oaxaca, Mexico*, pp. 439–500 (1986)

74. Reynolds, R.G.: An introduction to cultural algorithms. In: A.V. Sebalk, L.J. Fogel (eds.) *Proceedings of the Third Annual Conference on Evolutionary Programming*, pp. 131–139. World Scientific, River Edge, NJ (1994)

75. Reynolds, R.G.: Cultural algorithms: Theory and applications. In: D. Corne, M. Dorigo, F. Glover (eds.) *New Ideas in Optimization*, pp. 367–378. McGraw-Hill, Maidenhead, UK (1999)

76. Reynolds, R.G., Kohler, T.A., Kobti, Z.: The effects of generalized reciprocal exchange on the resilience of social networks: An example from the prehispanic Mesa Verde region. Computational and Mathematical Organization Theory **9**, 227–254 (2003)

77. Reynolds, R.G., Liu, D.: Multi-objective cultural algorithms. In: *IEEE Congress on Evolutionary Computation*, pp. 1233–1241 (2011)

78. Reynolds, R.G., Peng, B., Ali, M.Z.: The role of culture in the emergence of decision-making roles: An example using cultural algorithms. Complexity **13**(3), 27–42 (2008)
79. Rivera, D.C., Becerra, R.L., Coello Coello Carlos, A.: Cultural algorithms, an alternative heuristic to solve the job shop scheduling problem. Engineering Optimization **39**(1), 69–85 (2007)
80. Rosin, C.D., Belew, R.K.: New methods for competitive coevolution. Evolutionary Computation **5**, 1–29 (1997)
81. Rychtyckyj, N., Reynolds, R.G.: Using cultural algorithms to re-engineer large-scale semantic networks. International Journal of Software Engineering and Knowledge Engineering **15**(4), 665–694 (2005)
82. Saleem, S.M.: Knowledge-based solution to dynamic optimization problems using cultural algorithms. Ph.D. thesis, Wayne State University, Detroit, MI (2001)
83. Secker, A., Freitas, A., Timmis, J.: A danger theory inspired approach to web mining. In: J. Timmis, P. Bentley, E. Hart (eds.) *Artificial Immune Systems*. Lecture Notes in Computer Science, vol. 2787, pp. 156–167. Springer, Berlin, Heidelberg (2003). doi:10.1007/978-3-540-45192-1_16
84. Shi, Y.J., Teng, H.F., Li, Z.Q.: Cooperative co-evolutionary differential evolution for function optimization. In: L. Wang, K. Chen, Y. Ong (eds.) *Advances in Natural Computation*, Lecture Notes in Computer Science, vol. 3611, pp. 428–428. Springer, Berlin, (2005)
85. Simon, D.: *Evolutionary Optimization Algorithms: Biologically-Inspired and Population-Based Approaches to Computer Intelligence*. p. 624. Wiley (2013)
86. Simon, D.: Biogeography-based optimization. IEEE Transactions on Evolutionary Computation **12**(6), 702–713 (2008)
87. Sims, K.: Evolving 3D morphology and behavior by competition. Artificial Life **1**(4), 353–372 (1994)
88. Stanley, K.O., Miikkulainen, R.: Competitive coevolution through evolutionary complexification. Journal of Artificial Intelligence Research **21**(1), 63–100 (2004)
89. Storn, R.M., Price, K.V.: Differential evolution: A simple and efficient heuristic for global optimization over continuous spaces. Journal of Global Optimization **11**(4), 341–359 (1997)
90. Talbi, E.G.: *Metaheuristics: From Design to Implementation*, 1st edn. Wiley-Blackwell (2009)
91. Tan, K.C., Yang, Y.J., Goh, C.K.: A distributed cooperative coevolutionary algorithm for multiobjective optimization. IEEE Transactions on Evolutionary Computation **10**(5), 527–549 (2006)
92. Teng, N., Teo, J., Hijazi, M., Hanafi, A.: Self-adaptive population sizing for a tune-free differential evolution. Soft Computing **13**, 709–724 (2009)
93. Timmis, J., Andrews, P., Hart, E.: On artificial immune systems and swarm intelligence. Swarm Intelligence **4**(4), 247–273 (2010)
94. Timmis, J., Andrews, P., Owens, N., Clark, E.: An interdisciplinary perspective on artificial immune systems. Evolutionary Intelligence **1**(1), 5–26 (2008)
95. Timmis, J., Hone, A., Stibor, T., Clark, E.: Theoretical advances in artificial immune systems. Theoretical Computer Science **403**(1), 11–32 (2008)
96. Tylor, E.B.: *Primitive Culture*, vol. 2, 7th edition. Brentano's, New York (1924)
97. Ulutas, B.H., Kulturel-Konak, S.: A review of clonal selection algorithm and its applications. Artificial Intelligence Review **36**(2), 117–138 (2011)
98. Walker, A., Hallam, J., Willshaw, D.: Bee-havior in a mobile robot: The construction of a self-organized cognitive map and its use in robot navigation within a complex, natural environment. In: *Proceedings of ICNN'93, International Conference on Neural Networks*, vol. III, pp. 1451–1456. IEEE Press, Piscataway, NJ (1993)
99. Wallace, A.R.: *The Geographical Distribution of Animals* (two volumes). Adamant Media Corporation, Boston, MA (2005)
100. Weber, M.: Parallel global optimization, structuring populations in differential evolution. Ph.D. thesis, University of Jyvskyl (2010)
101. Wiegand, R.P.: An analysis of cooperative coevolutionary algorithms. Ph.D. thesis, George Mason University, Fairfax, VA (2004). AAI3108645

102. Wu, C., Zhang, N., Jiang, J., Yang, J., Liang, Y.: Improved bacterial foraging algorithms and their applications to job shop scheduling problems. In: *Proceedings of the 8th International Conference on Adaptive and Natural Computing Algorithms*, Part I, *ICANNGA '07*, pp. 562–569. Springer, Berlin, Heidelberg (2007)
103. Yang, X.S.: A new metaheuristic bat-inspired algorithm. In: J.R. González, D.A. Pelta, C. Cruz, G. Terrazas, N. Krasnogor (eds.) *Nature Inspired Cooperative Strategies for Optimization (NICSO 2010)*, vol. 284, Chap. 6, pp. 65–74. Springer, Berlin Heidelberg (2010)
104. Yang, X.S., Deb, S.: Cuckoo search via Lévy flights. In: *World Congress on Nature & Biologically Inspired Computing (NaBIC 2009)*, Coimbatore, India, IEEE Conference Publications, pp. 210–214. IEEE Press, Piscataway, NJ (2009)
105. Zaharie, D.: Influence of crossover on the behavior of differential evolution algorithms. Applied Soft Computing 9(3), 1126–1138 (2009). doi:10.1016/j.asoc.2009.02.012
106. Zhang, C., Yi, Z.: A danger theory inspired artificial immune algorithm for on-line supervised two-class classification problem. Neurocomputing **73**(7–9), 1244–1255 (2010). doi:10.1016/j.neucom.2010.01.005. http://www.sciencedirect.com/science/article/pii/S0925231210000573
107. Zheng, J., Chen, Y., Zhang, W.: A survey of artificial immune applications. Artificial Intelligence Review **34**, 19–34 (2010)

Chapter 10
Nature Inspires New Algorithms

Sébastien Aupetit and Mohamed Slimane

Nature modeling is a leading trend in optimization methods. While genetic algorithms, ant-based methods, and particle swarm optimization are well-known examples, there is a continuous emergence of new algorithms inspired by nature. In this chapter, we give a short overview of the most recent promising new algorithms.

For brevity, we adopt the following notation in this chapter. Except when explicitly stated otherwise, the search space S is continuous, has a dimension D, and is a Cartesian product of ranges $[l_i; u_i]$, that is to say $S = \times_{i=1}^{D}[l_i; u_i]$. The objective function to be minimized is $f : S \mapsto \mathbb{R}$. If a generated candidate solution is not in S, we suppose there is some algorithm able to take the candidate back into S. This can be done by cropping or any other suitable process. $\mathcal{U}(X)$ represents a function that returns an uniformly distributed random element from the set X. $\mathcal{R}(X \sim Y)$ has the same role, but the probability distribution is governed by Y instead of being uniform. To avoid dealing with implementation details and for readability of the algorithms, we assume that for a solution $x \in S$, the value $f(x)$ is computed only once and memorized for future use without any additional computing cost.

S. Aupetit (✉) · M. Slimane
Laboratoire Informatique (EA6300), Université François Rabelais Tours,
64 Avenue Jean Portalis, 37200 Tours, France
e-mail: aupetit@univ-tours.fr

M. Slimane
e-mail: slimane@univ-tours.fr

© Springer International Publishing Switzerland 2016
P. Siarry (ed.), *Metaheuristics*, DOI 10.1007/978-3-319-45403-0_10

10.1 Bees

There exist many optimization algorithms inspired by the behavior of bees [23]. Mating and foraging are the main behaviors exploited. Mating has led to the honeybee mating optimization (HBMO) method [17] and its numerous variants and improvements. The foraging behavior of bees has given birth to many algorithms, such as BeeHive [39], BeeAdHoc [40], Virtual Bee [43], and ABC [21]. The last is the one attracting most attention from researchers [23]. In the following, we describe the artificial bee colony (ABC) algorithm introduced by Karaboga [21].

10.1.1 Honeybee Foraging

Honeybees are social insects that live in a colony represented by a hive. In the colony, there are three kinds of bees: the queen, the drones, and the workers. The queen is the only fertile female bee in the hive and her only role is to ensure the survival of the colony by giving birth to bees in the hive. The drones are male bees, whose only role is to fertilize the queen during her mating flight. The workers are sexually undeveloped females and have equipment such as brood food glands, scent glands, wax glands, and pollen baskets. This equipment allows them to accomplish tasks such as cleaning the hive, feeding the brood, caring for the queen, guarding the hive, handling incoming nectar, and foraging for food (nectar, pollen, etc.).

Foraging relies on four components: food sources (nectar or pollen), scout bees, onlooker bees, and employed bees.

Food sources are evaluated by the bees using many criteria, such as distance from the hive, the availability of the food, and the case of extraction. Whatever criteria are used, they can be summarized as the "interest" of the food source for the hive. To ensure the survival of the hive, it is necessary to reduce the energy cost of the activity of the bees. To do so, the colony constantly adapts, and focuses its foraging efforts on food sources with the most interest (Fig. 10.1).

An employed bee has the task of exploiting food sources, by bringing back nectar or pollen to the hive. Back at the hive, it deposits its harvest and goes to a part of the hive commonly referred to as the dance floor (Fig. 10.2). This area contains chemicals [36] which attract the bees currently not exploiting any food source. Once on the dance floor, an employed bee which has found or exploited an interesting food source (or flower field) practices a round dance or a tail-wagging dance to give information to other bees about the flower field. This dance allows the bee to communicate the location (direction and distance) and the composition of the food source (through odor or nectar exchange). Onlooker bees interested in a food source use this information to exploit the food source. This recruitment allows the colony to minimize the energy cost of its foraging. If an employed bee comes back to the hive and has found that a food source does not have interest anymore, it does not dance on the dance floor. Either it becomes an onlooker bee and observes other bees' dances,

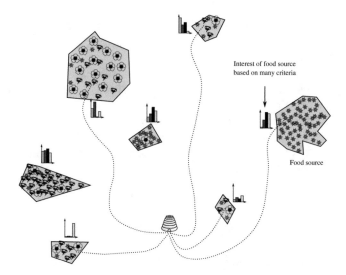

Fig. 10.1 Exploitation and evaluation of the interest of flower fields by bees

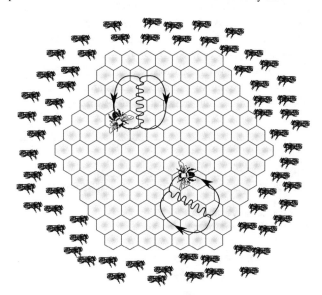

Fig. 10.2 Two employed bees dancing while being watched by many onlooker bees

or it becomes a scout bee and leaves the hive to discover a new food source. If this scout bee finds an interesting food source, it comes back to the hive as an employed bee and can decide whether or not it is interesting to dance. In a typical hive, about 5–10 % of the bees are scout bees [21].

The self-organizing and emergent properties of the foraging behavior of the bees originate mainly from the following properties:

- The more interesting a food source is, the more onlooker bees are recruited on the dance floor. This allows the colony to constantly adapt and focus its efforts on the best food sources.
- When a food source dries up or becomes less interesting, onlooker bees are no longer recruited and employed bees eventually abandon it. The colony replaces the food source with better ones.
- Scout bees ensure the regular discovery of new food sources. The bees can spread over many food sources, increasing the robustness of food provisioning.
- The bee dance allows bees to share information and to ensure recruitment based on the interest of the food sources.

10.1.2 Classical ABC Implementation

In the artificial bee colony (ABC) algorithm, the food sources are zones in the solution space S, usually represented as locations in S. Exploiting a food source consists in evaluating the interest of a location in its neighborhood. The colony is constituted of three kinds of bees: scout, onlooker, and employed bees. A scout bee becomes an employed bee when it chooses a food source. An employed bee becomes a scout bee when it abandons its food source. Let $\mathcal{N}$ be the number of food sources that can be exploited simultaneously by the hive, $\mathcal{N}_{\text{employed}}$ the number of employed bees, $\mathcal{N}_{\text{onlooker}}$ the number of onlooker bees, and $\mathcal{N}_{\text{scout}}$ the number of scout bees. The main steps of the ABC algorithm are summarized in Algorithm 10.1.

$\mathcal{N}_{\text{employed}}$ food sources are chosen
while *the stop criterion is not met* **do**
    Employed bees go out of the hive to exploit food sources
    Onlooker bees go out of the hive and split based on the interest of food sources to exploit them
    Employed bees eventually abandon some food sources
    Scout bees eventually search for new food sources
end

Algorithm 10.1: The main steps of the artificial bee colony (ABC) algorithm.

We denote by $\mathbb{S} = \{s_1, \ldots, s_{|\mathbb{S}|}\}$ the locations of the food sources and by $q : S \rightarrow \mathbb{R}^+$ the function measuring the interest (or quality) of a food source. To avoid dealing with implementation details and for readability of the detailed algorithms, we assume that for a solution $x \in S$, the value $q(x)$ is computed only once and memorized for future use without any additional computing cost. Finally, we denote by s^* the best solution (or food source location) found by the algorithm. In the following, we detail the each step of ABC.

10.1.2.1   Initial Choice of the Food Sources

By default, and without incorporating supplementary knowledge about the optimization problem, the initial choice of the food sources is the result of a uniform random sampling of the search space. For each food source, a failure counter (e_i) is maintained. Its value is set to 0 initially and when a better solution is found for the food source. Its value is increased when no improvement can be made by the exploitation of the food source. The initial choice of the food sources is described by Algorithm 10.2.

---

for $i = 1$ **to** $|\mathbb{S}|$ **do**
    $s_i \leftarrow \mathcal{U}(\mathbb{S})$
    $e_i \leftarrow 0$
$s^* \leftarrow \arg\min_{s \in \mathbb{S}} f(s)$

Algorithm 10.2: Initial choice of food sources.

10.1.2.2   Employed Bees Leave the Hive to Exploit Food Sources

Exploiting a food source requires the choice of a new solution in the neighborhood of the food source (s_i). In classical ABC, the new solution to be explored, $v_i = (v_{i,1}, \ldots, v_{i,D})'$ is obtained by mutating one coordinate of the location of the food source. The mutation is conducted according to the formula

$$v_{i,k} = s_{i,k} + \mathcal{U}([-1 : 1]) * (s_{i,k} - s_{n,k})$$

where s_i is the food source, s_n is another food source for the hive, randomly chosen, and k is the mutated coordinate, which is chosen randomly. Algorithm 10.3 details the process.

---

for $i = 1$ **to** $|\mathbb{S}|$ **do**
    $s_n \leftarrow \mathcal{U}(\mathbb{S} - \{s_i\})$ /* Choose the influencing food source */
    $k \leftarrow \mathcal{U}([\![1 : D]\!])$ /* Choose the modified coordinate */
    $v_i \leftarrow s_i$
    $v_{i,k} \leftarrow s_{i,k} + \mathcal{U}([-1 : 1]) * (s_{i,k} - s_{n,k})$ /* Mutate the solution */

Algorithm 10.3: Computation of new food sources.

If the interest $q(f(v_i))$ of the new solution v_i is higher than the interest $q(f(s_i))$ of the food source s_i, then s_i is replaced by v_i in the memory of the employed bee and the counter associated with the food source is reset to 0. In the other case, the

new solution is less interesting, so the counter is increased by 1. The update of the
food sources by employed bees is given in Algorithm 10.4.

---

for $i = 1$ **to** $|\mathbb{S}|$ **do**
    if $f(v_i) < f(s_i)$ **then**   /* The new solution is more interesting */
       $s_i \leftarrow v_i$
       $e_i \leftarrow 0$
       if $f(v_i) < f(s^*)$ **then**   /* The best solution is improved */
          $s^* \leftarrow v_i$
    else
       $e_i \leftarrow e_i + 1$ /* The new solution is worse */

Algorithm 10.4: New, more interesting food sources are memorized, and others
are forgotten.

10.1.2.3   Onlooker Bees Leave the Hive

Onlooker bees choose a food source based on their observation of the dance floor
and then based on the interest of the food sources. In ABC, this principle translates
into as a spreading of the onlookers over the food sources following the probability
distribution $\mathcal{P}$ derived from the interest of those sources. There exist many ways to
define this distribution. In classical ABC, the interest function q of a food source s
is defined by

$$
q(f(s)) = \begin{cases} \frac{1}{1+f(s)} & \text{if } f(s) \geq 0 \\ 1 + |f(s)| & \text{otherwise} \end{cases}
$$

such that $q(f(s))$ increases as $f(s)$ decreases. The probability for an onlooker to
choose the food source s_i is p_i, such that

$$
p_i = \frac{q(f(s_i))}{\sum\limits_{s \in \mathbb{S}} q(f(s))}.
$$

Each onlooker chooses a food source following $\mathcal{P}$ and exploits the food source as
an employed bee: a solution in the neighborhood is chosen, and the food source
and its counter are updated. Finally, the best ever food source found is memorized.
Implementation details are given in Algorithm 10.5.

```
/* Compute the probabilities from interests */
for i = 1 to |S| do
   p_i ← q(f(s_i)) / Σ_{s∈S} q(f(s))
/* Onlookers exploit the food sources */
for i = 1 to N_onlooker do
   x_i ← R([[1 : |S|]] ~ P)              /* s_{x_i} is the food source chosen */
                                          /* by the bee according to P */
   s_n ← U(S − {s_{x_i}})       /* Choose the influencing food source */
   k ← U([[1 : D]])                 /* Choose the modified coordinate */
   w_i ← s_{x_i}
   w_{i,k} ← s_{x_i,k} + U([−1 : 1]) * (s_{x_i,k} − s_{n,k})      /* Mutate the solution */
/* Update the food sources and their counters */
for i = 1 to N_onlooker do
   if f(w_i) < f(s_{x_i}) then  /* The new solution is more interesting */
      e_{x_i} ← 0
      s_{x_i} ← w_i
      if f(w_i) < f(s*) then         /* The best solution is improved */
         s* ← w_i
   else                                /* The new solution is worse */
      e_{x_i} ← e_{x_i} + 1
```

Algorithm 10.5: Exploitation of food sources by onlookers.

10.1.2.4 Dried-Up Food Sources are Abandoned and Scout Bees Work to Replace Them

In classical ABC, only a few employed bees are allowed to abandon a food source. For this purpose, the counter of the food source is checked. If its value is greater than or equal to a specified constant e_{Max}, then the bee becomes a scout. Each scout chooses a new food source in the search space and becomes an employed bee. After this step, the hive again has all its food sources. In classical ABC, there is at most $N_{scout} = 1$ scout bee. Implementation details are given in Algorithm 10.6.

10.1.3 Parameterization and Evolution of the Classical ABC Algorithm

In classical ABC, the whole algorithm requires very few parameters. The number of food sources is the number of employed bees, which constitutes half of the colony. We have

$$N_{employed} = N_{onlooker} = \frac{N}{2}.$$

```
n ← 0        /* The number of scout bees transformed into employed
bees */
C ← {i ∈ [[1 : |S|]]|eᵢ >= eₘₐₓ}        /* The candidate to be abandoned */
while n < 𝒩ₒₙₗₒₒₖₑᵣ and C ≠ ∅ do
    i ← arg max{eⱼ}        /* One of the most dried-up food sources */
          j∈C
    sᵢ ← 𝒰(𝒮)                            /* Choose a new food source */
    if f(sᵢ) < f(s*) then          /* The best solution is improved */
        └ s* ← sᵢ
    C ← {i ∈ 1..|S| |eᵢ >= eₘₐₓ}
    └ n ← n + 1
```

Algorithm 10.6: Dried-up food sources are abandoned and scout bees work to replace them.

Usually, only one food source can be abandoned in an iteration, and then $\mathcal{N}_{scout} = 1$. In experiments presented in [22], a maximum value e_{Max} of the failure counter of a food source equal to $D\mathcal{N}/2$ was found to give good results. The last parameter of ABC is the stop criterion. Usually, it is expressed as a maximum number of iterations.

Since its introduction, ABC has attracted a lot of interest and has been applied in many domains. In short, it has been shown that ABC performs as well as and sometimes better than many popular metaheuristics while requiring fewer settings. ABC was created to solve continuous problems, but many variants have been proposed for discrete, combinatorial, and multiobjective problems. Many improvements and hybridization with other metaheuristics have allowed researchers to consider ABC as one of the best optimization algorithms of the present moment. The architecture of ABC is also well suited to parallelization and therefore to the solution of large problems. For more details of ABC and a wider review of the ABC universe, we recommend the reader to read [23] and follow http://mf.erciyes.edu.tr/abc/. Algorithm 10.7 shows the complete ABC algorithm.

10.2   In Search of the Perfect Harmony

Music has been part of human civilization from the beginning. Throughout this time, humanity has sought to create perfect melodies. Usually, many musicians are required to play notes simultaneously in order to create an aesthetically pleasing chord or harmony. The search for such a harmony is done progressively by adjusting the notes until an aesthetic harmony is found. During the process, the musicians memorize the best harmonies and reuse them to make adjustments and to improvise new harmonies.

To improvise, musicians usually behave as follows. The choice of a new chord is dependent on the instrument (and on the musician). The first way to improvise consists in randomly choosing a note from a scale. The second possibility considers

```
for i = 1 to |S| do
    s_i ← U(S)
    e_i ← 0
s* ← arg min f(s)
        s∈S
while the stop criterion is not met do
    for i = 1 to |S| do
        s_n ← U(S − {s_i})           /* Choose the influencing food source */
        k ← U([[1 : D]])                /* Choose the modified coordinate */
        v_i ← s_i
        v_{i,k} ← s_{i,k} + U([−1 : 1]) * (s_{i,k} − s_{n,k})          /* Mutate the solution */
    for i = 1 to |S| do
        if f(v_i) < f(s_i) then      /* The new solution is more interesting */
            s_i ← v_i
            e_i ← 0
            if f(v_i) < f(s*) then            /* The best solution is improved */
                s* ← v_i
        else
            e_i ← e_i + 1               /* The new solution is worse */

    /* Compute the probabilities from interests */
    for i = 1 to |S| do
        p_i ←  q(f(s_i)) / ∑_{s∈S} q(f(s))

    /* Onlookers exploit the food sources */
    for i = 1 to N_onlooker do
        x_i ← R([[1 : |S|]] ∼ P)           /* s_{x_i} is the food source chosen */
                                           /* by the bee according to P */
        s_n ← U(S − {s_{x_i}})        /* Choose the influencing food source */
        k ← U([[1 : D]])                /* Choose the modified coordinate */
        w_i ← s_{x_i}
        w_{i,k} ← s_{x_i,k} + U([−1 : 1]) * (s_{x_i,k} − s_{n,k})          /* Mutate the solution */
    /* Update the food sources and their counters */
    for i = 1 to N_onlooker do
        if f(w_i) < f(s_{x_i}) then      /* The new solution is more interesting */
            e_{x_i} ← 0
            s_{x_i} ← w_i
            if f(w_i) < f(s*) then            /* The best solution is improved */
                s* ← w_i
        else                             /* The new solution is worse */
            e_{x_i} ← e_{x_i} + 1

    n ← 0   /* The number of scout bees transformed into employed bees
    */
    C ← {i ∈ [[1 : |S|]] | e_i >= e_Max}        /* The candidates to be abandoned */
    while n < N_onlooker and C ≠ ∅ do
        i ← arg max{e_j}       /* One of the most dried-up food sources */
                 j∈C
        s_i ← U(S)                             /* Choose a new food source */
        if f(s_i) < f(s*) then            /* The best solution is improved */
            s* ← s_i
        C ← {i ∈ 1..|S| | e_i >= e_Max}
        n ← n + 1
```

Algorithm 10.7: The artificial bee colony (ABC) algorithm.

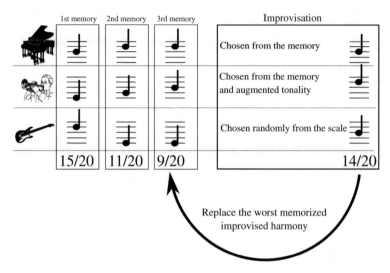

Fig. 10.3 Improvisation in harmony search

the notes played on that instrument in the more aesthetic memorized harmonies: a note is chosen, and a modification of the tonality is possibly applied. The resulting new chord is played and compared with the most aesthetic memorized harmonies. The process is repeated until the musicians are satisfied with the harmonies (see Fig. 10.3).

Harmony search [16], introduced by Geem et al., was derived from the iterative search for an aesthetic harmony described above. The solution vector represents the notes of the chord, and the objective function is used as an aesthetic measurement. By searching for a harmony with the best aesthetic value, the algorithm searches for a solution which minimizes the objective function. The main principles of the harmony search algorithm are described in Algorithm 10.8.

Initialize the memories
while *the stop criterion is not met* **do**
    Improvise a new harmony
    if *the new harmony is more aesthetic than the worst memorized*
    harmony **then**
        ⌊ Replace the worst memorized harmony with the new harmony

---

Algorithm 10.8: The harmony search algorithm.

10.2.1   Memory Initialization

A fixed set of slots is used for memorizing the best harmonies. Let $\mathbb{M} = \{m_1, \ldots, m_{|\mathbb{M}|}\}$ be the memory slots and their values. In the classical implementation, the memory slots are initialized through a sampling of the search space as described in Algorithm 10.9.

---

for $i = 1$ **to** $|\mathbb{M}|$ **do**
  $\lfloor \ m_i \leftarrow \mathcal{U}(\mathcal{S})$
  $s^* = \arg \min_{m \in \mathbb{M}} f(m)$

Algorithm 10.9: Initialization of memory slots.

10.2.2   Improvisation of a New Harmony

The exploration and exploitation capabilities of harmony search lie in the improvisation process. To improvise a new harmony, the memory may or may not be exploited. The improvisation is applied instrument by instrument, that is to say, coordinate by coordinate of the solution. The exploitation of the memory is applied with a probability of $\tau_{\text{memory}} \in]0 : 1[$. Otherwise, the search space is explored.

The exploitation process consists in choosing a note from the memorized aesthetic harmonies. For instrument j (dimension j), a note is chosen randomly and uniformly from the set of the notes of the instrument for the best-memorized harmonies $\{m_{1,j}, m_{2,j}, \ldots, m_{|\mathbb{M}|,j}\}$. The more a value is represented in the memorized harmonies, the greater the probability of choosing it is. The chosen value is modified with a probability $\tau_{\text{tonality}} \in]0 : 1[$. The modification consists in adding a uniformly generated random value from the range $[-\beta : \beta]$, where $\beta > 0$. In the literature, β is referred as the bandwidth or fret width, and controls the mutation applied to the improvisation process. When the memory is not used, the note is uniformly chosen from the scale, that is to say, in $[l_j : u_j]$, to ensure exploration of the search space. Table 10.1 summarizes the three possible outcomes of improvisation for an instrument, and their respective probabilities of occurring.

Many choices have been proposed for τ_{tonality} and β. When the search space is discrete, β is usually set to 1. When $\mathcal{S}$ is symbolic, the tonality adjustment is considered as an increase or a decrease in the tonality. For example, with musical notes, a tonality adjustment for E could be D or F. When $\mathcal{S}$ is continuous, many formulas have been devised and are still regularly being devised [2].

In the original definition of the harmony search algorithm by Geem et al. [16], τ_{memory}, τ_{tonality}, and β are fixed at the start of the algorithm. This setting was rapidly replaced by more elaborate strategies. The best-known strategy is that of the improved

Table 10.1 The three cases of improvisation for coordinate j, and their respective probabilities of occurring

Improvisation outcome	Probability
$m_{\mathcal{U}(\llbracket 1:|M|\rrbracket),j}$	$\tau_{memory} * (1 - \tau_{tonality})$
$m_{\mathcal{U}(\llbracket 1:|M|\rrbracket),j} + \mathcal{U}([-\beta : +\beta])$	$\tau_{memory} * \tau_{tonality}$
$\mathcal{U}([l_j : u_j])$	$1 - \tau_{memory}$

harmony search (IHS) algorithm [27]. In IHS, the probability of tonality adjustment is increased over time from $\tau_{tonality}^{min}$ to $\tau_{tonality}^{max}$ while the fret width is decreased exponentially from β^{max} to β^{min}. Denoting the improvisation step by $t \in [\![0 : T_{Max}]\!]$, we have

$$\tau_{tonality}(t) = \tau_{tonality}^{min} + \frac{\tau_{tonality}^{max} - \tau_{tonality}^{min}}{T_{Max}} t$$

and

$$\beta(t) = \beta^{max} \left(\frac{\beta^{min}}{\beta^{max}} \right)^{t/T_{Max}}$$

As stated earlier, if the solution obtained is not in the search space, it is taken back to $\mathcal{S}$ using any valid way, such as cropping of the values.

The improvisation of a new harmony is summarized in Algorithm 10.10.

---

for $j = 1$ **to** D **do**
  if $\mathcal{U}([0 : 1]) \leq \tau_{memory}$ **then**
    $i \leftarrow \mathcal{U}(\llbracket 1 : |M| \rrbracket)$ /* A memory slot is chosen as source */
    if $\mathcal{U}([0 : 1]) \leq \tau_{tonality}$ **then**
      $v_j \leftarrow m_{i,j} + \mathcal{U}([-\beta : \beta])$ /* Tonality is adjusted */
    else
      $v_j \leftarrow m_{i,j}$ /* Only the memory is exploited */
  else
    $v_j \leftarrow \mathcal{U}([l_j : u_j])$ /* A note is chosen from the scale */

Algorithm 10.10: Improvisation of a new harmony.

10.2.3 Updating of the Memory Slots

In the classical harmony search, the memory update process consists in replacing the least aesthetic harmony in the memory if and only if the new improvised harmony is more aesthetic. If the improvisation produces a less aesthetic harmony, it is ignored. Many other alternative strategies have been considered for the memory update, such as forbidding duplicate harmonies or maintaining a minimum diversity among the values of a dimension. The memory update process in the classical harmony search is given in Algorithm 10.11.

$$w \leftarrow \arg\max_{m \in \mathbb{M}} f(m) \qquad \text{/* The least aesthetic memorized harmony */}$$

if $f(v) < f(w)$ **then**

    $w \leftarrow v$ /* Replace the least aesthetic memorized harmony */

    if $f(v) < f(s^*)$ **then**   /* The best solution is improved */

        $s^* \leftarrow v$

Algorithm 10.11: Updating of the memory slots.

10.2.4   Parameterization and Evolution of the Classical Algorithm

The parameter settings are dependent on the optimization problem. However, experiments conducted over the years have shown some tendencies. Usually, the memory size is 30, but the use of a size of 100 can easily be found for some problems. τ_{memory} must have a high value, between 0.70 and 0.98, to allow proper exploitation of the memory. A value of 0.9 seems to be common. The fret width takes values lower than 0.5, with a mean of 0.3. β is dependent on the problem but is of the order of 1–10 % of the amplitude of the values ($u_j - l_j$ for a continuous range). Recent studies such as [12, 18, 41] have tried to devise self-adapting parameters for the optimization problem.

Harmony search has been hybridized with many other metaheuristics. More details can be found in [2, 14, 15]. The reader can also find a broader source of information about harmony search at http://www.hydroteq.com.[1] The complete harmony search algorithm is given in Algorithm 10.12.

10.3   The Echolocation Behavior of Microbats

Bats are mammals in the order Chiroptera. They are the only mammals that are able to fly and to sustain flight like birds. The Bats are the second largest order of mammals, with more than a thousand species. The order is usually divided into two parts: the megabats and the microbats. Unlike the megabats, the microbats are small and use

---

[1] Alternative link: https://sites.google.com/a/hydroteq.com/www/.

```
for i = 1 to |M| do
  ⌊ mᵢ ← 𝒰(𝒮)
s* ← arg min f(m)
       m∈M
while the stop criterion is not met do
  for j = 1 to D do
    if 𝒰([0 : 1]) ≤ τ_memory then
      i ← 𝒰(⟦1 : |M|⟧)        /* A memory slot is chosen as source */
      if 𝒰([0 : 1]) ≤ τ_tonality then
        │ vⱼ ← m_{i,j} + 𝒰([−β : β])     /* Tonality is adjusted */
      else
        ⌊ vⱼ ← m_{i,j}              /* Only the memory is exploited */
    else
      ⌊ vⱼ ← 𝒰([lⱼ : uⱼ])        /* A note is chosen from the scale */
  τ_tonality ← τ_tonality^min + (τ_tonality^max − τ_tonality^min)/T_Max · t    /* IHS parameter update */
  β ← β^max (β^min/β^max)^{t/T_Max}                /* IHS parameter update */
  w ← arg max f(m)   /* The least aesthetic memorized harmony */
       m∈M
  if f(v) < f(w) then
    │ w ← v  /* Replace the least aesthetic memorized harmony */
    if f(v) < f(s*) then              /* The best solution is improved */
      ⌊ s* ← v
```

Algorithm 10.12: The harmony search algorithm with the parameter update process of IHS.

echolocation. Besides being sighted, most echolocating microbats are insectivores and only use echolocation to hunt at night.

The echolocation capabilities of bats are due to a mutated gene, named *Prestin*, that allows the ears to perceive ultrasound. When a microbat emits ultrasound from its mouth and nose, the waves are reflected by obstacles and detected by the ears (Fig. 10.4). Using ultrasound, the microbat reconstitutes a 3D model of its environment. While moving, most bats modulate this ultrasound according to their movement, the hunting strategy that they are following, and the distance from their prey. The modulation consists in varying the loudness, frequency, and rhythm of the bursts of ultrasound in order to adjust the precision of the echolocation as needed. The hunting behavior of echolocating microbats led to the definition of the bat algorithm by Yang [46].

The bat algorithm is based on the hypothesis that only echolocation is required to hunt prey, and to perceive distances and the environment. It supposes that microbats move by flying. The solutions in the search space $\mathcal{S}$ are locations in the environment. The bat algorithm considers a colony of $\mathcal{N}$ microbats. At each time t, microbat i is located at $x_i \in \mathcal{S}$ and has a velocity v_i. During a move, each microbat emits ultrasound with a loudness $L_i \in [L_{min} : L_{max}]$ at a frequency $f_i \in [f_{min} : f_{max}]$. The ultrasound is emitted in bursts with a pulse rate $\tau_i \in [0 : 1]$. When the prey is near

Fig. 10.4 Ultrasound emitted by a microbat reflected from a prey organism

the bat, the pulse rate is high (τ_i increases) and the loudness is low (L_i decreases). In the opposite case, when the prey is far away, the pulse rate is low and the loudness is high in order to be able to see further. The main steps of the bat algorithm are given in Algorithm 10.13.

Initialize the position and velocity of the bats
Initialize the properties governing the emission of ultrasound by
the bats
while *the stop criterion is not met* **do**
   Move the bats
   Update the properties governing the emission of ultrasound by
   the bats
   Update the best ever known solution

Algorithm 10.13: Main steps of the bat algorithm.

10.3.1 Initialization Step

At initialization, the bats are usually spread uniformly over the search space. The initial velocity is zero. In most implementations, the loudness is bounded by $[0 : 1]$, that is to say, $L_{\min} = 0$ and $L_{\max} = 1$. Usually, the initial loudness L_i has a value around 0.5. In this case, the bat moves in a random direction half the time (see Sect. 10.3.2). Many other initial values can be used, such as random values for L_i or τ_i, or values more suited to specific features of the problem. The details of the initialization step are given in Algorithm 10.14, where $f : S \longmapsto \mathbb{R}$ is the objective function to be minimized and s^* is the best solution known.

$$\tau_{max} \leftarrow 0.5$$
for $i = 1$ **to** $\mathcal{N}$ **do**
$\quad x_i \leftarrow \mathcal{U}(\mathcal{S})$
$\quad v_i \leftarrow 0$
$\quad \tau_i \leftarrow \tau_{max}$
$\quad L_i \leftarrow 0.5$
$s^* \leftarrow \min_{i=1..\mathcal{N}} f(x_i)$

Algorithm 10.14: Initialization step for the bat algorithm.

10.3.2   Moves of the Bats

The moves of a bat obey some simple rules: either it continues moving in the same direction or it changes direction.

In the first case, similar principles to those in particle swarm optimization are used. The new velocity vector is obtained by adding the current velocity and an external velocity vector. In the classical bat algorithm, the external velocity vector is computed as the multiplication of the frequency f_i and the direction between the current location and the location of the best solution ever found. The frequency, uniformly generated in the range $[f_{min} : f_{max}]$, controls the speed of movement. To move the bat, the new velocity is added to the position. This movement exploits current knowledge to explore the search space.

In the second case, the new position depends on the position of another bat, randomly chosen. The position is perturbed in proportion to the mean value of the loudness of the ultrasound emitted by all bats. This behavior allows one to explore the search space.

The pulse rate governs the choice between the two strategies. The higher the pulse rate is, the more the bat exploits its current velocity and the knowledge about the best solution ever found. The lower the pulse rate is, the more random moves are allowed. This rate plays a similar role to the temperature in simulated annealing by controlling the balance between exploitation of knowledge and exploration of the search space. The move step of the algorithm is summarized in Algorithm 10.15.

for $i = 1$ **to** $\mathcal{N}$ **do**
$\quad$ **if** $\mathcal{U}([0 : 1]) > \tau_i$ **then**   /* Change direction */
$\quad\quad k \leftarrow \mathcal{U}(\llbracket 1 : \mathcal{N} \rrbracket)$
$\quad\quad x_i \leftarrow x_k + \frac{1}{\mathcal{N}} * \left(\sum_{j=1}^{\mathcal{N}} L_j \right) * \mathcal{U}([-1 : 1]^D)$
$\quad$ **else**   /* Keep direction */
$\quad\quad f_i \leftarrow \mathcal{U}([f_{min} : f_{max}])$
$\quad\quad v_i \leftarrow v_i + (x_i - s^*) * f_i$
$\quad\quad x_i \leftarrow x_i + v_i$

Algorithm 10.15: Moves of the bats.

10.3.3   Update of the Emission Properties of the Ultrasound

When the position is better than the best solution ever known, the emission properties are updated with a probability L_i. In this case, the loudness is reduced by a factor $\alpha \in]0:1[$ and the pulse rate τ_i is increased according to $\tau_{max}(1 - e^{-\gamma t})$, where t is the iteration number and $\gamma \in]0:1[$ is usually a constant. A progressive decrease in L_i leads to a decrease in the probability of τ_i being increased. An increase in τ_i diminishes the frequency of random moves, thereby increasing exploitation at the expense of exploration. The parameters α and γ are classically fixed at 0.9. These parameters control the convergence of the bat algorithm like the cooling factor in simulated annealing. The update process for the emission properties is given in Algorithm 10.16.

```
for i = 1 to N do
    if f(x_i) < f(s*) then              /* The best solution is improved */
        s* ← x_i
        if U([0 : 1]) < L_i then
            L_i ← α * L_i
            τ_i ← τ_max(1 − e^{−γt})
```

Algorithm 10.16: Update of the emission properties of the ultrasound.

10.3.4   Evolution of the Algorithm

The bat algorithm is quite recent. During the creation of the algorithm, Yang [46] tried to include best practices derived from many metaheuristics. Using specific parameters, the bat algorithm can be reduced to a simple form of particle swarm optimization or to a harmony search. In [46], Yang suggested that the bat algorithm was probably better than the others considered in that paper (ABC and harmony search).

Recently, the method has attracted attention and has led to many suggestions for improvement. In [10], the Doppler effect was incorporated to the moves. In [6], some new ideas, similar to those used in $(\mu + \mu)$-ES, were considered for the selection of new positions and the update of the emission properties. In [38], mutation mechanisms were added. In [32], the algorithm was adapted to binary search spaces. The study in [13] took constraints into account using penalty functions. Finally, in [25], a random walk of the chaotic Lévy flight type was added. These multiple improvements are only an extract of what has been proposed, and the reader can be sure that many others are to come. The complete bat algorithm is given in Algorithm 10.17.

$\tau_{max} \leftarrow 0.5$
for $i = 1$ **to** $\mathcal{N}$ **do**
   $x_i \leftarrow \mathcal{U}(\mathcal{S})$
   $v_i \leftarrow 0$
   $\tau_i \leftarrow \tau_{max}$
   $L_i \leftarrow 0.5$
$s^* \leftarrow \min_{i=1..\mathcal{N}} f(x_i)$
while *the stop criterion is not met* **do**
   for $i = 1$ **to** $\mathcal{N}$ **do**
      if $\mathcal{U}([0:1]) > \tau_i$ **then**   /* Change direction */
         $k \leftarrow \mathcal{U}([\![1:\mathcal{N}]\!])$
         $x_i \leftarrow x_k + \frac{1}{\mathcal{N}} * \left(\sum_{j=1}^{\mathcal{N}} L_j \right) * \mathcal{U}([-1:1]^D)$
      else   /* Keep direction */
         $f_i \leftarrow \mathcal{U}([f_{min} : f_{max}])$
         $v_i \leftarrow v_i + (x_i - s^*) * f_i$
         $x_i \leftarrow x_i + v_i$
   for $i = 1$ **to** $\mathcal{N}$ **do**
      if $f(x_i) < f(s^*)$ **then**   /* The best solution is improved */
         $s^* \leftarrow x_i$
         if $\mathcal{U}([0:1]) < L_i$ **then**
            $L_i \leftarrow \alpha * L_i$
            $\tau_i \leftarrow \tau_{max}(1 - e^{-\gamma t})$

Algorithm 10.17: The bat algorithm.

10.4 Nature Continues to Inspire New Algorithms

The algorithms based on bees, music improvisation, and bats are only three examples of nature-inspired metaheuristics. Based on physical properties or social behavior, from the tiniest bacteria to larger organisms such as the cuckoo, there exist many other metaheuristics. In the following, we give the reader some insights and starting points for studying them.

10.4.1 Bacterial Foraging Optimization

The foraging and movement behavior of bacteria led to the bacterial foraging optimization (BFO/BFOA) algorithm [9, 26, 31]. In this algorithm, the bacteria move in a solution space, and both the objective function and proximity to other bacteria are considered. Through successive moves, deaths, and births, by spreading of new bacteria, the population of bacteria searches for optima of the objective function.

10.4.2 Slime Mold Optimization

Dictyostelium discoideum, usually referred as slime mold, is a species of soil-living amoeba belonging to the phylum Mycetozoa. The primary diet of slime mold consists of bacteria and yeasts. Although it is unicellular, the amoeba is able to behave as a multicellular organism by the aggregation of many amoebas in order to survive. The behavior of these "social" amoebas inspired the slime mold optimization algorithm [3, 29, 30].

10.4.3 Fireflies and Glowworms

Fireflies and glowworms are insects in the order Coleoptera, more precisely from the Lampyridae family. These insects have the ability to emit light. The light is produced in the abdomen through a type of chemical reaction called bioluminescence from molecules of luciferin produced by the insect. While the insects' light can be used for attracting prey, it is mainly used for mate selection. Blinking of the light is used to attract a mate. Two algorithms are directly derived from the blinking phenomenon: the firefly algorithm and the glowworm swarm optimization algorithm.

The firefly algorithm was introduced by Yang [44, 45]. This algorithm considers many fireflies moving in the search space. Each firefly emits a blinking light whose intensity depends on the quality of the solution (the objective function). In each iteration, the fireflies perceive the blinking of the other fireflies. When the intensity of a remote firefly is stronger than its own intensity, an insect moves toward the remote insect. The velocity of the movement depends on the distance and the intensity. The process is repeated as needed. Although it is based on different principles, this algorithm is quite similar to a particular form of particle swarm optimisation.

Glowworm swarm optimization was defined by Krishnanand and Ghose [24, 42]. As in the firefly algorithm, the intensity of the emitted light depends directly on the quality of the position (the objective function). Each glowworm sees only the glowworms in its neighborhood and is attracted by a stronger light. There are three steps in each iteration of the algorithm: the light intensities are updated, the glowworms move, and the radius of the neighborhoods is updated. The update of the intensities consists in increasing or decreasing the intensity based on the quality of the solution. To move a glowworm, another glowworm in the neighborhood is chosen using a probability law depending on the difference in the light intensities of the glowworms. The first glowworm is then moved toward the second (chosen) one. Finally, the radius of the neighborhood is updated so that the number of glowworms in the neighborhood loosely matches a required value. The process is repeated as needed.

10.4.4 Termites

Although less known, termites have been used to solve network problems. The resulting algorithms are quite similar to ant algorithms such as ACO [1, 20, 28, 34, 35, 48].

10.4.5 Roach Infestation

Roach infestation served as a model for the roach infestation optimization algorithm [19].

10.4.6 Mosquitoes

In [11], the hunting behavior of mosquitoes inspired the mosquito host-seeking algorithm to solve the traveling salesman problem.

10.4.7 Wasps

Some specific features of wasps led to the wasp swarm optimization algorithm [7, 8, 33, 37].

10.4.8 Spiders

Social spiders have been used for region detection in images [5] and for security in wireless networks [4].

10.4.9 Cuckoo Search

The social behavior of animals such as the cuckoo (a bird of the Cuculidae family) has also inspired algorithms. Cuckoos are brood parasites, relying on other species to raise their young. The other bird species can have two reactions: either they discover the trick and destroy the cuckoo eggs, or they are oblivious to the trick. To increase the survival of the young, a cuckoo seeks to lay its eggs in nests where the survival rate is higher. This brood-parasitizing strategy led to the cuckoo search algorithm [47].

10.5   Conclusion

Throughout this chapter, we have tried to provide a short overview of algorithms inspired by nature by considering either physical or biological behavior. We saw that this inspiration can come from the tiniest organisms such as bacteria to complex organisms birds. Many algorithms have been created. Some have been successful, some less. Although this chapter is a very incomplete list of nature-inspired algorithms for optimization, we can be sure that inspiration has been, is, and will be fruitful for a long, time to come.

10.6   Annotated Bibliography

Reference [23]   This article is a relatively comprehensive overview of optimization algorithms derived from the behavior of bees.

Reference [21]   This article by Karaboga establishes the foundations of the artificial bee colony algorithm.

Reference [16]   This article introduces the fundamental principles behind the harmonic search algorithm.

Reference [46]   This article by Yang describes the principles of echolocation of bats and their use in the associated optimization algorithm.

References

1. Ajith, A., Crina, G., Vitorino, R., Martin, R., Stephen, W.: Termite: A swarm intelligent routing algorithm for mobilewireless ad-hoc networks. In: J. Kacprzyk (ed.) *Stigmergic Optimization*, vol. 31, pp. 155–184. Springer, Berlin, Heidelberg (2006). http://www.springerlink.com/index/ 10.1007/978-3-540-34690-6_7

2. Alia, O.M., Mandava, R.: The variants of the harmony search algorithm: An overview. Artificial Intelligence Review **36**(1), 49–68 (2011). doi:10.1007/s10462-010-9201-y

3. Becker, M., Wegener, M.: An optimization algorithm similar to the search of food of the slime mold *Dictyostelium Discoideum*. In: *IRAST International Congress on Computer Applications and Computational Science (CACS 2010)*, pp. 874–877 (2010)

4. Benahmed, K., Merabti, M., Haffaf, H.: Inspired social spider behavior for secure wireless sensor networks. International Journal of Mobile Computing and Multimedia Communications **4**(4), 1–10 (2012). doi:10.4018/jmcmc.2012100101

5. Bourjot, C., Chevrier, V., Thomas, V.: A new swarm mechanism based on social spiders colonies: From web weaving to region detection. Web Intelligence and Agent Systems **1**(1), 47–64 (2003). http://dl.acm.org/citation.cfm?id=965057.965061

6. Carbas, S., Hasancebi, O.: Optimum design of steel space frames via bat inspired algorithm. In: *10th World Congress on Structural and Multidisciplinary Optimization*, Orlando, FL (2013)

7. Cicirello, V.A., Smith, S.F.: Wasp-like agents for distributed factory coordination. Technical Report CMU-RI-TR-01-39, Robotics Institute, Carnegie Mellon University, Pittsburgh, PA (2001)
8. Cicirello, V.A., Smith, S.F.: Wasp-like agents for distributed factory coordination. Autonomous Agents and Multi-Agent Systems **8**(3), 237–266 (2004). doi:10.1023/B:AGNT.0000018807. 12771.60
9. Das, S., Biswas, A., Dasgupta, S., Abraham, A.: Bacterial foraging optimization algorithm: Theoretical foundations, analysis, and applications. In: J. Kacprzyk, A. Abraham, A.E. Hassanien, P. Siarry, A. Engelbrecht (eds.) *Foundations of Computational Intelligence*, vol. 3. Studies in Computational Intelligence, vol. 203, pp. 23–55. Springer, Berlin, Heidelberg (2009). http:// www.springerlink.com/index/10.1007/978-3-642-01085-9_2
10. Faritha Banu, A., Chandrasekar, C.: An optimized approach of modified BAT algorithm to record deduplication. International Journal of Computer Applications **62**(1), 10–15 (2013). doi:10.5120/10043-4627. http://research.ijcaonline.org/volume62/number1/pxc3884627.pdf
11. Feng, X., Lau, F.C.M., Gao, D.: A new bio-inspired approach to the traveling salesman problem. In: O. Akan, P. Bellavista, J. Cao, F. Dressler, D. Ferrari, M. Gerla, H. Kobayashi, S. Palazzo, S. Sahni, X.S. Shen, M. Stan, J. Xiaohua, A. Zomaya, G. Coulson, J. Zhou (eds.) *Complex Sciences*, vol. 5, pp. 1310–1321. Springer, Berlin, Heidelberg (2009). http://www.springerlink. com/index/10.1007/978-3-642-02469-6_12
12. Fourie, J., Green, R., Geem, Z.W.: Generalised adaptive harmony search: A comparative analysis of modern harmony search. Journal of Applied Mathematics **2013**, 1–13 (2013). doi:10. 1155/2013/380985. http://www.hindawi.com/journals/jam/2013/380985/
13. Gandomi, A.H., Yang, X.S., Alavi, A.H., Talatahari, S.: Bat algorithm for constrained optimization tasks. Neural Computing and Applications **22**(6), 1239–1255 (2012). doi:10.1007/ s00521-012-1028-9
14. Geem, Z.W.: *Recent Advances in Harmony Search Algorithm*. Studies in Computational Intelligence, vol. 270. Springer, Berlin (2010)
15. Geem, Z.W.: State-of-the-art in the structure of harmony search algorithm. In: Z.W. Geem (ed.) *Recent Advances in Harmony Search Algorithm*. Studies in Computational Inatelligence, vol. 270, pp. 1–10. Springer, Berlin, Heidelberg (2010). http://www.springerlink.com/index/ 10.1007/978-3-642-04317-8_1
16. Geem, Z.W., Kim, J.H., Loganathan, G.: A new heuristic optimization algorithm: Harmony search. Simulation **76**(2), 60–68 (2001). doi:10.1177/003754970107600201
17. Haddad, O.B., Afshar, A., Mario, M.A.: Honey-bees mating optimization (HBMO) algorithm: A new heuristic approach for water resources optimization. Water Resources Management **20**(5), 661–680 (2006). doi:10.1007/s11269-005-9001-3
18. Hasanebi, O., Erdal, F., Saka, M.P.: Adaptive harmony search method for structural optimization. Journal of Structural Engineering **136**(4), 419–431 (2010). doi:10.1061/(ASCE)ST.1943-541X.0000128
19. Havens, T.C., Spain, C.J., Salmon, N.G., Keller, J.M.: Roach infestation optimization. In: *Swarm Intelligence Symposium 2008 (SIS 2008)*, St. Louis, MO, pp. 1–7. IEEE (2008). doi:10. 1109/SIS.2008.4668317
20. Hedayatzadeh, R., Akhavan Salmassi, F., Keshtgari, M., Akbari, R., Ziarati, K.: Termite colony optimization: A novel approach for optimizing continuous problems. In: *18th Iranian Conference on Electrical Engineering (ICEE)*, pp. 553–558. IEEE (2010). doi:10.1109/ IRANIANCEE.2010.5507009
21. Karaboga, D.: An idea based on honey bee swarm for numerical optimization. Technical Report TR06, Engineering Faculty, Computer Engineering Department, Erciyes University, Kayseri, Turkey (2005)
22. Karaboga, D., Akay, B.: A comparative study of artificial bee colony algorithm. Applied Mathematics and Computation **214**(1), 108–132 (2009). doi:10.1016/j.amc.2009.03.090. http:// linkinghub.elsevier.com/retrieve/pii/S0096300309002860
23. Karaboga, D., Gorkemli, B., Ozturk, C., Karaboga, N.: A comprehensive survey: Artificial bee colony (ABC) algorithm and applications. Artificial Intelligence Review **42**(1), 21–57 (2012). doi:10.1007/s10462-012-9328-0

24. Krishnanand, K., Ghose, D.: Detection of multiple source locations using a glowworm metaphor with applications to collective robotics. In: *Proceedings of IEEE Swarm Intelligence Symposium 2005 (SIS 2005)*, pp. 84–91. IEEE (2005). doi:10.1109/SIS.2005.1501606

25. Lin, J.H., Chou, C.W., Yang, C.H., Tsai, H.L.: A chaotic Levy flight bat algorithm for parameter estimation in nonlinear dynamic biological systems. Journal of Computer and Information **2**(2), 56–63 (2012). www.AcademyPublish.org

26. Liu, Y., Passino, K.: Biomimicry of social foraging bacteria for distributed optimization: Models, principles, and emergent behaviors. Journal of Optimization Theory and Applications **115**(3), 603–628 (2002). doi:10.1023/A:1021207331209

27. Mahdavi, M., Fesanghary, M., Damangir, E.: An improved harmony search algorithm for solving optimization problems. Applied Mathematics and Computation **188**(2), 1567–1579 (2007). doi:10.1016/j.amc.2006.11.033. http://linkinghub.elsevier.com/retrieve/pii/S0096300306015098

28. Martin, H.R.: Termite: A swarm intelligent routing algorithm for mobile wireless ad-hoc networks. Ph.D. thesis, Faculty of the Graduate School of Cornell University (2005)

29. Monismith, D.R.: The uses of the slime mold lifecycle as a model for numerical optimization. Ph.D. thesis, Oklahoma State University (2008)

30. Monismith, D.R., Mayfield, B.E.: Slime mold as a model for numerical optimization. In: *Swarm Intelligence Symposium 2008 (SIS 2008)*, St. Louis, MO, pp. 1–8. IEEE (2008). doi:10.1109/SIS.2008.4668295

31. Muller, S., Marchetto, J., Airaghi, S., Kournoutsakos, P.: Optimization based on bacterial chemotaxis. IEEE Transactions on Evolutionary Computation **6**(1), 16–29 (2002). doi:10.1109/4235.985689

32. Nakamura, R.Y.M., Pereira, L.A.M., Costa, K.A., Rodrigues, D., Papa, J.P., Yang, X.S.: BBA: A binary bat algorithm for feature selection. In: *25th SIBGRAPI Conference on Graphics, Patterns and Images (SIBGRAPI 2012)*, pp. 291–297. IEEE (2012). doi:10.1109/SIBGRAPI.2012.47

33. Pinto, P.C., Runkler, T.A., Sousa, J.M.C.: Wasp swarm algorithm for dynamic MAX-SAT problems. In: B. Beliczynski, A. Dzielinski, M. Iwanowski, B. Ribeiro (eds.) *Adaptive and Natural Computing Algorithms*. LNCS, vol. 4431, pp. 350–357. Springer, Berlin, Heidelberg (2007). http://www.springerlink.com/index/10.1007/978-3-540-71618-1_39

34. Roth, M.: A framework and model for soft routing: The markovian termite and other curious creatures. In: M. Dorigo, L.M. Gambardella, M. Birattari, A. Martinoli, R. Poli, T. Stützle (eds.) *Ant Colony Optimization and Swarm Intelligence*. LNCS, vol. 4150, pp. 13–24. Springer, Berlin, Heidelberg (2006). http://www.springerlink.com/index/10.1007/11839088_2

35. Sharvani, G.S., Ananth, A.G., Rangaswamy, T.M.: Ant colony optimization based modified termite algorithm (MTA) with efficient stagnation avoidance strategy for MANETs. International Journal on Applications of Graph Theory in wireless Ad Hoc Networks and Sensor Networks **4**(2/3), 39–50 (2012). doi:10.5121/jgraphoc.2012.4204. http://www.airccse.org/journal/graphhoc/papers/4312jgraph04.pdf

36. Tautz, J.: *L'tonnante abeille*. De Boeck, Brussels (2009)

37. Theraulaz, G., Goss, S., Gervet, J., Deneubourg, J.L.: Task differentiation in *Polistes* wasp colonies: A model for self-organizing groups of robots. In: J.-A. Meyer, S.W. Wilson *From Animals to Animats: Proceedings of the First International Conference on Simulation of Adaptive Behavior*, pp. 346–355. MIT Press, Cambridge, MA (1990). http://dl.acm.org/citation.cfm?id=116517.116556

38. Wang, G., Guo, L., Duan, H., Liu, L., Wang, H.: A bat algorithm with mutation for UCAV path planning. Scientific World Journal **2012**, 1–15 (2012). doi:10.1100/2012/418946. http://www.hindawi.com/journals/tswj/2012/418946/

39. Wedde, H.F., Farooq, M., Zhang, Y.: BeeHive: An efficient fault-tolerant routing algorithm inspired by honey bee behavior. In: M. Dorigo, M. Birattari, C. Blum, L.M. Gambardella, F. Mondada, T. Stützle (eds.) *Ant Colony Optimization and Swarm Intelligence*. Lecture Notes in Computer Science, vol. 3172, pp. 83–94. Springer, Berlin, Heidelberg (2004). http://www.springerlink.com/index/10.1007/978-3-540-28646-2_8

40. Wedde, H.F., Farooq, M., Pannenbaecker, T., Vogel, B., Mueller, C., Meth, J., Jeruschkat, R.: BeeAdHoc: An energy efficient routing algorithm for mobile ad hoc networks inspired by bee behavior. In: *Proceedings of the 2005 Conference on Genetic and Evolutionary Computation, GECCO'05*, pp. 153–160. ACM, New York (2005). doi:10.1145/1068009.1068034

41. Worasucheep, C.: A harmony search with adaptive pitch adjustment for continuous optimization. International Journal of Hybrid Information Technology **4**(4), 13–24 (2011)

42. Wu, B., Qian, C., Ni, W., Fan, S.: The improvement of glowworm swarm optimization for continuous optimization problems. Expert Systems with Applications **39**(7), 6335–6342 (2012). doi:10.1016/j.eswa.2011.12.017. http://linkinghub.elsevier.com/retrieve/pii/S0957417411016885

43. Yang, X.S.: Engineering optimizations via nature-inspired virtual bee algorithms. In: J. Mira, J.R. Alvarez (eds.) *Artificial Intelligence and Knowledge Engineering Applications: A Bioinspired Approach: First International Work-Conference on the Interplay Between Natural and Artificial Computation, IWINAC'05*, Part II, pp. 317–323. Springer, Berlin, Heidelberg (2005). doi:10.1007/11499305_33

44. Yang, X.S.: Firefly algorithm, Lévy flights and global optimization. In: M. Bramer, R. Ellis, M. Petridis (eds.) *Research and Development in Intelligent Systems XXVI*, pp. 209–218. Springer, London (2010). http://www.springerlink.com/index/10.1007/978-1-84882-983-1_15

45. Yang, X.S.: *Nature-Inspired Metaheuristic Algorithms*, 2nd edn. Luniver Press, Frome, UK (2010)

46. Yang, X.S.: A new metaheuristic bat-inspired algorithm. In: J. Kacprzyk, J.R. González, D.A. Pelta, C. Cruz, G. Terrazas, N. Krasnogor (eds.) *Nature Inspired Cooperative Strategies for Optimization (NICSO 2010)*. Studies in Computational Intelligence, vol. 284, pp. 65–74. Springer, Berlin, Heidelberg (2010). http://www.springerlink.com/index/10.1007/978-3-642-12538-6_6

47. Yang, X.S., Deb, S.: Cuckoo search via Lévy flights. In: *World Congress on Nature & Biologically Inspired Computing 2009 (NaBIC 2009)*, pp. 210–214. IEEE (2009). doi:10.1109/NABIC.2009.5393690

48. Zungeru, A.M., Ang, L.M., Seng, K.P.: Performance of termite-hill routing algorithm on sink mobility in wireless sensor networks. In: Y. Tan, Y. Shi, Z. Ji (eds.) *Advances in Swarm Intelligence*. Lecture Notes in Computer Science, vol. 7332, pp. 334–343. Springer, Berlin, Heidelberg (2012). http://www.springerlink.com/index/10.1007/978-3-642-31020-1_39

Chapter 11
Extensions of Evolutionary Algorithms to Multimodal and Multiobjective Optimization

Alain Petrowski

11.1 Introduction

Real world problems can seldom be fully formalized. Many decisions depend on the image that a company desires to project, the policy it wants to adopt vis-à-vis its customers and its competitors, its economic or legal environment, etc. Its decisions regarding the design of a new product, its manufacture, and its launch depend on dialogue and negotiations with several players. All of these factors make it difficult to formalize such decision problems with the aim of solving them by means of a computer.

In the context of optimization, a problem can have several optimal solutions of equivalent value. This occurs when the objective function of an optimization problem is multimodal, i.e., when it has several solutions which are global optima. This also occurs in the field of multiobjective optimization, which consists in optimizing simultaneously several objectives, leading generally to making compromises between them.

Theoretically, one solution is enough. However, when factors that have not been formalized, i.e., have not been integrated into the constraints or objective functions of a problem, this is not adequate. It is therefore valuable to have a representative sample of the diversity of equivalent-value solutions so that a decision maker can choose the one that seems best.

This chapter is therefore devoted to the presentation of extensions of evolutionary algorithms to address multimodal and multiobjective problems.

A. Petrowski (✉)
Telecom SudParis, 91000 Evry, France
e-mail: Alain.Petrowski@telecom-sudparis.eu

© Springer International Publishing Switzerland 2016
P. Siarry (ed.), *Metaheuristics*, DOI 10.1007/978-3-319-45403-0_11

11.2 Multimodal Optimization

11.2.1 The Problem

Multimodal optimization consists in locating multiple global optima, and possibly the best local optima, of an objective function. Evolutionary algorithms are good candidates for achieving this task because they handle a population of solutions that can be distributed among various optima. We should note that there are methods to search for several optima, such as sequential niching [2], which do not require a population-based algorithm to succeed, but they show quite poor performance. This is why this section is entirely devoted to evolutionary methods. However, if a multimodal objective function is subjected to a standard evolutionary algorithm, experiments show that the population is stabilized on only one of the maxima of the fitness function (see Fig. 11.1a), and this is not necessarily a global maximum. For example, consider a function with two peaks of equal maximum value. An initial population is built in which the individuals are evenly located, around the two optima. After a few generations, the balance will be broken because of genetic drift, until the population is mainly localized around a single peak. The problem of multimodal optimization could be correctly solved if a mechanism that could stabilize subpopulations located on the highest peaks of the fitness function. This is *speciation*, which makes it possible to classify the individuals of a population into different *subpopulations*, and *niching*, which stabilizes subpopulations within *ecological niches* containing the optima of the objective function. There are several methods of speciation and niching. The most common and the most effective ones are described below.

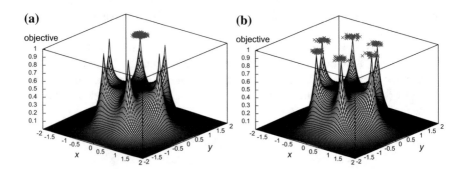

Fig. 11.1 **a** Selection without sharing: the individuals converge towards only one of the optima. **b** Selection with sharing: the individuals converge towards several optima

11.2.2 Niching with the Sharing Method

The concept of "sharing of limited resources within an *ecological niche*", suggested by Holland [16], constitutes one of the most effective approaches to creating and maintaining stable subpopulations around the peaks of the objective function with an evolutionary algorithm. The concept of an ecological niche originates from the study of population dynamics. It was formalized by Hutchinson in 1957 [19], who defined it as a hypervolume in an n-dimensional space, each dimension representing a living condition (e.g., quantity of food, temperature, or the size of the vital domain). An ecological niche cannot be occupied by several species simultaneously. This is the empirical principle of *competitive exclusion*. The resources within a niche being limited, the size of the population that occupies it stabilizes.

In 1987, Goldberg and Richardson [15] proposed an adaptation of this concept for genetic algorithms, which can be generalized to any evolutionary algorithm. The technique is known under the name of the *sharing method*. It is essential that a concept of dissimilarity between individuals be introduced. For example, if the individuals are bit strings, the Hamming distance may be appropriate. If they are vectors in $\mathbb{R}^n$, the Euclidean distance is a priori a good choice. The value of the dissimilarity is the criterion for deciding whether two individuals belong to the same niche or not. The method consists in attributing a *shared fitness* to each individual, which is equal to its raw fitness divided by a quantity that increases with the number of individuals resembling it. The shared fitness can be viewed as representing a quantity of a resource available to each individual in a niche. The selection is ideally proportional, so that the number of offspring of an individual is proportional to its shared fitness, although the method has also been employed with other selection models. Thus, for the same raw fitness, an isolated individual will definitely have more offspring than an individual that has many neighbors in the same niche. At equilibrium, the number of individuals located on each peak becomes proportional, to a first approximation, to the fitness associated with that peak. This appears to give rise to a stable *subpopulation* in each niche. The shared fitness of an individual i can be expressed as

$$\tilde{f}(i) = \frac{f(i)}{\sum_{j=1}^{\mu} \text{sh}(d(i, j))}$$

where sh is of the form

$$\text{sh}(d) = \begin{cases} 1 - \left(\frac{d}{\sigma_s}\right)^{\alpha} & \text{if } d < \sigma_s \\ 0 & \text{otherwise} \end{cases}$$

Here, sh is the sharing function; $d(i, j)$ is the distance between the individuals i and j, which depends on the representation chosen; σ_s is the *niche radius*, or dissimilarity threshold; α is the "sharpness" parameter; and μ is the population size.

Let us assume that α is chosen very large, tending towards infinity; then $(d/\sigma_s)^{\alpha}$ tends towards 0 and sh(d) is 1 if $d < \sigma_s$, or otherwise equal to 0. Then

$\sum_{j=1}^{\mu}$ sh $(d(i, j))$ is the number of individuals located within a ball of radius σ_s centered on the individual i. The shared fitness is thus, in this case, the raw fitness of the individual i divided by the number of its neighbors. This type of niching performs well on condition that the distances between the peaks are less than the niche radius σ_s. However, for a given optimization problem, barring a few rare cases, the distances between the peaks are not known a priori. Then, if the radius is selected to be too large, all optima cannot be discovered by the individuals of the population. An imperfect solution to this problem consists in defining niches as balls with a fuzzy boundary. Thus, the individuals j for which the distances to the individual i are close to σ_s have a weaker contribution to the value of sh($d(i, j)$) than the others. In this way, if unfortunately the niche (already presumably centered on a peak) contains another peak close to its boundary, it will be less probable that the latter peak will perturb the persistence of the presence of individuals on the central peak. The "sharpness" of the niche boundaries is controlled by the parameter α, which is assigned a default value of 1.

Now, let us consider the case where the radius σ_s is selected to be too small compared with the distances between the peaks. There will then be several niches for each peak. In theory, this presents no difficulty but in practice it implies putting many more individuals in the niches than necessary, and thus it will require a population size larger than necessary. This will cause wastage of computational resources. If the population is not of sufficient size, it is very much possible that all the global optima of the problem will not be discovered. Hence an accurate estimation of σ_s is of prime importance. For this reason, we shall present some suggestions for ways to come close to this objective.

Figures 11.1 shows the distributions of the individuals on the peaks of a multimodal function defined in $\mathbb{R}^2$ after convergence of an evolutionary algorithm, with and without sharing of the fitness function. The individuals are projected on to a plane at the height of the optimum, parallel to the x and y axes, so that they are more visible.

11.2.2.1 Genetic Drift and the Sharing Method

Let us assume that the individuals are distributed over all the global peaks of the fitness function after a sufficient number of generations. Let N be the population size and let p be the number of peaks; each peak will be occupied by a subpopulation of approximately N/p individuals. Assume also that the fitnesses of all the individuals are close to the fitness value for the global optima. When an equilibrium situation is reached, the subpopulations for the next generation will have approximately the same size. Consequently, each individual is expected to have a number of offspring close to unity. In this case, the effective number of offspring of an individual that is obtained by employing a stochastic selection technique can be zero, with a nonnegligible probability. Even with a sampling technique of minimal variance such as SUS selection, an individual can have zero or one effective offspring if the expected number of offspring is slightly lower than unity. Hence, there is a possibility, which becomes more significant for a small population, that a subpopulation covering a peak may disappear because of stochastic fluctuations. To reduce this possibility to an acceptable level, it is necessary to allot to each peak a high number of individuals, so that the sharing method requires a priori large population sizes.

11.2.2.2   Advantages and Difficulties of the Application of the Method

The basic sharing method possesses excellent stability if the population size is large enough to counter genetic drift. With the help of variation operators capable of ensuring good diversity, the distribution of the population after some generations does not depend on the initial population. The main difficulty of this method lies in the appropriate choice of the niche radius σ_s. Another drawback relates to the algorithmic complexity, which is given by $\mathcal{O}(\mu^2)$, where μ is the population size. As the method requires large population sizes, this can be seriously disadvantageous except when the calculation of the fitness function is very long. The basic sharing method is not compatible with elitism. Lastly, it is well suited to being used with a proportional selection technique. Various authors have proposed solutions to overcome these disadvantages. The long history of the sharing method and its effectiveness in terms of the maintenance of diversity make it, even today, the best known and the most often used niching technique.

11.2.3   Niching with the Deterministic Crowding Method

The first method of nitching by *crowding* was presented by De Jong in 1975 [7]. Like the sharing method, it utilizes a value of distance, or at least of dissimilarity, between individuals, but it operates at the level of the replacement operator. De Jong suggested that for each generation, the number of offspring should be of the order of ten times less than the number of parents. A higher value decreases the effectiveness of the method. A lower value would favor genetic drift too much. All the offspring find themselves in the population of the parents for the next generation, and hence the parents that they replace have to be chosen. The replacement operator selects a parent that must "die" for the offspring that resembles it closest. Nevertheless, the similarity comparisons are not systematic, and an offspring is compared only with one small sample of C_F parents randomly drawn from the population. C_F is the crowding factor. De Jong showed, for some test functions, that a value of C_F fixed at two or three gives interesting results. Hence the individuals tend to be distributed among the various peaks of the fitness function, thus preserving preexisting diversity in the population.

However, the method makes frequent replacement errors due to the low value of C_F, which is prejudicial to the niche effect. But a high value of C_F produces too strong a reduction of the selection pressure. Indeed, the parents which are replaced, being similar to the offspring, have almost the same fitnesses if the function is continuous.

Their replacement thus improves the fitnesses within the population very little. In contrast, the selection pressure is stronger if efficient offspring replace less efficient parents, i.e., if errors in replacement are made, which implies that C_F must be small.

In 1992, Mahfoud [23] proposed the *deterministic crowding method* as a major improvement over the method of De Jong. The main idea is that a pair of offspring e_1 and e_2, obtained after crossover and mutation, enters into competition only with its two parents, p_1 and p_2. There are two possibilities for replacement:

(a) e_1 replaces p_1 and e_2 replaces p_2;
(b) e_1 replaces p_2 and e_2 replaces p_1.

The choice (a) is selected if the sum of dissimilarities $d(p_1, e_1) + d(p_2, e_2)$ is less than $d(p_1, e_2) + d(p_2, e_1)$; otherwise, the choice (b) is made. Lastly, the replacement of a parent by an offspring is effective only if the parent is less efficient than the offspring. This can be described as a deterministic tournament. This implies that the method is elitist, because if the best individual is in the population of the parents and not in that of the offspring, it will not be able to disappear from the population in the next generation.

11.2.3.1 Advantages and Difficulties of the Application of the Method

Deterministic crowding does not require the determination of appropriate parameter values that depend on the problem, such as a niche radius. In fact, only the population size is significant and is chosen according to a very simple criterion: the larger the number of optima to be found, the larger the population. The number of calculations of distances to be carried out is of the order of the population size, which is lower by an order of magnitude compared with the sharing method. Deterministic crowding is a replacement operator that favors the best individuals. Thus, selection for reproduction may be absent, i.e., reduced to its simplest expression: a parent always produces only one offspring, irrespective of its fitness. In this case, the selection operators involve only computational dependencies between pairs of offspring and their parents. Thus parallelization of the method is both simple and efficient. All these qualities are interesting, but deterministic crowding does not reduce genetic drift significantly compared with an algorithm without niching. From this point of view, this method is less powerful than the sharing method. This implies that, if the peaks are occupied by individuals for a certain number of generations, the population will finally converge towards only one optimum. This disadvantage often leads us to the conclusion that methods with low genetic drift are preferred, even if their use is less simple.

11.2.4 The Clearing Procedure

The *clearing procedure* was proposed in 1996 by Petrowski [26]. This is based on limited resource sharing within ecological niches, according to the principle suggested by Holland, with the particularity that the distribution of resources is not equitable among the individuals. Thus the clearing procedure typically assigns all the resources of a niche to the best individual, referred to as the dominant individual. The other individuals in the same niche will not have anything. This means that only the dominant individual will be able to reproduce to generate a subpopulation for the next generation. The algorithm thus determines the subpopulations in which the dominant individuals are identified. The simplest method consists in choosing a distance d that is significant for the problem and to represent the niches with balls of radius σ_c centered on the dominants. The value of σ_c must be lower than the distance between two optima of the fitness function, so that they can be distinguished to maintain individuals on all of them. Thus the problem now consists in discovering all the dominant individuals in a population. The population is initially sorted according to decreasing fitness. A step of the algorithm is implemented in three phases to produce a niche:

1. The first individual in the population is the best individual. This individual is obviously a dominant individual.
2. The distances of all the individuals from the dominant one are computed. The individuals located at a distance closer than σ_c belong to a niche centered on the dominant individual. Hence, they are dominated and thus their fitnesses are assigned to zero.
3. The dominant and dominated individuals are virtually withdrawn from the population. The procedure is then reapplied, starting from step 1, to the new reduced population.

The operator has as many steps as the algorithm finds dominant individuals. These preserve the fitness which they obtained before the application of niching. The operator is applied just before application of proportional selection.

11.2.4.1 Elitism and Genetic Drift

The clearing procedure lends itself easily to implementing an elitist strategy: it suffices to preserve the dominant individuals from the better subpopulations to inject them into the population for the next generation. If the number of optima to be discovered is known a priori, the same number of dominant individuals is preserved. In the opposite case, one simple strategy, among others, consists in preserving in the population the dominant individuals whose fitness is better than the average fitness of the individuals in the population before clearing. Nevertheless, it is necessary to take precautions so that the number of individuals preserved is not too large compared with the population size.

If the dominant individuals have located the optima of the function in a given generation, elitism will maintain them indefinitely on the peaks. This algorithm is perfectly stable, unlike the methods discussed before. Genetic drift does not have a detrimental effect in this context! This enables us to reduce the required population sizes compared with other methods.

11.2.4.2   Niche Radius

Initially, the estimation of the niche radius σ_c follows the same rules as for the sharing method. Theoretically, it should be lower than the minimum distance between all the global optima considered in pairs, so that all of them will be discovered. However, the choice of too large a niche radius does not have the same effects as in the sharing method, where this situation leads to instabilities with an increased genetic drift. If this occurs in the clearing procedure, certain optima will be ignored by the algorithm, without disturbing its convergence towards those which have been located. Hence, the criterion for determination of the radius can be different. Indeed, the user of a multimodal optimization algorithm does not need to know all the global optima, which is impossible when there is an infinite number of them in a continuous domain, but only a representative sample of the diversity of these optima. Locating the global optima corresponding to instances of almost identical solutions will not be very useful. On the other hand, it is more interesting to determine instances of optimal solutions distant from each other in the search space. Thus, the determination of σ_c depends more on the minimum distance between the desired optimal solutions, a piece of information independent of the fitness function, than on the minimum distance between the optima, which depends strongly on the fitness and is generally unknown. If, however, a knowledge of all the global optima is required, there are techniques which enable estimation of the niche radius by estimating the width of the peaks. It is also possible to build niches which are not balls, with the help of an explicit speciation (see Sect. 11.2.5).

11.2.4.3   Advantages and Difficulties of Application of the Method

The principal quality of the method lies in its great resistance to the loss of diversity by genetic drift, especially in its elitist version [28]. Therefore it can work with relatively modest population sizes, which results in reduced computing time. The niche radius is a parameter which can be defined independently of the landscape of the fitness function, unlike the case for sharing method, and instead according to the desired diversity of the multiple solutions.

The clearing procedure requires about $\mathcal{O}(c\mu)$ distance calculations, where c is the number of niches and μ the population size. This number is less than in the sharing method, but higher than in the deterministic crowding method.

If it is found during the process of evolution that the number of dominant individuals is of the same order of magnitude as the population size, this indicates that:

- Either the population size is insufficient to discover the optima using the sampling step fixed by the niche radius.
- Or this step is too small, compared with the computational resources assigned to the solution of the problem. It is then preferable to increase the niche radius, so that the optima found are distributed as widely as possible in the entire search space.

The method performs unsatisfactorily with the condition of restricted mating using a restriction radius less than or equal to the niche radius (see Chap. 6, Sect. 6.4.2). In that case, crossover will be useless, because it will be applied only to similar individuals: the selected individuals, which are clones of the same dominant individual. To overcome this problem, there are at least two solutions. One solution is to carry out mutation at a high rate before the crossover, in order to restore diversity within each niche. The other is to increase the restriction radius. In the latter case, the effect of exploration due to the crossover becomes more significant. Indeed, it may be that between two dominant individuals around two peaks there are located some areas of interest that the crossover is likely to explore. But this can also generate a high rate of lethal crossovers, reducing the convergence speed of the algorithm.

11.2.5 Speciation

The main task of speciation is to identify the existing niches in a search space. So far, in our discussions, only one species can occupy a niche; it is then assumed that the individuals of a population that occupy it belong to a species or a *subpopulation*. Once determined by speciation, a subpopulation can be used in several ways. For example, it can be stabilized around a peak by employing a niching technique. Restricted mating can also be practiced inside subpopulations, which, in addition to the improvement due to the reduction in the number of lethal crossovers, conforms to the biological metaphor, which requires that two individuals of different species cannot mate and procreate.

The balls used in the techniques of niching described above can be viewed as niches created by an implicit speciation. The sharing method and the clearing procedure also perform satisfactorily if the niches are provided to them by the explicit, prior application of a speciation method. For that purpose, such a method must provide a partition of the population $S = \{S_1, S_2, \ldots, S_c,\}$ into c subpopulations S_i. It is then easy to apply, for example:

- niching by the sharing method, by defining the shared fitness as

$$\tilde{f}(i) = \frac{f(i)}{\text{card}(S_j)}, \quad \forall i \in S_j$$

for all subpopulations S_j;
- niching by the clearing procedure, by preserving the fitness of the best individual of any subpopulation S_j and forcing the fitnesses of other individuals to zero;
- restricted mating, which operates only between the individuals of any subpopulation S_j.

Moreover, an explicit speciation technique is compatible with elitism: since the individuals of a subpopulation are clearly identified, it is possible to preserve the best one from each subpopulation in a generation for the next generation.

11.2.5.1 Label-Based Speciation

In 1994, Spears proposed [30] a simple speciation technique using *tag-bits*, where an integer number belonging to a set $\mathbf{T} = \{T_1, T_2, \ldots, T_c\}$ is associated with each individual in a population. The value of the label T_i signifies the subpopulation $\mathbf{S}_i$ to which all the individuals labeled by T_i belong; c is the maximum number of subpopulations which can exist in the population. The method was so named because originally Spears had proposed his method within the framework of genetic algorithms, and the labels were represented by bit strings. During the construction of the initial population, the labels attached to each individual are drawn randomly from the set $\mathbf{T}$. During evolution, the labels can mutate, by selecting randomly a new value in $\mathbf{T}$. Mutation corresponds in this case to a *migration* from one subpopulation towards another. After some generations, the subpopulations are placed on the peaks of the fitness function because of the selective pressure. However, there is no guarantee that each peak containing a global optimum will be occupied by one and only one subpopulation. Some of them can be forgotten, while others can be occupied by several subpopulations. Hence the method is not a reliable one. It is quoted here because it is well known in the world of evolutionary computation.

11.2.5.2 Island Model

The *island model* is also a classical concept in the field of evolutionary computation. This model evolves several subpopulations $\mathbf{S}_i$ through a series of epochs. During each epoch, the subpopulations evolve independently of each other, over a given number of generations G_i. At the end of each epoch, the individuals move between the subpopulations during a phase of *migration*, followed by a possible phase of assimilation. The latter phase is employed to carry out operations of integration of the migrants into their host subpopulations, for example by stabilizing the sizes of the subpopulations. This iterative procedure continues until a user-defined termination criterion for the algorithm is satisfied. The migration does not take place arbitrarily, but according to a relation of neighborhood defined between the subpopulations. The proportion of migrating individuals is determined by migration rates chosen by the user.

Originally, this model was developed as a model for parallelization of a genetic algorithm. This enables it to be efficiently implemented in a distributed-memory multiprocessor computer, where each processing unit deals with a subpopulation [5]. It can be noticed that, logically, this process is similar to label-based speciation, with the mutation of the labels constrained by the neighborhood relations. Label mutation takes place only at the end of each epoch. Like label-based speciation, this method is lacking in reliability for the distribution of the subpopulations on the peaks of the fitness function. However, the fact that the subpopulations evolve independently during each epoch, offers the advantage of a more accentuated local search for optima.

11.2.5.3 Speciation by Clustering

During evolution, the individuals of a population tend to gather in the areas of the search space showing high fitness under the action of the selection pressure. These areas are good candidates for containing global optima. The application of a classical clustering method (e.g., the K-means algorithm or the LBG algorithm) partitions the search space as many areas as the number of accumulations of individuals that are detected. Each detected area corresponds to niche, and the individuals located there constitute a subpopulation [35]. The method is reliable with large population sizes because a niche can be identified only if it contains a large enough cluster. This number can be significantly reduced if the speciation algorithm exploits the fitness values of the individuals in each area, in order to recognize better the existence of possible peaks in those regions [27]. It is interesting to combine clustering-based speciation with an island model, in order to profit from the advantages of both methods: a reliable global search for the highest peaks, which occurs during the migration phases (exploration) and an improved local search for the optima (exploitation) during the epochs [3].

11.3 Multiobjective Optimization

Multiobjective or multicriteria optimization concerns the case of the simultaneous presence of several objectives, or criteria, often contradictory with each other. Let $\mathbf{f}(\mathbf{x})$ be a vector of c objectives associated with an instance of a solution $\mathbf{x}$ of a multiobjective optimization problem. Each of its components $f_i(\mathbf{x})$ is equal to the value of the ith objective for solution $\mathbf{x}$. Without loss of generality, we will consider in the following sections the case where all the objectives of a problem have to be minimized. When a problem does not conform to this condition, it is enough to change the signs of those objectives which must be maximized.

11.3.1 Problem Formalization

11.3.1.1 Pareto Dominance

Let us consider two vectors of objectives $\mathbf{v}$ and $\mathbf{u}$. If all the components of $\mathbf{v}$ are less than or equal to the components of $\mathbf{u}$, with at least one strictly lower component, then the vector $\mathbf{v}$ corresponds to a better solution than $\mathbf{u}$. In this case, it is said that $\mathbf{v}$ *dominates* $\mathbf{u}$ in the Pareto sense. In a more formal way, can be written as $\mathbf{v} \overset{P}{<} \mathbf{u}$:

$$\mathbf{v} \overset{P}{<} \mathbf{u} \iff \forall i \in \{1, \ldots, c\}, v_i \leq u_i \text{ and } (\exists j \in \{1, \ldots, c\}, v_j < u_j)$$

Figure 11.2 represents the relations of domination between six objective vectors in a two-dimensional space. c is dominated by a, b, and e. d is dominated by e. f is dominated by b, d, and e.

11.3.1.2 Pareto Optimum

The set of the objective vectors which cannot be dominated constitutes the optimal values of the problem in the Pareto sense. These vectors belong to the *Pareto front*, or *trade-off surface*, denoted by $\mathcal{P}$:

$$\mathcal{P} = \{\mathbf{f}(\mathbf{x}) | \mathbf{x} \in \mathbf{\Omega}, \nexists \mathbf{y} \in \mathbf{\Omega}, \mathbf{f}(\mathbf{y}) \overset{P}{<} \mathbf{f}(\mathbf{x})\}$$

The *Pareto-optimal set* $\mathbf{X}^*$ is defined as the set of the solutions in the search space $\mathbf{\Omega}$ whose objective vectors belong to the Pareto front:

$$\mathbf{X}^* = \{\mathbf{x} \in \mathbf{\Omega} | \mathbf{f}(\mathbf{x}) \in \mathcal{P}\}$$

11.3.1.3 Multiobjective Optimization Algorithms

Multiobjective optimization consists in building the Pareto-optimal set $\mathbf{X}^*$. However, $\mathbf{X}^*$ can contain an infinite number of solutions if the search space is continuous. Even if $\mathbf{\Omega}$ is finite, $\mathbf{X}^*$ must not be too large if a decision maker is to be able to exploit it effectively. Thus, it is expected that a multiobjective optimization algorithm should produce a set of nondominated solutions, not too large, such that they are as close as possible to the Pareto front, covering it as evenly and as completely as possible [8].

Fig. 11.2 Domination in the Pareto sense in an objective space of dimension 2

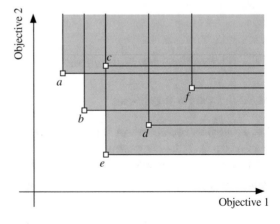

11.3.2 The Quality Indicators

There is a wide choice of multiobjective optimization algorithms, each with its preferred areas of application, which are rarely well characterized. Additionally, these algorithms often use parameters whose values can strongly influence the quality of the results, while their values are hard to estimate to best achieve the goals of the user. Too often, practitioners have no other way to select the best approach than comparing the results provided by several algorithms and parameter sets. It is therefore important that they have quality indicators to facilitate performance analysis of the approaches tested.

Many quality indicators have been proposed for multiobjective optimization [21, 41]. Three commonly used indicators are described below. The first two of them are described because they are mentioned in Sect. 11.3.5.3. The third indicator is described because of its good properties, although it requires high computational power.

11.3.2.1 The Generational Distance

This metric [32] gives the distance between the Pareto front and a set of n nondominated solutions. The expression for it is

$$D_p = \frac{(\sum_{i=1}^{n} d_i{}^p)^{1/p}}{n}$$

where d_i is the distance between the objective vector associated with solution i and the closest point to the Pareto front, and p is a constant, usually chosen equal to 2. $p = 1$ is also often used.

This metric has the advantage of a relatively low computational cost. However, the difficulty in using it is that, in most real-world problems, the Pareto front is not known in advance. In this case, if a lower approximation (for minimization of the objectives) is available, it may replace the Pareto front. Then, obviously, the value of D_p is no longer significant in itself, but it allows one to compare the results given by several optimizers. Another drawback of this metric is that it does not take into account the quality of the coverage of the Pareto front by the nondominated solutions obtained after a mutiobjective optimization.

11.3.2.2 The "Two-Set Coverage Metric", or "C-Metric"

This metric was proposed in [37]. Let **A** and **B** be two sets of objective vectors, and let C be a function of **A** and **B** to the interval $[0, 1]$:

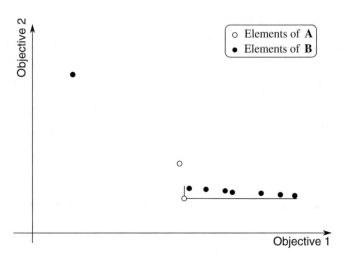

Fig. 11.3 $\mathcal{C}(\mathbf{A}, \mathbf{B}) = 7/8, \mathcal{C}(\mathbf{B}, \mathbf{A}) = 0$: according to the $\mathcal{C}$-metric, **A** should be better than **B**. When sets **A** and **B** have cardinalities that are too different and/or are not evenly distributed, the $\mathcal{C}$-metric can give unreliable results

$$\mathcal{C}(\mathbf{A}, \mathbf{B}) = \frac{|\{\mathbf{b} \in \mathbf{B} | \exists \mathbf{a} \in \mathbf{A}, \mathbf{a} \overset{P}{<} \mathbf{b} \vee \mathbf{a} = \mathbf{b}\}|}{|\mathbf{B}|}$$

This is the fraction of elements of **B** that are dominated by one or more elements of **A**. If $\mathcal{C}(\mathbf{A}, \mathbf{B}) = 1$, this means that all elements of **B** are dominated by those of **A**. In this case, $\mathcal{C}(\mathbf{B}, \mathbf{A}) = 0$ when **A** contains only nondominated vectors. If **A** and **B** contain only Pareto front elements with $\mathbf{a} \neq \mathbf{b}, \forall \mathbf{a} \in \mathbf{A}, \forall \mathbf{b} \in \mathbf{B}$, then $\mathcal{C}(\mathbf{A}, \mathbf{B}) = \mathcal{C}(\mathbf{B}, \mathbf{A}) = 0$. Usually, there is no simple relation between $\mathcal{C}(\mathbf{A}, \mathbf{B})$ and $\mathcal{C}(\mathbf{B}, \mathbf{A})$. Thus, a comparison of the qualities of two sets of nondominated objective vectors **A** and **B** with metric $\mathcal{C}$ requires one to calculate $\mathcal{C}(\mathbf{A}, \mathbf{B})$ and $\mathcal{C}(\mathbf{B}, \mathbf{A})$. We could say that **A** is better than **B** if $\mathcal{C}(\mathbf{A}, \mathbf{B}) > \mathcal{C}(\mathbf{B}, \mathbf{A})$. The metric is more reliable when $\mathcal{C}(\mathbf{A}, \mathbf{B})$ or $\mathcal{C}(\mathbf{B}, \mathbf{A})$ is close to one.

An advantage of this indicator is that it does not require a knowledge of the Pareto front $\mathcal{P}$. It also has a low computational cost and is not affected by differences in orders of magnitude between the objectives. This indicator gives results consistent with intuition when the sets **A** and **B** have similar cardinalities and if their distribution is even. When this is not the case, the indicator can be misleading (Fig. 11.3). In addition, care must be taken not to consider the relation "is better than" in the sense of the $\mathcal{C}$-metric as an order relation. Indeed, there may be configurations where, for three sets **A**, **B**, and **C**; $\mathcal{C}(\mathbf{A}, \mathbf{B}) < \mathcal{C}(\mathbf{B}, \mathbf{A}), \mathcal{C}(\mathbf{B}, \mathbf{C}) < \mathcal{C}(\mathbf{C}, \mathbf{B})$, and $\mathcal{C}(\mathbf{A}, \mathbf{C}) > \mathcal{C}(\mathbf{C}, \mathbf{A})$. This means that **C** would be better than **B**, **B** would be better than **A** and **A** would be better than **C**, which is inconsistent with the transitivity of an order relation [21, 41].

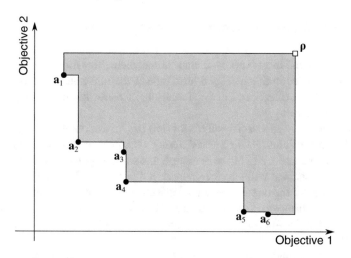

Fig. 11.4 The hypervolume for a two-objective minimization problem $v(\mathbf{A}, \boldsymbol{\rho})$ with $\mathbf{A} = \{a_1, \ldots, a_6\}$ is represented by the *gray* area

11.3.2.3 The Measure of the Hypervolume, or "$\mathcal{S}$-Metric"

Let $\boldsymbol{\rho} = (\rho_1, \ldots, \rho_c)$ be a reference point in the objective space and let $\mathbf{a} = (a_1, \ldots, a_c)$ be an element of a set $\mathbf{A}$ of nondominated objective vectors. $a_i \leq \rho_i$ is required when the objectives are to be minimized. $\boldsymbol{\rho}$ and $\mathbf{a}$ allow one to define a hyperrectangle whose edges are parallel to the coordinate axes of the objective space. The expression for its hypervolume is $v(\mathbf{a}, \boldsymbol{\rho}) = \prod_{i=1}^{c}(\rho_i - a_i)$.

The set $\mathbf{A}$ and point $\boldsymbol{\rho}$ define a hypervolume $v(\mathbf{A}, \boldsymbol{\rho})$ in the objective space by the union of the hyperrectangles associated with the elements of $\mathbf{A}$ (Fig. 11.4). $\boldsymbol{\rho}$ is chosen so that each of its coordinates is an upper bound of the coordinates of all points of $\mathbf{A}$ (for minimization of the objectives). The measure of the hypervolume $v(\mathbf{A}, \boldsymbol{\rho})$ is a good comparison indicator for nondominated sets because it is strictly monotonic according to the Pareto dominance relation [21]. Namely, if any element of a set $\mathbf{B}$ is dominated by at least one element of $\mathbf{A}$, then hypervolume $v(\mathbf{B}, \boldsymbol{\rho})$ is less than the hypervolume $v(\mathbf{A}, \boldsymbol{\rho})$. So far, the indicator $v(\mathbf{A}, \boldsymbol{\rho})$ is the only one that has this monotonicit property, which explains the interest being shown in it.

The maximum hypervolume is obtained when $\mathbf{A}$ is the Pareto front. This indicator is more significant than the "generational distance" or the $\mathcal{C}$-metric. Indeed, a value of $v(\mathbf{A}, \boldsymbol{\rho})$ near the maximum indicates that the nondominated vectors of a set $\mathbf{A}$ are close to the Pareto front, with a good-quality distribution.

The main drawback of this metric lies in its exponential computational overhead in the number of objectives. The reference point $\boldsymbol{\rho}$ must also be appropriately chosen. There are several approaches to calculating hypervolumes; [13] is one of the recent references in this field.

11.3.3 Multiobjective Evolutionary Algorithms

Unquestionably, the most employed class of metaheuristics for multiobjective optimization is that of evolutionary algorithms. Indeed, they are well suited for simultaneous searching for a collection of optimal solutions, because they deal with populations of solution instances.

The evolutionary approach requires a priori the implementation of an archive of the nondominated solutions discovered during a complete evolution. In fact, there is no guarantee that at the end of the evolution, the solutions that have approached the Pareto-optimal set best will have been preserved in the population. Thus, at the end of each generation, the population is copied into the archive, and then the dominated individuals are eliminated from the archive. However, the management of an archive could be useless for multiobjective optimization algorithms that implement a form of elitism.

Two types of evolutionary approaches are widely considered in the literature:

- Methods using a *Pareto ranking* to evaluate the fitness function.
- *Aggregation* (or *scalarization*) methods that transform a multiobjective optimization problem into a collection of single-objective problems. The resolution of each single-objective problem then gives a point for the Pareto front.

Some methods that are among the most currently used or are representative of the different types of approaches, or are milestones in the field, are described in the following sections.

11.3.4 Methods Using a Pareto Ranking

These methods were the first to show their efficiency in producing an even coverage of a Pareto front. The individuals of a population correspond to instances of solutions in the search space. An objective vector is evaluated and assigned to each of them. Then, each individual is given a scalar fitness value, computed from the objective vectors, such that the nondominated individuals will be selected more often than the others.

The even coverage of the Pareto front, or at least, its nearest nondominated set of solutions found, is obtained by using a mechanism for the preservation of diversity in the population, which may be a speciation/niching method (Sect. 11.2).

Dimensionality of the objective space. There is a difficulty in applying techniques based on Pareto dominance, related to the dimensionality of the objective space. The more objectives there are to optimize, the larger the Pareto front is, and the less likely it is that individuals will be dominated by others. In this case, if a maximum fitness is assigned to the nondominated individuals in order to favor their reproduction, then many individuals will have that fitness. This situation generates a low selection pressure, and thus slow convergence of the algorithm. Consequently, strategies using

Pareto dominance have to take this problem into account. Currently, the "Pareto ranking" approach makes it difficult to go beyond problems involving four objectives.

History of Pareto ranking methods. An initial approach of "Pareto ranking" was proposed by Goldberg in his well-known book [14]. However, he did not describe any concrete implementation of the algorithm, and obviously did not present any performance results. The idea, however, inspired many researchers in the following years. It gave birth to the first generation of multiobjective methods using a Pareto ranking, such as the MOGA (1993), NPGA (1994), and NSGA (1994) algorithms. These are presented below.

In the 2000s, these approaches were improved by the introduction of elitism, either by selection or by using a secondary population, which gave birth to a second generation of multiobjective methods. The algorithms NSGA-II (which is an improvement of the NSGA method), SPEA, and SPEA2 are presented below because they are in widespread use. Several other approaches of the same generation have been published such as PAES (Pareto archived evolution strategy) [22], MOMGA (Multiobjective messy genetic algorithm) [33], and its extension MOMGA-II [42].

11.3.4.1 Goldberg's Pareto Ranking

Calculation of the individual fitnesses. In the original proposal of Goldberg, the calculation is based on the ranking of the individuals according to the domination relation between the solutions which they represent. First of all, rank 1 is assigned to the nondominated individuals in the complete population: they belong to the nondominated front. These individuals are then fictitiously withdrawn from the population, and the new nondominated individuals are determined, that are assigned rank 2. It can be said that they belong to the rank-2 dominated front. One can proceed in this manner until all the individuals have been ranked. The fitness value of each individual is then calculated using a decreasing function of the rank of each individual in a way similar to the technique described in Sect. 6.3.3.5, keeping in mind to the need assign to each equally placed individual the same fitness.

Niching. Goldberg chose the sharing method (Sect. 11.2.2), possibly reinforced by restricted mating (Sect. 6.4.2). Goldberg did not specify whether the niching should be implemented in the search space or the objective space.

11.3.4.2 The "Multiple Objective Genetic Algorithm" (MOGA) Method

Fonseca and Fleming proposed the MOGA algorithm in 1993 [12], based on the approach suggested by Goldberg. When the fitnesses are evaluated, each individual is assigned a rank equal to the number of individuals that dominate it. Then a selection according to the rank is applied, in accordance with the ideas of Goldberg. The niching is carried out in the objective space, which allows an even distribution of the nondominated individuals in the population in the objective space, but not in the search space. This choice does not permit one to perform multimodal and multiobjective optimization at the same time. The niche radius σ_s should be calculated so that the distribution of μ individuals of the population is even over the whole Pareto front. Fonseca and Fleming proposed a method to estimate its value [12].

11.3.4.3 The "Niched Pareto Genetic Algorithm" Method

In Pareto ranking methods, the selection according to rank can be done by a tournament selection between ranked individuals. In 1994, Horn et al. [17] proposed the "niched pareto genetic algorithm" (NPGA) method, in which the tournaments are performed directly according to the relations of dominance, thus avoiding a computationally expensive preliminary ranking of the entire population. Applying a simple binary tournament (Sect. 6.3.4.2) is not satisfactory because of the low selection pressure in this context. To increase it, Horn et al. proposed an unusual type of binary tournament: the *Pareto domination tournament*.

Let two individuals x and y be drawn randomly from the population to take part in a tournament. Those are compared with a comparison sample γ, which is also drawn at random and contains t_{dom} individuals. The winner of the tournament is x if it is not dominated by at least one individual of γ and if y is dominated. The winner is y in the opposite case. If x and y are in the same situation, either both dominated or both nondominated, the winner of the tournament is the one with the fewest neighbors within a ball of radius σ_s in the objective space. This last operation has the effect of implementing a form of niching, with the aim of reducing the genetic drift which would be induced by the choice of a winner at random. A significant genetic drift would be harmful to an even distribution of nondominated individuals.

The parameters t_{dom} and σ_s are chosen by the user; t_{dom} is an adjustment parameter for the selection pressure. Horn et al. noticed in some case studies that if t_{dom} was too low, less than one percent of the population, there were too many dominated solutions and the solutions close to the Pareto-optimal set had less chance of being found. If it was larger than 20 %, premature convergences was frequent, owing to too high a selection pressure. A value of about 10 % was suitable for distributing the individuals near the Pareto front evenly. The parameter σ_s proves to be relatively robust. An estimate of its value is given in [12, 17].

The NPGA method has low computational complexity. It was one of the most used methods in the years following its publication. It was superseded by the elitist approaches proposed by various authors in the 2000s.

11.3.4.4 The "Non Dominated Sorting Genetic Algorithm" Method

The "non dominated sorting genetic algorithm" method was presented in 1994 by Srinivas and Deb [31] and was inspired directly by the idea of Goldberg. It uses the same Pareto ranking. On the other hand, it carries out niching in a way different from MOGA. Instead, the sharing method is applied, front by front, in the search space with a sharpness parameter α equal to 2. The niche radius has to be estimated by the user of the algorithm, which can be difficult.

The computational complexity of the Pareto ranking used by NSGA is high. To determine whether a solution is dominated or not, it needs to be compared, objective by objective, with all other solutions. Thus, the process takes μc comparisons of objectives, where μ is the size of the population and c is the number of objectives. So, $\mu^2 c$ comparisons are required to discover all the nondominated solutions of rank 1 in the population. The search to obtain nondominated individuals must be repeated for each domination rank. There are at most μ ranks in the population, which requires in the worst case $\mathcal{O}(\mu^3 c)$ comparisons to sort all the individuals according to their domination rank. This involves a need for high computing power in the case of large population sizes.

11.3.4.5 NSGA-II

The NSGA-II method [10] was introduced in 2002 as an improvement of NSGA in the following respects:

- the algorithmic complexity is reduced to $\mathcal{O}(\mu^2 c)$;
- the sharing method is replaced by a niching technique without parameters;
- it implements elitism (Sect. 6.3.6.4) to accelerate the convergence of the algorithm.

Reducing the algorithmic complexity of the Pareto ranking. The Pareto ranking is decomposed into two phases in NSGA-II: an initialization phase followed by a rank assignment phase. During the initialization phase, described in Algorithm 10.1, the following items are associated with each individual i of the population **P**:

- a domination counter α_i, giving the number of individuals that dominate i;
- the set of individuals $\mathbf{S}_i$ dominated by i.

The individuals for which α_i is zero constitute the set of all nondominated individuals of rank 1, denoted by $\mathcal{F}_1$. The constructions of $\mathbf{S}_i$ and computations of α_i for all individuals require $\mu^2 c$ comparisons.

The rank assignment phase (Algorithm 10.2) for all the individuals of the population follows the initialization phase. Assuming that set $\mathcal{F}_r$ of the rank r nondominated individuals has been built, it is possible to determine the rank $r + 1$ nondominated individuals as follows: for any individual i belonging to $\mathcal{F}_r$, the counters α_j of the individuals j dominated by i are decremented. The individuals j for which $\alpha_j = 0$ constitute the set $\mathcal{F}_{r+1}$. The complexity of this algorithm is also $\mathcal{O}(\mu^2 c)$. The fitness value of an individual is given by its rank, which the evolutionary algorithm tends to minimize.

Niching. The niching method uses a type of binary tournament selection (Sect. 6.3.4) specific to NSGA-II referred to as *crowded tournament*" selection. This tournament is designed to favor the selection of individuals with the same nondomination rank in sparsely populated areas in either the objective space or the search space Ω,

```
𝓕₁ ← ∅
for each individual i ∈ P do
    Sᵢ ← ∅
    αᵢ ← 0
    for each individual j ∈ P do
        if i <ᴾ j then
        |   Sᵢ ← Sᵢ ∪ {j}
        end

        else if j <ᴾ i then
        |   αᵢ ← αᵢ + 1
        end
    end
    if αᵢ = 0 then
    |   𝓕₁ ← 𝓕₁ ∪ {i}
    |   rᵢ ← 1 ;                              // rᵢ: nondomination rank of i
    end
end
```

Algorithm 11.1: Pareto ranking in NSGA-II: initialization.

```
r ← 1
while 𝓕ᵣ ≠ ∅ do
    𝓕ᵣ₊₁ ← ∅
    for each individual i ∈ 𝓕ᵣ do
        for each individual j ∈ Sᵢ do
            αⱼ ← αⱼ − 1
            if αⱼ = 0 then
            |   𝓕ᵣ₊₁ ← 𝓕ᵣ₊₁ ∪ {j}
            |   rⱼ ← r + 1
            end
        end
    end
    r ← r + 1
end
```

Algorithm 11.2: Pareto ranking in NSGA-II: rank assignment.

depending on the user's choice. The explanations in the following lines are related to the objective space. The adaptation to the search space is direct.

The crowded tournament method is based on a *crowded-comparison* operator, denoted by $\prec_n$. A *crowding distance* d_i is associated with each individual i. This represents an estimate of the distance between i and its neighbors in the space of the objectives. Let r_i be the nondomination rank of individual i. The crowded-comparison operator is defined as follows:

$$i \prec_n j \iff r_i < r_j \text{ or } (r_i = r_j \text{ and } d_i > d_j)$$

A crowded tournament between two individuals i and j selects i if $i \prec_n j$.

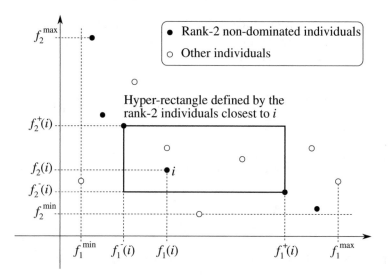

Fig. 11.5 Calculation of the crowding distance in a two-dimensional space of objectives for an individual i of rank 2: $d_i = \frac{(f_1^+(i) - f_1^-(i))}{(f_1^{max} - f_1^{min})} + \frac{(f_2^+(i) - f_2^-(i))}{(f_2^{max} - f_2^{min})}$

The designers of the method proposed that the crowding distance d_i should be calculated as follows. Let $f_m(i)$ be the value of objective m for individual i of $\mathcal{F}_r$ with a given value of r. We define:

- f_m^{max}: the maximum of objective m in the population.
- f_m^{min}: the minimum of objective m in the population.
- $f_m^+(i)$: the closest value to $f_m(i)$ in $\mathcal{F}_r$ such that $f_m^+(i) \geq f_m(i)$. For extreme individuals i where $f_m(i)$ is maximum in $\mathcal{F}_r$, $f_m^+(i)$ is set equal to ∞ for one of them and $f_m^+(i) = f_m(i)$ for the others, if any. This is useful for ensuring that extreme individuals can be selected with a sufficiently high probability to explore their neighborhoods with the variation operators (mutation and crossover).
- $f_m^-(i)$: the closest value to $f_m(i)$ in $\mathcal{F}_r$ such that $f_m^-(i) \leq f_m(i)$. For extreme individuals i where $f_m(i)$ is minimum in $\mathcal{F}_r$, $f_m^-(i)$ is set equal to $-\infty$ for one of them and $f_m^-(i) = f_m(i)$ for the others, if any.

The crowding distance d_i is expressed as

$$d_i = \sum_{m=1}^{c} \frac{f_m^+(i) - f_m^-(i)}{f_m^{max} - f_m^{min}}$$

Figure 11.5 shows an example of a calculation of the crowding distance for an individual i in a two-dimensional space of objectives.

Algorithm 10.3 describes the computation of d_i for the subpopulation $\mathcal{F}_r$ of nondominated individuals i with rank r. It allows one to deduce the computational complexity of the crowding-distance calculation which is $\mathcal{O}(c\mu \log \mu)$. This calculation follows the assignment of a rank to each individual with complexity equal to $\mathcal{O}(\mu^2 c)$. Consequently, the overall complexity of these two operations is $\mathcal{O}(\mu^2 c)$.

$l \leftarrow |\mathcal{F}_r|$ // l is the number of nondominated individuals of rank r
for *each individual $i \in \mathcal{F}_r$* **do**
$\quad | \quad d_i \leftarrow 0$
end
for *each objective m* **do**
$\quad \mathbf{T} \leftarrow \text{tri}(\mathcal{F}_r, m)$ // $\mathbf{T}$ is an array of individuals of $\mathcal{F}_r$ sorted according to objective m
$\quad d_{\mathbf{T}[1]} \leftarrow d_{\mathbf{T}[l]} \leftarrow \infty$
$\quad$ **for** $k = 2\,l - 1$ **do**
$\quad \quad | \quad d_{\mathbf{T}[k]} \leftarrow d_{\mathbf{T}[k]} + (f_m(\mathbf{T}[k+1]) - f_m(\mathbf{T}[k-1]))/(f_m^{\max} - f_m^{\min})$
$\quad$ **end**
end

Algorithm 11.3: Computation of crowding distances d_i for any individual i of $\mathcal{F}_r$.

Elitism. For a generation $g > 0$, the new generation is obtained by creating a population of children $\mathbf{Q}_g$ from the population $\mathbf{P}_g$ by applying in sequence the crowded tournament selection, crossover, and mutation operators (Fig. 11.6). The size of $\mathbf{Q}_g$ was chosen by the designers of the method to be equal to that of $\mathbf{P}_g$, i.e., μ. The Pareto ranking described above is applied to the union of $\mathbf{Q}_g$ and $\mathbf{P}_g$, which allows one to compute the nondomination ranks r_i of the individuals and to generate subpopulations $\mathcal{F}_r$. Parents and children participate in the same ranking, which implements elitism.

Environmental selection operator. The population $\mathbf{P}_{g+1}$ built by the environmental selection operator is initially composed of the individuals in the subpopulations $\mathcal{F}_1$ to $\mathcal{F}_k$, where k is the greatest integer such that the sum of the sizes of these subpopulations is less than or equal to μ. To complete $\mathbf{P}_{g+1}$ to μ individuals, the individuals in $\mathcal{F}_{k+1}$ are sorted with the comparison operator $\prec_n$ and the best solutions are inserted into the population $\mathbf{P}_{g+1}$ until it contains μ individuals.

The initial population. The initial population $\mathbf{P}_0$ is generated by a problem-dependent method if available, or else by construction of random individuals. Pareto ranking is then applied to $\mathbf{P}_0$ to calculate the initial fitness values of its individuals. This is different from what is done for the other generations, for which this ranking is applied to the union of $\mathbf{P}_g$ and $\mathbf{Q}_g$.

The generational loop. Figure 11.6 depicts the generational loop of NSGA-II. The two steps of the calculation of the composite fitness (r_i, d_i) for individual i are highlighted.

In conclusion. The NSGA-II method is recognized as being highly effective. Today, it is one of the reference methods for multiobjective evolutionary optimization.

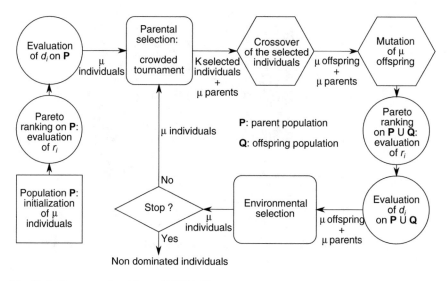

Fig. 11.6 The generational loop of NSGA-II

11.3.4.6 The "Strength Pareto Evolutionary Algorithm" (SPEA) Method

This method was presented in 1999 by Zitzler and Thiele [40]. Its originality lies in the utilization of the archive of nondominated solutions during the evolution of a population. It aims to intensify the search for new nondominated solutions, and thus to approach the Pareto front better by implementing a form of elitism (Sect. 11.2.4.1). Moreover, the authors of the method proposed a new niching technique without parameters, specifically dedicated to multiobjective optimization.

Only the operator for evaluation of the fitness of the individuals is specific to SPEA. Binary tournaments with replacement within the population are used to implement the parental selection operator. The fitness f_i of individual i is defined so that it is minimized. Thus, when two individuals i and j participate in a binary tournament, i is selected for reproduction if $f_i < f_j$.

Calculation of the fitnesses of the individuals. In each generation, the fitnesses of the individuals in population **P** and the archive **P'** are determined in the following way:

Stage 1. The fitness f_i of any individual i in **P'** is equal to its *strength* s_i:

$$f_i = s_i \quad \text{and} \quad s_i = \frac{\alpha_i}{\mu + 1}$$

where α_i is the number of solutions dominated by i in the population **P**, and μ is the size of **P**; s_i necessarily lies between 0 and 1.

Stage 2. The fitness f_j of any individual j in **P** is equal to the sum of the strengths of the individuals in **P'** that dominate it, added to unity:

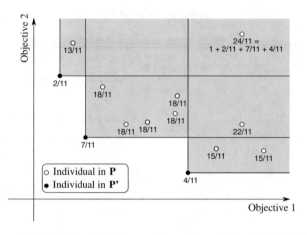

Fig. 11.7 Example of calculation of the fitness of the solutions in **P** and **P'** in the SPEA method

$$f_j = \left(1 + \sum_{i,i \overset{P}{<} j} s_i\right)$$

f_j is thus greater than or equal to 1, and consequently larger than the fitnesses of the solutions in **P'**.

Thus, an individual is less likely to be selected when many individuals in **P'** dominates it. Figure 11.7 illustrates the calculation of the fitness with the help of an example.

The jump in performance that SPEA provided compared with the best methods used before 1999 is mainly due to the elitism generated by the use of the archive **P'**.

11.3.4.7 The *"Strength Pareto Evolutionary Algorithm 2"* (SPEA2) Method

The "strength pareto evolutionary algorithm 2" method [39] was proposed in 2001 to improve SPEA in the following respects:

- The computation of individual fitnesses was modified to better guide the search towards the Pareto optimum by reducing the number of individuals with the same fitness value. In particular, the fitness values computed with SPEA2 take account of the local population densities in the objective space.
- The environmental selection operator was modified to better explore the neighborhood of the extreme points of the Pareto front, whereas the clustering operator used in the first version had the effect of removing these points.

Let **P'** be the parent population in a given generation and let **P** be the child population generated by the application in sequence of the operators of parental selection, crossover, and mutation to **P'**. In the terminology of SPEA2, **P'** is also referred to as the *archive* that holds the "best" nondominated individuals obtained during an evolution from the first generation. The meaning of "best" will be specified in the following paragraphs.

Fitness computation and niching. The calculation of the fitness values of the individuals in the population $\mathbf{P}' \cup \mathbf{P}$ is performed in two steps. The first step gives a *raw fitness* value to each individual from the Pareto dominance relations between the individuals. The second step estimates the population density in the vicinity of each individual, which is added to its raw fitness to give its effective fitness value.

The computation of the raw fitnesses requires the determination of the strength s_i associated with each individual i. This strength is the number of individuals dominated by i:

$$s_i = \text{card}(\{j \mid j \in \mathbf{P}' \cup \mathbf{P} \text{ and } i \overset{\text{P}}{<} j\})$$

The raw fitness b_i of each individual i is obtained by summing the strengths of the individuals j that dominate it:

$$b_i = \sum_{j \in \mathbf{P}' \cup \mathbf{P}, \, j \overset{\text{P}}{<} i} s_j$$

Figure 11.8 shows an example of the calculation of the strengths and raw fitnesses in the context of minimizing two objectives. The nondominated individuals have a raw fitness equal to 0. Conversely, individuals that are dominated by many other individuals have a high raw fitness value. This method of computation of b_i performs a form of niching. If there are regions of the objective space where the population is dense, the dominated individuals in these regions have high b_i values. These individuals then have little chance of being selected for reproduction. Conversely, if there is one individual dominated by a single nondominated individual in a region, its raw fitness will be $b_i = 1$, giving it more chance of being selected for reproduction. The computational complexity of calculating b_i for all individuals in the population is $\mathcal{O}((\alpha)^2)$, where $\alpha = \mu + \lambda$, with λ, being the size of offspring population $\mathbf{P}$, and μ, being size of the population $\mathbf{P}'$ of the parents.

However, especially when the number of objectives is large, it is possible that there may be few dominated individuals in the population. Most individuals would then have a fitness $b_i = 0$ and the search for the Pareto optimum would become almost a simple search at random. To avoid this phenomenon, a local density d_i of the population is estimated in the vicinity of each individual i according to an adapted version of the method of the kth closest neighbor used in statistics [29]:

$$d_i = \frac{1}{\delta_i^k + 2}$$

where δ_i^k is the distance of individual i from its kth closest neighbor. d_i is thus between 0 and 0.5. The usual value distances $k = \lfloor \sqrt{\lambda + \mu} \rfloor$, is chosen. The computation of d_i requires calculation of the distances from each individual i to all the others. δ_i^k

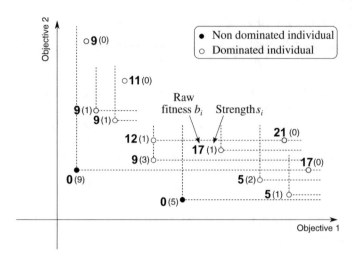

Fig. 11.8 An example of the assignment of raw fitness values b_i to individuals with SPEA2 for a two-objective minimization problem

is obtained after the sorting of these distances. The computational complexity of calculating d_i is $\mathcal{O}(\alpha^2 \log(\alpha))$, with $\alpha = \lambda + \mu$.

Finally, the fitness value of individual i is

$$f_i = b_i + d_i$$

Environmental selection operator. In generation g, this operator selects μ individuals in $\mathbf{P}'_g \cup \mathbf{P}$, whose size is $\lambda + \mu$, to build the population $\mathbf{P}'_{g+1}$ for the next generation. The α nondominated individuals in $\mathbf{P}'_g \cup \mathbf{P}$ constitute a subpopulation $\mathbf{Q}$. There are three cases:

- If $\mu = \alpha$, $\mathbf{P}'_{g+1} = \mathbf{Q}$.
- If $\mu > \alpha$, $\mathbf{Q}$ must be completed by $\mu - \alpha$ dominated individuals to constitute the population $\mathbf{P}'_{g+1}$. For this purpose, the dominated individuals in $\mathbf{P}'_g \cup \mathbf{P}$ are sorted according to increasing fitness. The first $\mu - \alpha$ sorted individuals are added to $\mathbf{Q}$ to constitute $\mathbf{P}'_{g+1}$.
- If $\mu < \alpha$, $\alpha - \mu$ individuals in $\mathbf{Q}$ must be removed from among those located in the most crowded regions of the objective space to constitute $\mathbf{P}'_{g+1}$. The method is iterative. In each step, an individual is removed from $\mathbf{Q}$. If there is only one individual i that has the shortest distance δ_i^1 to its first nearest neighbor, it is removed from $\mathbf{Q}$. If several individuals i have the same minimum values of δ_i^1 to δ_i^{k-1}, with only one for which δ_i^k is minimal, this individual is removed from $\mathbf{Q}$. If there are several individuals that have the same minimum values of δ_i^k for all k, one of them is removed from $\mathbf{Q}$. This happens especially when individuals have identical objective vectors. Formally, an individual i is chosen to be removed if $i \preccurlyeq_d j, \forall j \in \mathbf{Q}$, with

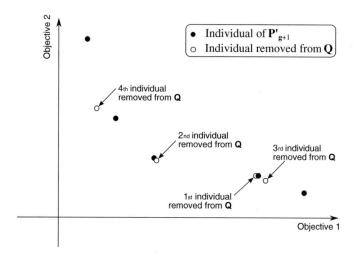

Fig. 11.9 Environmental selection operator: an example of construction of a population $\mathbf{P'}_{g+1}$ of five individuals from a population $\mathbf{Q}$ of nine individuals

$$i \preccurlyeq_d j \Leftrightarrow \forall\, 0 < k < \alpha : \delta_i^k = \delta_j^k \text{ or}$$
$$\exists\, 0 < k < \alpha : \left[(\forall\, 0 < l < k : \delta_i^l = \delta_j^l) \text{ and } \delta_i^k < \delta_j^k \right]$$

Fig. 11.9 illustrates this process with an example.

The computational complexity of the environmental selection operator is $\mathcal{O}(\alpha^2 \log(\alpha))$ an average, with $\alpha = \lambda + \mu$.

The generational loop. Figure 11.10 depicts the generational loop of SPEA2. The two stages of the computation of the fitnesses f_i of individuals i from the raw fitnesses b_i and densities d_i are highlighted.

In conclusion. The SPEA2 method is recognized for its efficiency. It is, with NSGA-II, one of the reference methods for evolutionary multiobjective optimization. The complexity of one generation is larger for SPEA2 than for NSGA-II, but the solutions obtained with SPEA2 are often more evenly distributed on the nondominated front. The computation times required by NSGA-II and SPEA2 have been compared on a set of standard test problems, (e.g., [9]). They results are summarized on pp. 95–100.

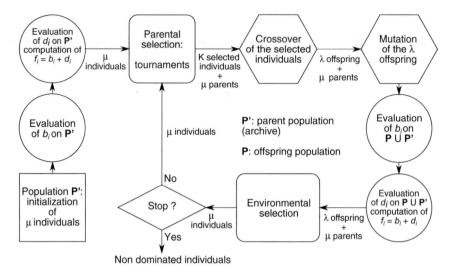

Fig. 11.10 The generational loop of SPEA2

11.3.5 Scalarization Methods

11.3.5.1 Scalarization of the Objectives

A simple method to obtain a nondominated solution, widely used in the field of multicriteria decision problems, consists in aggregating all the criteria, or objectives $f_i(\mathbf{x})$, Into a single criterion using a weighted summation. Thus, the problem is transformed by calculating an *aggregation* function of the objectives $G_1(\mathbf{x}|\mathbf{w})$ to be minimized, where:

$$G_1(\mathbf{x}|\mathbf{w}) = \sum_{i=1}^{c} w_i f_i(\mathbf{x})$$

For each weight vector $\mathbf{w} = (w_i)$ with $w_i > 0$ and $\sum_{i=1}^{c} w_i = 1$, there exists a Pareto-optimal solution. However, this linear approach does not allow one to obtain the Pareto-optimal solutions located on the concave parts of the Pareto front for the objectives a minimization problem, whatever the weight values. Indeed, such solutions cannot minimize a weighted sum of objectives (Fig. 11.11).

To prevent concave parts of the Pareto front being excluded from the search, the minimization of a weighted sum can be replaced by the minimization of the weighted Chebyshev distance $G_\infty(\mathbf{x}|\mathbf{w}, \boldsymbol{\rho})$ between the objective vector $\mathbf{f}(\mathbf{x})$ and a reference point $\boldsymbol{\rho}$, where

$$G_\infty(\mathbf{x}|\mathbf{w}, \boldsymbol{\rho}) = \max_{i=1}^{c} w_i |f_i(\mathbf{x}) - \rho_i|$$

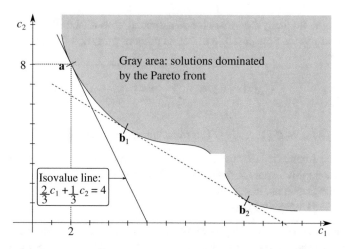

Fig. 11.11 Scalarization of the objectives by the weighted sum method: the minimum of $G_1(\mathbf{x}|\mathbf{w})$ is 4 when $\mathbf{w} = (2/3, 1/3)$ for the objective vector $\mathbf{a} = (2, 8)$. The points on the Pareto front located between $\mathbf{b}_1$ and $\mathbf{b}_2$ cannot minimize $G_1(\mathbf{x}|\mathbf{w})$ for all $\mathbf{w}$

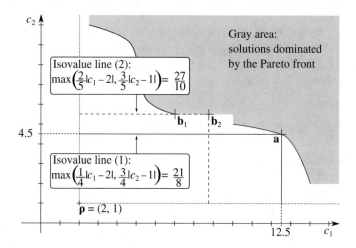

Fig. 11.12 Scalarization of the objectives with the Chebyshev distance from a reference point: the minimum of $G_\infty(\mathbf{x}|\mathbf{w}, \boldsymbol{\rho})$ is 21/8 for $\mathbf{w} = (1/4, 3/4)$ and $\boldsymbol{\rho} = (2, 1)$ for the objective vector $\mathbf{a} = (12.5, 4.5)$. The segment $]\mathbf{b}_1, \mathbf{b}_2]$ represents the dominated vectors that are optimal for $G_\infty(\mathbf{x}|\mathbf{w}, \boldsymbol{\rho})$ with $\mathbf{w} = (2/5, 3/5)$

Here, $\boldsymbol{\rho}$ is often chosen as the *ideal point* for which each coordinate ρ_i is the minimum of $f_i(\mathbf{x})$ regardless of the other criteria. Now, all the points of the Pareto front are reachable, provided that appropriate values are given to the weights w_i. However, it is possible that dominated solutions may also minimize such a distance, as shown in Fig. 11.12.

The weighted Chebyshev distance is a special case of the distances associated with L^p-norms. It is thus possible to define other aggregation functions

$$G_p(\mathbf{x}|\mathbf{w}, \boldsymbol{\rho}) = \sqrt[p]{\sum_{i=1}^{c} (w_i(f_i(\mathbf{x}) - \rho_i))^p} \qquad (11.1)$$

with $\rho_i \leq f_i(\mathbf{x})$, $\forall i \in \{1, \ldots, c\}$. The most common values of p are:

- $p = 1$: the Manhattan distance, the minimization of which is equivalent to minimizing a weighted sum (in the case of minimization problems);
- $p = 2$: the Euclidean distance,
- $p = \infty$: the Chebyshev distance.

The "scalarization" or "aggregation" methods presented above are simple and widely used. Despite the limitations of weighted sum methods on possible concave parts of a Pareto front, these methods can have better convergence properties towards the Pareto front than the Chebyshev distance method when the Pareto front is convex. There are also other scalarization methods, such as the boundary intersection approaches [6, 24].

To obtain several solutions approaching the Pareto-optimal set, the naive way would be to choose different weight vectors as many times as desired and to restart the optimization algorithm for each of them. However, such a method requires excessive computing power.

11.3.5.2 The "Steady-State ε-MOEA" Method

The Pareto ranking methods presented above use quite complex algorithms for fitness computation and diversity preservation. But they have the advantage of finding good-quality nondominated solutions, close to the Pareto-optimal set. The ε-MOEA method, as presented by its authors [9], aims to quickly find a set of nondominated solutions whose objective vectors are representative of the Pareto front. It is based on the notion of ε-*domination* combined with a scalarization (or aggregation) of the objectives.

$\varepsilon - domination$. For a minimization problem, an identification vector $\mathbf{B}(\mathbf{f}) = (B_1(f_1), \ldots, B_c(f_c))$ is associated with an objective vector $\mathbf{f} = (f_1, \ldots, f_c)$, such that

$$B_i(f_i) = \left\lfloor \frac{f_i - m_i}{\varepsilon_i} \right\rfloor \qquad (11.2)$$

where m_i is a lower bound of the values of the objective f_i, and ε_i is the tolerance is associated with objective i. This is a parameter of the method.

Definition. Let $\mathbf{f}, \mathbf{g} \in \mathbb{R}^c$ be two objective vectors; then $\mathbf{f}$ ε-dominates $\mathbf{g}$, denoted $\mathbf{f} \overset{\varepsilon}{<} \mathbf{g}$, if and only if

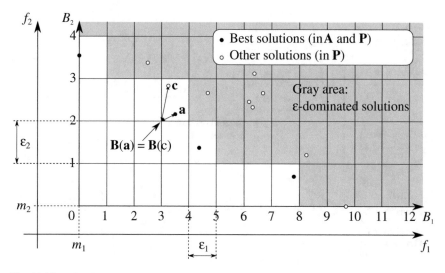

Fig. 11.13 ε-dominance between boxes, and preference relation in a box

$$\mathbf{B(f)} \overset{\mathrm{p}}{<} \mathbf{B(g)}$$

Each vector $\mathbf{B}$ defines a box in the space of objectives as a hyperrectangle that is the Cartesian product of the intervals $[B_i \varepsilon_i + m_i, (B_i + 1)\varepsilon_i + m_i[$ for $i \in \{1, \ldots, c\}$. Figure 11.13 represents the boxes defined by the vectors $\mathbf{B}$ as a grid in a plane generated by two objectives, and the regions ε-dominated by the set of the non-ε-dominated solutions. In this figure, $\mathbf{a}$ and $\mathbf{c}$ are associated with the same identification vector, $\mathbf{B(a)} = \mathbf{B(c)} = (3, 2)$.

The algorithm. The method uses two populations $\mathbf{A}$ and $\mathbf{P}$ evolving simultaneously. The population $\mathbf{P}$ contains the dominated and nondominated solutions obtained according to the diagram for a steady-state evolutionary algorithm (see Sect. 6.3.6.3) in which only one offspring is generated in each generation. The selection operators are specific, also involving the population $\mathbf{A}$.

Population $\mathbf{A}$ is an archive containing only the best solutions that are not ε-dominated and have been found since the beginning of an evolution. In addition, each box for $\mathbf{A}$ can contain a maximum of only one solution: the one that minimizes an aggregation function of the objectives. This is the niching mechanism of the ε-MOEA algorithm. $\mathbf{A}$ is initialized according to Algorithm 10.5, taking each solution in the initial population $\mathbf{P}$ as the argument $\mathbf{c}$ of the procedure.

Figure 11.14 summarizes the ε-MOEA algorithm. It can be noticed that the parental selections in $\mathbf{A}$ and $\mathbf{P}$ do not depend on one another and thus can be performed in parallel. The situation is the same for environmental selections. The parental selection in $\mathbf{P}$ is a domination tournament described by Algorithm 10.4. It gives a solution $\mathbf{r}$. The parental selection in $\mathbf{A}$ consists in randomly choosing a solution $\mathbf{a}$. $\mathbf{a}$ and $\mathbf{r}$ are then crossed and mutated to give a solution $\mathbf{c}$. The environmental selections which

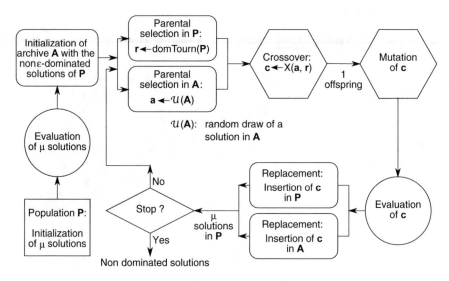

Fig. 11.14 The generational loop of the steady-state ε-MOEA algorithm

follow the application of the variation operators aim to include **c** in **A** and **P** when that is beneficial. These selection processes are detailed below.

```
p ← 𝒰(P)        // 𝒰(P): equiprobable drawing of a solution from P
q ← 𝒰(P)
p_f ← f(p)                              // f: multiobjective function
q_f ← f(q)
            p
if p_f < q_f then
 |  r ← p
end

            p
else if q_f < p_f then
 |  r ← q
end
else
 |  r ← 𝒰(p, q)                  // equiprobable drawing of p or q
end
return r
```

Algorithm 11.4: Function domTourn(**P**).

Replacement in archive **A**. Algorithm 10.5 describes the replacement operator for the archive **A**. This algorithm aims firstly to ensure that at any time this archive contains only solutions that are non-ε-dominated. Thus, if a solution **c** obtained after mutation is ε-dominated by one solution in **A** at least, then **c** is rejected.

Moreover, the replacement operator for the archive introduces a preference function between solutions in the same box so as to keep only one of them. The preference function is an aggregation function of the objectives.

The authors of the method proposed using $G_2(\mathbf{x}|\mathbf{w}, \boldsymbol{\rho})$ (Eq. (11.1)) with $\mathbf{w} = (1, \ldots, 1) = (1)$ and $\boldsymbol{\rho} = \mathbf{B}(\mathbf{x})$. If the insertion of a solution $\mathbf{c}$ leads to there being two non-ε-dominated solutions in a box defined by the identification vector $\mathbf{B}(\mathbf{a}) = \mathbf{B}(\mathbf{c})$, then only the solution that minimizes $G_2(\mathbf{x}|(1), \mathbf{B})$ is preserved (Algorithm 10.5).

This case is shown in Fig. 11.13, p. 31, for points $\mathbf{a}$ and $\mathbf{c}$. According to the figure, the Euclidean distance $G_2(\mathbf{a}|(1), \mathbf{B}(\mathbf{a}))$ between $\mathbf{a}$ and $\mathbf{B}(\mathbf{a})$ is less than $G_2(\mathbf{c}|(1), \mathbf{B}(\mathbf{c}))$. Therefore, $\mathbf{a}$ is preserved in the archive, and $\mathbf{c}$ is rejected.

Replacement in population $\mathbf{P}$. This operator has no particular specificity. It only needs to promote good solutions, keeping a constant size of $\mathbf{P}$. The authors of the method proposed the algorithm 10.6. If the solution $\mathbf{c}$ is not dominated by a solution in $\mathbf{P}$, it is inserted into the population. In this case, $\mathbf{c}$ preferably replaces one of the individuals in $\mathbf{P}$ that it dominates. If it does not dominate any individual, it replaces an individual in $\mathbf{P}$, randomly chosen.

```
c_f ← f(c)
c_ε ← (c_f − m)./ε                    // "./" : component-by-component division
B_c ← ⌊c_ε⌋                                    // according to equation (11.2)
rejected ← False
foreach a ∈ A do
   a_f ← f(a)
   a_ε ← (a_f − m)./ε
   B_a ← ⌊a_ε⌋
   if B_c ≺ᵖ B_a then
    |  A ← A \ {a}
   end
   if B_c = B_a then
      if G_2(c_ε|(1), B_c) < G_2(a_ε|(1), B_c) then
       |  A ← A \ {a}                 //  G_2(x|w, ρ): equation (11.1)
      end
      else
       |  rejected ← True
      end
   end
   if B_a ≺ᵖ B_c then
    |  rejected ← True
   end
end
if not rejected then
 |  A ← A ∪ {c}
end
```

Algorithm 11.5: Subroutine replacementArchive($\mathbf{A}$, $\mathbf{c}$).

Performance. Deb et al. [9] have performed comparisons between ε-MOEA, SPEA2, and NSGA-II for five two-objective test functions ZDT1, ZDT2, ZDT3, ZDT4, and ZDT6 proposed in 2000 by Zitler, Deb and Thiele [38] and five three-objective test functions DTLZ1, DTLZ2, DTLZ3, DTLZ4, and DTLZ5 proposed in 2002 by Deb,

```
cf ← f(c)
rejected ← False
D ← ∅                          // D: set of the solutions dominated by c
foreach p ∈ P do
  pf ← f(p)
  if cf <p pf then
  |  D ← D ∪ {p}
  end
  if pf <p cf then
  |  rejected ← True
  end
end
if not rejected then
  if D ≠ ∅ then
  |  r ← U(D)                  // equiprobable drawing of an element of D
  end
  else
  |  r ← U(P)
  end
  P ← P \ {r}
  P ← P ∪ {c}
end
```

Algorithm 11.6: Subroutine replacementPopulation(P, c).

Thiele, Laumanns, and Zitler [11]. For each test configuration, five evolutions with different initial populations were carried out. The results presented were averages of the results obtained for each test configuration.

The approximation to the Pareto front was satisfactory for all test functions and all methods, except for DTLZ4, for which the Pareto front was not approached in 50 % of the evolutions, whatever the method used, whether ε-MOEA, SPEA2, or NSGA-II.

The significant advantage that was found for ε-MOEA lies in the low computation times compared with the other two methods. According to [9], ε-MOEA was on average:

- **16** times faster than NSGA-II and **390** times faster than SPEA2 for the functions ZDT1, ZDT2, ZDT3, ZDT4, and ZDT6;
- **13** times faster than NSGA-II and **640** times faster than SPEA2 for the functions DTLZ1, DTLZ2, DTLZ3, DTLZ4, and DTLZ5.

However, the essential advantage of ε-MOEA lies in its efficiency in solving problems with four objectives or more. Wagner et al. [34] have compared the qualities of the Pareto front approximations for ε-MOEA, SPEA2, and NSGA-II, on the functions DTLZ1, and DTLZ2, for three to six objectives. Two quality metrics (Sect. 11.3.2) were used, including the hypervolume measure (Sect. 11.3.2.3). For these two metrics, the advantage of ε-MOEA is obvious beyond four objectives, both of the other methods being unable to approach the Pareto front.

Conclusion. ε-MOEA has proved to be an interesting approach, firstly because it obtains good approximations to the Pareto front, even for a relatively large number of objectives, and secondly because of its computation speed, compared with NSGA-II and SPEA2. This is due to the implementation of two effective mechanisms:

- one to preserve diversity (niching), by the distribution of the solutions in boxes forming a hypergrid;
- another to scalarize the objectives within each box.

However, the method is sensitive to the choice of the tolerance vector $\boldsymbol{\varepsilon}$, which is of critical importance to the quality of results.

11.3.5.3 MOEA/D: A Multiobjective Evolutionary Algorithm Based on Decomposition

The MOEA/D method [36] was chosen to be described in this chapter rather than others using a similar approach [20, 25] because of its simplicity and its good efficiency. It uses an approach to the multiobjective optimization problem of decomposition into a set of μ mono-objective subproblems $\mathcal{P}_i$, obtained by scalarization of the objectives. The subproblems are solved simultaneously and give for each of them an objective vector moving to the Pareto front. In this presentation, the weighted Chebyshev distance is chosen as the aggregation function. So, every subproblem $\mathcal{P}_i$ is a search for $\mathbf{x}_i^* \in \Omega$ that minimizes the objective function

$$G_\infty(\mathbf{x}_i | \mathbf{w}_i, \boldsymbol{\rho}) = \max_{j=1}^{c} w_{ij} \, |f_j(\mathbf{x}_i) - \rho_j|$$

where w_{ij} and ρ_j are given, and c is the number of objectives. The method uses a set of weight vectors $\mathbf{W} = \{\mathbf{w}_1, \ldots, \mathbf{w}_\mu\}$. These vectors remain constant during the search for the Pareto optimum. They are the ones that preserve the diversity of the solutions on the Pareto front, if they are properly chosen.

MOEA/D is justified under the assumption that if the weights $\mathbf{w}_k$ and $\mathbf{w}_l$ are neighbors, then the optimal solutions $\mathbf{x}_k^*$ and $\mathbf{x}_l^*$ are also neighbors. This is not always true, especially when the Pareto front is discontinuous. If this hypothesis about neighborhoods is not true, the convergence towards the Pareto front is found to be degraded.

Let $\mathbf{P}$ be a population of solutions with size μ. In a given generation, MOEA/D ensures that the solution $\mathbf{x}_i$ in $\mathbf{P}$ is the best that has been found for the vector $\mathbf{w}_i$ from the first generation. Algorithm 10.7 summarizes the MOEA/D approach.

Initialization. The weight vectors $\mathbf{w}_i$ are built so that they are uniformly distributed in the weight space. The method requires one to consider the ω nearest neighbors of each $\mathbf{w}_i$. To memorize the neighborhood relations, a set $\mathbf{V}_i$ with cardinality ω is built during the initialization for each $\mathbf{w}_i$. $\mathbf{V}_i$ contains the indices in $\mathbf{W}$ of the nearest

---

Input: μ, ω
Output: **P**

Initialize $\mathbf{W} = \{\mathbf{w}_1, \ldots, \mathbf{w}_\mu\}$
Initialize $\mathbf{V}_i$ of size ω, $\forall i \in \{1, \ldots, \mu\}$
Initialize $\mathbf{P} = \{\mathbf{x}_1, \ldots, \mathbf{x}_\mu\}$
Initialize ρ

// Generational loop
repeat
    for $i = 1$ μ **do**
        Reproduction: random selection of two indices $k \in \mathbf{V}_i$ and $l \in \mathbf{V}_i$

        $\mathbf{y} \leftarrow$ crossover$(\mathbf{x}_k, \mathbf{x}_l)$
        $\mathbf{y} \leftarrow$ mutation$(\mathbf{y})$
        $\mathbf{y} \leftarrow$ improvement$(\mathbf{y})$
        // Reference point ρ adjustment
        for $j = 1$ c **do**
            if $\rho_j > f_j(\mathbf{y})$ **then**
                $\rho_j \leftarrow f_j(\mathbf{y})$
            end
        end
        // Environmental elitist selection
        for *each index* $j \in \mathbf{V}_i$ **do**
            if $G_\infty(\mathbf{y}|\mathbf{w}_j, \rho) < G_\infty(\mathbf{x}_j|\mathbf{w}_j, \rho)$ **then**
                $\mathbf{x}_j \leftarrow \mathbf{y}$
            end
        end
    end
until *stopping criteria satisfied*

Algorithm 11.7: A simple version of the MOEA/D algorithm using the Chebyshev distance.

neighbors of $\mathbf{w}_i$, in the sense of the Euclidean distance; ω is a parameter that adjusts the capacities for exploration/exploitation of the algorithm. If it is too small, the search for the optimum of each subproblem is essentially local, reducing the probability of finding a global optimum. If it is too large, the solutions obtained after applying the variation operators are often poor, noticeably slowing down the convergence. The search for a suitable value for ω is a priori empirical. The experiments described by the authors of the method used values of ω between 10 and 20 for population sizes μ ranging from 100 to 500.

The population of solutions **P** is preferably intialized through a problem-dependent heuristic, if there exists one. Otherwise, the solutions $\mathbf{x}_i$ are generated randomly. A first estimate of the reference point ρ is computed using a method specific to the problem.

The generational loop. This is repeated until satisfaction of a stopping criterion defined as required by the user. During a generation and for each subproblem $\mathcal{P}_i$, two indices k and l are randomly selected in $\mathbf{V}_i$. The two corresponding solutions $\mathbf{x}_k$

and $\mathbf{x}_l$ give a new solution $\mathbf{y}$ after applying the variation operators of crossover and mutation. Since $\mathbf{x}_k$ and $\mathbf{x}_l$ are the best solutions found so far during the evolution for neighboring weight vectors $\mathbf{w}_k$ and $\mathbf{w}_l$, there is a relatively high probability that $\mathbf{y}$ is also of good quality for $\mathbf{w}_i$ and its nearest neighbors, if $\mathbf{x}_k$ and $\mathbf{x}_l$ are neighbors according to the hypothesis of continuity outlined above.

The operator of improvement that transforms the solution $\mathbf{y}$ obtained after crossover and mutation is optional. In the case of constrained optimization, it uses a problem-dependent heuristic that repairs unfeasible solutions to transform them into feasible solutions (see Sect. 12.4.1). Moreover, the improvement operator can also implement a heuristic local optimization, again specific to the problem, if it is useful for improving the convergence to the Pareto front.

The coordinates of the reference point must then be adjusted if some coordinates of $\mathbf{f}(\mathbf{y})$ are lower than those of the current reference point, for a problem of minimization of the objectives.

At the end of each generation, the environmental selection operator is applied to the subproblems $\mathcal{P}_i$. It replaces $\mathbf{x}_j$ in the population $\mathbf{P}$ with $\mathbf{y}$ provided that $\mathbf{y}$ is better than $\mathbf{x}_j$, according to the aggregation function $G_\infty(\mathbf{x}|\mathbf{w}_j, \boldsymbol{\rho})$, for all $j \in \mathbf{V}_i$. It is an elitist operator.

11.3.5.4 Comparisons Between MOEA/D and NSGA-II.

Algorithmic complexity. The complexity of a generation is given by the environmental selection operator, which is $\mathcal{O}(\mu\omega c)$, where c is the number of objectives; ω is less than μ. This complexity is lower than that of NSGA-II, which is $\mathcal{O}(\mu^2 c)$. The experiments that have been performed [36] show that MOEA/D spends 1.7–8.5 times less computing time than NSGA-II.

Experimental comparisons. Zhang and Li [36] have performed comparisons between MOEA/D and NSGA-II for five test problems with the two-objective functions ZDT1, ZDT2, ZDT3, ZDT4, and ZDT6 mentioned Sect. 11.3.5.2 and two test problems with the three-objective functions DTLZ1 and DTLZ2. The comparison metrics chosen were the *generational distance* and the *C-metric* (Sect. 11.3.2). The performance in the ZDT1 test was measured for population sizes $\mu = 100$ and $\mu = 20$. In the other tests, μ was 100 in the case of ZDT2 to ZDT6, and 300 for DTLZ1 and DTLZ2.

In the case $\mu = 20$, MOEA/D proved to be efficient, with a good approximation to the Pareto front, whereas NSGA-II failed. With the ZDT1 to ZDT6 tests and $\mu = 100$, both NSGA-II and MOEA/D provided a good approximation to the Pareto front, with a slight advantage to NSGA-II in 4 out of 10 tests. MOEA/D proved significantly better than NSGA-II for DTLZ1 and DTLZ2 functions with the *C*-metric. The authors of the study did not present results for problems involving more than four objectives, although it is for such functions that MOEA/D should really show its qualities. Table 11.1 summarizes the results reported.

Table 11.1 Scores of MOEA/D versus NSGA-II in 30 independent evolutions for each test problem, according to [36]. "4/5" in column 3, for example, means that MOEA/D is better than NSGA-II for 4 test problems out of 5. c is the number of objectives

Test problems	μ	Score of MOEA/D versus NSGA-II according to	
		C-metric	Generational distance
ZDT1 ($c = 2$)	20	1/1	1/1
ZDTx ($c = 2$)	100	4/5	2/5
DTLZx ($c = 3$)	300	2/2	2/2

MOEA/D also provides a gain in computing time compared with NSGA-II, especially for the problems DTLZ1 and DTLZ2, though this is less important than the gain observed for ε-MOEA (Sect. 11.3.5.2).

11.3.5.5 Scalarization Methods: Conclusion

The main advantage of scalarization methods lies in their ability to show good performance for problems involving many objectives (more than four) as long as they incorporate an effective diversity preservation mechanism. This advantage of algorithms based on scalarization of the objectives has been confirmed in [34] for the MSOPS method [18].

11.4 Conclusion

In this chapter, we have presented some possible answers to some highly important questions raised by modern optimization problems: how to obtain several diverse solutions, but of equivalent value, to facilitate taking finer decisions according to additional possible criteria which cannot be formalized?

Multiobjective evolutionary optimization is a rich field in which innovation is constantly occurring. Methods and approaches recognized to date for their effectiveness or their specific qualities have been presented. Those based on a Pareto ranking have shown their ability to find good-quality approximations to a Pareto front in reasonable computing times, as long as the size of the objective space is less than or equal to four. Although the methods based on scalarization of the objectives were disappointing in the past when applied to solving problems with few objectives, they can now show performance comparable to or even better than the Pareto ranking methods, especially for problems involving many objectives.

Other approaches are also the subject of active research, always with the aim of effectively addressing problems with many objectives. These include, for example, methods based on the use of quality indicators as a fitness function to be optimized, such as the hypervolume measure (Sect. 11.3.2.3) [1].

11.5 Annotated Bibliography

Reference [8]   The first reference book in the field of multiobjective optimization evolutionary algorithms.

Reference [4]   This book of more than 800 pages is another, more recent reference book in the field of multiobjective evolutionary optimization. It contains, in particular, thorough discussions and examples of applications. Many approaches to solving multiobjective problems with other metaheuristics are also presented. The book also addresses the problem of multicriteria decision making that follows multiobjective optimization to choose the *best* nondominated solutions from the perspective of the decision maker.

References

1. Bader, J., Zitzler, E.: Hype: An algorithm for fast hypervolume-based many-objective optimization. Evolutionary Computation **19**(1), 45–76 (2011)
2. Beasley, D., Bull, D.R., Martin, R.R.: A sequential niche technique for multimodal function optimization. Evolutionary Computation **1**(2), 101–125 (1993)
3. Bessaou, M., Petrowski, A., Siarry, P.: Island model cooperating with speciation for multimodal optimization. In: H.P. Schwefel, M. Schoenauer, K. Deb, G. Rudolph, X. Yao, E. Lutton, J.J. Merelo (eds.) *Parallel Problem Solving from Nature, PPSN VI, 6th International Conference*, Paris. Springer (2000)
4. Coello Coello, C.A., Lamont, G.B., Von Veldhuizen, D.A.: *Evolutionary Algorithms for Solving Multi-Objective Problems*. Genetic and Evolutionary Computation. Springer, New York (2006)
5. Cohoon, J.P., Hedge, S.U., Martin, W.N., Richards, D.: Punctuated equilibria: A parallel genetic algorithm. In: J.J. Grefenstette (ed.) *Genetic Algorithms and Their Applications: Proceedings of the second International Conference on Genetic Algorithms*, pp. 148–154. Lawrence Erlbaum Associates, Hillsdale, NJ (1987)
6. Das, I., Dennis, J.E.: Normal-boundary intersection: A new method for generating Pareto optimal points in multicriteria optimization problems. SIAM Journal of Optimization **8**(3), 631–657 (1998)
7. De Jong, K.A.: An analysis of the behavior of a class of genetic adaptive systems. doctoral dissertation, University of Michigan (1975)
8. Deb, K.: Multi-Objective Optimization Using Evolutionary Algorithms. John Wiley and Sons (2001)
9. Deb, K., Mohan, M., Mishra, S.: A fast multi-objective evolutionary algorithm for finding well-spread Pareto-optimal solutions. Technical Report KanGAL 2003002, Indian Institute of Technology Kanpur (2003)
10. Deb, K., Pratap, A., Agarwal, S., Meyarivan, T.: A fast and elitist multiobjective genetic algorithm: NSGA-II. IEEE Transactions on Evolutionary Computation **6**(2), 182–197 (2002). doi:10.1109/4235.996017
11. Deb, K., Thiele, L., Laumanns, M., Zitzler, E.: Scalable multi-objective optimization test problems. In: *Congress on Evolutionary Computation (CEC'2002)*, vol. 1, pp. 825–830 (2002)
12. Fonseca, C.M., Fleming, P.J.: Genetic algorithms for multiobjective optimization: Formulation, discussion and generalization. In: *Proceedings of the Fifth International Conference on Genetic Algorithms*, pp. 416–423. Morgan Kaufmann (1993)

13. Fonseca, C.M., Paquete, L., Lopez-Ibanez, M.: An improved dimension-sweep algorithm for the hypervolume indicator. In: *IEEE Congress on Evolutionary Computation (2006).* doi:10. 1109/CEC.2006.1688440
14. Goldberg, D.E.: *Genetic Algorithms in Search, Optimization and Machine Learning.* Addison-Wesley (1989)
15. Goldberg, D.E., Richardson, J.: Genetic algorithms with sharing for multimodal function optimization. In: J. Grefenstette (ed.) *Proceedings of the 2nd International Conference on Genetic Algorithms,* pp. 41–49. Erlbaum, Hillsdale, NJ (1987)
16. Holland, J.H.: *Adaptation in Natural and Artificial Systems,* 2nd edn. MIT Press (1992)
17. Horn, J., Nafpliotis, N., Goldberg, D.E.: A niched pareto genetic algorithm for multiobjective optimization. In: *Proceedings of the 1st IEEE Conference on Evolutionary Computation,* pp. 82–87. IEEE Press, Piscataway, NJ (1994)
18. Hughes, E.J.: Evolutionary many-objective optimisation: Many once or one many? In: *Congress on Evolutionary Computation'05,* pp. 222–227 (2005)
19. Hutchinson, G.E.: Concluding remarks, population studies: Animal ecology and demography. In: *Cold Spring Harbor Symposia on Quantitative Biology,* vol. 22, pp. 415–427 (1957)
20. Ishibuchi, H., Murata, T.: A multi-objective genetic local search algorithm and its application to flowshop scheduling. IEEE Transcations on Systems, Man, and Cybernetics, Part C: Applications and Reviews, **28**(3), 392–403 (1998)
21. Knowles, J., Corne, D.: On metrics for comparing nondominated sets. In: *Evolutionary Computation, 2002, CEC '02, Proceedings of the 2002 Congress,* vol. 1, pp. 711–716 (2002)
22. Knowles, J.D., Corne, D.W.: Approximating the nondominated front using the Pareto archived evolution strategy. Evolutionary Computation **8**(2), 149–172 (2000). doi:10.1162/ 106365600568167
23. Mahfoud, S.W.: Crowding and preselection revisited. In: R. Manner, B. Manderick (eds.) *Proceedings of Parallel Problem Solving from Nature,* pp. 27–36. Elsevier (1992)
24. Messac, A., Ismail-Yahaya, A., Mattson, C.: The normalized normal constraint method for generating the Pareto frontier. Structural and Multidisciplinary Optimization **25**(2), 86–98 (2003)
25. Paquete, L., Stiitzle, T.: A two-phase local search for the biobjective traveling salesman problem. In: C.M. Fonseca, P.J. Fleming, E. Zitzler, L. Thiele, K. Deb *Evolutionary Multi-Criterion Optimization.* Lecture Notes in Computer Science, vol. 2632, pp. 479–493. Springer (2003)
26. Petrowski, A.: A clearing procedure as a niching method for genetic algorithms. In: *IEEE 3rd International Conference on Evolutionary Computation (ICEC'96),* pp. 798–803 (1996)
27. Petrowski, A., Girod Genet, M.: A classification tree for speciation. In: *Congress on Evolutionary Computation (CEC99),* pp. 204–211. IEEE Press, Piscataway, NJ (1999)
28. Sareni, B., Krahenbuhl, L.: Fitness sharing and niching methods revisited. IEEE Transactions on Evolutionary Computation **2**(3), 97–106 (1998)
29. Silverman, B.W.: *Density Estimation for Statistics and Data Analysis.* Chapman and Hall, London (1986)
30. Spears, W.M.: Simple subpopulation schemes. In: *Proceedings of the Conference on Evolutionary Programming,* pp. 296–307. World Scientific (1994)
31. Srinivas, N., Deb, K.: Multiobjective function optimization using nondominated sorting genetic algorithms. Evolutionary Computation **2**(3), 221–248 (1994)
32. Van Veldhuizen, D.A.: Multiobjective evolutionary algorithms: Classifications, analyses, and new innovations. Ph.D. thesis, Wright Patterson AFB, OH (1999). AAI9928483
33. van Veldhuizen, D.A., Lamont, G.B.: Multiobjective optimization with messy genetic algorithms. In: *SAC (1)'00,* pp. 470–476 (2000)
34. Wagner, T., Beume, N., Naujoks, B.: Pareto-, aggregation-, and indicator-based methods in many-objective optimization. In: *Proceedings of the 4th International Conference on Evolutionary Multi-Criterion Optimization, EMO'07,* pp. 742–756. Springer, Berlin, Heidelberg (2007)

35. Yin, X., Germay, N.: A fast genetic algorithm with sharing scheme using cluster methods in multimodal function optimization. In: R.F. Albrecht, C.R. Reeves, N.C. Steele (eds.) *Proceedings of the International Conference on Artificial Neural Nets and Genetic Algorithms*, pp. 450–457. Springer (1993)
36. Zhang, Q., Li, H.: MOEA/D: A multiobjective evolutionary algorithm based on decomposition. IEEE Transactions on Evolutionary Computation **11**(6), 712–731 (2007)
37. Zitzler, E.: Evolutionary algorithms for multiobjective optimization: Methods and applications. Ph.D. thesis, ETH Zurich, Switzerland (1999)
38. Zitzler, E., Deb, K., Thiele, L.: Comparison of multiobjective evolutionary algorithms: Empirical results. Evolutionary Computation **8**(2), 173–195 (2000)
39. Zitzler, E., Laumanns, M., Thiele, L.: SPEA2: Improving the strength Pareto evolutionary algorithm for multiobjective optimization. In: *Evolutionary Methods for Design, Optimisation, and Control*, pp. 95–100. CIMNE, Barcelona (2002)
40. Zitzler, E., Thiele, L.: Multiobjective evolutionary algorithms: A comparative case study and the strength Pareto approach. IEEE Transactions on Evolutionary Computation **3**(4), 257–271 (1999)
41. Zitzler, E., Thiele, L., Laumanns, M., Foneseca, C.M., Grunert da Fonseca, V.: Performance assessment of multiobjective optimizers: An analysis and review. IEEE Transactions on Evolutionary Computation **7**(2), 117–132 (2003)
42. Zydallis, J.B., van Veldhuizen, D.A., Lamont, G.B.: A statistical comparison of multiobjective evolutionary algorithms including the MOMGA-II. In: E. Zitzler, L. Thiele, K. Deb, C.A. Coello Coello, D. Corne (eds.) *Evolutionary Multi-Criterion Optimization*. Lecture Notes in Computer Science,vol. 1993, pp. 226–240 (2001)

Chapter 12
Extension of Evolutionary Algorithms to Constrained Optimization

Sana Ben Hamida

12.1 Introduction

Optimization problems from the industrial world must often respect a number of constraints. These are expressed as a set of relationships that the variables of the objective function must satisfy. These relationships are usually presented as equalities and inequalities that may be very hard to deal with. The general nonlinear parameter optimization problem is then defined as

$$\text{optimize } f(\mathbf{x}), \mathbf{x} = (x_1, \ldots, x_n) \in \mathcal{F} \subseteq \mathcal{S} \subseteq \mathbb{R}^n,$$

subject to

$$\begin{cases} g_i(\mathbf{x}) \leq 0 \text{ for } i = 1, \ldots, q \text{ (inequality constraints)} \\ h_j(\mathbf{x}) = 0 \text{ for } j = q+1, \ldots, m \text{ (equality constraints)} \end{cases}$$

Here, $\mathcal{F}$ is the feasible region where f, g_i, and h_j are real-valued functions on $\mathbb{R}^n$, $\mathcal{S}$ is the search space, q is the number of inequality constraints, and $m - q$ is the number of equality constraints (in both cases, constraints may be linear or nonlinear). In all that follows, we consider only minimization problems. A maximization problem can be transformed to a minimization problem by inverting the objective function $f(\mathbf{x})$.

The search space $\mathcal{S} \subseteq \mathbb{R}^n$ is defined by the lower and upper bounds of the solution vectors,

$$l(i) \leq x_i \leq u(i) \text{ for } 1 \leq i \leq n.$$

The satisfaction of the set of constraints g_j, $(j = 1, \ldots, q)$, h_j, $(j = q+1, \ldots, m)$ defines the feasible region $\mathcal{F}$. Any solution belonging to $\mathcal{F}$ is a

S. Ben Hamida (✉)
Université Paris Ouest, 92000 Nanterre, France
e-mail: sbenhami@u-paris10.fr

© Springer International Publishing Switzerland 2016
P. Siarry (ed.), *Metaheuristics*, DOI 10.1007/978-3-319-45403-0_12

Fig. 12.1 A search space $\mathcal{S}$
and its feasible and
unfeasible parts

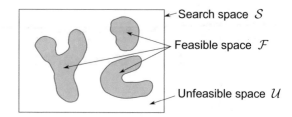

feasible solution, otherwise it is unfeasible. The search for a feasible optimum solution is all the more difficult when the size of the feasible space is small and/or its shape is complex (for example, $\mathcal{F}$ is a set of dispersed small areas as in Fig. 12.1).

The ratio $|\mathcal{F}|/|\mathcal{S}|$ can be used as a measure of difficulty for the problem [25]. The search for the optimum is often easier when it is inside the feasible region than when it is on its boundary. The latter case arises when one (or several) constraint(s) of the problem is (are) active at the optimum solution. This is often the case for real-world problems. Thus, constrained optimization problems require an exhaustive search of the feasible domain [12, 25].

There is not a standard evolutionary method to determine the global optimum of a constrained problem. The main question is: how to deal with unfeasible solutions? Two strategies have been devised in reply to this question. The first considers only feasible individuals, and therefore the objective function is computed only in the feasible domain. The second considers all individuals in the search space, but requires a special evaluation function for unfeasible individuals.

The choice of the right strategy depends on the nature of the problem: for example, whether the objective function is defined in the unfeasible domain or not.

A multitude of methods have been proposed using the two strategies, which can be classified into the following categories:

- penalty methods;
- methods based on the assumption of superiority of feasible individuals;
- methods based on the search for feasible solutions;
- methods based on preserving the feasibility of solutions;
- methods based on multiobjective optimization techniques;
- hybrid methods.

The papers [6, 20, 25] present comprehensive surveys of constraint-handling techniques for evolutionary algorithms published at different times (1996, 2002, and 2011, respectively). In this chapter, we present the basics and some reference methods for each category.

12.2 Penalization

Most constraint-handling methods are based on the concept of penalty functions, which penalize unfeasible solutions by adding a positive quantity to the objective function f (when the goal is to minimize f), in order to decrease the quality of such unfeasible individuals. The initial constrained problem is then converted into an unconstrained problem as follows:

$$\text{minimize } f(\mathbf{x}) + p(\mathbf{x})$$
$$p(\mathbf{x}) \begin{cases} = 0 \text{ if } \mathbf{x} \in \mathcal{F} \\ > 0 \text{ otherwise} \end{cases} \tag{12.1}$$

where $p(\mathbf{x})$ is the penalty function.

The design of the penalty function $p(\mathbf{x})$ is the main source of difficulty in penalty methods. Several techniques using different approaches and different heuristics have been proposed. The most popular approaches use measures of the constraint violations

$$p(\mathbf{x}) = F \left(\sum_{j=1}^{m} \alpha_j v_j^{\beta}(\mathbf{x}) \right) \tag{12.2}$$

where F is an increasing function; the positive real numbers α_j, $j = 1, \ldots, m$, are called the penalty coefficients; β is a parameter (often equal to 2); and $v_j(\mathbf{x})$ is the jth constraint violation (distance from the feasible region) defined as follows:

$$v_j(\mathbf{x}) = \begin{cases} \max(0, g_j(\mathbf{x})) \text{ for } 1 \le j \le q \text{ (inequality constraints)} \\ |h_j(\mathbf{x})| \quad\quad \text{ for } q+1 \le j \le m \text{ (equality constraints)} \end{cases} \tag{12.3}$$

The sum of the constraint violations $v_j(\mathbf{x})$ gives an estimate of the total violation of the solution $\mathbf{x}$, defined by

$$V(\mathbf{x}) = \sum_{j=1}^{m} v_j(\mathbf{x})$$

Using the constraint violation measure in the penalty function helps to distinguish the unfeasible individuals. Those which have a high violation measure are penalized more than those with low violation, which can guide the search towards the feasible space.

However, this measure is generally insufficient for choosing the penalty function, especially if the difference between the values of the objective function and the violation measures is very high (e.g. with an objective function on the order of 10^5 and a violation measure on the order of 10^{-1}). Penalty coefficients are then used to adjust the quantity to be added to the objective function. The main difficulty is to determine the appropriate value for each coefficient. If the penalty is too high or too low, then the problem might become very difficult for an evolutionary algorithm. With low values

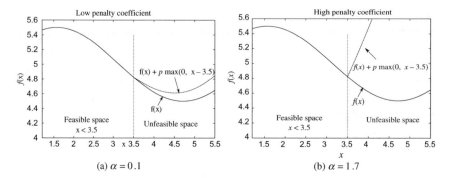

Fig. 12.2 Curves of the objective function $f(x) = 0.5 \sin(x) + 5$ and the fitness function $f(x) + \alpha \max(0, x - 3.5)$, with $\alpha = 0.1$ (**a**) and $\alpha = 1.7$ (**b**)

of penalty coefficients, the algorithm may produce many more unfeasible solutions than feasible solutions and then a lot of search time will be spent on exploring the unfeasible region. Otherwise, a large penalty discourages exploration of the unfeasible region and may increase the risk of premature convergence, especially if $\mathcal{F}$ is not convex or is disjoint.

In order to investigate the effect of the penalty coefficient on the performance of genetic algorithms, let us take a simple example. We consider the following problem:

$$\text{minimize } f(\mathbf{x}) = 0.5 \sin(x) + 5, \quad \text{with } 0 \le x \le 10$$

The problem is subject to a very simple single constraint: $x \le 3.5$. The optimum for this problem is $\mathbf{x}^* = 3.5$. The penalty function is defined as follows:

$$p(\mathbf{x}) = \alpha \max(0, x - 3.5)$$

Despite the simplicity of the problem, an evolutionary algorithm may fail to solve it if the value of α is not suitable. Figure 12.2a shows the curve of the objective function $f(\mathbf{x})$ and that of the evaluation function (fitness) $f(\mathbf{x}) + p(\mathbf{x})$ in the unfeasible region, with $\alpha = 0.1$. Clearly, the minimum of the evaluation function is $\mathbf{x}^* = 4.5$, which is an unfeasible solution. Thus, a penalty factor equal to 0.1 is too low to guarantee the feasibility of the solutions returned.

To overcome this problem, it is necessary to use a higher penalty coefficient. On the other hand, a very large value results in a sudden rejection of all unfeasible solutions in the population from the beginning of the evolution. In this case, the penalization may forbid any shortcuts across the unfeasible region and restrict the search to some parts of the feasible domain, which may lead to the failure of the method.

Figure 12.2b shows the curves of $f(\mathbf{x})$ and of $f(\mathbf{x}) + p(\mathbf{x})$ in the unfeasible area, with $\alpha = 1.7$. The slope of the curve of the fitness function for $x \le 3.5$ is very high, which induces weak production of solutions located in this part of the search space.

Since the optimum is on the boundary, the algorithm will have some difficulty in generating accurate solutions close to its position. In this case, there is always more chance of locating the optimum when the surrounding areas of the boundary are explored from both sides than from one side only.

Otherwise, if the feasible space $\mathcal{F}$ is nonconvex or disjoint, the presence of unfeasible solutions in the population improves the exploration capacity of the algorithm by spreading the population throughout the different parts of $\mathcal{F}$.

The choice of the penalization method must take into account the topological properties of the feasible space, the number of active constraints at the optimum, the ratio between the sizes of $\mathcal{F}$ and $\mathcal{S}$, and the types of the objective function and the constraints.

We distinguish four approaches to defining the penalty function: the static approach, the dynamic approach, the adaptive approach, and the self-adaptive approach. In the static approach, the penalty coefficients are parameters of the algorithm and their values are constant during the evolution. In the other approaches, their values are modified during the evolution, according to a predefined pattern in the dynamic approach, and depending on the historical and/or current status of the population in the adaptive and self-adaptive approaches.

The dynamic penalty function is generally an increasing function of the generation number in order to ensure the feasibility of the solutions at the end of the evolution, but the modification scheme is not simple to define and depends on the problem. The adaptive and self-adaptive methods modify the penalty coefficients according to some information extracted from the population, essentially the feasibility of the best solution or of a certain proportion of the population, and the distance from the feasible region.

12.2.1 "Death Penalty" Method

This is a quite simple method that just rejects unfeasible solutions from the population [1]. Although it does not need a penalty function, because the rejection takes place in the selection step, it can be viewed as a penalty method with an infinite penalty:

$$p(\mathbf{x}) = +\infty$$

The quality of the results given by this method depends strongly on:

- the ratio $|\mathcal{F}|/|\mathcal{S}|$, where $|\mathcal{F}|$ is the size of $\mathcal{F}$; note that the method cannot be applied when $\mathcal{F}$ has a null measure;
- the initialization scheme, which may cause instability of the method and increase the dispersal of the solutions returned since the search direction during evolution depends essentially on the starting population.

For these reasons, the "death penalty" method often has a very low performance.

12.2.2 Static Penalty Methods

The static penalty methods use user-defined values for the penalty coefficients α_j. The choice of values for these coefficients may be problematic because of the risks of overpenalization or underpenalization discussed above. In an attempt to avoid this risk, Homaifar et al. proposed in 1994 [13] a sophisticated method that defines a family of violation intervals for each constraint and then, for each interval, an appropriate penalty coefficient is defined. The method can be summarized in the following steps:

- For each constraint, create a number (l) of violation levels.
- For each constraint and for each violation level, define a penalty coefficient α_{ij} ($i = 1, 2, \ldots, l$, $j = 1, 2, \ldots, m$).
- The coefficients with the largest values are allocated to the highest violation levels.
- The initial population is generated randomly without taking into account the feasibility of the individuals.
- The population is evolved; each individual is evaluated with the following formula:

$$eval(\mathbf{x}) = f(\mathbf{x}) + \sum_{j=1}^{m} \alpha_{ij} v_j^2(\mathbf{x})$$

The main drawback of this method is the number of parameters to be defined before the evolution. For m constraints, the method requires a total of $m(2l + 1)$ parameters, where l is the number of violation levels defined for each constraint. This large number of parameters makes the quality of the results highly dependent on the values chosen.

12.2.3 Dynamic Penalty Methods

With the dynamic strategy, the values of the penalty coefficients are modified during the evolution according to a user-defined schedule—usually one in which they are increased—in order to ensure the generation of feasible individuals in the end.

Joines and Houk [14] proposed to evolve the penalties as follows:

$$p(\mathbf{x}) = (C \times t)^{\delta} \sum_{j=1}^{m} v_j^{\beta}(\mathbf{x}),$$

where t is the current generation and C, δ, and β are constant values, which are parameters of the method. A good choice of these parameters reported by Joines and Houck [14] is $C = 0.5$, $\delta = \beta = 2$.

This method requires much fewer parameters than methods based on static penalties, and gives better results thanks to the increasing selection pressure on the unfeasible solutions due to the term $(C \times t)^\delta$ in the penalty function. However, the factor $(C \times t)^\delta$ often increases very quickly, and the pressure becomes too strong. This affects the exploration process and reduces the chance of avoiding possible local optima.

Another approach based on dynamic penalties was proposed by Michalewicz and Attia [21] for their system GENOCOP II. GENOCOP II is the second version of the system GENOCOP ("GEnetic algorithm for Numerical Optimization of COnstrained Problems"). The latter had the handicap of being able to handle only linear constraints (see Sect. 12.5.1).

The algorithm begins by first distinguishing linear constraints LC and nonlinear constraints NC. It then builds an initial population, which has to satisfy the set LC. The feasibility of the population according to this set is maintained during the evolution thanks to some special operators in the GENOCOP system (see Sect. 12.5.1), which transform a feasible solution into an other feasible one.

To take the nonlinear constraints into account, Michalewicz and Attia were inspired by the cooling strategy of simulated annealing to define a penalization function

$$p(\mathbf{x}, \tau) = \frac{1}{2\tau} \sum_{j=1}^{m} v_j^{\ 2}(\mathbf{x})$$

where τ is the temperature of the system, which is a parameter of the algorithm. τ is decreased every generation according to a "cooling" scheme defined beforehand. Its goal is to increase the pressure on unfeasible individuals during the evolution. The algorithm stops when τ reaches a minimum temperature τ_f, which is also a parameter of the algorithm.

Experiments done by Michalewicz and Attia [21] showed that their algorithm may converge in a few iterations with a good choice of the "cooling" scheme, but it may give unsatisfactory results with other schemes.

12.2.4   Adaptive Penalty Methods

The main idea of the methods based on adaptive penalties is to introduce into the penalization function a component dependent on the state of the search process in a given generation. Thus, the weight of the penalty is adapted in every iteration, and it can be increased or decreased according to the quality of the solutions in the population. Numerous methods in this category have been proposed. Three are presented in this chapter: the methods of Hadj-Alouane and Bean [11], Smith and Tate [39], and Ben Hamida and Schoenauer [2, 3].

12.2.4.1 Method of Hadj-Alouane and Bean, 1992

With this method, the weight of the penalty depends on the quality of the best solution found in generation t. The penalty function is defined as follows:

$$p(\mathbf{x}) = \alpha(t) \sum_{j=1}^{m} v_j^2(\mathbf{x})$$

where $\alpha(t)$ is updated in each generation t as follows:

$$\alpha(t+1) = \begin{cases} (1/\beta_1) \cdot \alpha(t) & \text{if } \mathbf{x}_b \in \mathcal{F} \text{ over the last } k \text{ generations} \\ \beta_2 \cdot \alpha(t) & \text{if } \mathbf{x}_b \in (\mathcal{S} - \mathcal{F}) \text{ over the last } k \text{ generations} \\ \alpha(t) & \text{else} \end{cases}$$

where $\mathbf{x}_b$ is the best solution in the current population, and $\beta_1, \beta_2 > 1$ (with $\beta_1 \neq \beta_2$ to avoid cycles). In other words, this method decreases the value of the component $\alpha(t+1)$ in generation $t+1$ if all the best solutions over the last k generations were feasible, and increases its value in the opposite case (i.e., if all the best solutions were unfeasible). On the other hand, if during these k generations some of the best solutions were feasible and, at the same time, others were unfeasible, $\alpha(t+1)$ keeps the same value as $\alpha(t)$.

The aim of Hadj-Alouane and Bean was to increase the penalties only if they posed a problem for the search process; otherwise, they were reduced. However, the strategy of adaptation is based only on the state of the best individual over the last k generations. It does not consider the general state of the population.

12.2.4.2 Method of Smith and Tate, 1993

The adaptive penalty function proposed by Smith and Tate incorporates, as in the previous method, a component indicating the state of evolution of the search process, as well as a component indicating the degree of violation of the constraints. The first component depends on the quality of the best solution found during the evolution (up to the current iteration t). The second component is determined by the distance from the best unfeasible solutions to the feasible region. The purpose of this function is to expand the search space by introducing interesting unfeasible solutions (close to the feasible domain), which may facilitate the process of search when the optimum is located on the boundary of $\mathcal{F}$.

The penalty function is defined as follows:

$$p(\mathbf{x}) = (F_{\text{feas}}(t) - F_{\text{all}}(t)) \sum_{j=1}^{m} (v_j(\mathbf{x})/q_j(t))^k$$

where $F_{all}(t)$ is the value of the objective function (without penalty) of the best solution found during the evolution (up to the current generation t), $F_{feas}(t)$ is the evaluation of the best feasible solution found over the evolution, $q_j(t)$ is an estimate of the feasibility expansion threshold for each constraint, and k is a constant which allows adjustment of the "severity" of the penalty function.

Note that the thresholds $q_j(t)$ are dynamic; they are adjusted during the search process. For example, it is possible to define $q_j(t) = q_j(0)/(1 + \beta_j \cdot T)$ where $q_j(0)$ is the maximum threshold and β_j is a parameter to be set manually. However, Smith and Tate recommend that the technique for adjusting $q_j(t)$ should be changed according to the nature of the problem.

As in the method of Hadj-Alouane and Bean, the penalty function does not consider the general state of the population. Only the performance of the best feasible and unfeasible solutions is considered. Besides, this method has a further difficulty due to the choices that have to be made in the adjustment of the components $q_j(t)$.

12.2.4.3 Method of Ben Hamida and Schoenauer, 2000

Ben Hamida and Schoenauer [2] proposed the Adaptive Segregational Constraint Handling Evolutionary Algorithm (ASCHEA). The main idea of ASCHEA is to enhance the exploration around the boundaries of $\mathcal{F}$ by maintaining both feasible and unfeasible individuals in the population. In order to achieve this goal, ASCHEA relies on three main ingredients:

1. *A population-based adaptive penalty function* that uses global information about the population to adjust the penalty coefficients:

$$p(\mathbf{x}) = \alpha(t) \sum_{j=1}^{m} v_j(\mathbf{x}) \qquad (12.4)$$

   where $v_j(\mathbf{x})$ $j = 1, \ldots, m$ are the constraint violation measures (Eq. (12.3)). Increasing the value of the penalty coefficient $\alpha(t)$ in Eq. (12.4) clearly favors feasible individuals in subsequent selections, while decreasing it favors unfeasible individuals. Hence, in order to try to maintain a given proportion τ_{target} of feasible individuals, ASCHEA adopts the following strategy:

$$
\begin{array}{ll}
\text{if } (\tau_t > \tau_{target}) \ \alpha(t+1)(\mathbf{x}) = \alpha(t)/fact \\
\text{else} \qquad\qquad\quad \alpha(t+1)(\mathbf{x}) = \alpha(t) * fact
\end{array}
\qquad (12.5)
$$

   where τ_t denotes the proportion of feasible individuals in the current population and $fact > 1$ is a user-defined parameter.

2. *A constraint-driven recombination*, where in some cases feasible individuals can only mate with unfeasible individuals. It is known that, in many real-world problems, the constrained optimum lies on the boundary of the feasible domain (e.g., when one is minimizing some cost with technological constraints). In order to both

achieve better exploration of the boundary region and attract unfeasible individuals more rapidly toward feasible regions, ASCHEA uses a selection/seduction mechanism, which chooses the mates of feasible individuals to be unfeasible. However, to allow also the exploration of the feasible region, this mechanism is only applied when too few feasible individuals are present in the population

if $(0 < \tau_t < \tau_{\text{target}})$ **and** (x_1) is feasible

    select (x_2) among unfeasible individuals only

else select (x_2) according to fitness only

3. *A segregational selection* that distinguishes between feasible and unfeasible individuals. This starts by selecting without replacement feasible individuals, based on their fitness, until $\tau_{\text{select}} * \mu$ individuals have been selected (τ_{select} is a user-defined proportion and μ is the population size), or no more feasible individuals are available. The population is then filled using standard deterministic selection on the remaining individuals, based on the current penalized fitness. So, only a proportion τ_{select} of feasible individuals is considered superior to all unfeasible points.

The difficulty of satisfying constraints differs from one constraint to another, particularly in the case of active constraints. To adapt the search process to the difficulty of constraints and so ensure better exploration around the boundaries, Ben Hamida and Schoenauer proposed an improved version of ASCHEA [3], where each constraint is handled independently. Hence, the adaptive penalty $\alpha(t)$ is extended to several penalty coefficients $\alpha_j(t)$, one for each constraint j:

$$p(\mathbf{x}) = \sum_{j=1}^{m} \alpha_j(t) v_j(\mathbf{x})$$

Each coefficient is adapted, as for a single penalty (Eq. (12.5)), according to $\tau_t(j)$, which is the proportion of individuals satisfying the constraint. The idea is to have individuals on both sides (feasible and unfeasible) of the corresponding boundary.

Another component was introduced in the improved version of ASCHEA [3] to allow it to handle the equality constraints better. This component transforms the equalities $h_j(\mathbf{x}) = 0$ into two inequality constraints $-\epsilon_j(t) \leq h_j(\mathbf{x}) \leq \epsilon_j(t)$, where the value of ϵ_j is adjusted during the evolution. The objective is to start with a large feasible domain to enhance the process of exploration of the search space, and then to reduce it progressively in order to bring it as close as possible to the real domain with null measure at the end of the evolution, as illustrated in Fig. 12.3. In each generation t, unfeasible solutions are then *pushed* to the new feasible space $\mathcal{F}_{h_j}(t+1)$ thanks to the penalization and selection/seduction strategies of ASCHEA.

At the end of evolution, $\epsilon_j(t)$ takes very small values close to 0, which means that the equality constraint is satisfied numerically.

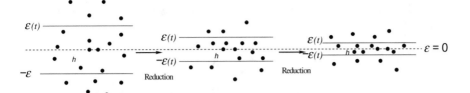

Fig. 12.3 Progressive reduction of the feasible domain $\mathcal{F}_{h_j}$ during evolution corresponding to an equality constraint h_j, using adaptive adjustment of ϵ_j in each generation t

Thanks to its components, ASCHEA has a great capacity for exploration and exploitation of the best feasible and unfeasible solutions, especially at the end of the evolution. It has given good results for several benchmarks, but remains costly in term of the number of evaluations.

12.2.5   Self-adaptive Penalty Methods

The main characteristic of the methods based on self-adaptive penalties is that they do not require additional parameters. In fact, the penalties are adapted solely on the basis of information from the population. Two classes of methods in this category have been proposed: those that consider the state of the population for several generations, such as the method of Coello [5], which was the first technique published in this category, and those that consider only the state of the population in the current generation, such as the methods of Farmani and Wright (2003) [9] and of Tessema and Yen (2006) [41]. Below, we present a method in each class.

12.2.5.1   Method of Coello, 1999

This is based on the principle of coevolution, where the population of penalties coevolves with the population of solutions. The population of solutions $P1$ is evaluated according to the following formula:

$$eval(\mathbf{x}) = f(\mathbf{x}) - \left(\alpha_1 \times \sum_{j=1}^{m} v_j(\mathbf{x})\right) + \alpha_2 \times \theta(\mathbf{x}))$$

where $\sum_{j=1}^{m} v_j(\mathbf{x})$ is the sum of the measures of the constraint violations by the solution $\mathbf{x}$, $\theta(\mathbf{x})$ is the number of constraints violated by $\mathbf{x}$, and α_1 and α_2 are two penalty coefficients.

A population $P2$ of vectors of penalty coefficients $A_j = (\alpha_1, \alpha_2)$ is maintained and coevolves in parallel with the population of solutions $P1$. Each vector A_j of $P2$ is used to evaluate all individuals of $P1$ for a given number of generations, after which the fitness of the corresponding vector A_j is computed by using the following formula:

$$\varphi(A_j) = \sum_{i=1}^{N} \frac{eval(x_i)}{N_f} + N_f$$

where $\varphi(A_j)$ is the average fitness of A_j, and N and N_f are the size of the population $P1$ and the number of feasible solutions in it, respectively. The best performance corresponds to the vectors A_j that allow one to generate more feasible solutions and solutions closer to the optimum.

The genetic operators are applied to the population $P2$ after the fitnesses of all vectors A_j have been computed. Thus, the penalty coefficients are adjusted automatically according to the information provided by the evolution of the population of solutions $P1$.

The major disadvantage of this method is its cost due to the large number of evaluations.

12.2.5.2 Method of Tessema and Yen, 2006

The SAPF (self-adaptive penalty function) method of Tessema and Yen [41] is based using on the distribution of the current population in the search space for the adjustment of the penalties. The algorithm of the SAPF method can be summarized in the following four steps:

1. Normalize the values of $f(\mathbf{x})$ for all solutions in the population according to the following formula:

$$\tilde{f}(\mathbf{x}) = \frac{f(\mathbf{x}) - \min_{\mathbf{x}} f(\mathbf{x})}{\max_{\mathbf{x}} f(\mathbf{x}) - \min_{\mathbf{x}} f(\mathbf{x})}$$

where $\min_{\mathbf{x}} f(\mathbf{x})$ and $\max_{\mathbf{x}} f(\mathbf{x})$ correspond to the fitness without penalty of the best and the worst solution, respectively, in the population.
2. Normalize the measures of constraint violations $v(\mathbf{x})$ for all the solutions in such a way that $\tilde{v}(\mathbf{x}) \in [0.1]$.
3. For each solution $\mathbf{x}$, compute the distance $d(\mathbf{x})$ as follows:

$$d(\mathbf{x}) = \begin{cases} \tilde{v}(\mathbf{x}) & \text{if all solutions are unfeasible} \\ \sqrt{\tilde{f}(\mathbf{x})^2 + \tilde{v}(\mathbf{x})^2} & \text{otherwise} \end{cases}$$

4. Evaluate the solutions with the following formula:

$$eval(\mathbf{x}) = d(\mathbf{x}) + (1 - r_t)\alpha_1(\mathbf{x}) + r_t\alpha_2(\mathbf{x})$$

where r_t is the proportion of feasibility of the population in generation t defined by the ratio of the number of feasible solutions to the size of the population, and α_1 and α_2 are two penalty functions defined as follows:

$$\alpha_1(\mathbf{x}) = \begin{cases} 0 & \text{if } r_t = 0 \\ \widetilde{v}(\mathbf{x}) & \text{otherwise} \end{cases}$$

$$\alpha_2(\mathbf{x}) = \begin{cases} 0 & \text{if } \mathbf{x} \text{ is feasible} \\ \widetilde{f}(\mathbf{x}) & \text{otherwise} \end{cases}$$

The steps described above allow the SAPF method to automatically adapt the penalties according to the distribution of the population in the search space while taking into account:

- the proportion of feasible solutions;
- the values of the objective function $f(\mathbf{x})$;
- the distances of unfeasible solutions from the feasible space $\mathcal{F}$.

This technique has given good results for several test cases [41]. However, it has low performance when the feasible region is too small or has a null measure, since the algorithm focuses more on the search for feasible solutions than on the optimization of the objective function.

12.2.6 Segregated Genetic Algorithm (SGGA)

The SGGA was proposed by Le-Riche et al. in 1995 [16] as a different approach to handling constraints by penalization that uses two different penalty functions at the same time. The first penalty applies weak penalties, while the second applies strong penalties. The aim is to overcome the problem of penalties that are too high or too low, discussed in the first part of Sect. 12.2. SGGA creates two groups of individuals from the population that coexist and cooperate. Each group is evaluated using one of the two penalty coefficients defined as parameters of the algorithm. The two groups are *segregated* during the selection step, where the individuals are sorted into a list using one of the two penalties. However, the variation operators are applied to a single ranked population that combines the two groups. The new population is made by selecting the best solutions from both lists. Two advantages result from this strategy:

1. The search space is explored by two different trajectories, one for each group. Also, thanks to hybridization of the two groups, the population can avoid local optima.
2. In constrained optimization problems, the global optimum is often on the boundary between the feasible and unfeasible areas. The hybridization between the two groups favors exploration of the boundaries and the global optimum is thus localized quite easily.

This algorithm has been tested [22] and has shown better performance than the static penalty method. However, the quality of the results remains sensitive to the choice of the penalty coefficients.

12.3 Superiority of Feasible Solutions

The approach of the superiority of the feasible individuals is based on the following heuristic rule: "any feasible solution is better than any unfeasible solution." This property is not guaranteed in the case of the penalty methods discussed above. The first method to use this heuristic rule was that of Powell and Skolnick published in 1993 [28]; then there was the method of *stochastic ranking* of Runarsson and Yao [31]; but the simplest to implement is that of Deb, published in 2000 [8].

12.3.1 Method of Powel and Skolnick

The method of Powell and Skolnick also uses penalty functions, but in a different way. The purpose is to map the fitness of all unfeasible solutions into the interval $(1, +\infty)$ and the fitness of all feasible solutions into the interval $[-\infty, 1)$ (for a minimization problem). To evaluate unfeasible solutions, in addition to the constraint violation measures, it uses the evaluation of feasible solutions as follows:

$$p(\mathbf{x}) = r \sum_{j=1}^{m} v_j(\mathbf{x}) + \theta(t, \mathbf{x})$$

where r is a constant parameter of the algorithm.

The component $\theta(t, \mathbf{x})$ is a function of the state of the population in the current generation t, and it has a great influence on the assessment of unfeasible individuals. It is defined by

$$\theta(t, \mathbf{x}) = \begin{cases} 0 & \text{if } \mathbf{x} \in \mathcal{F} \\ \max\{0, \delta\} & \text{otherwise} \end{cases}$$

with

$$\delta = \max_{\mathbf{y} \in \mathcal{F}} \{f(\mathbf{y})\} - \min_{\mathbf{y} \in (\mathcal{S} - \mathcal{F})} \left\{ f(\mathbf{y}) + r \sum_{j=1}^{m} v_j(\mathbf{y}) \right\}$$

With this additional heuristic, the performance of the unfeasible solutions depends on that of the feasible solutions: the best fitness of an unfeasible solution cannot be better than the worst fitness of any feasible solution ($\max_{\mathbf{x} \in \mathcal{F}} \{f(\mathbf{x})\}$).

This method requires the choice of a single parameter, r. The use of a small value allows the algorithm to explore the unfeasible region in parallel with the feasible domain, but if r is large, few unfeasible individuals survive in the population. Otherwise, the success of the method depends also on the topology of the feasible search space. Experimental results published in [24] indicate that for problems where $\mathcal{F}$ is too small, the method may fail.

12.3.2 Deb's Method

The method proposed by Deb [8] avoids the computation of the objective function in the unfeasible region. The proposed approach is based on the idea that in a constrained search, any individual must comply first with the constraints and then with the objective function. It uses a binary tournament selection, where two solutions are compared according to the following criteria:

1. Every feasible solution is better than every unfeasible solution.
2. From among two feasible solutions, the one with the best fitness is selected.
3. From among two unfeasible solutions, the one with the smallest violation measure is selected.

The evaluation of unfeasible solutions does not use penalty coefficients, but is instead based on constraint violation measures and the fitness of the feasible solutions:

$$eval(\mathbf{x}) = \begin{cases} f(\mathbf{x}) & \text{if } \mathbf{x} \in \mathcal{F} \\ f_{\max} + \sum_{j=1}^{m} v_j(\mathbf{x}) & \text{otherwise} \end{cases}$$

where $f_{\max}$ is the value of the objective function for the worst feasible solution in the population. To maintain diversity in the feasible domain, the method uses a niching technique applied during the selection step.

This method has the advantage of not requiring additional parameters. In addition, it does not compute the objective function in the unfeasible region. However, as with the previous method, it may fail if the ratio $|\mathcal{F}|/|\mathcal{S}|$ is too small.

12.3.3 Stochastic Ranking

Proposed by Runarsson and Yao in 2000 [31], this method introduces a new approach to creating a balance between the objective function and the penalty function, based on a stochastic ranking of the individuals, as described below.

If we assume that the solutions are evaluated using Eq. (12.1), then the penalty function is defined as follows:

$$p(\mathbf{x}) = r_t . \sum_{j=1}^{m} v_j(\mathbf{x}) = r_t \theta(\mathbf{x})$$

where r_t is the penalty coefficient and $\theta(\mathbf{x})$ is the violation measure.

To compare two adjacent solutions $\mathbf{x}_i$ and $\mathbf{x}_{i+1}$, the authors of the method introduced the concept of the critical penalty coefficient,

$$\tilde{r}_i = \frac{f(\mathbf{x}_{i+1}) - f(\mathbf{x}_i)}{\theta(\mathbf{x}_i) - \theta(\mathbf{x}_{i+1})} \quad \text{for } \theta(\mathbf{x}_i) \neq \theta(\mathbf{x}_{i+1})$$

For a given choice of $r_t > 0$, three types of comparisons are possible:

1. Comparison *dominated by the objective function*:
   $f(\mathbf{x}_i) \le f(\mathbf{x}_{i+1})$, $\theta(\mathbf{x}_i) \ge \theta(\mathbf{x}_{i+1})$, and $0 < r_t < \tilde{r}_i$.
2. Comparison *dominated by the penalty function*:
   $f(\mathbf{x}_i) \ge f(\mathbf{x}_{i+1})$, $\theta(\mathbf{x}_i) < \theta(\mathbf{x}_{i+1})$, and $0 < \tilde{r}_i < r_t$.
3. *Nondominated* comparison:
   $f(\mathbf{x}_i) < f(\mathbf{x}_{i+1})$, $\theta(\mathbf{x}_i) < \theta(\mathbf{x}_{i+1})$, and $\tilde{r}_i < 0$.

If $\bar{r}_t$ and $\underline{r}_t$ are the largest and smallest critical penalty coefficients, respectively, calculated from the adjacent individuals sorted according to the objective function, then it is necessary that $\underline{r}_t < r_t < \bar{r}_t$ so that the penalization is efficient. If $r_t < \underline{r}_t$, then all the comparisons are based only on the objective function: this is the case of *underpenalization*. On the other hand, if $r_t > \bar{r}_t$, then all the comparisons are based only on the penalty function: this is the case of *overpenalization*.

Finding a good strategy to adjust r_t in each generation, while avoiding *overpenalization* and *underpenalization*, is itself an optimization problem.

To overcome this difficulty, Runarsson and Yao proposed a stochastic ranking. They defined a probability P_f to decide whether to use the objective function or the penalty function for the comparison. Thus, two adjacent individuals $\mathbf{x}_i$ and $\mathbf{x}_{i+1}$, at least one of which is unfeasible, have a probability P_f of being compared according to their values of the objective function, and a probability $(1 - P_f)$ of being compared according to their constraint violation measures. If both individuals are feasible, $P_f = 1$.

This method was tested for a set on benchmark numerical problems. The best results were obtained with $P_f = 0.45$ [31].

Stochastic ranking has been developed further in [30] by using surrogate models for fitness approximations in order to reduce the total number of function evaluations needed during a search. The author of [30] found that the improved version provided the most competitive performance on a set of benchmark problems.

The simplicity of stochastic ranking has made it suitable for use in different domains. It has provided robust results for several different problems and benchmarks [17].

12.4 Searching for Feasible Solutions

The main goal of the methods in this category is to bring individuals into the feasible space $\mathcal{F}$. These methods can be divided into two subcategories: repairing unfeasible individuals and sampling the feasible space. A method is presented below for each subcategory.

12.4.1 Repair Methods: GENOCOP III

This is the third version of the GENOCOP system, proposed by Michalewicz and Nazhiyath in 1995 [24]. It is based on the idea of repairing unfeasible solutions (to make them feasible), and also uses some concepts of coevolution. This method incorporates the original GENOCOP system (described in Sect. 12.5.1) and extends it with an additional module that coevolves two separate populations. The first population, P_s includes points that satisfy the linear constraints of the problem, and these are called the search points. The feasibility of the points within P_s (with respect to the linear constraints) is maintained thanks to the special operators of the GENO-COP system (see Sect. 12.5.1). The second population, P_r includes points that satisfy all constraints of the problem (linear and nonlinear), and these are called reference points. The reference points r_i of P_r, being feasible, are evaluated directly with the objective function ($eval(\mathbf{r}) = f(\mathbf{r})$). However, the search points P_s, that are not feasible are repaired before they are evaluated. The repair process is described in Algorithm 12.1.

Let $\mathbf{s} \in P_s$ be a search point.
If $\mathbf{s} \in \mathcal{F}$, then $eval(\mathbf{s}) \leftarrow f(\mathbf{s})$,
else ($\mathbf{s} \notin \mathcal{F}$),

    repeat

        select a reference point $\mathbf{r}$ in P_r
        draw a random number a in the interval $[0, 1]$
        $\mathbf{z} \leftarrow a.\mathbf{s} + (1 - a).\mathbf{r}$
    while ($\mathbf{z}$ is unfeasible)

    $eval(\mathbf{s}) \leftarrow eval(\mathbf{z}) \leftarrow f(\mathbf{z})$
    replace $\mathbf{s}$ with $\mathbf{z}$ in P_s with a probability Q
    if $f(\mathbf{z}) < f(\mathbf{r})$,
    then replace $\mathbf{r}$ with $\mathbf{z}$ in P_r

Algorithm 12.1: Repair process in GENOCOP III

A search point $\mathbf{s}$ is replaced by a point $\mathbf{z}$ in the population P_s with a replacement probability Q. It should also be noted that there is an asymmetry between the evolutions of the two populations P_s and P_r: the application of the reproduction operator and the selection procedure to the population P_s is done in each generation, whereas it is only done every k generations for the population P_r, where k is a parameter of the method.

The coevolution strategy for the two populations is given by the main procedure of the GENOCOP III system, presented in Algorithm 12.2.

```
t ← 0
initialize  Ps(t),  Pr(t)
evaluate  Ps(t),  Pr(t)
while  (not stop condition ) do
     t ← t + 1
     select  Ps(t) from  Ps(t − 1)
     Reproduction of  Ps(t)
     evaluate  Ps(t)
     if  t mod  k = 0 then
          reproduction of  Pr(t)
          select  Pr(t) from  Pr(t − 1)
          evaluate  Pr(t)
end while
End
```

Algorithm 12.2: GENOCOP III algorithm

Note that the reproduction is done before the selection in the evolution process of P_r, owing to the low probability of generating a feasible offspring. Thus, the offspring are created first, and then the best feasible individuals among the parents and the offspring are selected to form the new population.

The advantages of GENOCOP III are that it does not evaluate the objective function in the unfeasible space and that it always returns a feasible solution. By contrast, the algorithm has great difficulty in creating the population of reference points if the ratio $|\mathcal{F}|/|\mathcal{S}|$ is very small. In particular, if the feasible region is not convex and if the population P_r has been initialized in a single component of $\mathcal{F}$, then the system will encounter difficulties in generating new feasible individuals in the other components of $\mathcal{F}$.

12.4.2 Behavioral Memory

This method was proposed by Schoenauer and Xanthakis in 1993 [36]. It is based on the concept of behavioral memory of the population: "the population contains not only information on the strict optimum, but also information on its behavior in the past."

The main purpose of this method is to sample the feasible space by processing the various constraints of the problem one by one and in a particular order. The algorithm begins with a random population. Then, for each constraint, it evolves the individuals until a certain percentage of the population becomes feasible for the constraint under consideration, while continuing to respect the previous constraints. There are $q + 1$ steps for q constraints to be satisfied. The population obtained at the end of each step is used as a starting point for the evolution for the next constraint. A linear order for processing the constraints must be defined. For the first q steps, the fitness in step i is a function $M(g_i(x))$ that is maximal when the constraint $g_i(x) \leq 0$ is satisfied. Individuals that do not satisfy the constraints g_1 to g_{i-1} are eliminated from the population by assigning them a null fitness.

The objective function is optimized in the last step using the death penalty method (see Sect. 12.2.1) for unfeasible points. In this step, the population may be located in a very small area of the search space, owing to the sequential processing of the constraints. This problem can be solved by using a niching procedure (see Sect. 11.2) to maintain diversity in each step.

This method has the advantage of avoiding evaluation of the objective function in the unfeasible region, but it can fail if the feasible domain is very small or disjointed. In addition, the sampling procedure requires the choice of a linear order for the treatment of the constraints of the problem. This choice greatly influences the quality of the results.

12.5 Preserving the Feasibility of Solutions

All the methods in this category have a common goal, which is to maintain the feasibility of the population. They use specific reproduction operators to generate feasible offspring from feasible parents (closed operators on $\mathcal{F}$).

12.5.1 GENOCOP System

The first version of GENOCOP ("GEnetic algorithm for Numerical Optimization of COnstrained Problems") was proposed in 1991 by Michalewicz and Janikow [23].

The system deals only with problems subject to linear constraints. It begins by eliminating the equality constraints by the elimination of a number of variables of the problem, which are replaced by linear combinations of the remaining variables. The inequality constraints are then reformulated by replacing the eliminated variables by these linear combinations. The remaining constraints, being linear, form a convex feasible space. Thus, it is quite easy to define closed operators that maintain the feasibility of the solutions.

For example, the arithmetic crossover of two feasible points $\mathbf{x}$ and $\mathbf{y}$ produces an offspring $\mathbf{z} = a\mathbf{x} + (1 - a)\mathbf{y}$, where $a = \mathcal{U}[0, 1]$ is a random number drawn uniformly in $[0, 1]$. It is then guaranteed that $\mathbf{z}$, in a convex domain, is always feasible.

Another crossover operator was added to GENOCOP, called heuristic crossover. This operator generates a child $\mathbf{z}$ from parents $\mathbf{x}$ and $\mathbf{y}$, selected such that the fitness $f(\mathbf{y})$ is better than $f(\mathbf{x})$, by applying the following rule:

$$\mathbf{z} = r \cdot (\mathbf{y} - \mathbf{x}) + \mathbf{x} \quad \text{where } r = \mathcal{U}[0, 1]$$

For the mutation operation, GENOCOP proceeds in two steps. It first determines the current domain $dom(x_i)$ for each component x_i of a solution vector x, which is a function of the linear constraints and the remaining values of the solution vector x. The new value of x_i is then taken from this domain.

This method has given good results for problems with a convex feasible space. However, it can be relatively expensive, as the heuristic crossover and uniform crossover may require several iterations before generating a feasible offspring.

12.5.2 Searching on the Boundary of the Feasible Region

In many cases, in constrained optimization problems, some constraints are active at the optimum. Thus the optimum is located on the boundary of the feasible space $\mathcal{F}$. Michalewicz and Schoenauer proposed an original approach which allows effective exploration of the boundary of the feasible region [33–35]. They introduced an evolutionary algorithm which starts from an initial population of points randomly selected on the boundary of $\mathcal{F}$. These solutions are then evolved while keeping their feasibility thanks to a set of "closed" genetic operators on $\mathcal{F}$.

The boundary is assumed to be a regular surface $\mathbf{S}$ of dimension $n - 1$ in the space $\mathbb{R}^n$. The operators used must be able to generate any point on the surface $\mathbf{S}$ (Fig. 12.4) and must respect the following conditions:

1. The crossover must be able to build the points in the neighborhood of both parents.
2. The mutation must be ergodic and must respect the principle of strong causality: a small change in the solution should cause a small change in the corresponding fitness.

Schoenauer and Michalewicz proposed several closed operators whose application depends on the type of surface of the boundary $\mathcal{F}$, such as specialized crossover and mutation operators for spherical and hyperboloidal surfaces [33, 34].

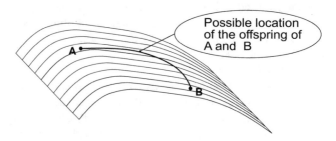

Fig. 12.4 Crossover operator on a surface

Fig. 12.5 Example of projection of points between the cube $[-1, 1]^n$ and the feasible space $\mathcal{F}$ (two-dimensional case)

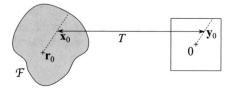

We can cite as an example the *curve-based operators*: given a curve joining two different points on the surface, a crossover operator can be defined by choosing one or two offspring on that curve (Fig. 12.4). We can also mention operators based on geodesic curves and plane operators based on curves resulting from the intersection of the surface **S** with two-dimensional planes.

This class of method has the advantage that it does not need to deal with unfeasible solutions, but it has the great disadvantage of being able to solve only those problems whose optimum is on the boundary of the feasible region. In addition, many difficulties may be encountered in the design of the genetic operators, which are specific to the problem to be solved.

12.5.3 "Homomorphous Mapping"

Proposed in 1999 by Koziel and Michalewicz [15], this method uses a set of decoders to transform a constrained problem into an unconstrained one. It evolves a population of encoded individuals, where each of them corresponds to a solution in the real search space. The following conditions must be satisfied to handle constraints with decoders:

1. For each solution $s \in \mathcal{F}$, there exists an encoded solution d.
2. Each encoded solution d corresponds to a feasible solution s.
3. All solutions in $\mathcal{F}$ must be represented by the same number of codes.
4. The encoding/decoding procedure T must not be too complex and should be fast in terms of computing time.
5. A small change in the encoded solution should generate a small change in the corresponding real solution.

"Homomorphous mapping" is a technique of encoding/decoding between any feasible search space $\mathcal{F}$ and an n-dimensional unit cube $[-1, 1]^n$ (Fig. 12.5). The encoded solution $\mathbf{y}_0 \in [-1, 1]^n$ for a point $\mathbf{x}_0 \in \mathcal{F}$ is obtained by a projection between the half-segment defined by the point $\mathbf{y}_0$ and the center of the cube **O**, and the half-segment defined by the point $\mathbf{x}_0$ and the reference point $\mathbf{r}_0 \in \mathcal{F}$. Thus, the encoded point $\mathbf{y}_0 \in \mathcal{F}$ corresponding to $\mathbf{x}_0$ is defined by $\mathbf{y}_0 = (\mathbf{x}_0 - \mathbf{r}_0) \cdot \tau$, where $\tau = (\|\mathbf{y}_M\|/\|\mathbf{x}_M - \mathbf{r}_0\|) \cdot \mathbf{y}_M$ is determined by a dichotomous search procedure.

This technique can only be applied for convex feasible spaces $\mathcal{F}$, but a generalization has been proposed for the case of a nonconvex space by introducing an additional encoding/decoding step. However, this generalization of the encoding technique may not respect the fifth condition for the validity of a decoder based on strong causality. The applicability of the method is therefore very limited.

12.6 Multiobjective Methods

The multiobjective approach relies on the idea of transforming the given constraints into additional objective functions to be minimized. Although the measure of violation of each constraint v_j $(j = 1, \ldots, m)$ can be handled as a separate objective in addition to the objective function f, the common approach considers the sum of the constraint violations as a second objective. Hence, the problem becomes a biobjective optimization problem. A second approach consists in transforming the constrained problem into an unconstrained multiobjective problem where the original objective function and each constraint are treated as separate objectives. The methods of the second approach can also be classified into subcategories: (1) those that use non-Pareto concepts (mainly based on multiple populations) and (2) those that use Pareto concepts (ranking and dominance) as selection criteria [7].

The first multiobjective method for constrained optimization was introduced by Parmee and Purchase in 1994 [27] for the optimization of gas turbine design with a heavily constrained search space. These authors used the multiobjective method proposed by Schaffer [32], called the "vector evaluated genetic algorithm" (VEGA), in which the aim is not to find the optimal solutions, but to search for feasible points to create a set of regions of $\mathcal{F}$ for a local search. The objective function is then optimized separately by a genetic algorithm using some special operators in order to help the algorithm to remain in the feasible region. This method was complex to implement, but the idea has inspired several researchers and has given birth to a large generation of multiobjective methods for handling constraints such as the method of Surry et al. (1995) [40], the method of Camponogara and Talukdar (1997) [4], the method of Ray et al. [29], the method of Coello (2002) [7], and the IDEA algorithm of Singh et al. (2008) [38].

In this chapter, three methods are presented from among those that we consider simple to implement.

12.6.1 Method of Surry et al.

The method proposed by Surry et al. in 1995 [40], called COMOGA ("Constrained Optimization by Multi-Objective Genetic Algorithms"), handles constraints as criteria of a multiobjective problem and simultaneously optimizes the objective function as an unconstrained optimization problem. To do this, all members of the search space S are labeled with some measure of their Pareto ranking R based on the constraint violations v_j (counting the number of individuals dominated by each solution).

Then, each solution is evaluated by both the Pareto rank and the value of the objective function f:

$$I_R(\mathbf{x}) = (R_{(v_1,\ldots,v_m)}(\mathbf{x}), f(\mathbf{x}))$$

The Pareto ranking is defined using the same sorting technique as in MOGA, proposed by Fonseca and Fleming in 1993 [10].

The environmental selection for the next generation proceeds in two steps. First, $p_{\text{cost}} \times N$ individuals are selected using a binary tournament selection based on the fitness f. Then, the rest of the individuals $((1 - p_{\text{cost}}) \times N)$ are selected linear according to their ranks R. To avoid convergence to an unfeasible solution, the value of p_{cost} is adapted dynamically according to the proportion of unfeasible solutions in the population, compared with a reference rate τ. The scheme of this method can be summarized in the following steps:

1. Compute the constraint violation measures v_j for all solutions.
2. Compute Pareto ranks R for all solutions using the violation measures v_j.
3. Compute the fitness f.
4. Select a proportion p_{cost} of solutions using f, and the rest in proportion to R.
5. Apply the crossover and mutation operators.
6. Adjust p_{cost}: if the proportion of feasible individuals is less than the reference rate τ, decrease p_{cost}: $p_{\text{cost}} \leftarrow (1 - \epsilon)p_{\text{cost}}$. Otherwise, increase p_{cost}: $p_{\text{cost}} \leftarrow 1 - (1 - p_{\text{cost}})(1 - \epsilon)$, where $0 < \epsilon \ll 1$.

The method was successfully applied to design a gas network (dealing with the provision and pipe type) [40], and it gave good results. However, it did not give the same degree of accuracy for other benchmark problems.

12.6.2 Method of Camponogara and Talukdar

Camponogara and Talukdar [4] suggested handling the problem of constrained optimization as a two-objective optimization problem. The first objective is the objective function f of the initial problem, and the second objective is an aggregation of the constraint violations:

$$\Theta(\mathbf{x}) = \sum_{j=1}^{m}(v_j(\mathbf{x}));$$

where $v_j(\mathbf{x})$ is obtained from the formula (12.3).

Once the problem has thus been redefined, a set of nondominated solutions is built. These solutions define a new direction of search d that tends to minimize all the objectives; $d = (\mathbf{x}_i - \mathbf{x}_j)/(|\mathbf{x}_i - \mathbf{x}_j|)$, where $\mathbf{x}_i \in S_i$ and $\mathbf{x}_j \in S_j$, and S_i and S_j are Pareto sets. A line search is then applied in the direction of search defined by d in order to create a better solution y which dominates $\mathbf{x}_i$ and $\mathbf{x}_j$.

This technique is simple to implement but it has some difficulties in preserving population diversity. Additionally, the use of a line search within a genetic algorithm adds some extra computational cost.

12.6.3 IDEA Method of Singh et al.

Singh et al. proposed the method IDEA, ("Infeasibility Driven Evolutionary Algorithm") method in 2008 [38]. Its idea is not only to consider the constraints as objectives but also to maintain the best unfeasible solutions in the population to try to approach the optimum from both the feasible and the unfeasible sides of the search space. IDEA transforms then the constrained problem into a biobjective optimization problem as follows:

$$\text{Minimize} \begin{cases} f(\mathbf{x}) \\ \Theta(\mathbf{x}) = \sum_{j=1}^{m} (R_j(\mathbf{x})) \end{cases}$$

IDEA assigns to each solution $\mathbf{x}$ in the population m ranks $R_j(\mathbf{x})$, corresponding to the m constraints of the problem, based on the violation measures v_j. For each constraint j, rank 0 corresponds to the solutions respecting this constraint, rank 1 corresponds to the solutions having the minimum violation measure, and the remaining solutions have ascending ranks according to the violation measure. $\Theta(\mathbf{x})$ is then the sum of the ranks R_j assigned to the solution $\mathbf{x}$. The Pareto rank is then used by the genetic operators in the same way as in the NSGA-II method.

In the replacement step, a proportion λ of the new population is selected from the the set of solutions with $\Theta(\mathbf{x}) > 1$. The goal is to keep the best unfeasible solutions during evolution.

Thanks to this additional component, IDEA has a better convergence ability than the other methods in the same category. The method showed high performance and fast convergence when applied by Singh et al. [37] to a problem of dynamic optimization.

12.7 Hybrid Methods

The general goal of the methods in this category is to separate the individuals from the constraints, which are handled using other heuristics or approaches while the objective function continues to be solved with an evolutionary algorithm. There are two ways to accomplish this separation. The first one is to handle the constraints by a deterministic procedure for numerical optimization combined with the evolutionary algorithm. In the second approach, the evolutionary algorithm creates this separation by including a different evolutionary approach to handle constraints.

For the first approach, we can cite as an example the method of Myung and Kim [26], which extend the evolutionary algorithm with Lagrange multipliers. As an example of the second approach, we can consider the method of Leguizamon and Coello-Coello [18], which uses ant colonies to explore the boundaries of the feasible domain. Several methods using the same approach have been published during the last decade. A more detailed description can be found in the book by Mezura-Montes [19].

12.8 Conclusion

This chapter has presented a set of approaches to handling constraints in an optimization problem using evolutionary algorithms. The basic ideas of these approaches vary from a simple penalization function to hybrid methods. The choice of an appropriate technique depends on several criteria, mainly related to the nature of the problem. Several questions need to be asked in this context, such as:

- Is the objective function defined in the infeasible domain? If it is not, several techniques cannot be applied, such as a large proportion of the penalization methods.
- Are there some active constraints at the optimum? If there are no active constraints at the optimum, none of the methods based on a search at the boundaries of the feasible region can be chosen.
- What are the types of constraints? For example, if at least one of the constraints is a nonlinear inequality, the methods which handle only linear constraints are excluded, such as the GENOCOP system.
- Is the ratio between the feasible space and the search space too small? If the problem has equality constraints or the ratio $|\mathcal{F}|/|\mathcal{S}|$ is too small, it is preferable to avoid certain methods that have demonstrated weak performance for this case, such as some approaches based on the superiority of the feasible solutions.

Some other criteria should also be considered in relation to the choice of method, such as the effectiveness demonstrated in the solution of benchmark problems. The performance of some approaches has been proved in several comparative studies, which can be an arguement for the choice of the corresponding technique. However, the effectiveness of a method is often dominated by two other selection criteria, which are the complexity and the difficulty of implementation.

In conclusion, there is not a general approach to handling constraints in evolutionary algorithms that is able to deal with any type of problem. This subject continues to be the focus of several research projects in the field.

12.9 Annotated Bibliography

Reference [19] This book is a collection of articles about recent research on handling constraints in evolutionary algorithms. It covers mainly multiobjective methods, the hybrid method, constrained optimization by immune systems, and differential evolution, as well as other recent studies and real applications in this field.

Reference [42] Yu and Gen's book has a special chapter on constrained optimization that presents and discusses some approaches to constraint handling, such as penalty functions and feasibility maintenance.

References

1. Back, T., Hoffmeister, F., Schwefel, H.P.: A survey of evolution strategies. In: *Proceedings of the Fourth International Conference on Genetic Algorithms*, pp. 2–9. Morgan Kaufmann (1991)
2. Ben-Hamida, S., Schoenauer, M.: An adaptive algorithm for constrained optimization problems. In: M. Schoenauer, K. Deb, G. Rudolph, X. Yao, E. Lutton, J. Merelo, H.P. Schwefel (eds.) *Proceedings of 6th Parallel Problem Solving From Nature (PPSN VI), Paris*. Lecture Notes in Computer Science, vol. 1917 pp. 529–538. Springer, Heidelberg (2000)
3. Ben-Hamida, S., Schoenauer, M.: ASCHEA: New results using adaptive segregational constraint handling. In: *Proceedings of the Congress on Evolutionary Computation 2002 (CEC'2002)*, vol. 1, pp. 884–889. IEEE Press, Piscataway, NJ (2002)
4. Camponogara, E., Talukdar, S.N.: A genetic algorithm for constrained and multiobjective optimization (1997)
5. Coello, C.A.C.: Self-adaptive penalties for GA-based optimization. In: *Proceedings of the Congress on Evolutionary Computation 1999 (CEC'99)*, vol. 1, pp. 573–580. IEEE Press, Piscataway, NJ (1999)
6. Coello, C.A.C.: Theoretical and numerical constraint handling techniques used with evolutionary algorithms: a survey of the state of the art. Computer Methods in Applied Mechanics and Engineering **191**(11–12), 1245–1287 (2002)
7. Coello, C.A.C., Mezura-Montes, E.: Handling constraints in genetic algorithms using dominance-based tournaments. In: I. Parmee (ed.) *Proceedings of the Fifth International Conference on Adaptive Computing in Design and Manufacture (ACDM'2002)*, Exeter, Devon, UK, vol. 5, pp. 273–284. Springer (2002)
8. Deb, K.: An efficient constraint handling method for genetic algorithms. Computer Methods in Applied Mechanics and Engineering **186**(2/4), 311–338 (2000)
9. Farmani, R., Wright, J.A.: Self-adaptive fitness formulation for constrained optimization. IEEE Transactions on Evolutionary Computation **7**(5), 445–455 (2003)
10. Fonseca, C.M., Fleming, P.J.: An overview of evolutionary algorithms in multiobjective optimization. Evolutionary Computation **3**, 1–16 (1995)
11. Hadj-Alouane, A.B., Bean, J.C.: A genetic algorithm for the multiple-choice integer program. Operations Research **45**, 92–101 (1997)
12. Hamida, S.B., Petrowski, A.: The need for improving the exploration operators for constrained optimization problems. In: *Proceedings of the Congress on Evolutionary Computation 2000 (CEC'2000)*, vol. 2, pp. 1176–1183. IEEE Press, Piscataway, NJ (2000)

13. Homaifar, A., Lai, S.H.Y., Qi, X.: Constrained optimization via genetic algorithms. Simulation **62**(4), 242–254 (1994)
14. Joines, J., Houck, C.: On the use of non-stationary penalty functions to solve nonlinear constrained optimization problems with GAs. In: D. Fogel (ed.) *Proceedings of the first IEEE Conference on Evolutionary Computation*, pp. 579–584. IEEE Press, Orlando, FL (1994)
15. Koziel, S., Michalewicz, Z.: Evolutionary algorithms, homomorphous mappings, and constrained parameter optimization. Evolutionary Computation **7**(1), 19–44 (1999)
16. Le-Riche, R.G., Knopf-Lenoir, C., Haftka, R.T.: A segregated genetic algorithm for constrained structural optimization. In: L.J. Eshelman (ed.) *Proceedings of the Sixth International Conference on Genetic Algorithms (ICGA-95)*, Pittsburgh, pp. 558–565. Morgan Kaufmann, San Mateo, CA (1995)
17. Leguizamón, G., Coello, C.A.C.: A boundary search based ACO algorithm coupled with stochastic ranking. In: *Proceedings of the IEEE Congress on Evolutionary Computation, CEC 2007, 25–28 September 2007*, Singapore, pp. 165–172 (2007). doi:10.1109/CEC.2007. 4424468
18. Leguizamón, G., Coello-Coello, C.: A boundary search based aco algorithm coupled with stochastic ranking. In: *2007 IEEE Congress on Evolutionary Computation (CEC'2007)*, pp. 165–172. IEEE Press (2007)
19. Mezura-Montes, E. (ed.): *Constraint-Handling in Evolutionary Optimization*. Springer, Berlin (2009)
20. Mezura-Montes, E., Coello, C.A.C.: Constraint-handling in nature-inspired numerical optimization: Past, present and future. Swarm and Evolutionary Computation **1**(4), 173–194 (2011)
21. Michalewicz, Z., Attia, N.F.: Evolutionary optimization of constrained problems. In: *Proceedings of the 3rd Annual Conference on Evolutionary Programming*, pp. 98–108. World Scientific (1994)
22. Michalewicz, Z., Dasgupta, D., Riche, R.L., Schoenauer, M.: Evolutionary algorithms for constrained engineering problems. Computers & Industrial Engineering Journal **30**(4), 851–870 (1996)
23. Michalewicz, Z., Janikow, C.Z.: Handling constraints in genetic algorithms. In: R.K. Belew, L.B. booker (eds.) *Proceedings of the Fourth International Conference on Genetic Algorithms (ICGA-91)*, San Diego, pp. 151–157. Morgan Kaufmann, San Mateo, CA (1991)
24. Michalewicz, Z., Nazhiyath, G.: Genocop III: A co-evolutionary algorithm for numerical optimization with nonlinear constraints. In: D.B. Fogel (ed.) *Proceedings of the Second IEEE International Conference on Evolutionary Computation*, pp. 647–651. IEEE Press, Piscataway, NJ (1995)
25. Michalewicz, Z., Schoenauer, M.: Evolutionary algorithms for constrained parameter optimization problems. Evolutionary Computation **4**(1), 1–32 (1996)
26. Myung, H., Kim, J.H.: Hybrid interior-lagrangian penalty based evolutionary optimization. In: V. Porto, N. Saravanan, D. Waagen, A. Eiben (eds.) *Proceedings of the 7th International Conference on Evolutionary Programming (EP98)*, San Diego. Lecture Notes in Computer Science, vol. 1447, pp. 85–94. Springer, Heidelberg (1998)
27. Parmee, I.C., Purchase, G.: The development of a directed genetic search technique for heavily constrained design spaces. In: I.C. Parmee (ed.) *Adaptive Computing in Engineering Design and Control-'94*, Plymoth, UK, pp. 97–102 (1994)
28. Powell, D., Skolnick, M.M.: Using genetic algorithms in engineering design optimization with non-linear constraints. In: S. Forrest (ed.) *Proceedings of the Fifth International Conference on Genetic Algorithms (ICGA-93)*, University of Illinois, pp. 424–431, Morgan Kaufmann, San Mateo, CA (1993)
29. Ray, T., Kang, T., Chye, S.K.: An evolutionary algorithm for constrained optimization. In: *Genetic and Evolutionary Computation Conference*, pp. 771–777 (2000)
30. Runarsson, T.: Approximate evolution strategy using stochastic ranking. In: G.G. Yen, S.M. Lucas, G. Fogel, G. Kendall, R. Salomon, B.T. Zhang, C.A.C. Coello, T.P. Runarsson (eds.) *Proceedings of the 2006 IEEE Congress on Evolutionary Computation*, Vancouver, pp. 745–752. IEEE Press, Piscataway, NJ (2006)

31. Runarsson, T.P., Yao, X.: Stochastic ranking for constrained evolutionary optimization. IEEE Transactions on Evolutionary Computation **4**(3), 284–294 (2000)
32. Schaffer, J.D.: Multiple objective optimization with vector evaluated genetic algorithms. In: *International Conference on Genetic Algorithms*, pp. 93–100 (1985)
33. Schoenauer, M., Michalewicz, Z.: Evolutionary computation at the edge of feasibility. In: H.M. Voigt, W. Ebeling, I. Rechenberg, H.P. Schwefel (eds.) *Proceedings of the Fourth Conference on Parallel Problem Solving from Nature (PPSN IV)*, pp. 245–254. Springer, Heidelberg (1996)
34. Schoenauer, M., Michalewicz, Z.: Boundary operators for constrained parameter optimization problems. In: T. Bäck (ed.) *Proceedings of the Seventh International Conference on Genetic Algorithms (ICGA-97)*, pp. 322–329. Morgan Kaufmann, San Francisco, CA (1997)
35. Schoenauer, M., Michalewicz, Z.: Sphere operators and their applicability for constrained optimization problems. In: V. Porto, N. Saravanan, D. Waagen, A. Eiben (eds.) *Proceedings of the 7th International Conference on Evolutionary Programming (EP98)*, San Diego. Lecture Notes in Computer Science, vol. 1447 pp. 241–250. Springer, Heidelberg (1998).
36. Schoenauer, M., Xanthakis, S.: Constrained GA optimization. In: S. Forrest (ed.) *Proceedings of the Fifth International Conference on Genetic Algorithms (ICGA-93)*, University of Illinois pp. 573–580, Morgan Kauffman, San Mateo, CA (1993)
37. Singh, H.K., Isaacs, A., Nguyen, T.T., Ray, T., Yao, X.: Performance of infeasibility driven evolutionary algorithm (idea) on constrained dynamic single objective optimization problems. In: *2009 IEEE Congress on Evolutionary Computation (CEC'2009)*, Trondheim, pp. 3127–3134. IEEE Press, Piscataway, NJ (2009)
38. Singh, H.K., Isaacs, A., Ray, T., Smith, W.: Infeasibility driven evolutionary algorithm (IDEA) for engineering design optimization. In: *Australasian Conference on Artificial Intelligence*, pp. 104–115 (2008)
39. Smith, A.E., Tate, D.M.: Genetic optimization using a penalty function. In: S. Forrest (ed.) *Proceedings of the Fifth International Conference on Genetic Algorithms (ICGA-93)*, University of Illinois pp. 499–503. Morgan Kaufmann, San Mateo, CA (1993)
40. Surry, P.D., Radcliffe, N.J., Boyd, I.D.: A multi-objective approach to constrained optimisation of gas supply networks: The COMOGA Method. In: T.C. Fogarty (ed.) *Evolutionary Computing. AISB Workshop*, Sheffield, U.K, Selected Papers. Lecture Notes in Computer Science, vol. 993 pp. 166–180. Springer (1995).
41. Tessema, B., Yen, G.G.: A self adaptative penalty function based algorithm for constrained optimization. In: *2006 IEEE Congress on Evolutionary Computation (CEC'2006)*, Vancouver, pp. 950–957. IEEE (2006)
42. Yu, X., Gen, M. (eds.): *Introduction to Evolutionary Algorithms*. Springer, London (2010)

Chapter 13
Methodology

Eric Taillard

13.1 Introduction

We will certainly disappoint those readers who have been patient enough to read
the present book up to here and who would know which metaheuristic they should
try first for solving a problem under their consideration. Indeed, this question is a
perfectly legitimate one, but we must confess that it is not possible to recommend
one specific technique or another. It has been seen that the weak theoretical results
known about metaheuristics are of almost no use in practice. In fact, in a sense, these
theorems state that to ensure that the optimum is correctly determined, it is required
to examine a number of solutions that is greater than the total number of solutions
of the problem. In other words, they recommend (trivially!) that one should use an
exact method if the optimum needs to be determined absolutely correctly. However,
the present chapter will make an attempt to draw up some guidelines for developing
a heuristic method based on the metaprinciples discussed earlier.

At the methodological level, it is of prime importance to use an adequate model
for solving the problem. The first question to ask is whether to treat the problem as
an optimization, a classification, or a multicriteria problem. The choice of the right
model is essentially intuitive, but a few general principles must be followed. The
first one is *divide and conquer*. The first part of this chapter treats decomposition
techniques, either the decomposition of a complex problem into a series of simpler
subproblems or the decomposition of a large-size problem into smaller subproblems.

Once the model has been chosen, the first natural attempt is to try to derive rules for
building a solution of adequate quality. This means evaluating whether greedy choices
appear to be adequate. If this is the case, a GRASP-based implementation can be
recommended, especially for someone who is only starting to work in metaheuristics.
Other building methods, such as artificial ant colonies are more difficult to tune. If a

E. Taillard (✉)
HEIG-VD, Route de Cheseaux 1, CP, 1401 Yverdon-les-Bains, Switzerland
e-mail: eric.taillard@heig-vd.ch

© Springer International Publishing Switzerland 2016   357
P. Siarry (ed.), *Metaheuristics*, DOI 10.1007/978-3-319-45403-0_13

greedy building method seems too hard to implement, the alternative choice is to start with a local search. Simulated annealing and tabu search can be recommended, as well as variable neighborhood search, because these methods have a limited number of parameters and are easy to tune.

When local searches focus too rapidly on bad-quality local optima, they must be hybridized with a learning level, in the spirit of what is called *adaptive memory programming*. One of the simplest of these adaptive memory methods is certainly GRASP with path relinking (GRASP-PR).

Finally, the design of an algorithm based on metaheuristic principles requires one to tune parameters and to make choices from among various algorithmic options. The second part of this chapter presents a few techniques for comparing iterative heuristics.

Following the same principles as we adopted in the chapter on tabu search, this illustration will be presented with the help of a particular optimization problem. The vehicle routing problem has been chosen for this specific purpose. In order to make the illustration as clear as possible, we limit ourselves to the simplest version of the problem, known as the *capacitated vehicle routing problem* (CVRP) in the literature, and to one of its subproblems, the traveling salesman problem, as well as an extension, the location–routing problem.

13.1.1 Academic Vehicle Routing Problem

An academic problem, which is a simplification of practical vehicle routing problems, can be described as follows. An unlimited set of vehicles, each one capable of carrying a volume V of goods, is required to deliver n orders to customers, starting from a unique depot, in such a way that the total distance traveled by the vehicles is minimized. Each order (or, as is commonly said, customer) i has a volume v_i ($i = 1, \ldots, n$). The direct distances d_{ij} between customers i and j ($i, j = 0, \ldots, n$), are known, with 0 representing the depot. The vehicles execute tours T_k ($k = 1, 2, \ldots$) that start from and finish at the depot. A variant of the problem imposes the additional constraint that the lengths of the tours must be bounded from above by a given value L. Figure 13.1 illustrates the shape of a solution obtained for a Euclidean problem instance considered in the literature [3], with 75 customers (marked by circles, whose area is proportional to the volume ordered) and a depot (marked by a black disk, whose area is proportional to the volume of the vehicles).

A solution of this problem can be viewed as a partition of the set of customers into a number of ordered subsets, the order defining the sequence in which each vehicle has to visit the customers constituting a tour. The vehicle routing problem has the traveling salesman problem as a subproblem; if one knows the set of customers to be serviced by a given tour, one has to find the tour of shortest length, which is a traveling salesman problem. The VRP is a subproblem of the *location–routing*

Fig. 13.1 Best solution known for a small academic vehicle routing problem with 75 customers. It has not yet been proved that this solution is an optimal one

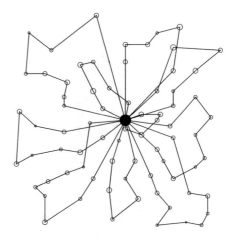

problem, where the position of the depot must be chosen (opening a depot has a given cost) and, simultaneously, finding vehicle tours. This means choosing the starting depot for each tour.

13.2 Decomposition Methods

13.2.1 Chain of Decomposition

The first reflex one has when facing a complex problem is to decompose it into a series of simpler subproblems. In the case of the location–routing problem, the first attempt may be to find clusters of customers that are close each other so that the sum of their delivery volumes is not greater than the capacity of a vehicle. Once these clusters have been determined, for instance by solving a p-median problem with capacity, an optimal tour visiting all customers in a cluster can be found. If the possible positions of the depots are not given, the median of each cluster can define these positions. Finally, the depot-opening costs are limited by connecting several tours to the same depot.

This technique is illustrated in Fig. 13.2 for a set of customers located in the islands of Corsica and Sardinia.

For a given problem, the decomposition into a series of subproblems may vary. For instance, for the location–routing problem, instead of having successively p-median $\longrightarrow$ traveling salesman $\longrightarrow$ depot location, an alternative would be to solve a p-median problem with a number p of centers equal to the final number of depots and to solve a complete VRP for each cluster of customers assigned to the same depot-center.

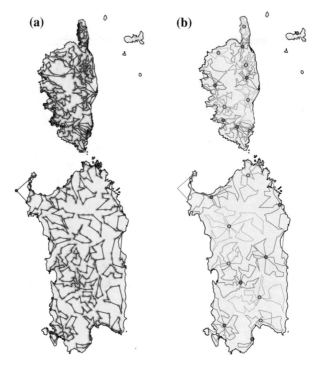

Fig. 13.2 Decomposition of a location–routing problem into a series of simpler problems. First, a *p*-median problem is solved to identify the groups of customers that would be logical to assign to a vehicle tour. Then, the depot-opening costs are limited by connecting several tours to the same depot

A third possibility that has been proposed by some other authors for the location–routing problem is to solve a traveling salesman problem on the whole set of customers and then to decompose this large tour into subpaths whose volume is compatible with the capacity of the vehicles. The depot positioning is finally chosen by solving a *p*-median problem.

Naturally, this technique of decomposition into a series of subproblems is a heuristic one. It is not necessarily efficient for every problem instance and requires good intuition from the designer to estimate the characteristics of good solutions.

13.2.2 Decomposition into Subproblems of Smaller Size

When solving large-size problem instances, a natural tendency is to proceed by decomposing the problem into independent subproblems. These subproblems can then be solved by employing an appropriate procedure. In this way, large-size problem instances can be approached efficiently, since the global complexity of the method grows very slowly, typically as $O(n)$ or $O(n\log(n))$, where n is the problem size.

However, implementing an a priori decomposition of a problem may induce low-quality solutions, since the subproblems will have been created more or less arbitrarily without considering the structure of the solutions. It is not easy to decompose a problem conveniently without having an intuition about the structure of good solutions. The idea behind POPMUSIC is to locally optimize parts of a solution *a posteriori*, once a global solution is known.

These local optimization procedures can be repeated until a local optimum— relative to a very special neighborhood—is obtained. POPMUSIC is an acronym for *Partial Optimization Metaheuristic Under Special Intensification Conditions* [23]. Several authors have proposed techniques that are slightly different from POPMUSIC. These techniques are sometimes less general and have been given different names such as LOPT (*Local OPTimizations* [20]), LNS (*Large Scale Neighborhood* [16]), shuffle, MIMAUSA [11], VNDS [10], and hybrid branch & bound tabu search.

More recently, several matheuristic methods sharing several similarities with POPMUSIC have been proposed. The advantage of the latter technique is that it has a single parameter that defines the size of the subproblems to be solved. Consequently, if a method is available that is able to solve efficiently subproblems *up to a given size*, a good value for the unique POPMUSIC parameter is easy to find.

For many combinatorial optimization problems, a solution S can be represented by a set of parts $s_1, \ldots, s_p$. For the vehicle routing problem, a part can be a tour, for example. The relations existing between each pair of parts may vary. For instance, two tours containing customers that are close to each other will have a stronger interaction than tours located in opposite directions relative to the depot.

The central idea of POPMUSIC is to build a subproblem with a *seed part*, s_i, and a given number $r < p$ of parts $s_{i_1}, \ldots, s_{i_r}$ which are specially related to the seed part s_i. These r parts build a subproblem R_i, smaller than the initial problem, that can be solved by an ad hoc procedure. If each improvement in subproblem R_i implies an improvement of the complete solution, then the framework for a local search can be defined. This local search is relative to a neighborhood that consists in optimizing subproblems. So, by storing a set O of those parts that have been used as seeds for building a subproblem and are unable to improve the complete solution, the search can be stopped as soon as all p parts constituting the complete solution have been contained in O. So, a special local search has been designed. This local search is parameterized by r, the number of parts constituting a subproblem. The method can be described as follows:

POPMUSIC(r)

1. Input: Solution S composed of parts $s_1, \ldots, s_p$
2. Set $O = \emptyset$
3. While $O \neq \{s_1, \ldots, s_p\}$ repeat

a. Select $s_i \notin O$
b. Create subproblem R_i composed of the r parts $s_{i_1}, \ldots, s_{i_r}$ that are the most related to s_i
c. Optimize R_i
d. If R_i has been improved, set $O \leftarrow O \backslash \{s_{i_1}, \ldots, s_{i_r}\}$, update S (as well as the set of parts).
   Else, set $O \leftarrow O \cup \{s_i\}$

This technique corresponds exactly to an improving method which, starting from an initial solution, stops as soon as a local optimum, relative to a very large neighborhood, is obtained. Hence, the method was named LOPT (local optimizations) in [20] and LNS (large neighborhood search) in [16].

In fact, the structure of the neighborhood so built contains all solutions s' that differ from s only by subproblem R_i, $i = 1, \ldots, p$. This means that the size of the neighborhood is defined by the number of solutions contained in the subproblems. This number is naturally very large and grows exponentially with the parameter r (the subproblem created for $r = p$ is the whole problem).

13.2.2.1 Parts

When a POPMUSIC-based intensification scheme is desired to be implemented, the first requirement is to define the meaning of a part of a solution. For vehicle routing problems, a tour (i.e., the set of orders delivered by the same vehicle) is perfectly convenient for defining a part. This approach was used in [14, 15, 17]. It is also possible to consider each customer as a part. This approach was used in [16]. If the problem instances are large enough and contain a relatively large number of tours, then considering a tour as a part has the advantage that the subproblems so defined are also vehicle routing problems. They can be solved completely independently.

13.2.2.2 Seed Part

The second point not precisely specified in the pseudocode of POPMUSIC is the way the seed part is selected. The simplest policy can be to systematically choose it at random. In the case of parallel optimization of subproblems, the seed parts can advantageously be chosen so that, as far as possible, the interactions between the subproblems are minimized.

13.2.2.3 Relations Between Parts

The definition of the relations between different parts is the third point that has to be discussed in the framework of POPMUSIC. Sometimes this relation is naturally defined. For example, if the parts are chosen as the customers of a vehicle routing problem, the distance between customers is a natural measure of the relation between parts. If the parts are defined as the tours of a vehicle routing problem, the notion of proximity is not so easy to define. In [15, 17], where Euclidean problems were treated, the proximity is measured by the center of gravity of the tours. The quantity ordered by each client is interpreted as a mass. Figure 13.3 illustrates the principle of the creation of a subproblem from a seed tour. Alvim and Taillard [1] have proposed a more general measure for the distance between tours by considering the minimum distance separating two customers belonging to different tours.

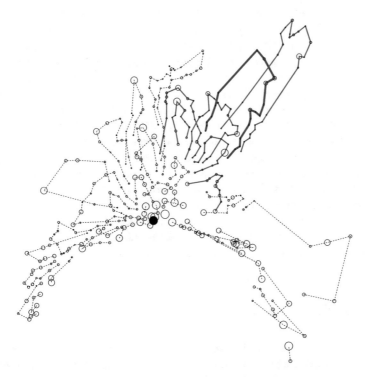

Fig. 13.3 Example of the definition of a subproblem for a vehicle routing problem. The seed part (tour) is drawn with a *thick line*, the tours most related to the seed tour by *normal lines*, and the tours that are not considered in the optimization of a subproblem by *dashed lines*. The routes from or to the depot are not drawn, so that the figure is not overloaded

13.2.2.4   Optimization Procedure

Finally, the fourth point not specified in the POPMUSIC framework is the procedure used for optimizing subproblems. In [15, 17], this procedure is a relatively basic tabu search. Shaw [16] uses an exact method based on constraint programming, making the whole method a matheuristic.

13.2.2.5   Complexity of POPMUSIC

An essential aspect when one is facing a large-size problem is the algorithmic complexity of the method. It is not practically possible to use an $O(n^2)$ algorithm when the number of entities in the problem is higher than 100 000 or an $O(n^3)$ algorithm if the size is higher than one thousand. Empirically, POPMUSIC repeats steps 3a to 3d a number of times that grows quasi-linearly with the problem size. The step requiring the most computational effort is the subproblem optimization (step 3c). For a fixed value of the parameter r, each of these optimizations takes a computational time that can be considered as constant. This means that steps 3a, 3b, and 3d can be performed globally in quasi-linear time if appropriate data structures are used.

The main difficulty with a POPMUSIC approach is in building an initial solution of adequate quality with a computational effort lower than $O(n^2)$. Alvim and Taillard [1] proposed a technique based on solving a kind of p-median problem with capacities to generate in $O(n^{3/2})$ an acceptable solution to a location–routing problem. Figure 13.4 illustrates the evolution of the computational time as a function of the problem size for the main steps of POPMUSIC. The subproblems defined in this reference are multidepot VRPs that are solved by a basic tabu search. In this figure, we see that building an initial solution has a higher complexity than the subproblem optimization, even if the computational effort needed for building an initial solution to an instance with 2 million customers is still moderate.

13.3   Problem Modeling

A key element for successfully solving a problem is to use an adequate model. First of all, the set S of feasible solutions must be defined. It may happen that the shape of this set is very complicated; i.e., without the definition of a very large neighborhood, it is impossible to generate all feasible solutions or, more precisely, it is not possible to reach an optimal solution starting from any feasible solution. In this case, to avoid the definition of an unmanageably large neighborhood (and therefore making the computational effort required to perform one iteration of local search

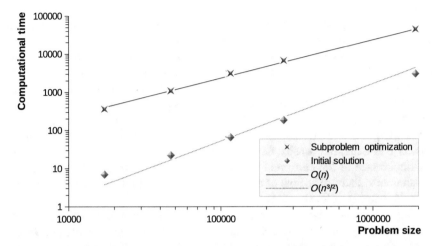

Fig. 13.4 Empirical complexity of POPMUSIC for a location–routing problem. The time for building the initial solution grows faster than the time needed for optimizing subproblems, which seems to be quasi-linear but which remains preponderant for instances with less than 2 million customers

prohibitive), the set of feasible solutions is extended, while penalizing solutions that violate constraints of the initial problem. Therefore, the problem is modified as follows:

$$\min_{s \in S^{\text{extended}}} f(s) + p(s)$$

where $S \subset S^{\text{extended}}$, $p(s) = 0$ for $s \in S$, and $p(s) > 0$ if $s \notin S$. This penalization technique, inspired by Lagrangian relaxation, is very useful in applications where finding a feasible solution is already difficult. For example, this is the case for school timetables, where the variety of constraints is impressive. Such a model is mandatory as soon as we have a *min max* objective, i.e., when we are searching for the minimum of a maximum, for instance if the longest tour of a VRP must be minimized.

In the CVRP, the number of vehicles can be chosen a priori and solutions where some customers are not delivered to can be accepted with some penalty. In this way, creating a feasible (but not operational) solution is a trivial job. The value of the penalty for not delivering an order can simply be the cost of a return trip between the depot and the customer.

The penalties can be modified during the search: If, during previous iterations, a constraint was systematically violated, then the penalty associated with the violation of that constraint can be increased. Conversely, if a constraint has never been violated, then the penalty associated with that constraint can be decreased. This technique has been used in the context of the CVRP [7]. This technique is very suitable if only one constraint is relaxed. If several constraints are simultaneously introduced into the objective, then it may happen that only nonfeasible solutions are visited. This is due to the fact that the different penalties associated with different constraints could vary in opposite phase in such a way that at least one constraint is always violated, the violated constraint changing during the search.

It is not always easy to model a problem, especially when the (natural) objective is to minimize a maximum. The choice of the function to minimize and the penalty function can be difficult. These functions must take a number of different values that is as large as possible over their domain of definition, in such a way that the search can be directed efficiently. How can the choice of a suitable move be made when a large number of solutions with the same cost exist in the neighborhood? To answer this issue, the penalty function may be chosen, for instance, by measuring the importance of the constraint violations rather just by counting the number of constraints violated. The goal of the penalties is to smooth the objective function to limit the number of local optima.

This last remark assumes a priori that a local search will be used. However, evolutionary algorithms and artificial ant methods do not relate to local searches, at least in their most elementary versions. But now, almost all efficient implementations inspired by these metaheuristics embed a local search, at least a simple improving method. A noticeable exception is the biased random-key genetic algorithms, where advanced population management seems to be sufficient. So, metaheuristics seem to be evolving toward a common framework that can be described by the higher level *adaptive memory programming* (AMP) template.

13.4 Population Management and Adaptive Memory Programming

A minute observation of recent implementations of evolutionary algorithms, scatter search, and artificial ant colonies reveals that all these techniques seem to be evolving toward the adaptive memory programming template [18, 22]. This framework is the following:

Adaptive Memory Programming

1. Initialize memory
2. Repeat, until a termination criterion is satisfied:

   a. Build a new solution with the help of the memory
   b. Improve the solution with a local search
   c. Update the memory with information carried by the new solution

Now, let us justify why various metaheuristics follow the same framework.

13.4.1 *Evolutionary or Memetic Algorithms*

In the case of evolutionary algorithms, the population of solutions can be considered as a form of memory. Indeed, some characteristics of the solutions—hopefully the best ones—are transmitted and improved, from one generation to the next. Recent implementations of evolutionary algorithms have replaced the "random mutation" metaphor by a more elaborate operator.

Instead of performing several local, random modifications to the solution obtained after the crossover operation, a search for a local optimum is initiated. Naturally, a more elaborate search can be executed, for example a tabu search or simulated annealing. In the literature, this type of method is called a "hybrid genetic algorithm" or "memetic algorithm" [12].

Another key element of memetic algorithms is an "intelligent" management of the population. A small population implies rapid convergence of a genetic algorithm. This is both an advantage—little effort is spent on generating bad-quality solutions— and a disadvantage—the solutions so obtained are not so good. To combine the advantages of a small and a large population, the idea is to divide the population into islands that evolve independently for a while. Periodically, one or more individuals among the best of an island migrate toward another island. This brings fresh blood to the population of the islands and avoids or strongly delays the convergence of the global population.

13.4.2 *Scatter Search*

Scatter search is almost as old as genetic algorithms, as the technique was originally proposed, completely independently, in 1977 [8]. However, the technique only started to gain prominence among academic communities by the end of the 1990s. In contrary to evolutionary algorithms, simulated annealing, and tabu search, this method has been used very little in the industrial world so far. Scatter search can be viewed as an evolutionary algorithm with the following specific characteristics:

1. Binary vectors are replaced by integer vectors.
2. The selection operator for reproduction may select more than two parent solutions.
3. The crossover operator is replaced by a convex or nonconvex linear combination.
4. The mutation operator is replaced by a repair operator that projects the newly created solution into the feasible solution space.

These characteristics may also be considered as generalizations of evolutionary algorithms which have been proposed and exploited later by various authors, especially [13]:

1. The use of crossover operators is different from the exchange of bits or subchains,
2. A local search is applied to improve the quality of solutions produced by the crossover operator,

3. More than two parents are used to create a child,
4. The population is partitioned with the help of classification methods instead of an elementary survival operator.

In scatter search, the production of new individuals from solutions in the population is a generalization of the crossover in evolutionary algorithms. In "pure" genetic algorithms, solutions of a problem are only considered in the form of a fixed-length chain of bits. For many problems, it is not natural to code a solution using a binary vector and, depending on the coding scheme chosen, a genetic algorithm may produce results of varying quality. In the initial versions of genetic algorithms, the main point was to choose an appropriate coding scheme, the other operators belonging to a standard set. In contrast, in scatter search a natural coding of solutions is advocated, implying the design of "crossover" operators (the generation of new solutions from those in the populations) strongly dependent on the problem to be solved.

Since crossover operators on naturally represented solutions do not necessarily lead to a feasible solution, repairing or improving operators must be designed. In scatter search, the population is managed by maintaining a *reference set* composed on the one hand of a few of the best solutions found by the search (elite solutions) and on the other hand of a few solutions that are as diverse as possible (scattered in the solution space). The template for scatter search is as follows:

1. Generate an initial population of solutions that are as scattered as possible. The solutions are not necessarily feasible, but they are repaired and improved with an appropriate operator.
2. While the population changes, repeat

    a. Select from the population a reference set composed of a few elite solutions and a few solutions as different as possible from the elite ones.
    b. Generate all possible subsets (with more than one solution) from the reference set.
    c. Combine all solutions in each subset into a tentative solution and repair/improve this tentative solution.
    d. Add all the new solutions to the reference set, which then constitutes the new population for the next step.

13.4.3 Ant Colonies

In the spirit of adaptive memory programming, the trails of pheromone in ant colonies can be considered as a form of memory. This memory is utilized to build new solutions, following specific rules for simulated ants or, expressed in other terms, by following a magic formula, a belief in the precepts of the designers of ant colony optimization. Initially, the process did not embed a local search. However, simulation experiments very soon revealed that the quality of the process was more efficient when a local search was incorporated. Unfortunately, the designer of ant colonies used to hide this component in the pseudocode of the metaheuristic in the form of a "daemon action," which might consist, potentially, of anything!

13.4.4 Vocabulary Building

Vocabulary building is a concept introduced in [9] in the context of tabu search, but the principles of this concept have certainly been used, under different names, by different authors. Vocabulary building can be conceived as a special type of GRASP or as an ant colony working with a memory called a dictionary. Here, instead of storing complete solutions in the memory, only fragments (or *words*) are memorized. These words are employed to build a *vocabulary*. A new solution (i.e., a *sentence*) is obtained by combining different fragments. In the context of the vehicle routing problem, a fragment—or a part of a solution, following the terminology of POPMUSIC—can be defined as a tour. Then the following procedure can be applied to build a new solution **s**, where M is a set of tours, each tour T being composed of a set of customers:

Building a New Solution

1. $\mathbf{s} = \emptyset$
2. Repeat, while $M \neq \emptyset$:

   a. Choose $T \in M$
   b. Set $\mathbf{s} \leftarrow \mathbf{s} \cup T$
   c. Set $M \leftarrow M \backslash T'$ $\forall T' \in M$ such that $T' \cap T \neq \emptyset$

3. If the solution **s** does not contain all customers, then complete it.
4. Improve the solution with a local search.

Thus, the idea is to build a solution by successively choosing tours belonging to a set of memorized tours. The chosen tours must not contain those customers which are already contained in the partially built solution.

In the case of vehicle routing problems, this technique was first applied in [15]. It succeeded in obtaining several best solutions of instances of benchmark problems in the literature. This method shows significant performance, particularly for the following reason: an elementary tabu search, embedded within the framework of POPMUSIC, is capable of finding a few of the tours in the best solutions known very rapidly. This is illustrated in Fig. 13.5. Therefore, a lot of computational effort can be spared by collecting already existing tours without having to build them from scratch.

13.4.5 Path Relinking

Path relinking [9] is another technique that works with a population of solutions, and was also proposed by Glover in the context of tabu search. The initial idea is to store

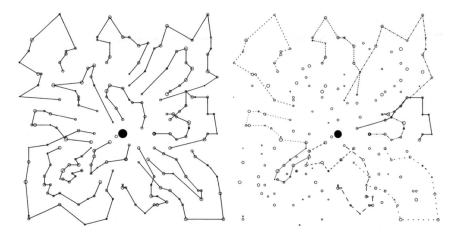

Fig. 13.5 An example of the utility of creating *words* (=tours) in vocabulary building. On the *left*, one of the best solutions known for a CVRP instance. On the *right*, a few of the tours found within a few seconds with a POPMUSIC-based search

a set of good solutions visited during a tabu search. All of the solutions are linked together by a path corresponding to the successive neighboring solutions visited by the tabu search. Path relinking tries to connect pairs of selected solutions by using another path, hoping that better-quality solutions are met with along this new path. An iteration of scatter search can be described as follows:

1. Select two solutions s_1 and s_2 from among those memorized
2. Repeat, while $s_1 \neq s_2$

    a. Consider all neighboring solutions of s_1 that allow one to get closer to s_2
    b. Retain the best neighboring solution; it becomes s_1

Variants of this template exist: it is also possible to try a path going from s_2 to s_1 or to simultaneously modify s_1 and s_2 and to stop at an intermediate solution.

Hence, there can be infinite number of possibilities for extending a technique. A bottom-up methodology seems relatively logical to follow. In fact, the addition of a level that increases the complexity of a method is not very difficult to implement. For example, modifying an improving method that will terminate at the first local optimum to transform it into a simulated annealing method takes only a few minutes to code, if the first version was developed without implementing any algebraic or software optimization. Nowadays, users even have several libraries at their disposal that allow them to embed a basic method into a more complex framework (see, for example, [2, 24], related to simplifying parallel implementations).

However, it is much more problematic to find suitable parameters (e.g., annealing scheme, type and duration of tabu conditions, penalty factors, intensification and diversification mechanisms in a tabu search, coding scheme, crossover operators, population size in an evolutionary algorithm,...). In order to find good parameters without performing elaborate numerical experiments, it is important to make use of statistical tests, sometimes relatively specific. This leads us directly to a point that has been quite neglected in the metaheuristics literature: the comparison of iterative heuristics.

13.5 Comparison of Heuristics

The implementation of a heuristic method for solving a complicated combinatorial problem necessitates that the designer considers several choices. Some of them may be relatively easy to justify, but others, such as the numerical tuning of parameters or the choice of a neighborhood, may be much more hazardous. When theory or intuition cannot support the researcher's choice, the researcher must justify their decision with the help of numerical experiments. However, it is all too often observed that these choices are not supported by scientific considerations. The present section discusses a few techniques for comparing improving heuristic techniques.

13.5.1 Comparing Proportions

The first question that needs to be clarified concerns the comparison of the success rates of two methods $\mathcal{A}$ and $\mathcal{B}$. Practically, experiments are conducted as follows: Method $\mathcal{A}$ is run n_a times and succeeds in solving the problem a times. Similarly, method $\mathcal{B}$ is executed n_b times and succeeds in solving the problem b times. So, the following question arises: is a success rate of a/n_a significantly higher than a success rate of b/n_b? A researcher who is a perfectionist will carry out a large number of experiments and work with a sufficiently large number of runs to conduct a standard statistical test based on the central limit theorem. Conversely, a less careful researcher will not conduct the 15 or 20 runs theoretically needed to validate their choice from among several options, but will assume, for instance, that if $\mathcal{A}$ has 5 positive results over 5 runs, it will certainly be better than $\mathcal{B}$ which has only 1 positive run over 4. Is the above conclusion correct or not? A nonparametric statistical test [21] shows that a success rate of $5/5$ is significantly higher—with a confidence level of 95 %—than a success rate of $1/4$. The contents of Table 13.1, which were originally presented in [21], provide, for a confidence level of 95 %, the pairs (a, b) for which a success rate greater than or equal to a/n_a is significantly better than a success rate less than or equal to b/n_b.

This table can be particularly useful for finding good parameters for a technique, both quickly and in a rigorous manner. A suitable procedure is to fix two different parameter sets (thus defining two different methods $\mathcal{A}$ and $\mathcal{B}$) and to compare the results obtained with both methods. In order to make proper use of Table 13.1, it is required to define what a success is (for instance, the fact that the optimal solution or a solution of a given quality has been found for a given problem instance) and,

Table 13.1 Pairs (a, b) for which a success rate $\geq a/n_a$ is significantly higher than a success rate $\leq b/n_b$, for a confidence level of 95 %

n_b	n_a								
	2	3	4	5	6	7	8	9	10
2	—	(3,0)	(4,0)	(5,0)	(5,0)	(6,0)	(7,0)	(7,0)	(8,0)
3	(2,0)	(3,0)	(3,0)	(4,0) (5,1)	(4,0) (6,1)	(5,0) (7,1)	(5,0) (8,1)	(6,0) (8,1)	(6,0) (9,1)
4	(2,0)	(3,1)	(3,0) (4,1)	(3,0) (5,1)	(4,0) (5,1) (6,2)	(4,0) (6,1) (7,2)	(5,0) (7,1) (8,2)	(5,0) (7,1) (9,2)	(5,0) (8,1) (10,2)
5	(2,0)	(2,0) (3,1)	(3,0) (4,2)	(3,0) (4,1) (5,2)	(3,0) (5,1) (6,2)	(4,0) (5,1) (7,2)	(4,0) (6,1) (7,2) (8,3)	(4,0) (6,1) (8,2) (9,3)	(5,0) (7,1) (9,2) (10,3)
6	(2,1)	(2,0) (3,2)	(2,0) (3,1) (4,2)	(3,0) (4,1) (5,3)	(3,0) (4,1) (5,2) (6,3)	(3,0) (5,1) (6,2) (7,3)	(4,0) (5,1) (7,2) (8,3)	(4,0) (6,1) (7,2) (9,3)	(4,0) (6,1) (8,2) (9,3) (10,4)
7	(2,1)	(2,0) (3,2)	(2,0) (3,1) (4,3)	(3,0) (4,2) (5,3)	(3,0) (4,1) (5,2) (6,4)	(3,0) (4,1) (6,3) (7,4)	(3,0) (5,1) (6,2) (7,3) (8,4)	(4,0) (5,1) (7,2) (8,3) (9,4)	(4,0) (6,1) (7,2) (9,3) (10,4)
8	(2,1)	(2,0) (3,3)	(2,0) (3,1) (4,3)	(2,0) (3,1) (4,2) (5,4)	(3,0) (4,1) (5,3) (6,4)	(3,0) (4,1) (5,2) (6,3) (7,5)	(3,0) (5,1) (6,2) (7,3) (8,5)	(3,0) (5,1) (6,2) (7,3) (8,4) (9,5)	(4,0) (5,1) (7,2) (8,3) (9,4) (10,5)
9	(2,2)	(2,1) (3,3)	(2,0) (3,2) (3,2)	(2,0) (3,1) (4,3) (5,5)	(3,0) (4,2) (5,3) (6,5)	(3,0) (4,1) (5,2) (6,4) (7,5)	(3,0) (4,1) (5,2) (6,3) (7,4) (8,6)	(3,0) (5,1) (6,2) (7,3) (8,4) (9,6)	(3,0) (5,1) (6,2) (7,3) (8,4) (9,5) (10,6)
10	(2,2)	(2,1) (3,4)	(2,0) (3,2) (4,5)	(2,0) (3,1) (4,3) (5,5)	(2,0) (3,1) (4,2) (5,4) (6,6)	(3,0) (4,1) (5,3) (6,4) (7,6)	(3,0) (4,1) (5,2) (6,3) (7,5) (8,6)	(3,0) (4,1) (5,2) (6,3) (7,4) (8,5) (9,7)	(3,0) (5,1) (6,2) (7,3) (8,4) (9,5) (10,7)

naturally, it is assumed that the runs are conducted independently of each other. The test works with a given problem instance and nondeterministic methods $\mathcal{A}$ and $\mathcal{B}$ (such as simulated annealing) or with problem instances randomly chosen from a set (for example, randomly generated instances of a given size). An online version of this statistical test is available at http://mistic.heig-vd.ch/taillard/qualopt/.

13.5.2   Comparing Iterative Optimization Methods

13.5.2.1   Computational Effort for a Given Target

When optimization techniques are robust enough to achieve a given goal with a high success rate, one can compare the computational effort needed to reach this target. A diagram used in the literature gives the cumulative probability of reaching the target as a function of the number of iterations performed by the method. The target is typically the best solution known for a problem instance. This instance is solved many times by every method. Multiple solutions of the same instance have a meaning only if the method embeds a random component, such as a simulated annealing, a GRASP, or a tabu search method starting with a random solution. Another possibility is to choose a set of problem instances with given characteristics (such as the problem size) that are randomly generated. This allows the comparison of deterministic methods.

A time-to-target diagram is given in Fig. 13.6. This figure provides the probability of obtaining the best solution known for a quadratic assignment problem (QAP) for tabu searches using different tabu list sizes and various diversification parameters.

This comparison technique is sometimes criticized: the target cannot be fixed by a heuristic method itself. Indeed, the best solution or the optimum must be found by another method. Therefore, it is hard to speak about the success rate, since the target to be reached is not clear. More precisely, the processes we are interested in have two objectives: in addition to optimizing the solution quality, we want to minimize the computational effort. This last objective can be chosen freely by the user, for instance by changing the number of iterations of a tabu search, by changing the number of generations of an evolutionary algorithm, or by changing the number of solutions built by an ant system. Moreover, these methods are often nondeterministic. Two runs on the same problem instance generally produce different solutions. The next subsubsection focuses on techniques for comparing the quality of solutions obtained as a function of the computational effort for nondeterministic iterative methods.

13.5.2.2   Comparing the Quality of Iterative Searches

Traditionally, the measure of quality of a method is the average value of the solutions it produces. The computational burden is measured in terms of CPU time consumed in seconds. However, neither of these measures is satisfactory. If the computational burden is fixed for two nondeterministic methods $\mathcal{A}$ and $\mathcal{B}$, and if it is desired to

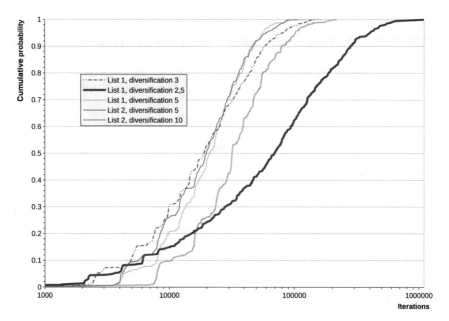

Fig. 13.6 Time-to-target diagram for parameter tuning. This figure is based on the number of iterations needed by a tabu search starting with a random solution to get the best solution to the QAP instance *tai40b*. Two different tabu durations were evaluated (*n* and 2*n*) and four different values for the diversification mechanism (forcing a move that was not performed during a number of iterations to be a multiple of 2.5, 3, 5, and 10 times the neighborhood size)

rigorously compare the quality of the solutions produced by these methods, both methods must be executed several times and a statistical test comparing the two methods must be conducted. Unfortunately, the distribution function of the quality of the solutions produced by a method is unknown and generally not Gaussian. Therefore it is not possible to use a standard statistical test unless large samples are available. This means that the numerical experiments must be repeated a large number of times—practically, this may correspond to many hundreds of times, contrary to the common belief that a sample size of 20 or 30 is large enough.

If quality can be measured by some method other than the average solution values obtained, interesting comparisons can be performed with very few runs. One of these nonparametric methods consists in ranking the set of all solutions obtained with methods A and B and computing the sum of the ranks obtained by one method. If this sum is lower than some value—which depends on the level of significance, which can be found in numerical tables—then one cannot exclude the possibility that a run of this method has a probability significantly higher than $1/2$ of obtaining a better solution than a run of the other method. In the literature [4], this test is known as the *Mann–Whitney test*.

Another statistical technique that is easy to undertake but that requires higher computational effort is the bootstrap technique. This allows one to calculate various reliable statistics with relatively small samples, typically a few dozen runs.

Let us suppose that an optimization algorithm has been run n times and the solution quality observed is $\mathbf{x} = (x_1, x_2, \ldots, x_n)$. The observations are assumed to be independent, with finite variance, but without making hypotheses about the distribution function, for instance its symmetry. Indeed, the solution values obtained by nondeterministic metaheuristics are not symmetrical, just because it is impossible to get values beyond the optimum. Moreover, researchers favor a small number of runs so that they can observe what happens for long runs. Under these circumstances, what is needed is a confidence interval for the statistics $s(\mathbf{x})$, such as the average or the median. This information can be obtained by resampling the observations a large number of times as follows:

- Generate B vectors $\mathbf{x}^b (b = 1, \ldots, B)$ of size n by choosing each component randomly, uniformly, and with replacement from among the observed values $(x_1, \ldots, x_n)$.
- Compute the value of the wanted statistics $s(\mathbf{x}^1), \ldots, s(\mathbf{x}^B)$ for the B vectors generated and order them by increasing value.
- It can be considered that $s(\mathbf{x})$ has a probability $1 - 2\alpha$ of belonging to the interval $[s_1, s_2]$ by taking the $s_1 = 100 \cdot \alpha$ and $s_2 = 100 \cdot (1 - \alpha)$ percentiles of the ordered values $s(\mathbf{x}^b)$.

Practically, one can choose $B = 2000$, $\alpha = 2.5\%$, $s_1 = 50$, and $s_2 = 1950$. This technique is very simple and is suitable for metaheuristics practitioners who are familiar with simulation. Note that this technique does not necessarily produce either the smallest interval or an interval centred on the chosen statistics. The reader can find more elaborate techniques in the books [5, 6].

Naturally, if iterative methods are to be compared using such a test, the test must be repeated each time with a fixed computational effort. In practice, as mentioned before, the computation time on a given computer is used to measure the computational effort. This is a relative measure as it depends on the hardware used, the operating system, the programming language, the compiler, etc. To make a more rigorous comparison, an absolute measure of the computational burden must be used. Typically, the evaluation of the neighboring solutions is the most demanding part of a metaheuristic-based method, such as simulated annealing, tabu search, evolutionary algorithms, or ant colony methods (provided that the latter two techniques are hybridized with a local search). Thus, it is often possible to express the computational burden not in seconds but in iteration numbers and to specify the theoretical complexity of one iteration. For instance, one iteration of the tabu search proposed in reference 13 of chap. 3 for the QAP has a complexity of $O(n^2)$. By making the code of their method available in the public domain, everyone can now express the computation burden of their own method for a given problem example in terms of the equivalent number of tabu search iterations. So, there is no necessity to provide a reference to a computation time relative to a given machine—which will very soon become obsolete.

Even if it is possible to use a general software package such as OpenOffice Calc to producing a diagram like that in Fig. 13.6, the work for the programmer of an iterative heuristic could be crippling. The STAMP software [19], available online on the site http://mistic.heig-vd.ch/taillard/qualopt, allows one to generate both types of diagrams respecting good practice. The first diagram that STAMP offers provides the average (or the median) of the solutions obtained as a function of the computational effort, expressed both as an absolute measure (number of iterations) and a relative measure (seconds). The chosen statistic is given with a confidence interval estimated by a bootstrap technique. An example of such a diagram is given in Fig. 13.7.

Finally, by concentrating on the information that is really needed—is method $\mathcal{A}$ significantly better than method $\mathcal{B}$?—the STAMP software allows one to generate a second type of diagram, which provides the probability that a given method is better than another as a function of the computational effort. By using this type of diagram, the area needed for drawing the the essential information is reduced by a large proportion. So, it is possible to draw many probability diagrams on the same figure, for example to compare many methods with each other for the same problem instance or to compare two methods solving different problem instances. This possibility is

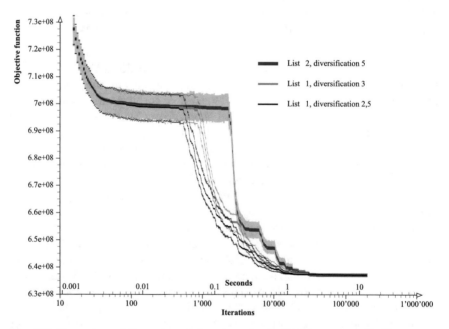

Fig. 13.7 Comparison of three tabu search variants for the QAP instance *tai40b*. The average solution values obtained by these variants are given as a function of the computational effort, measured in terms of both tabu iterations and time on a given machine. The average values are bounded by their 95 % confidence interval. This shows that between 1000 and 10 000 iterations, the difference observed between the two variants with tabu list type 1 could be fortuitous, and does not show that one of the variants is truly better than the other

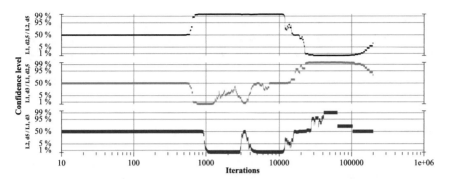

Fig. 13.8 Comparison of three tabu search variants on QAP instance *tai40b*. Each diagram provides the probability that a method is better (or worse) than another as a function of the computational effort. At a glance, we can see that between 1000 and 10 000 iterations the variant *List 1, diversification 2,5* is significantly better (at 99 % confidence level) than the variant *List 2, diversification 5,* while between 20 000 and 100 000 iterations, the reverse is true

illustrated in Fig. 13.8, where three tabu search variants are compared pairwise when they are run on a QAP instance in the literature.

At a glance, these diagrams provides much more information than a traditional numerical table. The main advantages are that they present comparisons for a continuum of computational effort and provide exactly the information wanted (is this method better than another?)

13.6 Conclusion

It is sincerely hoped that this chapter will guide researchers who are engaged in the design of a heuristic based on the techniques presented in the previous chapters. We are well aware of the fact that every practical problem will be a specific case and that our advice sometimes may not be judicious. For example, for the traveling salesman problem, one of the best heuristic methods available at present is a simple improving method, which is based on the appropriate neighborhood. For the p-median problem, one of the best methods is based on a POPMUSIC method that does not embed other metaheuristic principles such as tabu search or simulated annealing. Finally, we should mention that evolutionary algorithms and scatter search implementations do not necessarily embed ejection chains, partial optimization, or other metaheuristic principles.

However, in our opinion, researchers should be more careful concerning the methodology for comparing iterative heuristics. Indeed, in the literature, tables that formally contain no reliable information are too often presented, and their authors draw conclusions that are not supported by the experiments performed. This is why we hope that the last part of this chapter, where a comparison of improving heuristics is presented, will lead to research topics that will gain increasing importance in the near future.

References

1. Alvim, A.C.F., Taillard, É.D.: POPMUSIC for the world location-routing problem. EURO Journal on Transportation and Logistics **2**(3), 231–254 (2013). URL http://mistic.heig-vd.ch/taillard/articles.dir/AlvimTaillard2013.pdf
2. Cahon, S., Melab, N., Talbi, E.G.: Paradiseo: A framework for the reusable design of parallel and distributed metaheuristics. Journal of Heuristics **10**(3), 357–380 (2004)
3. Christofides, N., Mingozzi, A., Toth, P.: The vehicle routing problem. In: N. Christofides, A. Mingozzi, P. Toth, C. Sandi (eds.) *Combinatorial Optimization*, pp. 315–338. Wiley (1979)
4. Conover, W.J.: *Practical Nonparametric Statistics*, 3rd edn. Wiley, Weinheim (1999)
5. Davison, A.C., Hinkley, D.: *Bootstrap Methods and Their Application*, 5th edn. Cambridge University Press (2003)
6. Efron, B., Tibshirani, R.J.: *An Introduction to the Bootstrap*. Chapman and Hall (1993)
7. Gendreau, M., Hertz, A., Laporte, G.: A tabu search heuristic for the vehicle routing problem. Management Science **40**, 1276–1290 (1994)
8. Glover, F.: Heuristics for integer programming using surrogate constraints. Decision Sciences **8**(1), 156–166 (1977)
9. Glover, F., Laguna, M.: *Tabu Search*. Kluwer, Dordrecht (1997)
10. Hansen, P., Mladenović, N.: An introduction to variable neighborhood search. In: S. Voß, S. Martello, I.H. Osman, C. Roucairol (eds.) *Meta-heuristics: Advances and Trends in Local Search Paradigms for Optimization*, pp. 422–458. Kluwer, Dordrecht (1999)
11. Mautor, T., Michelon, P.: MIMAUSA: A new hybrid method combining exact solution and local search. In: *2nd International Conference on Metaheuristics*, Sophia-Antipolis, France, p. 15 (1997)
12. Moscato, P.: Memetic algorithms: A short introduction. In: D. Corne, F. Glover, M. Dorigo (eds.) *New Ideas in Optimisation*, pp. 219–235. McGraw-Hill, London (1999)
13. Mühlenbein, H., Gorges-Schleuter, M., Krämer, O.: Evolution algorithms in combinatorial optimization. Parallel Computing **7**, 65–88 (1988)
14. Rochat, Y., Semet, F.: A tabu search approach for delivering pet food and flour in Switzerland. Journal of the Operational Research Society **45**, 1233–1246 (1994)
15. Rochat, Y., Taillard, E.D.: Probabilistic diversification and intensification in local search for vehicle routing. Journal of Heuristics **1**(1), 147–167 (1995)
16. Shaw, P.: Using constraint programming and local search methods to solve vehicle routing problems. Technical report, ILOG S.A., Gentilly, France (1998)
17. Taillard, E.D.: Parallel iterative search methods for vehicle routing problems. Networks **23**, 661–673 (1993)
18. Taillard, E.D.: Programmation à mémoire adaptative et algorithmes pseudo-gloutons : nouvelles perspectives pour les méta-heuristiques. Thèse d'habilitation à diriger des recherches, Université de Versailles, France (1998)
19. Taillard, E.D.: Principes d'implémentation des méta-heuristiques. In: J. Teghem, M. Pirlot (eds.) *Optimisation approchée en recherche opérationnelle*, pp. 57–79. Lavoisier, Paris (2002).
20. Taillard, E.D.: Heuristic methods for large centroid clustering problems. Journal of Heuristics **9**(1), 51–73 (2003)
21. Taillard, E.D.: A statistical test for comparing success rates. In: *Metaheuristic International Conference MIC'03*, Kyoto, Japan (2003)

22. Taillard, E.D., Gambardella, L.M., Gendreau, M., Potvin, J.Y.: Adaptive memory programming: A unified view of meta-heuristics. European Journal of Operational Research **135**(1), 1–16 (1998)
23. Taillard, E.D., Voß, S.: POPMUSIC—partial optimization meta-heuristic under special intensification conditions. In: C. Ribeiro, P. Hansen (eds.) *Essays and Surveys in Metaheuristics*, pp. 613–629. Kluwer, Dordrecht (2002)
24. Voß, S., Woodruff, D.L.: *Optimization Software Class Libraries*. OR/CS Interfaces Series. Kluwer, Dordrecht (2002)

Chapter 14
Optimization of Logistics Systems Using Metaheuristic-Based Hybridization Techniques

Laurent Deroussi, Nathalie Grangeon and Sylvie Norre

In the postwar years, the development of operaftional research provided companies with tools to deal with their logistical problems in a quantitative way. For a long time, these problems were split into unrelated subproblems, each subproblem often being tackled separately. This is mainly due to the fact that the subproblems considered, such as the localization problem, planning problem, scheduling problem, and transportation problem, are generally NP-hard problems and their computational complexity remains a significant issue for many researchers. Nevertheless, in an increasingly competitive industrial environment, companies continue to have a strong demand for decision aid tools to provide a global view of their organization.

The aim of this chapter is to present the challenges of providing such a view, to understand the consequences in terms of logistics system modeling, and to say something about new optimization techniques.

This chapter is organized as follows.The first part describes logistics systems in general and supply chains in particular. In this part, the concepts of horizontal and vertical synchronization are developed to allow a comprehensive vision of supply chains. We also show that metaheuristic-based hybridization techniques are very relevant to the characteristics of logistics systems.The second part is devoted to hybridization techniques of metaheuristics with optimization methods and metaheuristics with evaluation models. In the last part, we present some issues related to synchronization, as well as some hybridization methods proposed in the literature.

L. Deroussi (✉) · N. Grangeon · S. Norre
Laboratoire LIMOS, IUT d'Allier, Avenue Aristide Briand CS 82235,
03101 Montlucon Cedex, France
e-mail: laurent.deroussi@univ-bpclermont.fr

N. Grangeon
e-mail: nathalie.grangeon@univ-bpclermont.fr

S. Norre
e-mail: sylvie.norre@univ-bpclermont.fr

© Springer International Publishing Switzerland 2016
P. Siarry (ed.), *Metaheuristics*, DOI 10.1007/978-3-319-45403-0_14

14.1 Logistics Systems

14.1.1 Definitions and General Considerations

According to Ganeshan and Harrison [17] a supply chain "is a network of facilities and distribution options that performs the functions of procurement of materials, transformation of these materials into intermediate and finished products, and the distribution of these finished products to customers." This definition, chosen from among many others, defines a supply chain as a network of physical entities (sites, organizations, or actors) crossed by physical flows, information flows, and financial flows. It integrates a set of activities from raw material procurement to final consumption.

In this chapter, we use the term "logistics system" to refer to as any set of physical entities interconnected by a logistics network, in which both material and nonmaterial flows may occur. So, a logistics system can represent both a global supply chain and a part of it (by focusing, for instance, on entities in the same organization or on a site). Internal logistics represents the set of flows passing through the network. Procurement logistics (or inbound logistics) includes inflows (from component suppliers in any tier), whereas distribution logistics (or outbound logistics) includes outflows (customers, wholesalers, retailers, end consumers). Figure 14.1 presents an example of a supply chain.

Forrester's work on systems dynamics highlighted the fact that the efficiency of an organization arises from the coordination of its components [15]. The concept of supply chain management was proposed by Oliver and Webber in 1982 [34].

A very large number of definitions of supply chain management have been presented in Wolf [49]. Of these, we shall use the one proposed by Simchi-Levi et al. [42], which describes supply chain management as a set of approaches used to integrate effectively the actors taking part in the manufacturing process (suppliers,

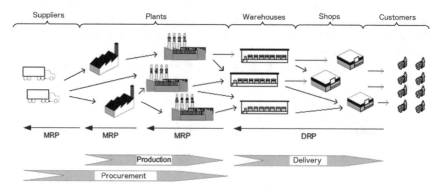

Fig. 14.1 A supply chain. MRP, material requirement planning; DRP, distribution resource planning

plants, warehouses, stores, …) so as to manufacture and dispatch goods in the right quantity, in the right place, and at the right time, with the objective of minimizing a set of costs while ensuring quality of service.

14.1.2 Integrated View of Supply Chain

Optimizing a single component of a logistics system may have a positive or negative impact on the global performance of the system. Thus, it is important to consider the system as a whole by integrating inbound and outbound logistics. There are several types of integration:

Functional. The smooth running of a logistics system implies the coordination of many activities (facility location, logistics network design, transportation of goods, warehouse management, inventory management, production logistics, product design and product life cycle, the information system, procurement logistics, distribution logistics, …). The MRP (material requirement planning) concept, also called "net requirements calculation," was born in the 1970s from the need to synchronize the quantities of raw materials and semifinished products in order to satisfy demands expressed by consumers. We speak of synchronization of physical flows [35].

Temporal. Wight [48] proposed the MRPII (manufacturing resources planning) as a development of MRP, taking capacities (procurement, production, storage, distribution, financial) into account. This approach is based on the definition of a hierarchical structure with five levels, each of them working on a temporal horizon with its own level of data precision. These levels are strategic planning, sales and operations planning (S&OP), the master production schedule (MPS), net requirements calculation, and shop floor control (SFC).

Geographical. Originally, MRPII was a monosite approach. However, current logistics systems are mostly multisite, which implies making decisions in terms of facility location, transportation of goods (procurement, production, and distribution), lead times, …. Thomas and Lamouri considered the concept of supply chain management as an extension of the MRPII approach [46].

Kouvelis et al. defined coordination as any action or approach that leads the actors in a logistics system to act in a way that improved the running of the system as a whole [20]. The coordination of the various actors constitutes a great challenge for operational research, whether in a centralized view (actors are grouped together in the same organization, which takes decisions for the whole group) or in a decentralized view (each actor is empowered with respect to decision-making). Schmidt and Wilhelm [41]describe logistics network models which may address each of the three decision levels, namely strategic, tactical, and operational [41]. The strategic level (long-term) covers decisions about logistics network design and, in particular, the facility location problem (FLP). The tactical level (mid-term) describes flow management policies, including for instance, lot-sizing problems. The operational level

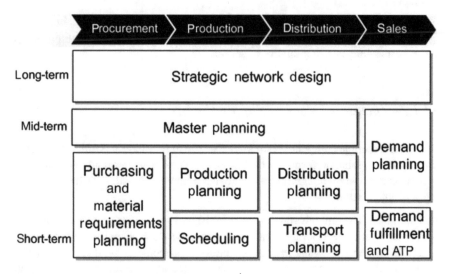

Fig. 14.2 Problems linked to supply chain planning [28]. ATP, available-to-promise

(short-term) concerns control of the supply chain and covers scheduling problems (flow-shop problem, job-shop problem, …). Schmidt and Wilhelm conclude that each level interacts with the others and that an approach unifying the three levels was necessary to design and manage a competitive logistics network.

Lemoine [23] defined the concepts of horizontal and vertical synchronization, which gather together the two previous examples. Horizontal synchronization addresses the difficulties of synchronization between the entities of a supply chain (for instance, a plan for a production site may be not feasible because of procurement constraints). Vertical synchronization consists in planning decisions in time. The levels of the MRPII approach are recomputed at various frequencies, and this may induce desynchronization between them. It is not certain that a modification made at a given level will be consistent with the other levels.

Figure 14.2 details the problems linked to supply chain planning and shows the need for concepts of synchronization to achieve better flow coordination.

14.1.3 Difficulties of Performance Optimization in a Supply Chain

The process of adopting a global view of a logistics system and integrating synchronization constraints allows one to optimize its performance and make it more competitive. Some difficulties have to be overcome, however. These difficulties are linked to:

Model design.   A logistics system is hard to model; actors, entities, activities, and interactions between entities must be defined. Management rules may be complex or hard to establish. Knowledge and data gathering may be a long and difficult task.

Algorithmic complexity.   Most of the classical models, regardless the decision level, are NP-hard problems. We have mentioned only a few of them, but it becomes necessary to combine them when one aims to achieve horizontal or vertical synchronization.

Size of systems studied.   The large size of logistics systems (in terms of number of actors, products, ...) often makes them hard to solve.

Consideration of uncertainties.   A high decision level results in greater uncertainties. The tactical level concerns a relatively long time horizon, generally from two to five years. For such a horizon, there may be major uncertainties concerning demand or the economic environment. It is important that a system can be adapted and remain efficient when facing uncertainties. Snyder [43] presented a review of the state of the art in the consideration of uncertainities in facility location problems.

Model precision.   A logistics system contains a huge quantity of data. It is necessary to aggregate the data according to the decision level considered and the objectives. For instance, S&OP works on product families, whereas MPS considers only products.

Competitiveness evaluation.   The performance criteria are generally costs (transport, storage, production, ...) and consumer service ratings. Apart from the fact that they may be difficult to evaluate, they are often conflicting.

Risk management.   This includes machine breakdown at an operational level, management of the maintenance of production units, and the study of the reactivity of the system to natural disasters.

14.1.4 Decision Support System

The performance of a logistics system is measured as its ability to manage flows passing through it, whether they are physical, informational, or financial. One of the keys is data sharing between actors in the system. Each actor must be able to read, at any time, all the information they need to take the best possible decisions. This is one of the major role of the information system, which is increasingly gathering data using tools such as ERP (enterprise resource planning). If these tools allow one to manage information flows, however, they are often difficult to use to take decisions. This is the very issue of business intelligence (BI), defined by Krmac [21] as the set of tools that helps an enterprise to better understand, analyze, explore, and forecast what happens in the enterprise and its environment. Figure 14.3 shows the interactions between these tools. ETL (extract-transport-load) tools allow one to extract data from many sources and format them (validation, filtering, transformation,

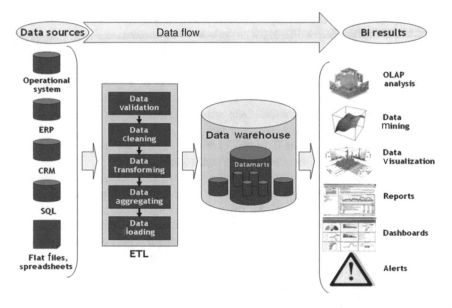

Fig. 14.3 Decision support system [21]

aggregation) and store them in a data warehouse. These data are then available to be used by analysis and decision aid tools such as those presented in this chapter.

14.1.5 Reason for Interest in Metaheuristics

We have pointed out some difficulties that have to be overcome to optimize a logistics system. A whole supply chain is composed of a complex network of sites and organizations that have interconnected activities but aim at various and contradictory objectives. Lourenço [24] has pointed out the major role that metaheuristics can play in decision aid tools for supply chains. Metaheuristics have good qualities for solving the very complex problems that arise in supply chain management. The elements outlined were the following:

- These methods are generally simple, easy to implement, and robust and have already proven successful in hard optimization problems.
- Their modular nature leads to short implementation and maintenance times, which give them advantages compared with other techniques for industrial applications.
- They are able to manipulate a large amount of data, rather than aggregating data or simplifying a model to obtain a solvable problem but with only a partial representation of reality.
- They are able to manage uncertainties, by building many scenarios rather than offering an exact solution to a model with estimates of many data items.

A global problem could be considered as composed of many subproblems, each one being an NP-hard problem, in order to optimize one or more performance indicators in the presence of uncertainties in the data. But, at the present time, there is no model that can tackle the whole complexity of a logistics system. Instead, decision aid tools are generally developed with a precise purpose and an appropriate vision of the system (choice of horizon, data precision, one or more performance criterion, …), by making simplifying assumptions. However, it seems essential to ensure the consistency of proposed solutions, whether for other actors or for other time scales.

14.2   Hybridization Techniques

There is no doubt that metaheuristics can play an important role in the integration of the complexity of logistics systems but it is equally clear that metaheuristics alone are not sufficient. That is why we wish to highlight metaheuristic-based hybridization techniques in this section.

14.2.1   Generalities

Optimization methods allow one to optimize the running of a system while minimizing (or maximizing) one or more performance criteria. Methods for combinatorial optimization problems are usually split into two categories: exact methods and approximate methods. Exact methods can provide optimal solutions and prove their optimality. They include techniques from integer linear programming (ILP) such as branch-and-bound, branch-and-cut, and Lagrangian relaxation. Approximate methods are used whenever an optimal solution cannot be obtained (because of the size of the instance, the impossibility of modeling the problem by a linear model, the time allocated for solving it, …). Among the approximate methods, we find metaheuristics based mostly on local searches. Optimization methods are well suited for tackling the algorithmic complexity of the systems studied.

In some cases, the performance criterion for the system is not easy to compute. It is then necessary to run a performance evaluation model (a deterministic or stochastic simulation model, or a markovian model). For these systems, Norre [33] defined the notion of functional and structural complexity. Norre also introduced the notion of dual complexity (Fig. 14.4) and proposed a combination of an optimization method and a performance evaluation model to solve problems linked to this dual complexity. In the following, a "method" will represent either an optimization method or an evaluation model.

In the previous section, we have shown that the logistics systems we want to study are characterized by two elements:

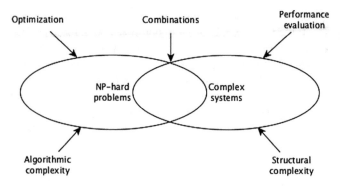

Fig. 14.4 Dual complexity

- On one hand, a wish to support an integrated view as part of horizontal or vertical synchronization, which may lead one to consider the logistics system as a combination of many optimization problems.
- On the other hand, the aim of improving the competitiveness of the system. The performance must be evaluated by taking into account sometimes contradictory criteria in the presence of uncertainties.

The techniques for hybridization between a metaheuristic and another method, whether an optimization or performance evaluation method, can be organized into three categories:

Sequential linking.   (A → B) (Fig. 14.5). Method A and method B are used sequentially. Method A solves a part of the problem (for instance for a given subset of variables). The other part of the problem is solved by method B. A classical example is the use of an optimization method to determine a feasible solution to the problem, and then a metaheuristic for optimizing this solution.

Sequential and iterative linking.   (A ⇆ B) (Fig. 14.6). Method A and method B are used in a sequential and iterative way. The result of method B is an input to method A, which allows one to iterate the solution process.

Hierarchical linking.   (A ↓ B) (Fig. 14.7). The methods are used according a "master–slave" scheme. For instance, method A may build one or more solutions, which are evaluated or optimized by method B.

These three techniques may be combined to obtain more elaborate hybridization methods. For instance, (A → ((B ↓ C) ⇆ D)) means that a hierarchical linking between methods B and C follows method A and is sequentially linked with method D.

In this section, we consider two types of metaheuristic-based hybridization methods: metaheuristic/optimization-method hybridization, which is well suited when a problem can be decomposed into subproblems, and metaheuristic/performance-evaluation-method hybridization, which is useful when performance criteria are hard to evaluate.

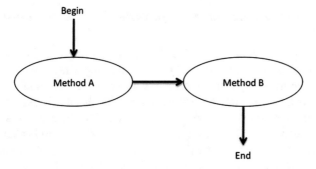

Fig. 14.5   Principle of sequential linking of two methods

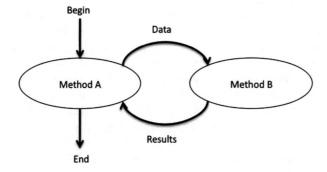

Fig. 14.6   Principle of sequential and iterative linking of two methods

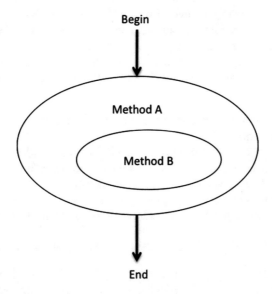

Fig. 14.7   Principle of hierarchical linking two methods

14.2.2 Metaheuristic/Optimization-Method Hybridization

Blum et al. [6] have noted that an increasing number of published metaheuristics are not strictly in agreement with the paradigm of a single traditional metaheuristic. Instead, they combine algorithmic elements which come from optimization methods from domains other than those of metaheuristics. Such approaches are defined by Blum et al. as hybrid metaheuristics. Hybrid metaheuristics appeared nearly two decades ago. Since then, they have proved their efficiency in solving hard optimization problems. We first present hybridization between two metaheuristics before talking about hybridization with another optimization method.

The metaheuristic/metaheuristic hybridization technique consists in combining two metaheuristics. The aim is to design a win–win method. A good example is a hybridization (Pop ↓ Ind) between a population algorithm (for instance, an evolutionary algorithm or a particle swarm optimization) and an individual-based method (for instance, a local search, simulated annealing, or a tabu search). Such a hybridization takes advantage of the exploratory nature of the population algorithm and the ability of an individual-based method to intensify the search in a promising area of the search space. Many examples of such hybridization exists in the literature, most of them combine a metaheuristic with a local search (Meta ↓ LS). The hybridization (genetic algorithm ↓ LS) is a technique often used in the literature and is known as memetic algorithms [29] or genetic local search [27]. The hybridization (simulated annealing ↓ LS) is known as C-L-O (chained local optimization) [25] or SALO (simulated annealing local optimization) [10] and is part of the set of iterated local searches [24] in which an acceptance criterion follows the simulated annealing process. Talbi [45] proposed a taxonomy of hybrid methods based essentially on the degree of encapsulation of one technique in another and the degree of parallelization.

In recent years, many approaches have combined a metaheuristic with another optimization method. Several classifications have been proposed in the literature [11, 19, 37]. For example, Dumitrescu and Stützle [11] split hybridization techniques into five categories:

- those which use exact methods to explore large-size neighboring systems into a local search algorithm;
- those which run several replications of a local search algorithm and exploit the information contained in good-quality solutions to define a subproblem with a reduced size which can be solved by a exact method;
- those which exploit bounds in greedy algorithms;
- those which guide a local search with information obtained by relaxing an ILP model;
- those which solve exactly some specific subproblems with a hybrid metaheuristic.

Fernandes and Lourenço presented a mapping of hybrid methods according to the problems considered [13]. Among problems about logistics systems, many references concern logistics network design (p-median problem), vehicle-routing problems (the travelling salesman problem (TSP) and the vehicle routing problem (VRP)), planning problems (the lot-sizing problem), and scheduling problems (the flow-shop and job-shop problems, etc.).

Constraint programming (CP) is a programming paradigm in which relations between variables are stated in the form of constraints. The search is based on constraint propagation, which reduces the set of possible values for the variables. Unlike metaheuristics, CP is known to be an efficient technique for decision problems but not for optimization problems. Hybridization of these two techniques is a good idea when one aims to profit from their respective advantages. Two strategies are possible, according to the optimization method driving the hybrid method. The first is a metaheuristic in which constraint programming is used as an efficient tool to explore a large neighborhood. The second is a tree search algorithm in which a metaheuristic is used to improve nodes or to explore neighboring paths. The article [14] and the book [47] are two suggestions for introductory reading on the subject. This hybrid technique has been successfully used on vehicle routing problems [8] and scheduling problems [5].

14.2.3   Metaheuristic/Performance-Evaluation-Method Hybridization

Performance evaluation models take into account the functional and structural complexity of logistics systems. Their use is particularly suitable when:

- Defined performance indicators cannot be computed by simple analytical functions as complex rules define the running of the system. It is then necessary to simulate the running of the system to evaluate its performance.
- Some data are described by distribution functions and it is necessary to run the model many times to know its robustness.

In this part, we will focus our discussion on simulation models. The terms "optimization by simulation" and "joint simulation/optimization approach" are often used in the literature. Optimization components based on evolutionary algorithms, scatter search, simulated annealing, and tabu search are included into? discrete event simulation software, as can be seen in [2, 16]. The aim of the resulting hybridization technique (simulation model ↓ optimization method) is that the optimization method provides solutions that are evaluated by the discrete event simulation software. Fu [16] discussed other types of links existing between optimization methods and simulation models. In the context of supply chain management, many studies have shown an interest in hybridization. Abo-Hamad and Arisha gave a recent review of the state of the art [1]. Figure 14.8, taken from that article shows the interactions between the optimization components and the simulation model. The Simulation model allows one to manage uncertainties and system complexity.

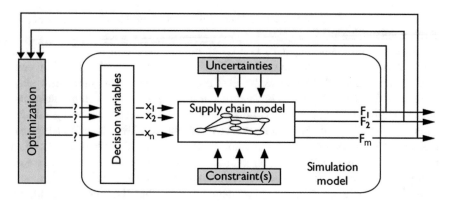

Fig. 14.8 Example of optimization/simulation-model hybridization for a supply chain [1]

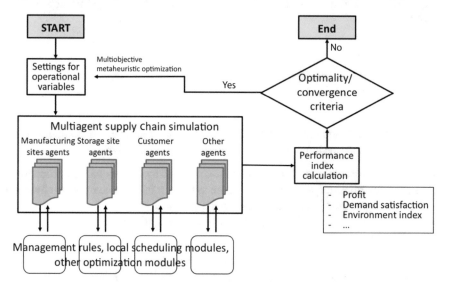

Fig. 14.9 Example of metaheuristic/simulation model hybridization for a decentralized supply chain [26]

Mele et al. [26] used this hybrid technique to a supply chain with a decentralized approach. Each actor of the chain was represented by an agent and all the agents were integrated into a simulation model. The model was combined (Fig. 14.9) with a genetic algorithm for the optimization part. More recently, a similar approach has been proposed by Nikolopoulou and Ierapetritou [32] with ILP.

14.3 Application to Supply Chain Management

14.3.1 Preamble

We have emphasized the importance of considering a logistics system as a whole. Griffis et al. [18] noted that being able to take into account many problems simultaneously is one of the major interesting features of metaheuristics for the study of logistics systems study (Griffis et al. use the term "hybrid problems"). These authors gave the following examples:

- *The location routing problem* consists in opening a subset of depots, assigning customers to them, and determining vehicle routes to minimize a total cost including the cost of open depots, the fixed costs of the vehicles used, and the total cost of the routes.
- *The inventory routing problem* consists in the distribution of a single product from a single facility to a set of customers over a given planning horizon. Each customer consumes the product at a given rate and has the capability to maintain a local inventory of the product up to a (maximum value) maximum.
- *The vehicle routing problem* is concerned with determining the optimal set of routes for a fleet of vehicles to deliver to a given set of customers.
- *The multilevel logistics network design problem* relates to the establishment of supply, warehousing, and distribution infrastructure. It encapsulates procurement, value-added, and postponement activities and inventory control policies.

The three first problems define a vertical synchronization with two decision levels: one about the logistics network design (choice of site location, choice of supply, delivery frequency) and the other about the route design. Joint problem solving allows one to obtain better results than solving the problems separately. The last example defines a horizontal synchronization between the activities of the supply chain. Griffis et al. considered this problem as a hybrid problem made up of a combination of many network design problems, one for each level (choice of production location and choice of distribution infrastructure, for instance).

In addition, many other combinations of problems are interesting to study in the context of supply chain management. Among them, we mention the following examples of horizontal synchronization:

- tactical planning: for the study of multisite lot-sizing problems;
- multisite scheduling: which takes into account product transport between sites;
- end product distribution: transportation sharing.

Examples of vertical synchronization include:

- tactical planning: synchronization between sales and operations planning and production planning;
- scheduling: synchronization between predictive and reactive scheduling (offline and online scheduling).

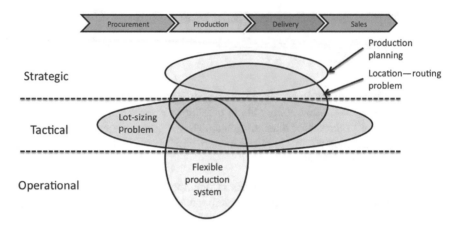

Fig. 14.10 Problems and synchronization types chosen for discussion

The methods implemented to solve these problems are generally based on a decomposition into basic problems. An optimization method is associated with each basic problem. We have the three categories defined in the previous section:

- *Sequential linking.* This technique can be used for a vertical synchronization problem where decisions taken at a high level may have an impact on lower level. The solution obtained by the first method is an input to the second method.
- *Sequential and iterative linking.* The previous scheme is iterated. Data are transmitted by the second method to the first one, restarting the process. In this technique, the methods are considered at the same level. The difficulty of this approach is to define the data transmitted from one method to the other.
- *Hierarchical linking.* In contrast to iterative methods, this combination induces a master slave relationship between the methods. During its execution, the first method calls the second method to solve a subproblem.

The combinations of problems is a first step in allowing an integrated view of a supply chain, in order to take decisions. For this reason, we shall highlight some of them (production planning, the location routing problem, the lot-sizing problem and the flexible production system) and present for each of them some metaheuristic-based hybrid methods proposed in the literature. Figure 14.10 shows the types of synchronization associated with each problem.

14.3.2 *Production/Distribution Planning*

Suon et al. [44] covered an international two-echelon production/distribution problem. This is a strategic planning problem which aims to define the movement of goods within a logistics network from tier-1 suppliers to the end customers.

The aim is to plan the manufacture of product types (total number N). The logistics network considered is composed of production zones (total number PZ), sales zones (total number SZ), and distribution links (total number DL) between the production and sales zones. We set $o_{k,u} = 1$ if distribution link k, $k = 1$, DL, begins at production zone u, $u = 1$, PZ, and we set $d_{k,v} = 1$ if distribution link k, $k = 1$, DL, ends at sales zone v, $v = 1$, SZ.

Each sales zone forecasts its requirements by type of product ($fd_{i,v}$ for type of product i, $i = 1$, N, and sales zone v, $v = 1$, SZ). Many production technologies (total number PT) available in the production zones are required to manufacture one type of product. A production zone may not offer all production technologies, and it may not manufacture certain products. Some production zones may be used for several types of products, whereas others may be dedicated to only one type of product. $xc_{i,t}$ is the ratio between the number of a product of type i, $i = 1$, N, and that of the reference product for production technology t, $t = 1$, PT. Each production technology t, $t = 1$, PT, for u, $u = 1$, PZ, has a minimum production capacity $cap\_min_{t,u}$, which represents the breakeven point of the installed industrial equipment, and it has a maximum production capacity $cap\_max_{t,u}$.

The problem is to determine the quantity of each type of product manufactured in each production zone and the delivery method to the sales zones, showing the quantity assigned to each distribution link.

The objective is to minimize the global delivery costs of the supply chain for all product types and distribution links. $sc_{i,u}$ represents the supply charge for the bill of material for a type of product i, $i = 1$, N, manufactured by production zone u, $u = 1$, PZ. $fc_{t,u}$ and $vc_{t,u}$) represent the fixed and variable charge respectively, for production technology t, $t = 1$, PT, and production zone u, $u = 1$, PZ; $tc_{i,k}$ and $dr_{i,k}$ are the unit transportation charge and duty rate, respectively, for type of product i, $i = 1$, N, and distribution link k, $k = 1$, DL.

The variables are:

$P_{i,u}$ the number of products of type i, $i = 1$, N, manufactured by production zone u, $u = 1$, PZ;

$Y_{i,k}$ the number of products of type i, $i = 1$, N, assigned to distribution link k, $k = 1$, DL;

$mc_{i,u}$ the unit manufacturing cost for a product of type i, $i = 1$, N, and production zone u, $u = 1$, PZ;

$dc_{i,k}$ the unit delivery cost for a product of type i, $i = 1$, N, and distribution link k, $k = 1$, PZ.

The problem is

$$\min z = \sum_{i=1}^{N} \sum_{k \in DL} Y_{i,k}.dc_{i,k} \qquad (14.1)$$

under the following constraints:

$$\sum_{k \in DL} d_{k,v}.Y_{i,k} = fd_{i,v}, \quad \forall i \in N, \ \forall v \in SZ \tag{14.2}$$

$$\sum_{i \in N} xc_{i,t}.P_{i,u} \leq cap\_max_{t,u}, \quad \forall t \in PT, \ \forall u \in PZ \tag{14.3}$$

$$\sum_{i \in N} xc_{i,t}.P_{i,u} \geq cap\_min_{t,u}, \quad \forall t \in PT, \ \forall u \in PZ \tag{14.4}$$

$$\sum_{k \in DL} o_{k,u}.Y_{i,k} = P_{i,u}, \quad \forall i \in N, \forall u \in PZ \tag{14.5}$$

$$mc_{i,u} = \sum_{\substack{t \in PT/ \\ xc_{i,t}>0}} \left(\frac{fc_{t,u}}{\displaystyle\sum_{\substack{i' \in N/ \\ xc_{i',t}>0}} P_{i',u}} + xc_{i,t}.vc_{t,u} \right) \quad \forall i \in N, \forall u \in PZ \tag{14.6}$$

$$dc_{i,k} = tc_{i,k} + dr_{i,k}.\left(\sum_{u \in PZ} o_{k,u}.(sc_{i,u} + mc_{i,u}) \right) \quad \forall i \in N, \forall k \in DL \tag{14.7}$$

$$P_{i,u} \geq 0, \quad \forall i \in N, \forall u \in PZ \tag{14.8}$$

$$Y_{i,k} \geq 0, \quad \forall i \in N, \forall k \in DL \tag{14.9}$$

The constraint (14.2) concerns the forecast demand. The constraint (14.5) specifies that no storage is allowed in either zone; all manufactured goods must be delivered. The constraints (14.3) and (14.4) concern production technology capacities. The constraint (14.6) computes the unit manufacturing cost. The constraint (14.7) computes the unit delivery cost. The manufactured and delivered quantities are nonnegative according to the constraints (14.8) and (14.9).

The problem is modeled by linear constraints and a nonlinear objective function. The nonlinearity is due to the fixed manufacturing costs, economies of scale, and duty costs. Moreover, this objective function is nonconvex (as was proven by a counterexample).

To tackle the nonlinearity, the problem was decomposed into two subproblems: the first one concerns the determination of the quantity to be manufactured in each production zone, and the second deals with distribution of the quantities determined to the sales zones. This decomposition comes from the fact that the second model is a classical transportation problem which can be modeled as a linear model. A hybrid metaheuristic was proposed (Fig. 14.11), with an iterated local search for the manufactured quantities ($P_{i,u}$) and a linear program for solving the transportation model. The proposed method can be denoted (ILS $\leftrightarrows$ LP).

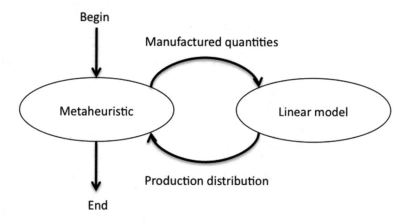

Fig. 14.11 Hybrid metaheuristic proposed in [44]

14.3.3   Location–Routing Problem

This synchronization problem is one of the oldest and most studied problems. The location–routing problem combines two NP-hard problems: The facility location problem (FLP) and the vehicle routing problem (VRP). The aim is to determine facility locations from among many potential locations, to assign customers to an open facility, and to solve a vehicle routing problem. The objective is to minimize a set of costs, i.e., the facility-opening costs, vehicle costs, and travel costs.

Let $V = I \cup J$ be a set of vertices, where I denotes the set of potential depot nodes and J the set of customers to be delivered to. A capacity W_i and an operating cost O_i are associated with each depot $i \in I$. Each customer $j \in J$ has a demand d_j. The travel cost for edge (i, j) is denoted by $c_{i,j}$. K denotes a set of available vehicles with capacity Q. F is a fixed cost per vehicle used.

The variables are the following:

y_i = 1 if depot $i \in I$ is opened, 0 otherwise;
$f_{i,j}$ = 1 if depot $i \in I$ delivers to customer $j \in J$, 0 otherwise;
$x_{i,j,k}$ = 1 if edge $(i, j) \in V^2$ is traversed by vehicle $k \in K$.

The problem is

$$\min z = \sum_{i \in I} O_i y_i + \sum_{i \in V} \sum_{j \in V} \sum_{k \in K} c_{i,j} x_{i,j,k} + \sum_{i \in I} \sum_{j \in J} \sum_{k \in K} F x_{i,j,k} \qquad (14.10)$$

under the following constraints:

$$\sum_{k \in K} \sum_{i \in V} x_{i,j,k} = 1, \quad \forall j \in J \qquad (14.11)$$

$$\sum_{j \in J} \sum_{i \in V} d_j x_{i,j,k} \le Q, \quad \forall k \in K \tag{14.12}$$

$$\sum_{j \in V} x_{i,j,k} - \sum_{j \in V} x_{j,i,k} = 0, \quad \forall k \in K, \forall i \in V \tag{14.13}$$

$$\sum_{i \in I} \sum_{j \in J} x_{i,j,k} \le 1, \quad \forall k \in K \tag{14.14}$$

$$\sum_{i \in S} \sum_{j \in S} x_{i,j,k} \le |S| - 1, \quad \forall S \subset J, \forall k \in K \tag{14.15}$$

$$\sum_{u \in J} x_{i,u,k} + \sum_{u \in V \backslash j} x_{u,j,k} \le 1 + f_{i,j}, \quad \forall i \in I, \forall j \in J, \forall k \in K \tag{14.16}$$

$$\sum_{j \in J} d_j f_{i,j} \le W_i y_i, \quad \forall i \in I \tag{14.17}$$

$$x_{i,j,k} = \{0, 1\}, \quad \forall i \in V, \forall j \in V, \forall k \in K \tag{14.18}$$

$$y_i = \{0, 1\}, \quad \forall i \in V \tag{14.19}$$

$$f_{i,j} = \{0, 1\}, \quad \forall i \in V, \forall j \in V \tag{14.20}$$

The objective defined in Eq. (14.10) is to minimize a sum of three terms: operating costs, travel costs, and vehicle costs. The constraint (14.11) guarantees that every customer belongs to one and only one route and that each customer has only one predecessor in that route. Capacity constraints are satisfied through the inequalities (14.12) and (14.17). The constraints (14.13) and (14.14) ensure the continuity of each route and a return to the depot of origin. The constraint (14.15) is a subtour elimination constraint. The constraint (14.16) specifies that a customer can be assigned to a depot only if a route linking them is opened. Finally, the constraints (14.18) to (14.20) state the Boolean nature of the decision variables.

This model is based on a CPLP (capacitated plant location problem) model and a VRP model. The CPLP is a monoperiod location problem. Once defined, the network structure cannot change over time. Further work will combine a multiperiod model with the VRP.

Nagy and Salhi [30] have reviewed the state of the art for the LRP. These authors indicate that for the specific problems for which exact methods are efficient, most of the proposed methods are hybrid methods based on decomposition into two subproblems, namely the FLP and VRP.

Prins et al. [36] proposed a two-phase iterative method. The principle of this method is to alternate between a depot location phase and a routing phase, exchanging information about the most promising edges. In the first phase, the routes and their customers are aggregated into supercustomers, leading to a facility location problem, which is then solved by Lagrangian relaxation of the assignment constraints. In the second phase, the routes obtained from the resulting multidepot vehicle routing problem are improved using a granular tabu search heuristic. At the end of each global iteration, information about the edges most often used is recorded to be used in the next phases. This method can be denoted $(LR) \leftrightarrows ((TS) \downarrow (LS))$.

Boccia et al. [7] tackled a two-echelon LRP. The first echelon was composed of large-capacity depots, generally far from the customers, and the second echelon contained satellite locations of less capacity. The problem was split into two monoechelon LRPs, each being split into two subproblems: a capacitated facility location problem and a multidepot VRP. These authors proposed a tabu search combining an iterative approach for the monoechelon problems and a hierarchical one for each of the subproblems $((TS)\downarrow(TS)) \leftrightharpoons ((TS)\downarrow(TS))$.

14.3.4 Multiplant Multiproduct Capacitated Lot-Sizing Problem

Considering a medium-term horizon (from 6 to 18 months), lot-sizing problems are aimed at determining quantities of products to be manufactured, with the objective of minimizing costs (production, setup, and inventory holding) while satisfying a given demand for each period. The setup costs are generally an estimate of the productivity loss due to a changeover in production, which may require adjustment of a production line. Capacity constraints ensure that, for each period, the production capacity is not exceeded. There are many production sites. The following model includes many product types in order to consider a bill of materials and calculate net requirements.

We give the mathematical model proposed by Sambasivan and Yahya [40]. The data are the following.

M denotes the set of production sites, N the set of product types, and T the set of periods. $d_{i,j,t}$ represents the demand for product i and site j in period t. $P_{j,t}$ is the production capacity of site j during period t. $M_{i,j,t}$, $V_{i,j,t}$, and $H_{i,j,t}$ represent the production costs, setup costs, and inventory holding costs, respectively, for product i and site j during period t. $r_{j,k,t}$ is the unitary transportation cost from site j to site k. $u_{i,j}$ represents the production rate, and $s_{i,j}$ the lead time for product i and site j.

The decision variables are:

$x_{i,j,t}$ quantity of product $i \in I$ manufactured by site $j \in M$ during period $t \in T$;
$I_{i,j,t}$ quantity of product $i \in N$ holded by site $j \in M$ during period $t \in T$;
$w_{i,j,k,t}$ quantity of product $i \in N$ transported from $j \in M$ to $k \in M$ during period $t \in T$;
$z_{i,j,t}$ $=1$ if there is a production setup for product $i \in N$ and site $j \in M$ during period $t \in T$, 0 otherwise.

The problem is

$$\min z = \sum_{i \in N} \sum_{j \in M} \sum_{t \in T} \left(M_{i,j,t} x_{i,j,t} + V_{i,j,t} z_{i,j,t} + H_{i,j,t} I_{i,j,t} + \sum_{k \in M \setminus \{j\}} r_{j,k,t} w_{i,j,k,t} \right)$$

$$(14.21)$$

under the following constraints:

$$I_{i,j,t} = I_{i,j,t-1} + x_{i,j,t}$$

$$-\sum_{k \in M \setminus \{j\}} w_{i,j,k,t} + \sum_{l \in M \setminus \{j\}} w_{i,l,j,t} - d_{i,j}, \quad \forall i \in N, \forall j \in M, \forall t \in T \qquad (14.22)$$

$$x_{i,j,t} \leq \left(\sum_{j \in M} \sum b = t^T d_{i,j,b} \right) z_{i,j,t}, \quad \forall i \in N, \forall j \in M, \forall t \in T \qquad (14.23)$$

$$\sum_{i \in N} \left(\frac{x_{i,j,t}}{u_{i,j}} + s_{i,j} z_{i,j,t} \right) \leq P_{j,t}, \quad \forall j \in M, \forall t \in T \qquad (14.24)$$

$$x_{i,j,t} \geq 0, \, I_{i,j,t} \geq 0, \quad \forall i \in N, \forall j \in M, \forall t \in T \qquad (14.25)$$

$$w_{i,j,k,t} \geq 0, \quad \forall i \in N, \forall j \in M, \forall k \in M \setminus \{j\} \qquad (14.26)$$

$$z_{i,j,t} \in \{0, 1\}, \quad \forall i \in N, \forall j \in M, \forall t \in T \qquad (14.27)$$

The objective function encodes the goal of the optimization, which is the minimization of the total cost, i.e., the production, setup, inventory, and transfer costs. The constraint (14.22) refers to the inventory balance for the quantity of item i during period t at plant j. The constraint (14.23) ensures that if item i is produced at plant j in period t then the setup of the plant has to be considered. The constraints (14.24) ensures that the available capacity is not exceeded. Finally, the constraints (14.25) to (14.27) impose the nonnegativity of the variables x, I, and w, and ensure that the variables z are binary.

Nascimento et al. [31] proposed a hybridization of GRASP and Path-relinking ((GRASP) ⇆ (PR)). GRASP [12] is a multistart metaheuristic similar to an iterated local search. GRASP typically consists of iterations made up of successive constructions of a greedy randomized solution and subsequent iterative improvements of it through a local search. Initially, path-relinking has been proposed for use with tabu search but has been successfully hybridized with genetic algorithms [38] and GRASP [39]. This technique is a way of exploring trajectories between elite solutions. The fundamental idea behind this method is that good solutions to a problem should share some characteristics. The hybridization consists in keeping a memory of the set of elite solutions and building new solutions by connecting elite solutions to those generated by GRASP.

For this kind of problem, there have been many studies using techniques such as Lagrangian relaxation (for production capacities and costs) or constraint programming. Metaheuristics have been used less because it is not easy to define neighboring systems for lot-sizing problem. Increasing or decreasing even slightly the quantity produced by a site during a period may impact upstream and downstream periods. An example of a neighboring system was detailed by Lemoine [23]. New types of hybridization between a metaheuristic and constraint programming seem to be promising ideas for that kind of problem.

14.3.5 Flexible Manufacturing System

This section presents a study of a logistics system reduced to a single production site: a flexible manufacturing system (FMS). In a supply chain, an FMS is used to manufacture products. An FMS is a fully automated system with production cells (referred to as machines) interconnected by a transportation system. The most commonly used transportation systems are automated guided vehicles. FMSs are known to be expensive and hard to manage, but they are flexible, which means that they can be adapted to fluctuations in demand. There is a huge literature about FMSs. We recommend [22] as a first read.

One of the advantage of FMSs is that they include problems similar to those found in multisite logistics systems. The facility layout problem concerns the layout of production cells in a workshop so as to minimize the physical flows. Other problems concern transportation system design, the layout of loading/unloading spots fleet sizing, offline scheduling (vehicles use a predefined route), and online scheduling (vehicles determine their route in real time). These problems are generally tackled separately because of their difficulty.

Deroussi and Gourgand studied vertical synchronization between a layout problem and a scheduling problem in an FMS [9]. This study was done in a context of workshop reorganization (tactical level) in which production zones and transportation network were not modified. Only permutations of machines inside production zones were possible. The problem considered can be modeled as a quadratic assignment problem.

The problem was modeled as a job-shop scheduling problem. Here, M denotes the set of machines and L the set of production zones (as the objective is to assign the machines to the production zones, we clearly have $|L| = |M|$). O is the set of tasks, and $o_{i,j} \in O$ is the ith task in the jth job. For each job, a fictitious task at the beginning of the routing corresponds to the input of the job into the workshop. O^+ is the set of all tasks (real and fictitious). The machine type required by task $o_{i,j} \in O$ is denoted $\mu_{i,j}$, and $\tau_{m,\mu_{i,j}} \in \{0, 1\}$ is a compatibility matrix for the machines and machine types. Finally, t_{l_1,l_2} is the matrix of transport times between zones l_1 and l_2.

The decision variables are:

$x_{m,l}$ = 1 if machine $m \in M$ is assigned to zone $l \in L$, 0 otherwise;
$y_{o_{i,j},l}$ = 1 if operation $o_{i,j} \in O^+$ is assigned to zone $l \in L$, 0 otherwise.

The problem is

$$\min z = \sum_{o_{i,j} \in O} \sum_{l_1 \in L} \sum_{l_2 \in L} t_{l_1,l_2} y_{o_{j,i-1},l_1} y_{o_{i,j},l_2} \tag{14.28}$$

under the following constraints:

$$\sum_{m \in M} x_{l,m} = 1, \quad \forall l \in L \tag{14.29}$$

$$\sum_{l \in L} x_{l,m} = 1, \quad \forall m \in M \tag{14.30}$$

$$\sum_{l \in L} y_{o_{ji},l} = 1, \quad \forall o_{i,j} \in O^+ \tag{14.31}$$

$$y_{o_{i,j},l} \leq \sum_{m \in M} \tau_{m,\mu_{i,j}} x_{m,l} \quad \forall o_{i,j} \in O^+, \forall l \in L \tag{14.32}$$

$$x_{m,l} \in \{0, 1\}, \quad \forall m \in M, \forall l \in L \tag{14.33}$$

$$y_{o_{i,j},l} \in \{0, 1\}, \quad \forall o_{i,j} \in O^+, \forall l \in L \tag{14.34}$$

The objective function minimizes the sum of transportation times (14.28). The constraints (14.29) and (14.30) ensure a bijection between the set of machines and the set of zones. The constraint (14.31) assigns a production zone to each task, whereas the constraint (14.32) guarantees that each task is assigned to a compatible machine.

This model does not allow loaded-vehicle transports to be taken into account. Yet [3, 4] underlined that empty-vehicle transports are as expensive as loaded-vehicle transports and it is important to consider them. One difficulty is that empty-vehicle transports are dependent on the transport sequence and are difficult to estimate except in specific cases. Deroussi and Gourgand [9] therefore proposed a hybrid meta-heuristic taking transportation times into account. The first step consists in solving exactly the quadratic assignment problem presented above. The second step takes into account empty transportation times by using an approach similar to GRASP. Solutions are built using an ant colony optimization paradigm. The assignment obtained during the first step allows one to define probabilities to build new assignments. New assignments are evaluated by solving a job-shop problem with transport (with a joint schedule of production means and transport). The technique used was an iterated local search combined with a discrete event simulation model. The results show that even for small size instances (five production zones) the assignment obtained in the first step can be improved by more than 50 %. The proposed method can be denoted $((ILP) \rightarrow ((ACO)\downarrow(ILS \leftrightarrows simul)))$.

14.4 Conclusion

Logistics systems in general and supply chains in particular are complex systems composed of many actors, each of them with its own concerns, but they must collaborate if one is to have a system as efficient as possible. In this chapter, we have shown the complexity that becomes apparent in the study of such systems, and suggested ways in which these issues can be resolved. In doing so, we have given reasons why metaheuristics are of interest to researchers. Indeed, these optimization methods offer tools to answer many specific features of logistics systems.

We have also explained how horizontal and vertical synchronization are relevant. For these kinds of problem, hybridization techniques are often obvious solutions. We have introduced the notions of sequential linking, iterative linking, and hierarchical linking for combining a metaheuristic with another optimization method or a performance evaluation model. Although the importance of synchronization in logistics system has been recognized for a long time by many researchers, the research field is wide open. Following the emergence of problems in fields such as inverse logistics, green logistics, and risk management, logistics systems now include new activities, new management rules, and new performance indicators, thereby enhancing further studies.

Let us hope that scientific research will remain very active in this domain during the coming years.

References

1. Abo-Hamad, W., Arisha, A.: Simulation–optimisation methods in supply chain applications: A review. Irish Journal of Management **1**, 95–124 (2010)
2. April, J., Glover, F., Kelly, J.P., Laguna, M.: Practical introduction to simulation optimization. In: Proceedings of the 2003 Winter, Simulation Conference, vol. 1, pp. 71–78 (2003)
3. Asef-Vaziri, A., Laporte, G., Ortiz, R.: Exact and heuristic procedures for the material handling circular flow path design problem. European Journal of Operational Research **176**, 707–726 (2007)
4. Asef-Vaziri, A., Hall, N.G., George, R.: The significance of deterministic empty vehicle trips in the design of a unidirectional loop flow path. Computers & Operations Research **35**, 1546–1561 (2008)
5. Beck, J.C., Feng, T.K., Watson, J.P.: Combining constraint programming and local search for job-shop scheduling. INFORMS Journal on Computing **23**(1), 1–14 (2011)
6. Blum, C., Puchinger, J., Raidl, G., Roli, A.: Hybrid metaheuristics in combinatorial optimization: a survey. Applied Soft Computing **11**, 4135–4151 (2011)
7. Boccia, M., Crainic, T.G., Sforza, A., Sterle, C.: A metaheuristic for a two echelon location-routing problem. In: P. Festa, *Experimental Algorithms*. Lecture Notes in Computer Science, vol. 6049, pp. 288–301. Springer, Berlin, Heidelberg (2010)
8. De Backer, B., Furnon, V., Shaw, P., Kilby, P., Prosser, P.: Solving vehicle routing problems using constraint programming and metaheuristics. Journal of Heuristics **6**(4), 501–523 (2000)
9. Deroussi, L., Gourgand, M.: A scheduling approach for the design of flexible manufacturing systems. In P. Siarry (ed.) *Heuristics: Theory and Applications*, pp. 161–222. Nova (2013)
10. Desai, R., Patil, R.: Salo: Combining simulated annealing and local optimization for efficient global optimization. In: *Proceedings of the 9th Florida AI Research Symposium (FLAIRS-'96)*, pp. 233–237 (1996)
11. Dumitrescu, I., Stützle, T.: Combinations of local search and exact algorithms. In: *EvoWorkshops*, pp. 211–223 (2003)
12. Feo, T., Resende, M.: A probabilistic heuristic for a computationally difficult set covering problem. Operations Research Letters **8**, 67–71 (1989)
13. Fernandes, S., Lourenço, H.: Hybrids combining local search heuristics with exact algorithms. In:*V Congreso Espanol sobre Metaheuristicas, Algoritmos Evolutivos y Bioinspirados*, pp. 269–274 (2007)
14. Focacci, F., Laburthe, F., Lodi, A.: Local search and constraint programming. International Series in Operations Research and Management Science **57**, 369–404 (2003)
15. Forrester, J.: *Industrial Dynamics*. Technical report, MIT Press, Cambridge, MA (1961)

16. Fu, M.C.: Optimization for simulation: Theory vs. practice. INFORMS Journal on Computing **14**(3), 192–215 (2002)
17. Ganeshan, R., Harrison, T.: *An Introduction to Supply Chain Management*. Technical report, Penn State University, Department of Management Science and Information System Operations. Prentice Hall (1995)
18. Griffis, S., Bell, J., Closs, D.: Metaheuristics in logistics and supply chain management. Journal of Business Logistics **33**, 90–106 (2012)
19. Jourdan, L., Basseur, M., Talbi, E.G.: Hybridizing exact methods and metaheuristics: A taxonomy. European Journal of Operational Research **199**(3), 620–629 (2009)
20. Kouvelis, P., Chambers, C., Wang, H.: Supply chain management research and production and operations management: Review, trends, and opportunities. Production and Operations Management **15**(3), 449–469 (2006)
21. Krmac, E.V.: Intelligent value chain networks: Business intelligence and other ICT tools and technologies. In: S. Renko (ed.) *Supply Chain Management: New Perspectives*. InTech (2011)
22. Le-Anh, T.: Intelligent control of vehicle-based internal transport systems. Ph.D. thesis, Erasmus University, Rotterdam, The Netherlands (2005)
23. Lemoine, D.: Modèles génériques et méthodes de résolution pour la planification tactique mono-site et multi-site. Ph.D. thesis, Blaise Pascal University, France (2008)
24. Lourenço, H.: Supply chain management: An opportunity for metaheuristics. Technical report, Pompeu Fabra University, Barcelona (2001)
25. Martin, O., Otto, S.: Combining simulated annealing with local search heuristics. Annals of Operations Research **63**, 57–75 (1996)
26. Mele, F.D., Espuna, A., Puigjaner, L.: Supply chain management through a combined simulation–optimisation approach. Computer Aided Chemical Engineering **20**, 1405–1410 (2005)
27. Merz, P., Friesleben, B.: Genetic local search for the TSP: New results. In: *Proceedings of the 1997 IEEE International Conference on Evolutionary Computation*, Indianapolis, pp. 159–164. IEEE Press (1997)
28. Meyr, H., Wagner, M., Rohde, J.: Structure of advanced planning systems. In: H. Stadtler, C. Kilger (eds.) *Supply Chain Management and Advanced Planning - Concepts, Models, Software and Case Studies*. Springer, Berlin (2002)
29. Moscato, P.: On evolution, search, optimization, genetic algorithms and martial arts: Towards memetic algorithms. In: *Caltech Concurrent Computation Program*, C3P Report, vol. 826 (1989)
30. Nagy, G., Salhi, S.: Location-routing: Issues, models and methods. European Journal of Operational Research **177**, 649–672 (2007)
31. Nascimento, M.C., Resende, M.G., Toledo, F.: GRASP heuristic with path-relinking for the multi-plant capacitated lot sizing problem. European Journal of Operational Research **200**(3), 747–754 (2010)
32. Nikolopoulou, A., Ierapetritou, M.G.: Hybrid simulation based optimization approach for supply chain management. Computers & Chemical Engineering (2012)
33. Norre, S.: Heuristique et métaheuristiques pour la résolution de problèmes d'optimisation combinatoire dans les systèmes de production. Ph.D. thesis, Blaise Pascal University, France (2005)
34. Oliver, R., Webber, M.D.: Supply-chain management: Logistics catches up with strategy. Outlook **5**, 42–47 (1982)
35. Orlicki, J.: *Material Requirements Planning*. McGraw-Hill, London (1975)
36. Prins, C., Prodhon, C., Ruiz, A., Soriano, P., Calvo, R.W.: Solving the capacitated location-routing problem by a cooperative Lagrangean relaxation-granular tabu search heuristic. Transportation Science **41**(4), 470–483 (2007)
37. Puchinger, J., Raidl, G.R.: Combining metaheuristics and exact algorithms in combinatorial optimization: A survey and classification. In: J. Mira, J.R. Álvarez (eds.) Artificial Intelligence and Knowledge Engineering Applications: A Bioinspired. Lecture Notes in Computer Science, vol. 3562, pp. 41–53. Springer, Berlin, Heidelberg (2005)

38. Reeves, C.R., Yamada, T.: Genetic algorithms, path relinking, and the flowshop sequencing problem. Evolutionary Computation **6**(1), 45–60 (1998)
39. Resende, M., Ribeiro, C.: GRASP with path-relinking: Recent advances and applications. In: T. Ibaraki, K. Nonobe, M. Yagiura (eds.) *Metaheuristics: Progress as Real Problem Solvers*. Operations Research/Computer Science Interfaces Series, pp. 29–63. Springer (2005)
40. Sambasivan, M., Yahya, S.: A Lagrangean-based heuristic for multi-plant, multi-item, multi-period capacitated lot-sizing problems with inter-plant transfers. Computers & Operations Research **32**(3), 537–555 (2005)
41. Schmidt, G., Wilhelm, W.: Strategic, tactical and operational decisions in multi-national logistics networks: A review and discussion of modeling issues. International Journal of Production Research **38**(7), 1501–1523 (2000)
42. Simchi-Levi, D., Kaminsky, P., Simchi-Levi, E.: *Designing and Managing the Supply Chain; Concepts, Strategies and Case Studies*. Irwin/McGraw-Hill (2000)
43. Snyder, S.: Facility location under uncertainty: A review. IIE Transactions **38**(7), 537–554 (2006)
44. Suon, M., Grangeon, N., Norre, S., Gourguechon, O.: A hybrid metaheuristic for a strategic supply chain planning problem with procurement–production–distribution activities and economy of scale. In: *3rd International Conference on Information Systems, Logistics and Supply Chain: Creating Value Through Green Supply Chains ILS 2010*, Casablanca, Morocco, April 14–16 2010 (2010)
45. Talbi, E.: A taxonomy of hybrid metaheuristics. Journal of Heuristics **8**, 541–564 (2002)
46. Thomas, A., Lamouri, S.: Flux poussés: MRP et DRP. Techniques de l'ingénieur **AGL1**(AG5110), 1–12 (2000)
47. Van Hentenryck, M., Michel, L.: *Constraint-Based Local Search*. MIT Press (2009)
48. Wight, O.: *Manufacturing Resource Planing: MRP II*. Oliver Wight (1984)
49. Wolf, J.: *The Nature of Supply Chain Management Research*. Springer Science (2008)

Chapter 15
Metaheuristics for Vehicle Routing Problems

Caroline Prodhon and Christian Prins

15.1 Introduction

The basic problem of vehicle routing is a classical operations research problem, known to be NP-hard [64] and called the *vehicle routing problem* (VRP) or *capacitated VRP* (CVRP). It consists in determining a least-cost set of routes from a depot for a fleet of capacitated vehicles in order to meet the demands of a set of customers. Figure 15.1 depicts a typical example of a solution to this problem.

The theoretical research on the applications related to the CVRP and its variants make them one of the most studied classes of combinatorial optimization problems. Dantzig and Ramser [26] introduced the CVRP in 1959 under the name of the *truck dispatching problem*, to model a real problem of distribution of gasoline to service stations. Since this seminal work, the body of models and solution methods related to the CVRP and its extensions has experienced strong growth. For instance, in 2009 Eksioglu et al. [31] established a typology based on more than one thousand articles. Today, a search on Google Scholar with the keywords "vehicle routing problem" returns more than 13,000 references. Industrial applications have not been neglected: a survey of commercial VRP software [92] revealed 22 products, widely spread across various industries. An article by Laporte [60] summarized the impressive achievements of fifty years of research.

Despite this abundant activity, the current exact methods have been limited to problems with 100 customers [4]. They recently reached 200 customers [93], but real cases can involve up to 1,000 clients. Metaheuristics are the methods of choice

C. Prodhon (✉) · C. Prins
ICD-LOSI, UMR CNRS 6281, Université de Technologie de Troyes,
12 Rue Marie Curie, CS 42060, 10004 Troyes Cedex, France
e-mail: caroline.prodhon@utt.fr

C. Prins
e-mail: christian.prins@utt.fr

© Springer International Publishing Switzerland 2016
P. Siarry (ed.), *Metaheuristics*, DOI 10.1007/978-3-319-45403-0_15

Fig. 15.1 A typical CVRP
solution

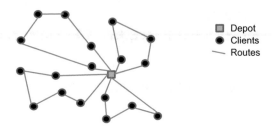

for dealing with realistic cases, and it can even be said that vehicle routing problems
constitute a successful application area for this class of algorithms.

This chapter defines, in Sect. 15.2, the basic vehicle routing problem and recalls
its main variants. Section 15.3 presents some constructive heuristics and stresses the
importance of some concepts of local search that are widely used in routing problems. Section 15.4 describes representative applications of the main metaheuristics
to vehicle routing. Section 15.5 details an approach based on splitting giant tours,
which results in efficient algorithms for various routing problems. An example of
an application of this technique is given in Sect. 15.6. Finally, Sect. 15.7 closes the
chapter.

15.2 Vehicle Routing Problems

15.2.1 Basic Version

The basic CVRP is defined in general, on a complete undirected graph $G = (V, E)$.
The set of nodes V is composed of a depot (node 0), where a fleet of identical vehicles
of capacity Q is based, and n customers with requests q_i for a product, $i = 1, 2, \ldots, n$.
Each edge $[i, j]$ in the set E represents an optimal path between nodes i and j in the
real road network. Its cost c_{ij}, often a distance or travel time, is precomputed. The aim
is to determine a set of routes of minimum total cost visiting every customer exactly
once. A route is a cycle starting and ending at the depot, performed by one vehicle,
and whose total load does not exceed Q. Depending on the authors, the number of
vehicles can be fixed or free, a service time s_i is sometimes defined for each client,
and the routes can be limited by a maximum distance or travel time L.

The CVRP is NP-hard because the single-route case (when the total demand fits
one vehicle, i.e., $\sum_{i=1}^{n} q_i \leq Q$) is the *traveling salesman problem* (TSP), which is
known to be NP-hard in the strong sense. In fact, it is particularly hard because it
combines a *bin packing problem* (the assignment of customers to vehicles) and a
sequencing problem (the TSP) for each vehicle. Several classical formulations as
integer linear programs are available [43, 130]. The challenge in these models is to
avoid the formation of subtours, i.e., cycles which do not include the depot.

The following model is probably the simplest one. The depot becomes two nodes 0 and $n + 1$, used to begin and end each route, and each edge $[i, j]$ is replaced by two opposite arcs (i, j) and (j, i). The binary variable x_{ij}^k is equal to 1 if arc (i, j) is traversed by vehicle k:

$$\min \sum_{k} \sum_{(i,j)} c_{ij} \cdot x_{ij}^k \tag{15.1}$$

$$\sum_{j \neq i} \sum_{k} x_{ij}^k = 1 \qquad \forall i \neq 0, n+1 \tag{15.2}$$

$$\sum_{j \neq i} x_{ji}^k = \sum_{j \neq i} x_{ij}^k \qquad \forall i \neq 0, n+1 \ \forall k \tag{15.3}$$

$$\sum_{i \neq 0, n+1} \sum_{j \neq i} q_i \cdot x_{ij}^k \leq Q \qquad \forall k \tag{15.4}$$

$$t_i^k + s_i + c_{ij} \leq t_j^k + M(1 - x_{ij}^k) \qquad \forall i \tag{15.5}$$

$$x_{ij}^k \in \{0, 1\} \qquad \forall (i, j) \ \forall k \tag{15.6}$$

$$t_i^k \geq 0 \qquad \forall i \ \forall k \tag{15.7}$$

The objective function (15.1), to be minimized, is the total cost of the routes. The constraint (15.2) guarantees that each customer is served, while the constraint (15.3) ensures route continuity: the same vehicle arrives at a customer and leaves it. Vehicle capacity is respected through the constraint (15.4). The variable t_i^k denotes the start of service of vehicle k at customer i. In Eq. (15.5), this variable serves to prevent subtours: if vehicle k traverses arc (i, j) ($x_{ij}^k = 1$), the term containing the large positive constant M is zero and the constraint means that the start of service at j can occur only once i has been served and the vehicle has moved from i to j. If arc (i, j) is not used by the vehicle, the constraint holds trivially.

The CVRP belongs to the family of *node routing problems*, in which tasks are associated with nodes of a network. There exist also *arc routing problems*, in which tasks must be performed on arcs or edges, such as in municipal refuse collection, where garbage must be picked up in every street. The problem equivalent to the CVRP in arc routing is called the CARP (*capacitated arc routing problem*): its definition is similar, except that a demand q_{ij} is now defined for each network edge, for example an amount of waste to be collected. Recent surveys of arc routing problems can be found in [21, 61].

15.2.2   Variants of the Classical Vehicle Routing Problem

Although the CVRP still attracts researchers [53, 76, 86], many variants are now being investigated. First, additional attributes or constraints may affect customers:

- The vehicle routing problem with *time windows* (VRPTW), in which customers can only be served within a given time interval $[e_i, l_i]$, [38, 50, 63, 131, 136].
- In *pickup and delivery problems* (PDPs), the goal is to execute a set of requests for transportation (i^+, i^-) from a pickup node i^+ to a delivery node i^-, in contrast to the CVRP where all goods are distributed from the depot [27, 42, 109, 116, 140, 141]. The *dial-a-ride problem* (DARP) denotes a PDP dedicated to passenger transportation, with time windows and service quality criteria, as in on-demand transportation systems [91, 118].
- The *team orienteering problem* (TOP) is inspired by routing problems for repair technicians, where serving a customer induces a profit. The aim is to determine a set of trips that maximizes the total profit collected, subject to a given time limit which hinders serving all customers [58, 70, 71].
- In problems with split deliveries (the *split-delivery VRP*, SDVRP), each customer can be visited several times to receive its request, which can lead to solutions using fewer vehicles [8].

Complications related to vehicles are very common in practice, as shown by the three following examples:

- The *heterogeneous fleet VRP* (HFVRP) considers several vehicle types, each type being defined by the number of vehicles available, a fixed utilization cost, and a cost per distance or time unit [14, 29, 85, 122].
- *Truck and trailer routing problems* (TTRPs) involve complex routes, where each truck can temporarily drop its trailer at a customer to reach other customers that cannot be accessed with a complete vehicle [72, 137].
- Problems involving vehicles with compartments, for example refrigerated or at room temperature (the *multicompartment VRP*, MC-VRP) [32].

Among other extensions, we have to mention some related to the type of network considered, the structure of the routes, the planning horizon, and the optimization criteria:

- The routes can originate from different depots in the *multidepot VRP* (MDVRP) [2, 56]. In the *two-echelon VRP* (VRP-2E), primary routes deliver to satellite depots from a main depot, and then secondary routes serve customers from these satellites [47, 52].
- In the *open VRP* (OVRP), met in some car rental contracts, vehicles are not required to return to the depot after having completed their service [68, 73].
- *Periodic vehicle routing problems* (PVRPs) are defined on a multiperiod horizon in which each customer must be visited according to a given frequency, as in waste collection [15, 139].
- In the *cumulative CVRP* (CCVRP), which arises in disaster logistics, the cost of a route is the sum of arrival times at visited nodes, which corresponds to the average rescue time after division by the number of stops. Computing the cost variations when a solution is modified is not trivial [87].

Finally, one can combine strategic, tactical, or operational decisions, and even merge a routing problem with another optimization problem. For instance, the routes must replenish customer stocks in the *inventory routing problem* (IRP) [74, 96]. The *integrated production–distribution problem* (IPDP) adds one production site to an IRP, which requires synchronization between a production planning problem and a VRP [13]. In the *location-routing problem* (LRP), the depots must be selected from among potential sites and the routes designed from open depots [20, 88, 103, 127]. Building routes while determining a feasible loading induces difficult problems (*VRP with two/three-dimensional loading constraints*—2L-VRP and 3L-VRP) [12, 30, 65, 66, 114].

We can go even further, by combining the above problems. For example, the *periodic location-routing problem* generalizes the LRP to a multiperiod planning horizon [1, 107]. This aggregation results in increasingly general problems, called *rich vehicle routing problems* [45]. In 2007, Hasle and Kloster [46] presented several rich vehicle routing problems emerging in industrial applications.

In summary, a large family of problems with a common structure exists beyond the CVRP. The challenges faced by metaheuristics are to produce results of good quality in an acceptable computation time, but these metaheuristics must also be easy to code and maintain, have few parameters, and be easily adaptable to the diversity of constraints encountered in real applications [25, 134].

15.3 Basic Heuristics and Local Search Procedures

The basic heuristics and local search procedures are important components of metaheuristics for vehicle routing problems that deserve a specific section. Then, basic heuristics are required to provide initial solutions quickly, while the local search procedures bring intensification.

15.3.1 Basic Heuristics

Basic heuristics are still widely used in commercial VRP software to quickly find feasible solutions of good quality. A review can be found in the paper by Laporte and Semet [62], who distinguish between constructive and two-phase methods.

The simplest constructive heuristic is the one called the *nearest-neighbor heuristic*: starting from the depot, a route is progressively extended by visiting the nearest unrouted customer, from among those compatible with the residual capacity of the vehicle. When no more customers can be added, the vehicle comes back to the depot and a new route is initialized. Other constructive methods such as the Clarke and Wright algorithm [19] rely on route mergers. In the initial solution, each customer is visited by one dedicated route. Then, the heuristic evaluates the possible mergers (concatenations) of two routes and executes the one with the largest saving. Insertion heuristics, such as the method proposed by Mole and Jameson [82], are also popular; these build a solution using successive insertions of customers, guided by weighted insertion costs.

The idea of the two-phase methods is to reduce the problem to a TSP. *Cluster-first, route-second* methods begin by creating groups of customers (clusters) whose total demand fits the vehicle capacity, and then solve a TSP for each cluster. Gillett and Miller's heuristic [41] is a good example, where clusters are defined as angular sectors centered on the depot. Fisher and Jaikumar's method [35] solves a generalized assignment problem in the clustering phase. The petal heuristic [5] builds a large number of routes and then solves a set partitioning problem to extract a subset of routes visiting each customer exactly once.

Conversely, *route-first, cluster-second* heuristics [7, 101] relax the vehicle capacity constraints to solve a TSP. The route obtained, covering all customers, is often called a *giant tour*. This is transformed into a VRP solution by a splitting procedure that cuts the giant tour into capacity-feasible routes.

15.3.2 Local Search

15.3.2.1 Classical Moves

An improvement procedure, or local search, starts from one initial solution s (often obtained by a constructive heuristic) and considers a subset $N(s)$ of solutions close to s in terms of structure, called the *neighborhood* of s. This neighborhood is inspected to find a better solution, s'. The strategy can be to look for the best improvement or to stop searching the neighborhood at the first improvement. Once s' has been detected, it becomes the incumbent solution and the process is repeated. In this way, the input solution is progressively improved until a local optimum is reached for the neighborhood considered.

In practice, $N(s)$ is implicitly defined by a transformation $s \to s'$ called a *move*, instead of being generated in extenso. The simplest moves have been defined for the TSP, and they can be applied to each route in the CVRP. For instance, one customer can be removed and then reinserted at a different position (*node relocation*), or the positions of two customers can be swapped (*node exchange*). The neighborhoods defined by these two moves can be browsed in $O(n^2)$. The k-opt moves [69], which are more efficient, remove k edges from a route and reconnect the subsequences obtained with k other edges. Since all possible moves are tested in $O(n^k)$ for n customers, the 2-opt and 3-opt moves are often used to keep the complexity low.

Or [89] proposed the Or-opt move, which relocates a string of 1 to λ consecutive customers, while Osman [90] introduced the λ-*interchange*, which exchanges two strings that have at most λ customers each (the two strings may have different lengths). As these two neighborhoods can be searched in $O(\lambda n^2)$ and $O(\lambda^2 n^2)$, respectively, $\lambda = 3$ is often selected to limit running times. λ-interchanges are particularly interesting: if one of the strings can be empty and if each string can be reversed during the reinsertion, these moves include node relocations, node exchanges, and 2-opt and Or-opt moves as particular cases.

Figure 15.2 depicts the 2-opt and λ-interchange moves. The key point is that one is able to evaluate the cost variation in constant time for each move. Thus, the 2-opt move on one route replaces arcs (u, x) and (v, y) by (u, v) and (x, y), inducing a cost variation $\Delta = c_{uv} + c_{xy} - c_{ux} - c_{vy}$, which is computable in $O(1)$.

15.3.2.2 Feasibility Tests

All the moves described above can be generalized to two routes, but it becomes more difficult to evaluate their feasibility or cost variation in $O(1)$. For instance, consider the 2-opt move applied to two routes T_1 and T_2 in Fig. 15.2. In this version, called 2-opt*, arcs (u, x) and (v, y) are replaced by (u, y) and (v, x) (there exists also a variant where the arcs are replaced by (u, v) and (x, y)). The neighborhood sizes are in $O(n^2)$ for the two versions. Let $C(T, i, j)$ and $W(T, i, j)$ be the cost and the load, respectively, for a route T between two nodes i and j (inclusive), and let $C(T)$ and $W(T)$ be the total cost and the total load for the route. After the move, the capacity constraints must still be satisfied for each route:

$$W(T_1, 0, u) + W(T_2) - W(T_2, 0, v) \leq Q \qquad (15.8)$$
$$W(T_2, 0, v) + W(T_1) - W(T_1, 0, u) \leq Q \qquad (15.9)$$

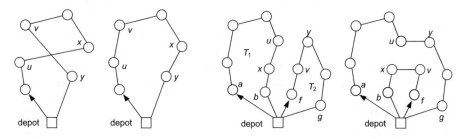

| 2-opt move on one route | 2-opt* move on two routes |

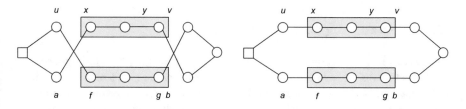

Osman's λ-interchange move

Fig. 15.2 Examples of 2-opt and λ-interchange moves

If an additional constraint on the maximum length L of a route must be satisfied (the driver's maximum working time, for example), it is also necessary to check the conditions

$$C(T_1, 0, u) + c_{uy} + C(T_2) - C(T_2, 0, y) \leq L \tag{15.10}$$

$$C(T_2, 0, v) + c_{vx} + C(T_1) - C(T_1, 0, x) \leq L \tag{15.11}$$

If W and C are calculated for each move by browsing the routes using $O(n)$ loops, the whole neighborhood is searched in $O(n^3)$ instead of $O(n^2)$. A way to perform feasibility tests in $O(1)$ is to *precalculate* interesting values. For the 2-opt* move, each route T in the input solution can be scanned to calculate $W(T, 0, u)$ and $C(T, 0, u)$ for each node u. After these precomputations, which cost $O(n)$ in total, the neighborhood can be searched in $O(n^2)$, since each feasibility test can now be done in $O(1)$. If an improving move is found and applied to the incumbent solution, W and C need to be updated only for the two modified routes, which can be done in $O(n)$.

Time windows constitute another frequent complication factor. For instance, in the VRP with time windows (VRPTW), inserting a customer into a route delays subsequent visits, which may violate the time windows. Once again, it is possible to check feasibility with a loop in $O(n)$, but can we reach $O(1)$? Kindervater and Savelsbergh [55] proposed to precompute the maximum delay (*push forward*) that is allowed at each node without violating any time windows.

To understand how this is done, consider a route $T = (T_1, T_2, \ldots, T_r)$. If $[e_k, l_k]$ denotes the time window for the start of service at customer T_k, t_k the arrival time at this customer, and s_k its service time, the customer must be serviced before the end of its time window ($t_k + s_k \leq l_k$), but it is possible to arrive before the beginning of the window ($t_k < e_k$) and wait. Hence, the maximum delay for the last customer T_r is $S_r = l_r - s_r - t_r$, and, for $k = r - 1, r - 2, \ldots, 1$, we have $S_k = \min(S_{k+1}, l_k - s_k - t_k)$. The delays for the whole set of customers can be computed in $O(n)$ at the beginning of the local search, which allows feasibility tests in $O(1)$ for various moves. For instance, inserting one customer i between nodes T_{k-1} and T_k delays the start of service at T_k by $\theta = C_{T_{k-1}, T_i} + s_i + C_{T_i, T_k} - C_{T_{k-1}, T_k}$, which is allowed only if $\theta \leq S_k$. The upper part of Fig. 15.3 gives an example with three customers: we obtain $S_1 = 2$ for the first customer.

15.3.2.3   General Approach of Vidal et al.

Vidal et al. [135] have proposed an even more general approach to handling precomputations for what they call *timing problems*: given a set of tasks, a set of constraints to be satisfied, and an optimization criterion, how does one determine the optimal starting times and how does one reoptimize them quickly when the sequence is modified? These problems are widespread in scheduling and vehicle routing problems. Vidal et al. noticed that all these moves can be expressed as concatenations of task

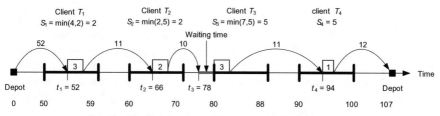

Example of trip with 4 clients in the VRP with time windows (VRPTW):
calculation of possible delay S_k to arrive at client k without violating time windows afterwards

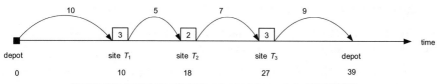

Example of trip to visit 3 sites affected by a disaster in the cumulative VRP (CCVRP):
The cost of the trip is the sum of arrival times at the sites 10 + 18 + 27 = 55 (average rescue time 55/3)

Fig. 15.3 Complications in the VRPTW and CCVRP (see text)

sequences. For each criterion Z used during the local search, they proposed to pre-calculate $Z(\sigma)$ for any sequence of nodes σ contained in the routes of the incumbent solution, using two main operators:

- an initialization operator that evaluates $Z(\sigma)$ if σ contains a single node;
- an operator that computes Z when two sequences σ and τ are concatenated $(\sigma \oplus \tau)$.

In practice, a matrix Z is prepared, in which Z_{ij} gives the value associated with the sequence delimited by nodes i and j (inclusive), if that sequence exists in a route. A route is coded by a list of nodes, beginning and finishing at the depot. Each node i is examined and, for each given i, each node j located after i. The initialization operator allows one to evaluate Z_{ii}, and then the second operator is called to provide the value of Z_{ij} for each node j, until the end of the route. In most cases, the two operators can be implemented in $O(1)$, so the matrix Z can be precalculated in $O(n^2)$.

This approach will now be illustrated by two examples, the CVRP and the CCVRP already introduced in Sect. 15.2.2. The length (number of nodes) of a sequence σ is denoted $|\sigma|$, σ_i is the node at the ith position, and $\sigma_{i,j}$ represents the subsequence from nodes σ_i to σ_j (inclusive).

In the simple case of the CVRP, the total demand $Q(\sigma)$ and the duration $D(\sigma)$ for each sequence σ within the routes is evaluated. If σ contains a single node i, the initialization operator sets $Q(\sigma) = q_i$ and $D(\sigma) = 0$. For two sequences σ and τ, the concatenation operator defines $Q(\sigma \oplus \tau) = Q(\sigma) + Q(\tau)$ and $D(\sigma \oplus \tau) = D(\sigma) + c(\sigma_{|\sigma|}, \tau_1) + D(\tau)$. Using this information, it becomes easy to check the capacity and maximum-duration constraints for any move. For example, if a string of customers

τ is inserted after σ_i in a route σ, the new route sequence is $\sigma_{1,i} \oplus \tau \oplus \sigma_{i+1,|\sigma|}$ and its load and duration can be derived in $O(1)$ from the precomputed values.

Now, consider the CCVRP as a more complicated example. The cost of a route is the sum of the arrival times at the customers, as depicted in the lower part of Fig. 15.3, and the cost of the return to the depot is ignored. Silva et al. [120] have shown that in the single-route case, called the *cumulative TSP* or *minimum latency problem*, the ad hoc quantities to be precalculated are:

- $D(\sigma)$, the total duration of visiting the nodes of σ, starting from σ_1;
- $C(\sigma)$, the cost (sum of arrival times) assuming a departure at time 0;
- $W(\sigma)$, the extra cost if the departure is delayed by one unit of time.

From these quantities, the total duration and the cost of any sequence generated during a move of the local search can be deduced:

- if $|\sigma| = 1$, then $D(\sigma) = C(\sigma) = 0$, and $W(\sigma) = 1$ for a customer and 0 for the depot;
- $D(\sigma \oplus \tau) = D(\sigma) + c(\sigma_{|\sigma|}, \tau_1) + D(\tau)$;
- $C(\sigma \oplus \tau) = C(\sigma) + W(\tau) \times [D(\sigma) + c(\sigma_{|\sigma|}, \tau_1)] + C(\tau)$
- $W(\sigma \oplus \tau) = W(\sigma) + W(\tau)$.

15.3.2.4 Very Constrained Problems

Very constrained problems, with tight time windows for instance, raise difficulties: the initial heuristics can fail to find a feasible solution and the local search may waste time on testing unfeasible moves. In the first case, a randomized constructive heuristic can be called several times until a feasible solution is returned. The second case can be improved by choosing ad hoc moves. For instance, a 2-opt move in one route reverses a subsequence of customers and has a high probability of violating time windows, in contrast to node relocation or exchange moves.

Another trick [24] consists in relaxing complicating constraints and adding a penalization for their violation to the objective function. By doing so, new paths between feasible solutions are created in the solution space, but searching this extended space may require more time.

Consider for instance a VRPTW solution S with p routes $T_1, T_2, \ldots, T_p$, a vehicle of capacity Q, a time window $[e_i, l_i]$ and a service time s_i for each customer i. The arrival time at customer i is t_i, and the load of route T_k is W_k. A possible penalized objective function adds the violations of vehicle capacity and time windows to the true solution cost $C(S)$ (the total cost of traversed arcs) as follows:

$$CP(S) = C(S) + \sum_{i=1}^{n} \alpha \cdot \max(0, t_i + s_i - l_i)^2 + \sum_{k=1}^{p} \beta \cdot \max(0, W_k - Q)^2 \quad (15.12)$$

The value of using squared expressions is that they accept small violations more easily than large ones; the coefficients α and β are used to change the relative weights of the two types of violation. At the end of the local search, a solution with $CP(S) = C(S)$ will be fully feasible. In fact, a solution may be acceptable even if there are still small violations. For instance, in waste collection, trucks have a compactor and the vehicle capacity can be slightly exceeded, while in distribution, many customers will tolerate small delays (*soft time windows*) when they are not too frequent. However, to be fair, local search, even with penalization, is not well suited to very constrained problems: constraint programming is certainly more effective.

15.3.2.5   Acceleration Techniques

For large-scale problems, even a neighborhood search in $O(n^2)$ can take too much time. Several techniques exist to reduce the computational time. A simple one consists in evaluating *a restricted number K of randomly selected moves*. This number can be fixed or proportional to the neighborhood size; for instance, $K = \sqrt{|N(s)|}$.

Lists of neighbors are also frequently used. For a given node i, its list of neighbors $LN(i)$ contains the $n - 1$ nodes $j \neq i$, sorted in increasing order of the costs c_{ij}. The lists of neighbors for all nodes can be precomputed in $O(n^2 \log n)$ at the beginning of the local search. A threshold θ is selected, for instance $n/10$ or $\sqrt{n}$, and then the only moves evaluated for each node i are the ones which add one arc (i, j) such that j belongs to the first θ nodes of $LN(i)$.

To illustrate the concept, consider a 2-opt move in one of the routes shown in Fig. 15.2, which adds an arc (u, x). In a fast implementation, a main loop inspects each node u while an inner loop tests each node x from among the first θ nodes of $LN(u)$. The choice of these two nodes is enough to define the move, since v and y are the successors of u and x, respectively, in the route.

The idea behind the restricted list of neighbors is that the presence of expensive arcs is unlikely in good solutions. Nevertheless, counterexamples can easily be found. Therefore, it is prudent to vary θ dynamically and even to try $\theta = n - 1$ from time to time, to browse the neighborhood completely.

Vertex marking was introduced under the name of *don't look bits* by Bentley in 2-opt moves for the TSP [9]. The principle is as follows: If none of the moves involving one given node is able to improve the incumbent solution, this node can be ignored in the subsequent iterations. Each node can be marked or unmarked. At the beginning of the search, they are all marked and are stored in an active list L. Then, each marked node x is inspected in the order of the list, to evaluate all moves involving edges incident to x. If these moves are unfruitful, x is unmarked and removed from L. Otherwise, all nodes concerned in the move (the extremities of inserted and removed edges) are marked and appended to L, unless they are already in the list. The local search ends when the list is empty. Compared with a traditional local search, which scans the whole neighborhood in each iteration, this version is faster since L contains only the nodes involved in recent successful moves. The other nodes are forgotten until an improving move involving incident edges is discovered. Muyldermans [83] described in detail an efficient implementation of a local search procedure for the CARP, which combines edge marking and lists of neighbors.

Finally, Irnich et al. [51] have proposed an approach called *sequential search* to accelerate the local search for the CVRP. The main idea relies on the decomposition of each move into partial moves, most of them being pruned by computing bounds on partial gains. Combined with lists of neighbors, this technique is very powerful, but its implementation is not trivial, since an ad hoc decomposition must be found for each move. Moreover, the approach becomes complicated when constraints such as time windows are added.

15.3.2.6 Complex Moves

The literature contains local searches based on very elaborate moves, such as the GENIUS method [39], cyclic transfers [126], and ejection chains [110, 111]. An example of an ejection chain is an attempt to move a customer to another route. If the capacity of the target route is violated, one of its customers can be ejected into a third route, etc. Obviously, the number of successive ejections must be limited to avoid excessive running times.

Large-neighborhood search (LNS) considers neighborhoods whose cardinality is not polynomial in n, while avoiding a complete exploration. The moves can be decomposed into elementary actions and an improving sequence of actions determined by computing a least-cost path in an auxiliary graph [33]. Another approach [119] consists in partially destroying the solution and then repairing it (*ruin and recreate moves*). In the CVRP for instance, k customers can be removed randomly and reinserted into the routes to minimize cost variation [94]. Funke et al. [37] reviewed the state of the art for most local search operators for vehicle routing problems and proposed a unified representation that can handle many complex constraints, including resource constraints.

15.4 Metaheuristics

The local search heuristics for routing problems have evolved into metaheuristics, which encompass various techniques to escape from local optima and achieve better results in reasonable computation times compared with exact algorithms. We now present these methods, distinguishing two classical categories: path methods, which determine a sequence of solutions tracing a path in the solution space, and population or agent-based methods, operating on a set of solutions. Vehicle routing problems being extremely combinatorial, all truly effective metaheuristics include local search procedures. The exceptions are simulated annealing, basic versions of genetic algorithms, ant colony optimization, and particle swarm techniques.

15.4.1 Path Methods

Simulated annealing is now little used for routing problems, although it was one of the first metaheuristics published for the CVRP, with an article from 1993 in which Osman introduced λ-interchange moves [90]. From time to time, effective implementations can be found, such as those of Lin and Yu [70, 71] for the team orienteering problem and that of Lin et al. [72] the TTRP with time windows. Deterministic variants of simulated annealing have been more successful: Li et al. [67] have proposed a *record-to-record travel* method for the CVRP. This method is well suited to parallel implementations [44].

Variable neighborhood search (VNS) and its easier variant *variable neighborhood descent* (VND) are fast, compact, and simple metaheuristics which explore successive neighborhoods with growing cardinalities. The idea behind them is that a local optimum for one neighborhood is not necessarily a local optimum for the others. VNS and VND are often used to replace a classical local search in another metaheuristic. Iterated local search methods integrating a VND have been proposed for the CVRP [17] and the CARP with split deliveries [8]. Effective VNSs are available, for example, for the open VRP [36], the multidepot VRP [56], the inventory routing problem [74, 96], and the CARP [48, 95]. These methods are outperformed by more sophisticated metaheuristics but, owing to their speed, they are often the only candidates for large problems [57].

Like the above metaheuristics, the *greedy randomized adaptive search procedure* (GRASP) [34] is not very efficient on routing problems. The reason probably derives from its independent iterations, which generate a solution using a randomized greedy heuristic and improve it via a local search. Although Marinakis has proposed a basic GRASP for the CVRP [76], the most efficient versions embed additional components. The path relinking technique has been employed in a GRASP for the LRP [105], the two-echelon LRP [88], and the CARP [132], while Qu and Bard [109] used a large-neighborhood search as an improvement procedure in a GRASP for a pickup and delivery problem.

Iterated local search (ILS) [75] and *guided local search* (GLS) [54] are very effective metaheuristics for vehicle routing problems. They generate a sequence of local optima by alternating a local search and a perturbation. ILS directly perturbs the solution, while GLS perturbs the edge costs so that a local optimum is no longer optimal with the modified costs. Two excellent examples are an ILS for the heterogeneous fleet VRP [122] and a GLS for the CARP [11].

In the 1990s, *tabu search* methods were the most effective metaheuristics for vehicle routing problems. Vehicle capacity and time windows were often relaxed to allow one to minimize a penalized objective function, as in Sect. 15.3.2.4. Successful versions are available for many problems, including the CVRP [6, 111, 129], the VRPTW [24], the HFVRP [14], and the CVRP and HFVRP with two-dimensional loading [65, 66]. Although these algorithms often involve classical moves, Rego and Roucairol [111] implemented ejection chains,while Toth and Vigo [129] designed a *granular tabu search* (GTS), in which the moves are restricted to a small fraction of the edges (the cheapest ones), which is dynamically adjusted during the algorithm.

Pisinger and Röpke [94] developed an effective metaheuristic able to solve several vehicle routing problems, such as the CVRP and PDP. This based on the ruin and recreate moves described in Sect. 15.3.2.6. Their *adaptive large-neighborhood search* (ALNS) includes several partial destruction and repair operators, implemented as heuristics. In each iteration, a randomly selected pair of operators is applied to the current solution, using probabilities updated by a learning process. This approach is conceptually simple but the codes can be long, since ten heuristics are typically employed. Successful ALNS metaheuristics have been published recently for the VRP-2E and the LRP [47], the real-time VRPTW [50], and the DARP [91].

Path relinking builds a path in the solution space between two existing solutions A and B. To do so, the first solution is gradually transformed into the second, for example by performing elementary modifications like the moves in a local search. In practice, the intermediate solutions are of poor quality and must be improved by calling a local search. This technique, rarely used alone, is mainly used to reinforce another metaheuristic such as GRASP [105] or tabu search [49].

There exist transitional forms between path methods and population-based metaheuristics. For example, a tabu search can be reinforced using an *adaptive memory* which records fragments of solutions to perform periodic intensifications [68, 113, 123, 124]. One can also maintain a pool of high-quality solutions in a GRASP and apply periodic stages of path relinking, as implemented by Villegas et al. for the TTRP [137]. Souffriau et al. even designed a metaheuristic for the TOP completely based on path relinking [121].

Another transitional form is *evolutionary local search* (ELS) [138]. This is actually an ILS, where each iteration generates p child solutions by applying a perturbation operator and a local search procedure to the current solution. This solution is replaced with the best child in the case of an improvement. ILS corresponds to the particular case $p = 1$. In our opinion, ELS is not a true population-based method, since the set of children is not retained. We describe in Sect. 15.6 a family of recent and very effective ELSs, which relaxes vehicle capacity, constraints, generates TSP tours, and applies a splitting procedure to convert these tours into feasible solutions for the original problem [29, 30, 99].

15.4.2 Population or Agent-Based Methods

We distinguish between the so-called evolutionary or true population methods, where new solutions are generated by combining solutions stored in a population, and agent-based methods (ant colony and particle swarm optimization), in which search agents are guided by a cooperation mechanism.

Genetic algorithms (GAs) appeared a short time after the first metaheuristics for vehicle routing (simulated annealing and tabu search), but with mixed performance, except for the VRPTW [97, 125]. The author of [125], employed chromosomes representing complete solutions, for instance lists of customers in successive routes, separated by copies of the depot node. Crossovers based on this encoding, such as

RBX [97], can produce children in which some routes violate vehicle capacity. The problem is easily solved by relocating customers on other routes, but the genetic transmission of good patterns from parents to children is degraded. Another explanation for these not entirely satisfactory results was the lack of a local search.

Good results were obtained on the CVRP from 2003 onwards via *memetic algorithms* (MAs), a family of genetic algorithm complemented by a local search applied with a certain probability to the offspring. Berger and Barkaoui [10] opened the way but still employed complete solutions as chromosomes. The problem of capacity violations was bypassed by Baker and Ayechew [3]. Here, each chromosome defines a partition of customers into clusters. It is decoded by solving a TSP for each cluster, using a constructive heuristic followed by a local search with 2-opt and λ-interchange moves. Prins [98] took the route-first cluster-second option, relaxing the capacity of vehicles in order to use chromosomes without route delimiters, similar to those used for the TSP. Prins called these chromosomes *giant tours*. A procedure named *Split*, described in Sect. 15.5, derives an optimal (subject to the sequence) solution to the CVRP from each chromosome. One advantage is that classical crossovers designed for TSP such as LOX and OX can be reused.

The MA proposed by Prins was the first to outperform tabu search methods. Then other effective memetic algorithms based on giant tours were designed for various vehicle routing problems, including the CARP [59], the multiperiod LRP [108], the multicompartment VRP [32], a production–distribution problem [13], and the cumulative CVRP [87]. Nagata and Bräysy [84] proposed the first effective MA without giant tours for the CVRP, using a sophisticated crossover operator called *edge assembly crossover* [84]. Some more recent MAs have been designed to solve several rich problems [133, 134, 136]. The latter reference describes the current best metaheuristic for 20 VRP variants.

Evolutionary strategies (ESs) make the population evolve by mutation and local search without combining solutions by crossover. This method, which recalls evolutionary local search, was applied by Mester and Bräysy to the VRPTW [79]. These same authors then developed a more powerful metaheuristic, which alternated between a GLS and an ES, for instances of the CVRP and VRPTW with more than 1000 customers [80].

Scatter search (SS) is an evolutionary method that is seldom used; it works with a small population combining high-quality solutions and diversified solutions relative to those solutions. In each iteration, a recombination operator, similar to the crossover of GAs but often deterministic, is systematically applied to each pair of parents and the resulting solutions are improved using local search. This metaheuristic is aggressive but often displays substantial computational times. A few good examples exist for the VRPTW [115], the multiperiod CARP [18], and the PDP with stochastic travel times [141].

Memetic algorithms with population management (MA|PM) are the missing link between memetic algorithms and scatter search. These are incremental memetic algorithms based on a distance measure $d(s, e)$ in the solution space between any two solutions s and e. The distance from a child e to the current population P is defined as $D(P, e) = \min\{d(s, e) \mid s \in P\}$. After the local search, this child is accepted into the population if $D(P, e) \leq \Delta$, where Δ is a fixed or dynamically varying threshold controlling the diversity. Prins [100] and Prins et al. [106] showed that population management improves memetic algorithms significantly for the heterogeneous fleet VRP and the CARP, respectively.

Ant colony optimization (ACO) is well adapted to problems where the construction of a solution is equivalent to finding a path in a graph. For routing problems, an ant can for example build successive routes by selecting successive arcs, using a nearest-neighbor heuristic biased by pheromone deposits. Reimann et al. [112] had the good idea to use ants to build routes by successive insertions, starting from loops on the depot. Their algorithm, enhanced by a local search, gives very good results on the CVRP. Santos et al. [117] developed another ACO algorithm with local search, which is one of the two current best metaheuristics for the CARP.

Particle swarm optimization (PSO) has only recently been applied to vehicle routing, and only hybrid versions have proved so far to be competitive with other metaheuristics. Chen et al. [16] proposed a PSO algorithm for the CVRP that assigns customers to vehicles, the routes being derived by a simulated annealing step applied to each vehicle. Marinakis and Marinaki [78] achieved better results but at the cost of a complex hybridization, combining PSO, GRASP, and path relinking. Two PSO algorithms have been developed for the CVRP with stochastic demands [77, 81].

15.4.3   Evolution of the Field, and Trends

A large number of metaheuristics are now available to solve various vehicle routing problems. Major trends emerge when reading the surveys published periodically on the subject. The most efficient algorithms until the early 2000s were tabu search methods. Cordeau and Laporte [23] identified ten of the most effective tabu search methods in 2002. In 2005, a study by Cordeau et al. [22] detected a turning point: the nine best metaheuristics still included three tabu search methods [24, 124, 129], but there were already three evolutionary algorithms [10, 79, 106] and one ant colony algorithm [112]. In 2008, a review by Gendreau et al. [40] confirmed this tendency. Currently, the most successful metaheuristic frameworks for the majority of routing problems are evolutionary local search, memetic algorithms, and adaptive large-neighborhood search.

Another direction is the development of hybrid methods. In fact, the best meta-heuristics already combine several components. For example, the best evolutionary algorithms for the CVRP are memetic algorithms and ELS, which all include a local search procedure [84, 106, 133]. This local search is sometimes replaced by a VND, a VNS, or an LNS to enhance intensification.

Hybridization does not forbid one to combine a metaheuristic and an exact method, which gives a *matheuristic*. A common technique is to generate a large number of good routes with a metaheuristic and then to solve a set, covering problem whose columns correspond to these routes (see, e.g., [15] for the PVRP). The cooperative method of Prins et al. for the LRP [103] alternates cyclically between the solution of a facility location subproblem, via Lagrangian relaxation, and a granular tabu search that optimizes the routes from selected depots. Labadie et al. [58] relax the TOP with time windows to solve an assignment problem via linear programming and used reduced costs to guide a granular VNS.

Parallel metaheuristics started to spread, at least in the academic world, with the multiplication of multicore PCs, powerful graphics cards (GPUs), and grid computing. Recently, this has led to a revival of tabu search methods because they are well suited to this kind of implementation [25, 53].

Nevertheless, many effective metaheuristics can still handle only a single routing problem. This lack of genericity is an obstacle to their incorporation into commercial software. The study of methods capable of solving several variants with a unique algorithm is still little developed. Such methods include the universal tabu search algorithm UTSA of Cordeau et al. [24], the large-neighborhood search of Pisinger and Röpke [94], the hybrid GA proposed by Vidal et al. [133] for the CVRP, the MDVRP, and the PVRP, and another GA by the same authors for various problems with time windows [136]. Recently, Vidal et al. [134] designed a more general algorithm, UHGS (unified hybrid genetic search), that is at least as good as the best published metaheuristics on 29 VRP variants.

15.5 The Split Approach

15.5.1 Principle and Advantages

As stated in Sect. 15.3.1, one possible constructive heuristic to solve vehicle routing problems consists in building a giant tour covering all customers, coded as a sequence of nodes, and then splitting this tour to get routes respecting vehicle capacity [7]. Less intuitive than the opposite cluster-first route-second approach [41], this route-first cluster-second principle was considered as a curiosity for a long time, before its integration into efficient metaheuristics [59, 98].

Since then, this technique has become very popular and can be found in more than 70 metaheuristics for a wide variety of routing problems; see [102] for a recent survey. Figure 15.4 illustrates its principle.

This approach has the following advantages:

- As observed by Beasley [7], the second phase (*cluster*) is equivalent to the computation of a shortest path in an auxiliary graph, as shown in the sequel.
- When the method is used in a metaheuristic, the smaller search space of giant tours is explored in the first phase (*route*), instead of the space of VRP solutions.
- The partition into routes can be done optimally, subject to the order defined by the giant tour. Conversely, one can show that there exists at least one "optimal" giant tour, i.e., one that yields an optimal solution to the original routing problem after splitting. Hence, the original problem can be solved without loss of information by searching the space of giant tours.

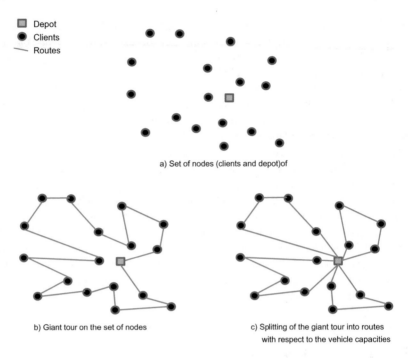

a) Set of nodes (clients and depot)of

b) Giant tour on the set of nodes

c) Splitting of the giant tour into routes
with respect to the vehicle capacities

Fig. 15.4 Illustration of the *route-first cluster-second* approach

- The construction of the giant tour can cope, at least partially, with many constraints concerning customers (precedence, time windows, ...), while the splitting phase can deal with constraints related to vehicles (capacity, working time, assignment to depots, ...), giving a very flexible framework for solving rich vehicle routing problems.
- Finally, the Split approach leads to state-of-the-art solution methods, competing with the best metaheuristics published.

Most metaheuristics based on the route-first cluster-second paradigm consist in alternating between an indirect representation of solutions to the routing problem at hand, the *genotype* (the giant tour), and a complete representation, the *phenotype* (the set of routes). The genotype defines a route in which capacity constraints are relaxed, corresponding to a Hamiltonian cycle on the set of nodes (customers and depot). The main structure of the metaheuristic (GRASP, GA, ...) searches the space of genotypes. The splitting procedure is used to decode the genotypes and evaluate the associated phenotypes. Intensification can easily be done by calling a local search. The alternation is generally completed by concatenating the trips of the resulting solutions and removing the depot nodes, which gives a new giant tour.

15.5.2   Split Algorithm

Starting from a giant tour defined as a sequence $T = (T_1, T_2, \ldots, T_n)$ of n customers, the Split algorithm begins by building an auxiliary graph $H = (X, U)$. X is a set containing $n + 1$ nodes, numbered from 0 to n. The arcset U contains one arc $(i - 1, j)$ for each subsequence of customers $(T_i, T_{i+1}, \ldots, T_j)$ which can be visited by a vehicle, i.e., $\sum_{k=i}^{j} q(T_k) \leq Q$. This arc is weighted by the cost of the corresponding route $cost(i, j) = c(0, T_i) + \sum_{k=i}^{j-1}(c(T_k, T_{k+1})) + c(T_j, 0)$, where $c(i, j)$ denotes the cost of arc (i, j) (distance or time). The optimal splitting is then obtained by computing a shortest path from node 0 to node n in H.

Figure 15.5 illustrates this principle using a small example with six customers. The giant tour considered is $T = (a, b, c, d, e, f)$. The values in parentheses indicate the demands. The associated auxiliary graph is given for this tour, assuming a vehicle capacity $Q = 15$. For instance, the arc ab represents a route visiting customers a and b, with load 12 and cost 10. The arc abc is not included, because the total demand of these three customers (16) exceeds the vehicle capacity. The arcs of the shortest path (thick lines) correspond to the trips in the optimal splitting, represented on the right: three routes with a cost of 27.

The shortest path in the auxiliary graph can be calculated using Bellman's algorithm for directed acyclic graphs. Algorithm 15.1 shows the compact version, called Split by Prins [98], in which the auxiliary graph is not generated explicitly. Two nested loops examine each subsequence $(T_i, T_{i+1}, \ldots, T_j)$ of customers and calculate its total demand (*load*) and the cost of the corresponding route (*cost*). If the load exceeds the capacity of a vehicle, the subsequence is rejected. Otherwise, for

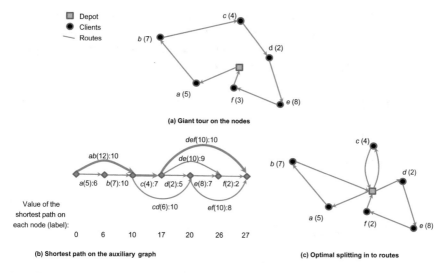

Fig. 15.5 Illustration of the Split algorithm

each node i, let V_i denote a label storing the cost of the shortest paths from node 0 to node i in H (see Fig. 15.5). The associated arc $(i-1, j)$ is implicitly created and the label V_j of node j is immediately updated if it can be improved, i.e., when $V_{i-1} + cost(i, j) < V_j$. At the end, the cost of the best possible CVRP solution for the given giant tour is given by the label of the last node, V_n.

```
V_0 ← 0; P_0 ← 0
for i ← 1 to n do
 |  V_i ← ∞
end
for i ← 1 to n do
 |  j ← i; load ← 0
 |  repeat
 |   |  load ← load + q(T_j)
 |   |  if i = j then
 |   |   |  cost ← c(0, T_i) + s(T_i) + c(T_i, 0)
 |   |  else
 |   |   |  cost ← cost − c(T_(j−1), 0) + c(T_(j−1), T_j) + s(T_j) + c(T_j, 0)
 |   |  end
 |   |  if load ≤ Q and V_{i−1} + cost < V_j then
 |   |   |  V_j ← V_{i−1} + cost
 |   |   |  P_j ← i − 1
 |   |  end
 |   |  j ← j + 1
 |  until j > n or load > Q;
end
```

Algorithm 15.1: Split algorithm

In Algorithm 15.1, a vector P is introduced to store the predecessor of each node j on the shortest path from 0 to j. This provides a quick and easy way to extract the CVRP solution S; see Algorithm 15.2. In this small procedure, S is encoded as a list of routes, and each route as a list of customers.

```
S ← ∅; j ← n
repeat
 |  route ← ∅
 |  for k ← P_j + 1 to j do
 |   |  add customer T_k at the end of list route
 |  end
 |  add list route at the beginning of list S
 |  j ← P_j
until j = 0;
```

Algorithm 15.2: Algorithm to extract the solution after Split

Note that Split evaluates each subsequence in $O(1)$, since $load$ and $cost$ are updated when j is incremented, instead of recomputing them completely using a loop from i to j. Hence, the complexity is proportional to the number of arcs in H (the number of feasible subsequences), i.e., $O(n^2)$ in the worst case. If, owing to vehicle capacity the average number of customers per route b is smaller than n, then the number of arcs in H and the complexity become $O(nb)$. Therefore, Split is very fast in practice and can be called frequently in a metaheuristic.

15.5.3 Integration into Heuristics and Metaheuristics

The simplest utilization of Split is in constructive heuristics. For instance, any exact or heuristic algorithm for the TSP can be recycled to build a giant tour which is then cut by Split to give routes. Such heuristics can be randomized to generate several giant tours, split them, and finally return the best solution obtained [101].

Split can be included in metaheuristics, where the metaheuristic explores the space of giant tours and Split is called to evaluate each tour. The most basic implementation is probably a GRASP: for instance, a greedy randomized heuristic samples the set of giant tours, each tour is decoded by Split, and the resulting solution is improved by a local search. Powerful memetic algorithms have also been developed. Here, chromosomes are encoded as giant tours, crossovers generate new giant tours, but the fitness calculation is done by applying Split and a local search procedure to the offspring [59, 87, 104, 106]. Various constraints can be added [101].

Another option is to alternate between the two search spaces. Starting from a giant tour T, Split is applied to obtain a complete solution S, which is then improved by a local search. If a giant tour T' is built by concatenating the lists of customers in S and discarding the depot nodes used as delimiters, it will differ from T provided at least one improving move has been executed during the local search. Hence, a new call to Split results in a new solution S' at least as good as S, and a simple repetition of this cycle already gives good results. Figure 15.6 illustrates this principle.

The implementation of this alternation in a metaheuristic leads to very powerful methods for various vehicle routing problems. As an example, the following section presents the GRASP×ELS method of Prins [99].

15.6 Example of a Metaheuristic Using the Split Approach

15.6.1 General Principle of GRASP×ELS

The method described here is a hybrid between two metaheuristics: a GRASP and an evolutionary local search. The principles of these two metaheuristics were presented in Sect. 15.4.1. Recall that a GRASP samples the local optima of the problem. Each iteration builds a solution, using a greedy randomized heuristic GRH, and then improves it by calling a local search procedure LS. Randomization is often based

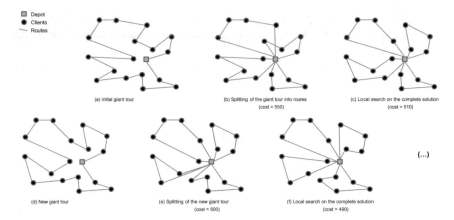

Fig. 15.6 Alternation between giant tours, complete solution, and local search

on a *restricted candidate list* (RCL). In each iteration of *GRH*, the α best decisions are placed in the RCL and one is chosen at random. If $\alpha = 1$, *GRH* becomes deterministic. If all possible decisions are possible in the RCL, the heuristic reduces to a kind of random walk and yields low-quality solutions. Usually $\alpha \in \{2, 3\}$ leads to solutions combining quality and diversity.

GRASP can be hybridized with a variety of techniques such as path relinking. Its successive solutions are independent (there is no memory), and it is worthwhile to explore the search space between them. The hybridization described here is done with an ELS.

As already mentioned, ELS generalizes ILS by generating in each iteration several child solutions instead of a single one. It requires three components: a constructive heuristic *CH*, an improvement procedure *LS*, and a randomized perturbation procedure *RP*. An initial solution *S* is built by *CH* and improved by *LS*. Then, each iteration generates a given number *nc* of child solutions by taking a copy of *S* and applying *RP* and *LS*. The incumbent solution is replaced by the best child if there is an improvement.

One advantage of this hybridization is that it strengthens the improvement phase of GRASP by replacing the local search by an ELS. Compared with an isolated ELS, another advantage is that the GRASP superstructure feeds the successive ELSs with diversified solutions.

15.6.2 Application to the Capacitated Vehicle Routing Problem

The key point for the effectiveness of the method when applied to the CVRP is combining the hybrid GRASP×ELS method and a cyclic alternation between giant

tours and CVRP solutions, as described in Sect. 15.5. The GRASP begins each iteration by building one giant tour using the simple nearest-neighbor heuristic described
in Sect. 15.3.1. This heuristic is easily randomized by selecting possible candidates
from an RCL. For a giant tour temporarily ending at customer i, the RCL may contain, for instance, the two or three unrouted customers closest to i. The resulting giant
tour is transformed by the Split algorithm into a CVRP solution S, which is in turn
improved by a local search. The routes of S are concatenated to give a new giant tour
T. The pair (S, T) is finally given to the ELS.

Each of the nc child solutions is obtained by applying to a copy T' of T a mutation
operator which exchanges the positions of two randomly selected customers. In fact,
the operator can perform p exchanges and p can be adjusted to control diversification. Initially, and after each improvement of the current best solution, p is set to a

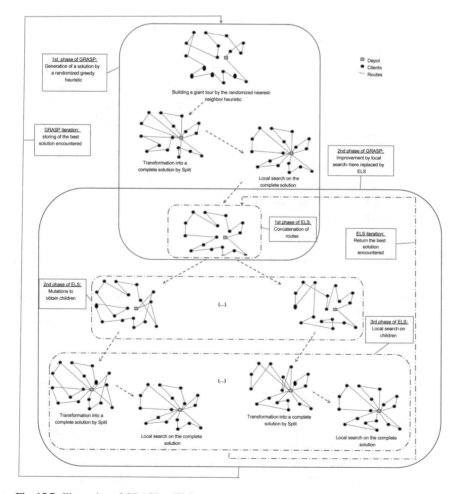

Fig. 15.7 Illustration of GRASP × ELS

minimum value $p_{min} = 1$. After each ELS iteration without improvement the incumbent solution, p is incremented but without exceeding a maximum value p_{max}, for instance 3. The perturbed giant tour is converted by Split into a CVRP solution and improved by local search. The local search used involves classical moves, including 2-opt, Or-opt, and λ-interchange. These moves are applied to each route and each pair of routes.

Once the nc children have been generated, the best one S' is compared with the parent solution S. If the latter has been improved on, S is replaced by S' ($S \leftarrow S'$) and the routes of S are concatenated to yield a new giant tour T, which provides the pair (S, T) for the next iteration of ELS. Figure 15.7 summarizes the method.

15.7 Conclusion

This chapter has shown that metaheuristics have been quite successful on vehicle routing problems. As with many other NP-hard optimization problems, the basic versions of the classic metaheuristics have been completely overtaken. The trend is to use increasingly hybridized algorithms, all including an improvement procedure. Parallel implementations and the exact solution of subproblems via metaheuristics can be used to reinforce this artillery. One negative aspect is the proliferation of metaheuristics for very similar variants of the same vehicle routing problem. Although a general problem solver does not yet exist in the field of vehicle routing, a few recent metaheuristics are able to deal with rich problems which contain many classical academic problems as particular cases. Such methods rely on flexible solution encodings and local search procedures that are able to handle many constraints simultaneously, but also on software engineering techniques such as the design of generic and reusable components.

15.8 Annotated Bibliography

We recommend the following references as good entry points to the rich literature on vehicle routing problems:

Reference [21]   This annotated bibliography is the latest on arc routing problems such as those found in waste collection.

Reference [24]   This article is a good example of a generic tabu search, able to handle several routing problems with a single code.

Reference [28]   The GRASP $\times$ ELS method based on giant tours described in Sect. 15.6 is generalized in this article to address the location–routing problem and heterogeneous fleet vehicle routing problems.

Reference [43]   This book provides a good overview of the field of vehicle routing problems, with exact methods, heuristics, and case studies.

Reference [45]   This special issue of the *Central European Journal of Operational Research* contains a selection of articles on "rich vehicle routing problems" and introduces this term for the first time.

Reference [46]   This comprehensive survey helps one to make the transition between academic vehicle routing problems and those encountered in industrial applications (large-scale problems in particular).

Reference [60]   This article is of interest because it highlights the main steps and key publications in the research on vehicle routing problems, from the first constructive heuristic published in 1964 until 2009.

Reference [62]   Although it is not dedicated to metaheuristics, this article describes very well various constructive heuristics that are still widely used in commercial software and to initialize metaheuristics. These algorithms are based on simple principles that we recommend one should discover before reading more complex references on metaheuristics.

Reference [98]   This article presents an efficient, simple to understand memetic algorithm for the CVRP, based on chromosomes encoded as giant tours and the Split approach.

Reference [128]   This summary book on routing problems is older than [43] but is still useful and complementary. A recent second edition has been completely rewritten to reflect recent developments [130].

Reference [134]   In the spirit of Cordeau et al. [24], this publication describes a hybrid genetic algorithm able to solve multiple versions of the VRP.

References

1. Albareda-Sambola, M., Fernández, E., Nickel, S.: Multiperiod location-routing with decoupled time scales. European Journal of Operational Research **217**(2), 248–258 (2012)
2. Aras, N., Aksen, D., Tekin, M.T.: Selective multi-depot vehicle routing problem with pricing. Transportation Research, Part C: Emerging Technologies **19**(5), 866–884 (2011)
3. Baker, B., Ayechew, M.: A genetic algorithm for the vehicle routing problem. Computers & Operations Research **30**, 787–800 (2003)
4. Baldacci, R., Christofides, N., Mingozzi, A.: An exact algorithm for the vehicle routing problem based on the set partitioning formulation with additional cuts. Mathematical Programming **115**, 351–385 (2008)
5. Balinski, M., Quandt, R.: On an integer program for a delivery program. Operations Research **12**, 300–304 (1964)
6. Barbarasoglu, G., Ozgur, D.: A tabu search algorithm for the vehicle routing problem. Computers & Operations Research **26**, 255–279 (1999)
7. Beasley, J.: Route-first cluster-second methods for vehicle routing. Omega **11**, 403–408 (1983)
8. Belenguer, J., Benavent, E., Labadi, N., Prins, C., Reghioui, M.: Lower and upper bounds for the split delivery capacitated arc routing problem. Transportation Science **44**(2), 206–220 (2010)
9. Bentley, J.L.: Fast algorithms for geometric tsp. ORSA Journal on Computing **4**, 387–411 (1992)

10. Berger, J., Barkaoui, M.: A new hybrid genetic algorithm for the capacitated vehicle routing problem. Journal of the Operational Research Society **54**, 1254–1262 (2003)
11. Beullens, P., Muyldermans, L., Cattrysse, D., Van Oudheusden, D.: A guided local search heuristic for the capacitated arc routing problem. European Journal of Operational Research **147**, 629–643 (2003)
12. Bortfeldt, A.: A hybrid algorithm for the capacitated vehicle routing problem with three-dimensional loading constraints. Computers & Operations Research **39**(9), 2248–2257 (2012)
13. Boudia, M., Prins, C., Ould-Louly, A.: A memetic algorithm with dynamic population management for an integrated production–distribution problem. European Journal of Operational Research **195**, 703–715 (2009)
14. Brandão, J.: A tabu search algorithm for the heterogeneous fixed fleet vehicle routing problem. Computers & Operations Research **38**(1), 140–151 (2011)
15. Cacchiani, V., Hemmelmayr, V., Tricoire, F.: A set-covering based heuristic algorithm for the periodic vehicle routing problem. Discrete Applied Mathematics (forthcoming) (2013)
16. Chen, A.L., Yang, G.K., Wu, Z.M.: Hybrid discrete particle swarm optimization algorithm for capacitated vehicle routing problem. Journal of Zhejiang University Science A **7**(4), 607–614 (2006)
17. Chen, P., Huang, H., Dong, X.: Iterated variable neighborhood descent algorithm for the capacitated vehicle routing problem. Expert Systems with Applications **37**, 1620–1627 (2010)
18. Chu, F., Labadi, N., Prins, C.: A scatter search for the periodic capacitated arc routing problem. European Journal of Operational Research **169**, 586–605 (2006)
19. Clarke, G., Wright, J.: Scheduling of vehicles from a central depot to a number of delivery points. Operations Research **12**, 568–581 (1964)
20. Contardo, C., Hemmelmayr, V., Crainic, T.G.: Lower and upper bounds for the two-echelon capacitated location-routing problem. Computers & Operations Research **39**(12), 3185–3199 (2012)
21. Corberan, A., Prins, C.: Recent results on arc routing problems: an annotated bibliography. Networks **56**(1), 50–69 (2010)
22. Cordeau, J.F., Gendreau, M., Hertz, A., Laporte, G., Sormany, J.: New heuristics for the vehicle routing problem. In: A. Langevin, D. Riopel (eds.) *Logistics Systems - Design and Optimization*, pp. 279–298. Springer (2005)
23. Cordeau, J.F., Laporte, G.: Tabu search heuristics for the vehicle routing problem. Technical Report G-2002-15, GERAD (2002)
24. Cordeau, J.F., Laporte, G., Mercier, A.: A unified tabu search heuristic for vehicle routing problems with time windows. Journal of the Operational Research Society **52**, 928–936 (2001)
25. Cordeau, J.F., Maischberger, M.: A parallel iterated tabu search heuristic for vehicle routing problems. Computers & Operations Research **39**(9), 2033–2050 (2012)
26. Dantzig, G.B., Ramser, J.H.: The truck dispatching problem. Management Science **6**(1), 80–91 (1959)
27. D'Souza, C., Omkar, S., Senthilnath, J.: Pickup and delivery problem using metaheuristics techniques. Expert Systems with Applications **39**(1), 328–334 (2012)
28. Duhamel, C., Lacomme, P., Prodhon, C.: Efficient frameworks for greedy split and new depth first search split procedures for routing problems. Computers & Operations Research **38**(4), 723–739 (2011)
29. Duhamel, C., Lacomme, P., Prodhon, C.: A hybrid evolutionary local search with depth first search split procedure for the heterogeneous vehicle routing problems. Engineering Applications of Artificial Intelligence **25**(2), 345–358 (2012)
30. Duhamel, C., Lacomme, P., Quilliot, A., Toussaint, H.: A multi-start evolutionary local search for the two-dimensional loading capacitated vehicle routing problem. Computers & Operations Research **38**(3), 617–640 (2011)
31. Eksioglu, B., Vural, A.V., Reisman, A.: The vehicle routing problem: A taxonomic review. Computers & Industrial Engineering **57**(4), 1472–1483 (2009)
32. El Fallahi, A., Prins, C., Wolfler-Calvo, R.: A memetic algorithm and a tabu search for the multi-compartment vehicle routing problem. Computers & Operations Research **35**(5), 1725–1741 (2008)

33. Ergun, Ö., Orlin, J.B., Steele-Feldman, A.: Creating very large scale neighborhoods out of smaller ones by compounding moves. Journal of Heuristics **12**(1-2), 115–140 (2006)
34. Feo, T., Resende, M.: A probabilistic heuristic for a computationally difficult set covering problem. Operations Research Letters **8**, 67–71 (1989)
35. Fisher, M., Jaikumar, R.: A generalized assignment heuristic for vehicle routing. Networks **11**, 109–124 (1981)
36. Fleszar, K., Osman, I., Indi, K.: A variable neighborhood search algorithm for the open vehicle routing problem. European Journal of Operational Research **195**, 803–809 (2009)
37. Funke, B., Grünert, T., Irnich, S.: Local search for vehicle routing and scheduling problems: review and conceptual integration. Journal of Heuristics **11**, 267–306 (2005)
38. Garcia-Najera, A., Bullinaria, J.A.: An improved multi-objective evolutionary algorithm for the vehicle routing problem with time windows. Computers & Operations Research **38**(1), 287–300 (2011)
39. Gendreau, M., Hertz, A., Laporte, G.: New insertion and post-optimization procedures for the traveling salesman problem. Operations Research **40**, 1086–1094 (1992)
40. Gendreau, M., Potvin, J.Y., Bräysy, O., Hasle, G., Løkketangen, A.: Metaheuristics for the vehicle routing problem and its extensions: A categorized bibliography. In: B. Golden, S. Raghavan, E. Wasil (eds.) *The Vehicle Routing Problem: Latest Advances and New Challenges*, pp. 143–170. Springer (2008)
41. Gillett, B., Miller, L.: A heuristic algorithm for the vehicle dispatch problem. Operation Research **22**, 340–349 (1974)
42. Goksal, F.P., Karaoglan, I., Altiparmak, F.: A hybrid discrete particle swarm optimization for vehicle routing problem with simultaneous pickup and delivery. Computers & Industrial Engineering **65**(1), 39–53 (2013)
43. Golden, B., Raghavan, S., Wasil, E. (eds.): *The Vehicle Routing Problem, Latest Advances and New Challenges*. Springer, New York (2008)
44. Groër, C., Golden, B., Wasil, E.: A parallel algorithm for the vehicle routing problem. INFORMS Journal on Computing **23**, 315–330 (2011)
45. Hartl, R.F., Hasle, G., Janssens, G.K.: Special issue on rich vehicle routing problems. Central European Journal of Operations Research **14**(2), 103–104 (2006)
46. Hasle, G., Kloster, O.: Industrial vehicle routing. In: G. Hasle, K.A. Lie, E. Quak (eds.) *Geometric Modelling, Numerical Simulation, and Optimization*, pp. 397–435. Springer, Berlin, Heidelberg (2007)
47. Hemmelmayr, V.C., Cordeau, J.F., Crainic, T.G.: An adaptive large neighborhood search heuristic for two-echelon vehicle routing problems arising in city logistics. Computers & Operations Research **39**(12), 3215–3228 (2012)
48. Hertz, A., Mittaz, M.: A variable neighborhood descent algorithm for the undirected capacitated arc routing problem. Transportation science **35**(4), 425–434 (2001)
49. Ho, S., Gendreau, M.: Path relinking for the vehicle routing problem. Journal of Heuristics **12**, 55–72 (2006)
50. Hong, L.: An improved LNS algorithm for real-time vehicle routing problem with time windows. Computers & Operations Research **39**(2), 151–163 (2012)
51. Irnich, S., Funke, B., Grünert, T.: Sequential search and its application to vehicle-routing problems. Computers & Operations Research **33**, 2405–2429 (2006)
52. Jepsen, M., Spoorendonk, S., Ropke, S.: A branch-and-cut algorithm for the symmetric two-echelon capacitated vehicle routing problem. Transportation Science **47**(1), 23–37 (2013)
53. Jin, J., Crainic, T.G., Løkketangen, A.: A parallel multi-neighborhood cooperative tabu search for capacitated vehicle routing problems. European Journal of Operational Research **222**(3), 441–451 (2012)
54. Kilby, P., Prosser, P., Shaw, P.: Guided local search for the vehicle routing problem. In: S. Voss, S. Martello, I. Osman, C. Roucairol (eds.) *Metaheuristics: Advances and Trends in Local Search Paradigms for Optimization*, pp. 473–486. Kluwer (1999)
55. Kindervater, G.A.P., Savelsbergh, M.W.P.: Vehicle routing: handling edge exchanges. In: E.H.L. Aarts, J.K. Lenstra (eds.) *Local Search in Combinatorial Optimization*, pp. 337–360. John Wiley & Sons (1997)

56. Kuo, Y., Wang, C.C.: A variable neighborhood search for the multi-depot vehicle routing problem with loading cost. Expert Systems with Applications **39**(8), 6949–6954 (2012)
57. Kytöjoki, J., Nuortio, T., Bräysy, O., Gendreau, M.: An efficient variable neighborhood search heuristic for very large scale vehicle routing problems. Computers & Operations Research **34**(9), 2743–2757 (2007)
58. Labadie, N., Mansini, R., Melechovský, J., Wolfler Calvo, R.: The team orienteering problem with time windows: An LP-based granular variable neighborhood search. European Journal of Operational Research **220**(1), 15–27 (2012)
59. Lacomme, P., Prins, C., Ramdane-Chérif, W.: Competitive memetic algorithms for arc routing problems. Annals of Operations Research **131**, 159–185 (2004)
60. Laporte, G.: Fifty years of vehicle routing. Transportation Science **43**(4), 408–416 (2009)
61. Laporte, G., Corberan, A.: *Arc Routing Problems–Methods and Applications. Society for Industrial and Applied Mathematics*, Philadelphia, PA (2014)
62. Laporte, G., Semet, F.: Classical heuristics for the capacitated VRP. In: P. Toth, D. Vigo (eds.) *The Vehicle Routing Problem*, pp. 109–128. Society for Industrial and Applied Mathematics, Philadelphia, PA (2001)
63. Lei, H., Laporte, G., Guo, B.: The capacitated vehicle routing problem with stochastic demands and time windows. Computers & Operations Research **38**(12), 1775–1783 (2011)
64. Lenstra, J.K., Kan, A.: Complexity of vehicle routing and scheduling problems. Networks **11**(2), 221–227 (1981)
65. Leung, S.C., Zhang, Z., Zhang, D., Hua, X., Lim, M.K.: A meta-heuristic algorithm for heterogeneous fleet vehicle routing problems with two-dimensional loading constraints. European Journal of Operational Research **225**(2), 199–210 (2013)
66. Leung, S.C., Zhou, X., Zhang, D., Zheng, J.: Extended guided tabu search and a new packing algorithm for the two-dimensional loading vehicle routing problem. Computers & Operations Research **38**(1), 205–215 (2011)
67. Li, F., Golden, B., Wasil, E.: Very large scale vehicle routing: New problems, algorithms, and results. Computers & Operations Research **32**(5), 1165–1179 (2005)
68. Li, X., Leung, S.C., Tian, P.: A multistart adaptive memory-based tabu search algorithm for the heterogeneous fixed fleet open vehicle routing problem. Expert Systems with Applications **39**(1), 365–374 (2012)
69. Lin, S., Kernighan, B.: An effective heuristic algorithm for the traveling salesman problem. Operations Research **21**, 498–516 (1973)
70. Lin, S.W.: Solving the team orienteering problem using effective multi-start simulated annealing. Applied Soft Computing **13**(2), 1064–1073 (2013)
71. Lin, S.W., Yu, V.F.: A simulated annealing heuristic for the team orienteering problem with time windows. European Journal of Operational Research **217**(1), 94–107 (2012)
72. Lin, S.W., Yu, V.F., Lu, C.C.: A simulated annealing heuristic for the truck and trailer routing problem with time windows. Expert Systems with Applications **38**(12), 15,244–15,252 (2011)
73. Liu, R., Jiang, Z.: The close–Dopen mixed vehicle routing problem. European Journal of Operational Research **220**(2), 349–360 (2012)
74. Liu, S.C., Chen, A.Z.: Variable neighborhood search for the inventory routing and scheduling problem in a supply chain. Expert Systems with Applications **39**(4), 4149–4159 (2012)
75. Lourenço, H.R., Martin, O.C., Stützle, T.: Iterated local search. In: M. Gendreau, J. Potvin (eds.) *Handbook of metaheuristics*, 2nd edn, International Series in Operations Research and Management Science, Vol. 146, pp. 363–397. Springer (2010)
76. Marinakis, Y.: Multiple phase neighborhood search-GRASP for the capacitated vehicle routing problem. Expert Systems with Applications **39**(8), 6807–6815 (2012)
77. Marinakis, Y., Iordanidou, G.R., Marinaki, M.: Particle swarm optimization for the vehicle routing problem with stochastic demands. Applied Soft Computing **13**(4), 1693–1704 (2013)
78. Marinakis, Y., Marinaki, M.: A hybrid multi-swarm particle swarm optimization algorithm for the probabilistic traveling salesman problem. Computers & Operations Research **37**(3), 432–442 (2010)

79. Mester, D., Bräysy, O.: Active guided evolution strategies for large scale vehicle routing problems with time windows. Computers & Operations Research **32**, 1593–1614 (2005)
80. Mester, D., Bräysy, O.: Active-guided evolution strategies for large-scale capacitated vehicle routing problems. Computers & Operations Research **34**(10), 2964–2975 (2007)
81. Moghaddam, B.F., Ruiz, R., Sadjadi, S.J.: Vehicle routing problem with uncertain demands: An advanced particle swarm algorithm. Computers & Industrial Engineering **62**(1), 306–317 (2012)
82. Mole, R.H., Jameson, S.R.: A sequential route-building algorithm employing a generalized savings criterion. Operational Research Quarterly **27**, 503–511 (1976)
83. Muyldermans, L.: Routing, districting and location for arc traversal problems. Ph.D. dissertation, Catholic University of Leuven, Belgium (2003)
84. Nagata, Y., Bräysy, O.: Edge assembly-based memetic algorithm for the capacitated vehicle routing problem. Networks **54**, 205–215 (2009)
85. Naji-Azimi, Z., Salari, M.: A complementary tool to enhance the effectiveness of existing methods for heterogeneous fixed fleet vehicle routing problem. Applied Mathematical Modelling **37**(6), 4316–4324 (2013)
86. Nazif, H., Lee, L.S.: Optimised crossover genetic algorithm for capacitated vehicle routing problem. Applied Mathematical Modelling **36**(5), 2110–2117 (2012)
87. Ngueveu, S., Prins, C., Wolfler Calvo, R.: An effective memetic algorithm for the cumulative capacitated vehicle routing problem. Computers & Operations Research **37**, 1877–1885 (2010)
88. Nguyen, V.P., Prins, C., Prodhon, C.: Solving the two-echelon location routing problem by a GRASP reinforced by a learning process and path relinking. European Journal of Operational Research **216**(1), 113–126 (2012)
89. Or, I.: Traveling salesman-type combinatorial optimization problems and their relation to the logistics of regional blood banking. Ph.D. dissertation, Northwestern University, Evanston, IL (1976)
90. Osman, I.: Metastrategy simulated annealing and tabu search algorithms for the vehicle routing problem. Annals of Operations Research **41**, 421–451 (1993)
91. Parragh, S.N., Schmid, V.: Hybrid column generation and large neighborhood search for the dial-a-ride problem. Computers & Operations Research **40**(1), 490–497 (2013)
92. Partyka, J., Hall, R.: On the road to connectivity. OR/MS Today **37**(1), 42–49 (2010)
93. Pecin, D., Poggi, M., Pessoa, A., Uchoa, E.: Improved branch-and-cut-and-price for capacitated vehicle routing. Operations Research (forthcoming)
94. Pisinger, D., Röpke, S.: A general heuristic for vehicle routing problems. Computers & Operations Research **34**, 2403–2435 (2007)
95. Polacek, M., Doerner, K., Hartl, R., Maniezzo, V.: A variable neighborhood search for the capacitated arc routing problem with intermediate facilities. Journal of Heuristics **14**(5), 405–423 (2008)
96. Popović, D., Vidović, M., Radivojević, G.: Variable neighborhood search heuristic for the inventory routing problem in fuel delivery. Expert Systems with Applications **39**(18), 13,390–13,398 (2012)
97. Potvin, J.Y., Bengio, S.: The vehicle routing problem with time windows. Part II: Genetic search. INFORMS Journal on Computing **8**, 165–172 (1996)
98. Prins, C.: A simple and effective evolutionary algorithm for the vehicle routing problem. Computers & Operations Research **31**, 1985–2002 (2004)
99. Prins, C.: A GRASP × evolutionary local search hybrid for the vehicle routing problem. In: F. Pereira, J. Tavares (eds.) *Bio-Inspired Algorithms for the Vehicle Routing Problem.* Studies in Computational Intelligence, vol. 161, pp. 35–53. Springer, Berlin, Heidelberg (2009)
100. Prins, C.: Two memetic algorithms for heterogeneous fleet vehicle routing problems. Engineering Applications of Artificial Intelligence **22**, 916–928 (2009)
101. Prins, C., Labadie, N., Reghioui, M.: Tour splitting algorithms for vehicle routing problems. International Journal of Production Research **47**, 507–535 (2009)

102. Prins, C., Lacomme, P., Prodhon, C.: Order-first split-second methods for vehicle routing problems: A review. Transportation Research, Part C **40**, 179–200 (2014)
103. Prins, C., Prodhon, C., Ruiz, A., Soriano, P., Wolfler Calvo, R.: Solving the capacitated location-routing problem by a cooperative Lagrangean relaxation–granular tabu search heuristic. Transportation Science **41**(4), 470–483 (2007)
104. Prins, C., Prodhon, C., Wolfler Calvo, R.: A memetic algorithm with population management (*MA | PM*) for the capacitated location-routing problem. In: J. Gottlicd, G.R. Raidl (eds.) *Evolutionary Computation in Combinatorial Optimization*. Lecture Notes in Computer Science, vil. 3906, pp. 183–194. Springer (2006)
105. Prins, C., Prodhon, C., Wolfler Calvo, R.: Solving the capacitated location-routing problem by a GRASP complemented by a learning process and a path relinking. 4OR **4**(3), 221–238 (2006)
106. Prins, C., Sevaux, M., Sörensen, K.: A genetic algorithm with population management (*GA | PM*) for the CARP. In: *Tristan V (5th Triennal Symposium on Transportation Analysis)*. Le Gosier, Guadeloupe (2004)
107. Prodhon, C.: A hybrid evolutionary algorithm for the periodic location-routing problem. European Journal of Operational Research **210**(2), 204–212 (2011)
108. Prodhon, C., Prins, C.: A memetic algorithm with population management (*MA | PM*) for the periodic location-routing problem. In: C. Blum, M.J.B. Aguilera, A. Roli, M. Sampels (eds.) Hybrid Metaheuristics. Studies in Computational Intelligence, vol. 114, pp. 43–57. Springer (2008)
109. Qu, Y., Bard, J.F.: A GRASP with adaptive large neighborhood search for pickup and delivery problems with transshipment. Computers & Operations Research **39**(10), 2439–2456 (2012)
110. Rego, C.: A subpath ejection method for the vehicle routing problem. Management Science **44**(10), 1447–1459 (1998)
111. Rego, C., Roucairol, C.: A parallel tabu search algorithm using ejection chains for the vehicle routing problem. In: I.H. Osman, J.P. Kelly (eds.) *Meta-Heuristics: Theory and Applications*, pp. 661–675. Kluwer, Boston (1996)
112. Reimann, M., Doerner, K., Hartl, R.: D-ants: Savings based ants divide and conquer the vehicle routing problem. Computers & Operations Research **31**(4), 563–591 (2004)
113. Rochat, Y., Taillard, E.: Probabilistic diversification and intensification in local search for vehicle routing. Journal of Heuristics **1**(1), 147–167 (1995)
114. Ruan, Q., Zhang, Z., Miao, L., Shen, H.: A hybrid approach for the vehicle routing problem with three-dimensional loading constraints. Computers & Operations Research **40**(6), 1579–1589 (2013)
115. Russell, R., Chiang, W.: Scatter search for the vehicle routing problem with time windows. European Journal of Operational Research **169**, 606–622 (2006)
116. Sahin, M., Cavuslar, G., Oncan, T., Sahin, G., Aksu, D.: An efficient heuristic for the multi-vehicle one-to-one pickup and delivery problem with split loads. Transportation Research Part C: Emerging Technologies **27**, 169–188 (2013)
117. Santos, L., Coutinho-Rodrigues, J., Current, J.: An improved ant colony optimisation based algorithm for the capacitated arc routing problem. Transportation Research Part B: Methodological **44**(2), 246–266 (2010)
118. Schilde, M., Doerner, K., Hartl, R.: Metaheuristics for the dynamic stochastic dial-a-ride problem with expected return transports. Computers & Operations Research **38**(12), 1719–1730 (2011)
119. Schrimpf, G., Schneider, J., Stamm-Wilbrandt, H., Dueck, G.: Record breaking optimization results using the ruin and recreate principle. Journal of Computational Physics **159**(2), 139–171 (2000)
120. Silva, M., Subramanian, A., Vidal, T., Ochi, L.: A simple and effective metaheuristic for the minimum latency problem. European Journal of Operational Research **221**, 513–520 (2012)
121. Souffriau, W., Vansteenwegen, P., Vanden Berghe, G., Van Oudheusden, D.: A path relinking approach for the team orienteering problem. Computers & Operations Research **37**(11), 1853–1859 (2010)

122. Subramanian, A., Penna, P.H.V., Uchoa, E., Ochi, L.S.: A hybrid algorithm for the heterogeneous fleet vehicle routing problem. European Journal of Operational Research **221**(2), 285–295 (2012)
123. Tarantilis, C.: Solving the vehicle routing problem with adaptive memory programming methodology. Computers & Operations Research **32**(9), 2309–2327 (2005)
124. Tarantilis, C., Kiranoudis, C.: Bone route: An adaptive memory-based method for effective fleet management. Annals of Operations Research **115**, 227–241 (2002)
125. Thangiah, S.: Vehicle routing with time windows using genetic algorithms. In: L. Chambers (ed.) *Application Handbook of Genetic Algorithms: New Frontiers*, pp. 253–277. CRC Press (1995)
126. Thompson, P., Psaraftis, H.: Cyclic transfer algorithms for multivehicle routing and scheduling problems. Operations Research **41**, 935–946 (1993)
127. Ting, C.J., Chen, C.H.: A multiple ant colony optimization algorithm for the capacitated location routing problem. International Journal of Production Economics **141**(1), 34–44 (2013)
128. Toth, P., Vigo, D. (eds.): The *Vehicle Routing Problem*. Society for Industrial and Applied Mathematics, Philadelphia, PA (2001)
129. Toth, P., Vigo, D.: The granular tabu search (and its application to the vehicle routing problems). INFORMS Journal on Computing **15**(4), 333–346 (2003)
130. Toth, P., Vigo, D.: *Vehicle Routing: Problems, Methods and Applications*, 2nd edn. SIAM, Philadelphia (2014)
131. Ursani, Z., Essam, D., Cornforth, D., Stocker, R.: Localized genetic algorithm for vehicle routing problem with time windows. Applied Soft Computing **11**(8), 5375–5390 (2011)
132. Usberti, F., França, P., França, A.: GRASP with evolutionary path-relinking for the capacitated arc routing problem. Computers & Operations Research (forthcoming) (2011)
133. Vidal, T., Crainic, T.G., Gendreau, M., Lahrichi, N., Rei, W.: A hybrid genetic algorithm for multidepot and periodic vehicle routing problems. Operations Research **60**(3), 611–624 (2012)
134. Vidal, T., Crainic, T.G., Gendreau, M., Prins, C.: A unified solution framework for multiattribute vehicle routing problems. Technical Report of 2013-22, CIRRELT, Montréal, Canada (2012)
135. Vidal, T., Crainic, T.G., Gendreau, M., Prins, C.: Timing problems and algorithms: time decisions for sequences of activities. Networks **65**(2), 102–128 (2015)
136. Vidal, T., Crainic, T.G., Gendreau, M., Prins, C.: A hybrid genetic algorithm with adaptive diversity management for a large class of vehicle routing problems with time-windows. Computers & Operations Research **40**(1), 475–489 (2013)
137. Villegas, J.G., Prins, C., Prodhon, C., Medaglia, A.L., Velasco, N.: A GRASP with evolutionary path relinking for the truck and trailer routing problem. Computers & Operations Research **38**(9), 1319–1334 (2011)
138. Wolf, S., Merz, P.: Evolutionary local search for the super-peer selection problem and the p-hub median problem. In: T. Bartz-Beielstein, M. Blesa Aguilera, C. Blum, B. Naujoks, A. Roli, G. Rudolph, M. Sampels (eds.) *Hybrid Metaheuristics*, Lecture Notes in Computer Science, vol. 4771, pp. 1–15. Springer (2007)
139. Yu, B., Yang, Z.Z.: An ant colony optimization model: The period vehicle routing problem with time windows. Transportation Research Part E: Logistics and Transportation Review **47**(2), 166–181 (2011)
140. Zachariadis, E.E., Kiranoudis, C.T.: A local search metaheuristic algorithm for the vehicle routing problem with simultaneous pick-ups and deliveries. Expert Systems with Applications **38**(3), 2717–2726 (2011)
141. Zhang, T., Chaovalitwongse, W., Zhang, Y.: Scatter search for the stochastic travel-time vehicle routing problem with simultaneous pick-ups and deliveries. Computers & Operations Research **39**(10), 2277–2290 (2012)

Chapter 16
Applications to Air Traffic Management

Nicolas Durand, David Gianazza, Jean-Baptiste Gotteland,
Charlie Vanaret and Jean-Marc Alliot

16.1 Introduction

Air traffic management (ATM) is an endless source of challenging optimization problems. Before discussing applications of metaheuristics to these problems, let us describe an ATM system in a few words, so that readers who are not familiar with such systems can understand the problems being addressed in this chapter. Between the moment passengers board an aircraft and the moment they arrive at their destination, a flight goes through several phases: pushback at the gate, taxiing between the gate and the runway threshold, takeoff and initial climb following a Standard instrument departure (SID) procedure, cruise, final descent following a standard terminal arrival route (STAR), landing on the runway, and taxiing to the gate. During each phase, the flight is handled by several air traffic control organizations: airport ground control, approach and terminal control, and en-route control. These control organizations provide services that ensure safe and efficient conduct of flights, from departure to arrival.

These services are provided by human operators. In order to share the tasks among several operators, the airspace is divided into several airspace sectors, each monitored by one or two air traffic controllers. Within this sectorized airspace, aircraft fly on a network of predefined routes, occasionally deviating from their route when instructed

N. Durand (✉) · D. Gianazza · J.-B. Gotteland · C. Vanaret
Laboratoire MAIAA (Ecole Nationale de l'Aviation Civile), Toulouse, France
e-mail: durand@recherche.enac.fr

D. Gianazza
e-mail: gianazza@recherche.enac.fr

J.-B. Gotteland
e-mail: gottelan@recherche.enac.fr

J.-M. Alliot
Institut de Recherche en Informatique de Toulouse, Toulouse, France
e-mail: alliot@irit.fr

© Springer International Publishing Switzerland 2016   439
P. Siarry (ed.), *Metaheuristics*, DOI 10.1007/978-3-319-45403-0_16

so by a controller, in order to avoid collisions with other aircraft. The design of the route network and sector boundaries must satisfy contradicting objectives: each flight should follow the most direct route from origin to destination; however, the overall traffic must be organized so as to be manageable by human controllers. The latter objective requires, for example, that each sector contains only a relatively small number of crossing points between routes, so that controllers can have a clear mental picture of the traffic of which they are in charge. In addition, there should be enough room around each crossing point to allow for lateral maneuvering of conflicting aircraft. As we shall see in this chapter, airspace sectorization and route network design are themselves challenging optimization problems that can be formulated and addressed in many different ways.

The smooth and efficient operation of an ATM system relies on an efficient organization of the system that is subject to many constraints. Large portions of airspace are the responsibility of various managerial units (air traffic control centers). The air traffic controllers working in these centers are trained and qualified to work in specific geographic areas. Consequently, airspace sectors are grouped by qualification zones, also called *functional airspace blocks* in this chapter. The staff operating a given functional airspace block follow a duty roster where several teams relays with each other to provide air navigation services to airspace users 24 hours a day, 7 days a week, all year long. In a control room, a controllers' working position can handle one or several airspace sectors belonging to the same functional airspace block. Several questions arise concerning the optimization of the ATM system regarding these organizational and operational issues. At the strategic level, how can the functional airspace blocks be optimized in order to balance the workload and minimize coordination? In daily operations, how does one allocate sectors to controllers' working positions in order to optimally balance the workload among the open working positions, while preventing overloads?

It is not always possible to avoid overloads only by reassigning airspace sectors to different working positions. One must sometimes enforce traffic regulation measures, for example, by rerouting some flights, or by allocating takeoff slots to departing flights. These measures smooth the traffic demand, so that it does not exceed the capacity of the ATM system to handle this traffic. The resulting slot allocation and flight rerouting problems are challenging constrained optimization problems of large size that must be addressed on a continental scale.

The core of the air traffic controllers' activity is to facilitate the flow of traffic through the sectors that they are responsible for, while avoiding collisions between aircraft. To satisfy this essential safety constraint, they must resolve conflicts between trajectories. Such conflicts may occur at any time during a flight, during taxiing, takeoff, climb, cruise, descent, or landing. The underlying constrained optimization problem is to minimize the deviations from the nominal trajectories while maintaining horizontal and vertical separations between conflicting aircraft. Conflicts related to runway occupancy can only be resolved by optimizing the landing and takeoff sequences. When aircraft are taxiing, conflict resolution can be achieved by choosing different paths or by making aircraft wait on some taxiways. An additional constraint may then occur: some flights must respect their takeoff slots. For airborne aircraft, the air traffic controller can order pilots to make different types of maneuvers: horizontal deviations, vertical maneuvers, a modified rate of climb or descent, or speed adjustments.

In this quick description of an ATM system, several optimization problems have been introduced. These problems are complex and not always easy to formulate explicitly for the actors, in the system, for several reasons. First, all these problems are related to one another, and ideally they should all be answered at once. For example, one can avoid airspace congestion by smoothing the traffic (e.g., by delaying departing flights), but one can also address this by dynamically reassigning airspace sectors to working positions or by addressing both problems simultaneously. This gives us three different formulations for the same general problem (airspace congestion). Second, ATM relies on complex systems involving many actors from different domains, operating with different temporal horizons. Airlines, air navigation service providers, and airports conduct different activities in the short, medium, and long term. Finally, these activities are subject to many uncertainties: predicting an aircraft trajectory is difficult because of errors generated by uncertainties in the weather, the pilot's intentions, and aircraft parameters. Before departure, missing luggage or passengers can generate unexpected delays in takeoffs. Dealing with uncertainties requires complex models that must be robust and reactive.

Modeling ATM problems is a difficult task in this context: if the model is too simple, it cannot handle realistic hypotheses; if it is too complex, it becomes impossible to optimize. Furthermore, when problems are correctly modeled, they are often hard to solve by exact methods, because of their huge size.

For all these reasons, metaheuristics are generally good candidates for answering many ATM optimization problems. We will see with some examples that they can sometimes be less efficient than exact methods, and with some other examples that they are the best methods known.

In this chapter, we present several examples of applications grouped by theme: route network optimization, airspace optimization, takeoff slot allocation, airport traffic optimization, and en-route conflict resolution. For each example, we give details of the model chosen, explain the complexity of the problem, describe the metaheuristics used, and present alternative methods when they exist.

16.2 Air Route Network Optimization

The air route network, as it exists today, is the result of successive modifications that have been made over time, taking into account some geographic and technical constraints. In the recent past, every air route was defined as a sequence of segments starting and ending at *waypoints*, which had to be located at the geographic coordinates of ground-based radionavigation aids. This is not the case anymore, as modern navigation systems can handle waypoints located almost anywhere. However, there remain other constraints on the positioning of the nodes of the network.[1] Typically, the crossing point of two intersecting routes should not be too close to a sector boundary, so that there is enough room for lateral maneuvers in the vicinity of this crossing point.

The continuous increase in overall traffic since the beginning of commercial aviation has often led people to rethink and redesign the route network, on a local or global scale, and even to propose new concepts for the operation of air routes and airspace sectors. An example of such a new concept, which has been proposed several times for air traffic but has not been implemented yet, is to define 3D routes dedicated to the most important traffic flows. This concept is similar to that of highways for ground traffic that accommodate flows of car traffic between large cities.

Optimizing the air route network is a problem that can be formulated and addressed in several ways. Let us list a few of them:

- *Node and edge positioning.* The route network can be seen as a planar graph in dimension 2 in which the edges must not cross.
- *Node positioning only.* Starting from an initial network (e.g., a regular grid), one can move the nodes in order to optimize a given criterion related to the routing of the traffic flow, while maintaining the planar property of the graph.
- *Optimal positioning of 2D routes* for the largest traffic flows.
- In dimension 3, *optimal placement of separated "3D tubes"* for the largest origin–destination flows.

16.2.1 Optimal Positioning of Nodes and Edges Using Geometric Algorithms

In current operations, air traffic controllers resolve conflicts occurring within the airspace volumes (sectors) of which they are in charge. The airways followed by aircraft must take this sectorization constraint into account: crossing points should not be near sector boundaries, and there must be enough space around each crossing point to allow for lateral maneuvers. In addition, the network must be designed so as to minimize trajectory lengthening when compared with the direct routes. Ideally, large traffic flows should be deviated less from their direct routes than small flows.

The horizontal projection of an air route network can be seen as a planar graph whose nodes are the intersections between the routes, and whose edges are route segments between crossing points. The objective, when building such a network, is to position the nodes and edges so as to satisfy a constraint on the distances between nodes while minimizing the trajectory lengthenings for aircraft flying on the network.

The method to address this problem that we are now going to present is not a metaheuristic. It consists in applying first a clustering method to the crossing points

---

[1] The waypoints are considered here as the nodes of the air route network. Note that a dual representation, where route segments are nodes and waypoints are edges, is also possible.

between direct routes, and then a geometric triangulation algorithm to build route segments joining the barycenters of the clusters. This method was introduced by Mehadhebi [72] (see also [42], (in French) for the application of a similar method). It is not aimed at finding a global optimum for the positioning problem. However, the method can build a network satisfying the node separation constraint, and the solutions are of good quality, by construction, because the method is applied to an initial situation where the routes are direct, from origin to destination. As such, this method could be used as a baseline in future work for trying to apply metaheuristics to the node and edge positioning problem. This is why it is worth mentioning here.

The aim of the clustering method is to position the nodes of the network taking account of the traffic demand, so that they satisfy a minimum separation distance between nodes. For this purpose, the crossing points between direct routes are first computed, using for example a sweep line geometric algorithm. Then, the crossing points are clustered according to proximity criteria, so that the barycenters of the clusters are at least a distance d_1 apart, and the points that are closer to a barycenter than a distance d_2 belong to the corresponding cluster. Typically, a variant of the k-means method can be used to compute the clusters. In computing the barycenter, weights related to the traffic flows passing through the crossing points can be used. Such a weighting of the crossing points avoids moving crossing points with heavy traffic too much. Figures 16.1 and 16.2 illustrate this clustering process applied to French airspace.

Once the network nodes have been computed, the edges are positioned so that they do not cross (otherwise the graph would not be planar), using a geometric triangulation method. Figures 16.3 and 16.4 show the results obtained by applying

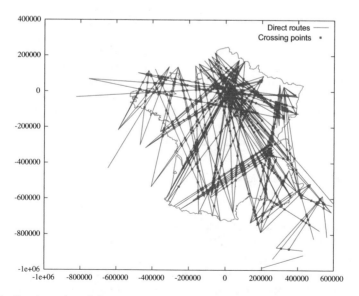

Fig. 16.1 Crossing points of direct routes with traffic flows above 10 flights per day

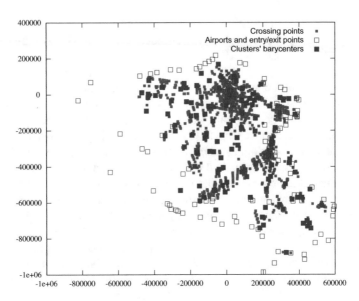

Fig. 16.2 Clustering process applied to crossing points

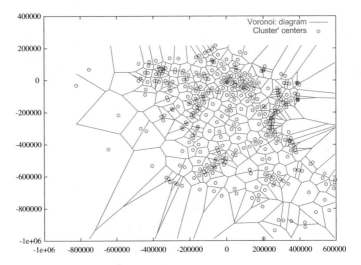

Fig. 16.3 Voronoi diagram of the barycenters of the clusters

the S. Fortune algorithm [39] to the barycenters of the clusters of crossing points. This algorithm computes both a Delaunay triangulation of the set of points and its dual graph, a Voronoi diagram.

Each polygonal cell of the Voronoi diagram is such that the points inside that cell are closer to the barycenter of the cell (i.e., a network node) than to any other barycenter. This interesting side effect of this geometric method allows us to associate

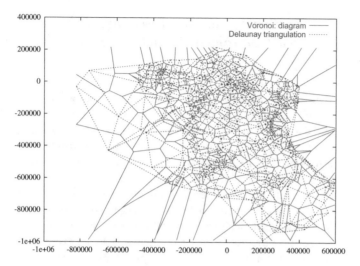

Fig. 16.4 Delaunay triangulation of the barycenters of the clusters

a cell of airspace with each node of the network. The area of this cell gives an indication of how much room is available in the vicinity of the node for the lateral maneuvers of conflicting aircraft.

In [72], Mehadhebi used the areas of the cells to measure the density of conflicts when building a network, in order to avoid excessive densities in a given airspace. For each crossing point, the density was obtained by computing the ratio of a number quantifying the conflicts[2] at that crossing point and the area of the Voronoi cell associated with the crossing point. In a dense area, moving the crossing points further apart has the effect of increasing the cell areas, thus decreasing the density. The optimization method used by Mehadhebi was not described in detail in [72], but it seems to be an iterative method that locally smooths the density in congested areas.

Once the full network (nodes and edges) has been defined, the flights have to choose a path in this network, from the departure airport to the destination airport. These paths must take into account a constraint on the angle between successive route segments: for any route to be actually flown by an aircraft, the angle between successive segments must not be too acute. This constraint was handled differently in [72], where it was satisfied as best as possible in the clustering phase, and in [42], where it was examined afterwards, when searching for the shortest path in the network for each flight.

---

[2]This quantification of conflicts can be done, for example, using the number of conflicts at the crossing point weighted by the difficulty of each conflict.

16.2.2   Node Positioning with Fixed Topology, Using a Simulated Annealing or Particle Swarm Optimization Algorithm

In [81], Riviere focused on a different problem, where the network topology has already been fixed, and where only the node positioning problem is addressed. Starting from an initial regular grid over the European airspace, he used simulated annealing [65] to modify this grid, minimizing the sum of trajectory lengthenings between origin and destination (Fig. 16.5). This optimization process takes account of a minimum distance that must be maintained between crossing points.

The evaluation of the trajectory-lengthening criterion requires the computation of the shortest paths in the network between all origin–destination pairs. This was done using the Floyd–Warshall algorithm, taking account of a constraint on the angle between two successive route segments: this angle should not exceed 90°.

As the objective function being minimized requires the computation of the shortest paths in the network, the gradient of the objective function cannot be computed and gradient descent methods cannot be used. One must instead use derivative-free methods, and metaheuristics such as the simulated annealing method used in [81] or the particle swarm optimization method used in [16] (which will be described later on) are a good option.

Starting from an initial point, the simulated annealing algorithm explores the search space by randomly choosing another point in the neighborhood of the current point. The move is accepted if the new point improves the objective function. It can also be accepted if it does not, with a probability that decreases with the number of iterations (according to the annealing scheme). In the route network design problem, a point in the search space is a route network, and a local move in the neighborhood of the current point is a random change in this network.

In more recent work [16], Cai et al. used an approach similar to that of Riviere [81], but for the Chinese airspace and with a formulation as a multiobjective optimization

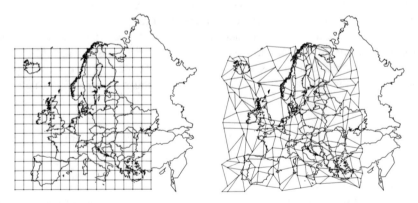

Fig. 16.5 Air route network found by simulated annealing (*right*), starting from an initial regular grid (*left*)

problem. Two criteria were minimized in this work. The first was related to the trajectory lengthenings, as in [81]. The second one, taken from [83], was the sum over all crossing points of the average number of potential conflicts per unit of time.

The metaheuristic used in [16] was a hybrid method combining a variant of particle swarm optimization (CLPSO; "Comprehensive Learning Particle Swarm Optimization," introduced in [69]) and an ad hoc method relying on local moves of the crossing points to improve the optimized criteria.

In its canonical version, the particle swarm optimization algorithm iteratively moves a population of particles, characterized by their positions and velocities, in the search space, memorizing the best positions found by each particle. Each particle is moved in the direction of its velocity vector. After each move, the speed vector is updated, combining several directions, namely, the current velocity vector (i.e., the inertia of the particle), the direction to the best position found by the particle, and the direction to the best position found by the whole swarm (or a subset of the population). The CLPSO variant uses all the best positions found by the particles to update the velocity vector, in order to avoid premature convergence toward a local minima.

The hybrid algorithm proposed in [16] is similar to CLPSO, except that a local optimization is performed after updating the particles' positions and velocities. For each particle (i.e., an air route network), the local optimization tries to move each node so as to improve the chosen criteria, considering the relative positions of the nodes and the traffic flows on the edges connected to each node.

Cai et al. compared their hybrid method with the simulated annealing proposed by Riviere [81], applied to the Chinese airspace. The simulated annealing approach minimized only one of the two criteria chosen by the authors, so the comparison of the Pareto fronts was naturally to the advantage of the multi-objective particle swarm optimization algorithm.

The results were also compared with the current route network in China, showing significant improvements. The method proposed by Cai et al. is being integrated into a program used to modify the air route network in China.

16.2.3 Defining 2D Corridors with a Clustering Method and a Genetic Algorithm

Xue and Kopardekar [91] proposed a method for positioning a limited number of 2D routes (or "corridors") to accommodate the largest flows over the territory of the United States. The aim was not to build a network for all of the traffic, but only for the large flows. How these corridors would be handled, concerning for example the entry and exit procedures and how to resolve conflicts at the crossing points of these corridors, was not detailed in the publication. The work focused on how to position these corridors, considering proximity criteria for the origin–destination flows.

There are many ways to specify an air traffic flow, for example by choosing an origin and a destination, or by considering the flow through a given sector, through a specific airspace sector boundary, or over a waypoint, etc. In their publication, Xue and Kopardekar considered aircraft trajectories as great circles on the Earth's surface, and a flow was defined as a group of such great circles that are close to one another.

To cluster these great circles according to a proximity criterion, Xue and Kopardekar transformed the direct trajectories from departure to arrival into a set of points in a dual space, using a Hough transform. In this dual space, each trajectory was represented by a pair (ρ, θ), where ρ is the shortest distance between the trajectory and a reference point, and θ is the angle between a reference direction and the line perpendicular to the trajectory passing through the reference point. Xue and Kopardekar then used a basic clustering technique, of a kind usually applied in image processing, to aggregate the trajectories. By placing a grid with a step size $(\Delta\rho, \Delta\theta)$ over the set of dual points, they simply counted the number of points in each cell and determined the cells of highest density.

This method allowed them to find groups of trajectories that were geographically close to one another. In the dual representation of the largest flows, the points in the cells with the highest densities were replaced by a single corridor (a point in the dual space). As a first approximation, they took the barycenter of the points (trajectories in the initial space).

One drawback of this representation in the dual space is that the arrival and departure points in the original space are lost in the transformation. One cannot directly measure the trajectory lengthening in the dual space for aircraft flying in the corridors computed by this method. The additional distance flown by the aircraft is a very important cost criterion for airline operators.

A genetic algorithm [54, 74] was then used to refine the approximate solution found by the above method. This algorithm iterated on a population of individuals, following a Darwinian process of selection (according to a fitness criterion), crossover, and mutation. An individual here was a set of barycenters (representing corridors in the initial space). This was encoded as a collection of coordinates (ρ, θ) in the dual space. The initial population was built from the approximate solution found by the first method. The fitness criterion was the sum of the trajectory lengthenings in the initial space, for all flights flying in the corridors.

With 200 elements in the population, 200 generations, a crossover probability of 0.8, and a mutation probability of 0.2, the proportion of flights flying in the corridors with no more than a 5 % trajectory lengthening increased from 31 % for the initial solution to 44 % for the best solution found by the genetic algorithm.

16.2.4 Building Separate 3D Tubes Using an Evolutionary Algorithm and an A* Algorithm

In the studies we have presented so far on the optimization of the air route network, there was no attempt to avoid intersecting routes (or corridors) while building the network. Crossing points were actually part of the planar graph representation that was used in the geometric methods [42, 72], where defining the nodes and edges of such a planar graph was the objective, as well as in the metaheuristic approaches [16, 81], where the aim was to position the network nodes, starting from an initial regular grid. The optimal positioning of 2D corridors in [91] also allowed corridors to intersect.

For the network structure actually to be beneficial in decreasing the number of conflicts between aircraft flying through it, one must introduce a vertical segregation of traffic flows. This vertical segregation can be introduced locally at the crossing points, or by considering origin–destination flows, or for each flight, depending on the direction of the route it follows. Graph coloration methods, which will not be described here, can be used to assign different flight levels to crossing flows [8, 67]. However, such methods only consider cruising flights. Descending or climbing aircraft are not taken into account.

Another approach, proposed by Gianazza et al. [42, 43, 49–51] is to build separate 3D tubes for the largest origin–destination flows. A 3D tube, as illustrated in Fig. 16.6, is a volume computed from the envelope of the minimum and maximum climb or descent profiles of all aircraft flying in the tube. Vertical and horizontal distance buffers are added to this envelope to take account of the standard vertical and horizontal separations.

The idea is that aircraft flying in such 3D tubes would be sequenced at the departure point and would ensure self-separation from other aircraft in the same tube. They would have priority over the rest of the traffic. The 3D tubes would be built so as not to intersect, thus ensuring there would be no conflicts between flights in the main traffic flows.

The aim is to assign one 3D tube to each flow of sufficient importance. A flow is defined here by two points (origin, destination) and a cruising flight level (the requested flight level, denoted by RFL). A variant of the k-means method is used to cluster the flights into origin–destination–RFL flows. As a consequence, there might be several flows for a given origin–destination pair, corresponding to several cruising flight levels.

The 3D tubes must be as short as possible. The tubes assigned to different origin–destination pairs must not intersect. For tubes having the same origin and destination with different cruising flight levels, the initial climb and final descent are common to them (and considered as the same tube for these phases). The possible lateral and vertical deviations of a 3D tube, which might be introduced to avoid other 3D tubes, are shown in Fig. 16.7. A 3D tube associated with an origin–destination–RFL flow

Fig. 16.6 Example of a 3D tube, with only one cruising flight level

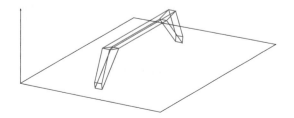

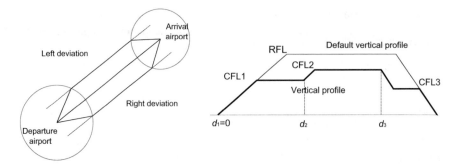

Fig. 16.7 Possible lateral and vertical deviations

is completely defined by a discrete choice from among several options for the 2D route, and by a sequence of pairs (d_k, CFL_k), where d_k is the distance along the route at which a vertical deviation toward the flight level CFL_k begins. (CFL stands for "cleared flight level.")

This constrained optimization problem is highly combinatorial. To solve it, Gianazza et al. used an evolutionary algorithm hybridized with an A^* algorithm. The evolutionary algorithm iterates on a population of elements where each individual is a full network of 3D tubes, applying selection, crossover, and mutation operators to this population of networks. The fitness of an individual is assessed by computing a triangular matrix C, where the diagonal elements i are the costs of the deviations of 3D tube number i from the most direct trajectory (the direct route between origin and destination, at requested flight level). These diagonal elements are the costs to be minimized. The nondiagonal elements contain the constraint violations. An element (i, j) of the matrix C, with $i < j$, contains the number of intersections of 3D tubes i and j. Denoting the number of constraint violations for tube i, by $f(i)$, the fitness criterion $\mathcal{F}$ is expressed as follows:

$$\mathcal{F} = \begin{cases} 1 + \frac{n}{1+\sum_i C_{ii}} & \text{if } \sum_i f(i) = 0 \\ \frac{1}{\sum_i f(i)} & \text{if } \sum_i f(i) > 0 \end{cases}$$

The fitness criterion to be maximized by the evolutionary algorithm is less than 1 when intersections of 3D tubes remain, and greater than 1 when all 3D tubes are separated. In the latter case, the fitness increases when the lateral or vertical deviation decreases. This raw fitness is scaled, using a sigma truncation scaling. A clusterized sharing operator is then applied, which modifies the fitness landscape so as to avoid premature convergence toward local optima. An elitist strategy is employed, preserving the best element of each cluster in the population when its fitness is close enough to the fitness of the best element. Apart from the best elements, the pool of parents is selected using the principle of stochastic remainder without replacement. The crossover and mutation operators are applied according to chosen probabilities.

The crossover operator is similar to the one proposed by Durand et al. [31, 35]. This operator is specifically designed for partially separable objective functions. It remains efficient when large problems are being addressed, as shown in [28]. This specific crossover operator requires one to define a *local fitness* for each gene (here a 3D tube) of each individual (here a full network) in the population. The local fitness chosen here is $f_k = -f(k)$, the negative of the number of constraint violations for flow k. The crossover itself is either a standard barycentric crossover (with probability $\frac{2}{3}$ in [43]) or a deterministic crossover (with probability $\frac{1}{3}$). In the deterministic crossover, the first descendant inherits gene k of parent p_1 and the second inherits gene k of parent p_2 when $f_k(p_1) = f_k(p_2)$. When the local fitnesses differ, both descendants inherit the best gene.

The mutation is where the hybridization with the A^* algorithm takes place. A gene (3D tube) is selected for mutation, preferentially, one picks a tube with a bad local fitness if $\mathcal{F} < 1$ (when there remain constraint violations) or with a high deviation cost if $\mathcal{F} \geq 1$ (when all tubes are separated). The mutation operator replaces the chosen tube with a new one, computed using an A^* algorithm. If no solution is found by the A^* algorithm, one of the parameters defining the chosen tube is randomly modified: the route choice, entry or exit flight levels, if any, or one of the cruising flight levels. For these last parameters, we have a choice (with equiprobability) from among several possibilities: add a new cruising flight level, remove an existing flight level, or modify one by changing the associated distance d_j or the level value CFL_j. As the A^* algorithm is relatively costly in computation time, it can be replaced (with a chosen probability) by a greedy method.

In [43], the two variants of the hybrid evolutionary algorithm (with A^* in the mutation, or with A^* and a greedy method) were compared with nonhybrid evolutionary algorithms (the canonical algorithm, with or without a bias in the selection for the mutated elements, or with the crossover operator for partially separated problems). The comparison was done on two test cases, one with 10 3D tubes and the other with 40 tubes. The results were improved when the hybrid method and the specific operators, were used. Figure 16.8 shows the evolution of the fitness criterion of the best element in the population for the two test cases where the origin and destination were located on a circle. The algorithm was run with 350 elements in the population, with a crossover probability of 0.6 and a mutation probability of 0.05.

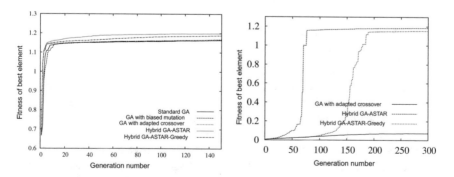

Fig. 16.8 Comparison of different algorithms on test cases with 10 3D tubes (*left*) and 40 tubes (*right*) (GA, genetic algorithm.)

Fig. 16.9 Solution by an A^*
algorithm, on a test case with
10 3D tubes

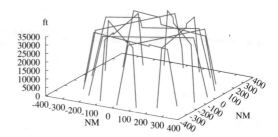

The hybrid evolutionary algorithm was also compared with a standalone A^* algo-rithm. In this case, the standalone A^* algorithm was applied successively to each 3D tube, building a new 3D tube that avoided the previous ones. The drawbacks of this approach are first that it does not aim to find a global optimum, and second that the solution found depends on the order in which we compute the tubes. Figure 16.9 shows a solution found by the A^* algorithm alone for the problem with 10 tubes. The fitness criterion $\mathcal{F}$ of this solution is 1.1674, which is less than the results found by the variants of the evolutionary algorithm, for which the average value over 10 runs was always above 1.19. For the problem with 40 tubes, the A^* algorithm was not able to find solutions satisfying the separation constraints.

To conclude on the construction of 3D tubes for the main traffic flows, the results presented in [43] show that using a metaheuristic to address this problem gives good results, and even better results when this metaheuristic is hybridized with an exact best-first tree-search method such as the A^* algorithm. The results of application of this hybrid method to real traffic over France and Europe [42] confirm these results, but they also show the limits of the concept. Building 65 separate 3D tubes over Europe, for flows with more than 20 flights per day, captured only 6 % of the overall traffic. This is due to the fact that the flows are built by considering departure and arrival airports. To improve the concept, one needs first to clusterize airports that are geographically close, as was done in [85], and then define 3D tubes between these clusters.

16.3   Airspace Optimization

In the previous section, we presented several approaches to building a route network, or independent "tubes," for the principal flows. In Sect. 16.2.1, the modeling of the partitioning of airspace into sectors with Voronoï cells for which a density criterion can be calculated was briefy described. This could be a way to build simultaneously a route network and partitioning of an airspace into sectors.

In this section, we suppose that the route network has been defined, and we focus on three problems related to the definition and management of airspace sectors. In the first problem, we want to define the sector edges so that we minimize different criteria such as the workload due to the coordination of flights crossing sector boundaries or the workload related to trajectory monitoring and conflict resolution within the sector boundaries.

In the second problem, elementary sectors have been defined, and we want to optimize the functional airspace blocks[3] in order to balance the traffic between blocks and limit the flows between the blocks.

In the third problem, we try to dynamically optimize the daily management of an airspace block: the problem is to group sectors in order to balance the controllers' workload, to avoid overloads and respect various operational constraints.

In the following sections, we give some examples of solutions using metaheuristics for these three problems.

16.3.1 Airspace Sectorization

The control sectors have evolved as traffic has increased, but they are still manually defined by human experts, mostly air traffic controllers. It is worth asking whether the partition of airspace into sectors is optimal regarding the evolution of traffic. The problem is difficult, because the model must be able to take into account various shapes of sectors, but remain simple enough to be solved. Delahaye presented a simple model for sectors in the horizontal plane in his Ph.D. thesis [20] (see also [21, 22]) (he did not consider the vertical dimension). In this model, n control sectors are characterized by n centers of a Voronoi diagram representing the limits of the sectors (see Fig. 16.10).

The main advantage of this model is that a sector is defined by a single point. However, different sets of points can define the same Voronoi diagram. This is the case for the example in Fig. 16.10, where the triplets (C_0, C_1, C_2) and (C_0', C_1', C_2') define the same sectorization. This is also the case for the triplets (C_1, C_2, C_0), (C_2, C_0, C_1) and for every permutation of the triplet (C_0, C_1, C_2) which gives the same result. Another issue with this model is that it only produces convex sectors, whereas real sectors are not always convex. Delahaye optimized the airspace sectorization with a classical evolutionary algorithm as described in [54, 58]:

- A vector of reals represents the coordinates of the class centers used to build the Voronoi diagram.
- The optimized criteria take into account the coordination workload (the number of aircraft flying from one sector to another), the monitoring workload (the number of aircraft inside the sector), and the resolution workload (the number of pairwise conflicts inside a sector). The objective function is aimed at balancing these three criteria while respecting constraints such as:

---

[3]A functional airspace block is a set of sectors in which several teams of controllers are qualified. Airspace blocks are independently managed by these different teams, which work in relays with one another. Several sectors in the same airspace block can be merged and controlled by the same pair of controllers. However, two sectors from different airspace blocks cannot be merged.

Fig. 16.10 Sector modeled
by class centers

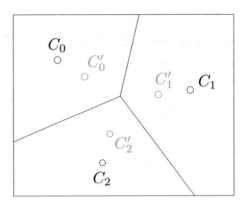

- the time spent by an aircraft in a sector should be longer than some minimum time;
- the routes followed by aircraft should not cross too close to the border of the sector.

An analytical expression that summarizes all these criteria is not possible; only a simulation can measure the quality of a sectorization. Metaheuristics are a good option in such a case because the objective function can be seen as a black box.

- The crossover operator identifies the class centers closest to both parents (which is a minimization problem) and applies a classical arithmetic crossover operation on these pairs.
- The mutation operator randomly moves one or several class centers in a defined neighborhood.

After his Ph.D. thesis, Delahaye proposed improved models in order to handle nonconvex sectors [26]. He also added the vertical dimension to his model in order to make it more realistic [24, 25]. Kicinger and Yousef [64] also proposed an evolutionary algorithm combined with an elementary cell aggregation heuristic in order to partition the airspace into sectors. Xue [90] introduced an approach using a Voronoi diagram optimized with an evolutionary algorithm, applied to the American airspace. In 2009, Zelinski [92] compared three methods for defining sectors, one based on traffic flow aggregation, another based on Voronoi diagrams optimized with evolutionary algorithms, and a third one using integer linear programming. Experiments showed the advantages and drawbacks of each method, but none really outperformed the others.

16.3.2 Definition of Functional Airspace Blocks

In Europe, the airspace structure follows the national borders of the different states. Nowadays, more than 60 control centers cover the airspace of around 40 member

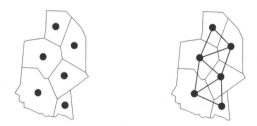

Fig. 16.11 Graph model for an airspace partition

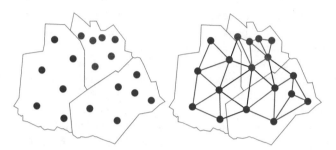

Fig. 16.12 Three functional blocks and the corresponding sectors

states of the European Organization for the Safety of Air Navigation (Eurocontrol). In the context of FABEC,[4] the problem is to reorganize the control centers in order to simplify the global structure. Among the numerous criteria that Eurocontrol has defined, three are quantifiable and could lead to a better balance of the distribution of centers:

- airspace blocks must minimize flows on their borders;
- important flows must take place inside the blocks;
- traffic must be balanced between different airspace blocks.

In his Ph.D. thesis, Bichot [11] modeled the problem as a graph partitioning problem. Here, the vertices of the graph are the sectors, and the edges are the flows connecting the sectors. The edge weights are the mean numbers of aircraft in the flows connecting the sectors.

Figure 16.11 shows a graph modeling a five-sector problem. Figure 16.12 shows a partition of the airspace into three functional blocks and the associated graph.

The minimization criterion chosen by Bichot was a normalized cut ratio criterion corresponding to the sum of the flows entering or exiting the functional blocks divided by the sum of the internal flows. He added a balance constraint: the weight of a block must not exceed k times the mean weight of every block. After showing the problem was NP-complete [12], Bichot tested several different classical algorithms on real

---

[4]Functional Airspace Block Central Europe.

recorded data (several months of European traffic), and compared them with two metaheuristics and also established with an innovative metaheuristic named "fusion–fission."

16.3.2.1 Simulated Annealing Algorithm

A simulated annealing algorithm requires a starting point. Bichot used a random configuration based on a percolation algorithm to build the starting point. He supposed that the graph was known, as well as the vertices and edges. The number of blocks was also fixed. A percolation algorithm simulates the movement of fluids through porous materials. Bichot defined as many sources of fluid as the desired number of functional blocks. Each source of fluid was a sector that was the kernel of the functional airspace block to which all other sectors were progressively linked. A detailed explanation of the algorithm is given in [12]. With this starting point, Bichot used a standard simulated annealing algorithm: in every step, a sector was randomly chosen in a functional airspace block and linked to another airspace block. The algorithm was divided into two phases. During the first phase of the algorithm, the control temperature was still high and the chosen sector was linked to a block with a low cut ratio. During the second phase, the control temperature was lower, and the chosen sector was linked to a neighboring block. The temperature adjustment and the time at which the algorithm switched to the second phase seem to have been chosen empirically.

16.3.2.2 Ant Colony Algorithm

In order to apply ant colony optimization to the functional airspace block partitioning problem, Bichot introduced a model in which one ant colony represented one block. Each block was the territory of one colony. The different colonies competed to get sectors and deposit their pheromones. More concretely, a sector belonged to the colony that had the largest amount of pheromones on it. After each ant movement, the value of the new state was calculated. If the ant movement decreased the criterion, the new partition was accepted, otherwise the partition was accepted with a probability following a rule similar to the simulated annealing method. This approach, like the previous one, requires one to adjust many parameters.

16.3.2.3 A Fusion–Fission Method

In his Ph.D. thesis [11], Bichot introduced a heuristic called "fusion–fission," by analogy with nuclear fusion and fission. For the fusion part, the idea is to merge two functional airspace blocks sharing the largest amount of traffic (as shown in Fig. 16.13). For the fission part, the principle is to divide the largest airspace block

Fig. 16.13 Fusion of two blocks

Fig. 16.14 Fission of the largest functional airspace block

into two blocks (see Fig. 16.14). Bichot refined his method by allowing some sectors to move from one block to another according to the cut ratio minimization criterion.

In [12], Bichot et al. showed that this last approach seemed more efficient and easier to apply than the simulated annealing and ant colony approaches. He also compared fusion–fission with classical graph-partitioning methods.

16.3.2.4 Comparison of Fusion–Fission and Classical-Graph Partitioning Methods

Bichot and Durand [13] compared two classical graph-partitioning algorithms (the Scotch and Graclus algorithms) with the fusion–fission approach and showed that the latter was more efficient than the Scotch and Graclus algorithms, but also much more time-consuming. Table 16.1 compares the normalized cut criterion, the balance between block sizes, and the maximum number of sectors per functional airspace block for the three algorithms. It also gives the values of the criteria for the existing partition of French airspace.

Figures 16.15 and 16.16 show the existing and optimized functional blocks for two flight levels (16 000 and 36 000 feet). The optimized partition divides the French airspace into only five blocks, instead of six for the existing partition. This result could provide an argument in favor of a partition with more blocks in the lower airspace and fewer blocks in the higher airspace.

Table 16.1 Partitions of french airspace

Algorithm	Ncut	Balance	Max number of sectors
Fusion–Fission	1.09	1.14	26
Scotch	1.18	1.20	30
Graclus	1.28	1.52	38
Existing partition	1.64	1.50	31

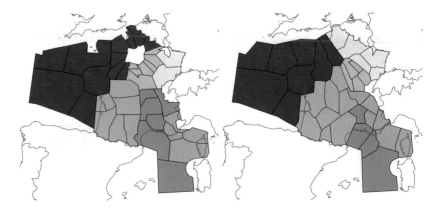

Fig. 16.15 Existing French functional airspace blocks (*left*, 16 000 feet; *right*, 36 000 feet)

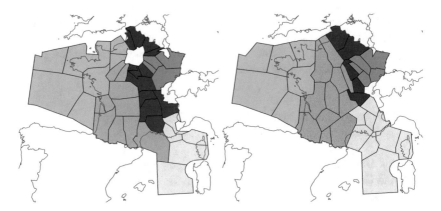

Fig. 16.16 Optimized French functional airspace blocks (*left*, 16 000 feet; *right*, 36 000 feet)

16.3.3 Prediction of ATC Sector Openings

We have seen in Sect. 16.3.1 how to define the airspace sector boundaries, given the air routes and traffic flows. In Sect. 16.3.2, we have seen how to group these airspace sectors into functional blocks, each placed under the responsibility of an air traffic control center. Operations such as sectorization and the definition of functional airspace blocks are in fact a strategic redesign of the whole airspace, which should be done well in advance before daily operations take place.

In this section, we focus on real-time or pretactical operations, assuming that the airspace sector geometry is fixed and that sectors have already been allocated to functional airspace blocks, as the result of a strategic design of the airspace. We consider a set of airspace sectors belonging to an air traffic control center (or a functional airspace block). In the daily operations of a control room, airspace sectors are dynamically assigned to air traffic controllers' working positions. The group of airspace sectors assigned to a working position is called an *air traffic control (ATC) sector*.

Figures 16.17 and 16.18 illustrate the partitioning of an airspace into ATC sectors using a toy example with five airspace sectors, denoted by numbers, and a list of acceptable groups denoted by letters.

The partitioning may change several times during the day, depending on the workload perceived by the controllers. Figure 16.19 shows a few other possible partitions that could be used instead of the partition presented in Fig. 16.17. Some operational constraints must also be taken into account: the duty roster, the maximum number of working positions that can be opened, and the list of possible groups that can actually be operated as ATC sectors (as already illustrated in Fig. 16.18).

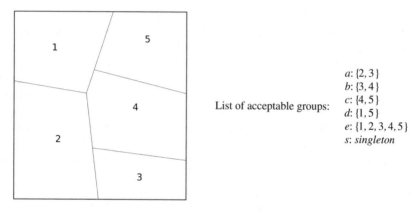

List of acceptable groups:

a: $\{2, 3\}$
b: $\{3, 4\}$
c: $\{4, 5\}$
d: $\{1, 5\}$
e: $\{1, 2, 3, 4, 5\}$
s: *singleton*

Fig. 16.17 A toy example of airspace sectors belonging to the same functional block

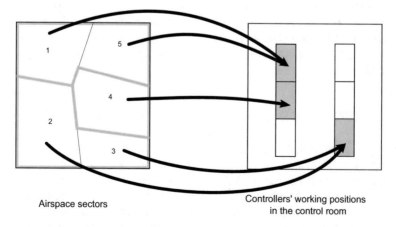

Airspace sectors

Controllers' working positions
in the control room

Fig. 16.18 Assignment of airspace sectors to controllers' working positions

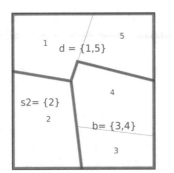 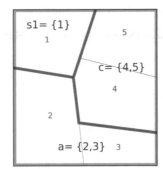

Fig. 16.19 Other possible partitions of the airspace

The primary objective of this dynamic partitioning of the set of elementary airspace sectors into ATC sectors is to avoid overloads, as these may threaten the overall safety of the flights controlled in the ATC sectors affected. When an ATC sector becomes overloaded, some of its airspace sectors are transferred to another working position (a new one, or one that is already open but underloaded) when this is possible. When such reassignments are not possible, one must enforce traffic regulation measures such as delaying departing flights or rerouting aircraft. Overloads must be anticipated with enough look-ahead time, so that regulation measures can be taken early enough. A secondary objective, which might sometimes come into contradiction with the primary objective of avoiding overloads, is to be as cost-efficient as possible by opening as few ATC sectors as possible and by avoiding under-loads.

Currently, this reassignment of airspace sectors to controllers' working positions is quite efficient for the purpose of sharing workload among ATC sectors in real time. However, we still lack prediction tools that would allow control room managers and flow management operators to anticipate how workload and airspace partitioning could evolve in the next few hours. Such tools require two things: a reliable estimation of the future workload in any given ATC sector, and an algorithm that can compute an optimal partition of the airspace into ATC sectors according to the predicted workload.

16.3.3.1 Difficulty of the Problem and Possible Approaches

The problem of optimal partitioning of airspace is highly combinatorial: the total number of candidate partitions is equal to the Bell number. However, taking operational constraints into account, such as restricting oneself to a list of acceptable groups of airspace sectors, reduces the number of sector combinations to be explored.

For relatively small and sufficiently constrained problem instances, exact tree-search methods that exhaustively explore (or discard) all possible partitions of the airspace into ATC sectors might be tractable. For larger instances, where the functional airspace block considered is made up of a large number of airspace sectors, or for less constrained problems with a larger number of acceptable sector groups, such methods are likely to be unsuccessful. In such cases, an optimal or nearly optimal partition can be searched for using a metaheuristic.

16.3.3.2 Using a Genetic Algorithm

In [47, 48], Gianazza and Alliot used a genetic algorithm [54, 74] to build an optimal partition of the airspace into ATC sectors. This metaheuristic approach was compared with two tree-search methods (a *depth-first* branch and bound search and a *best-first* search) on airspace sectors belonging to the five French en-route control centers.

In this approach, each element of the population is a sector configuration, i.e., a partition of the set of airspace sectors for the chosen control center. In each iteration, the genetic algorithm selects a pool of parents. Randomly chosen parents are then recombined, using crossing and mutation operators. The resulting offspring is added to the new population, which is completed by randomly picking individuals from the pool of parents. This completion is done so that the fittest individuals have a greater chance of being chosen. Several refinements exist for the selection, crossing, and mutation operations, with for example the application of scaling and sharing operators to the raw fitness. A description of these refinements can be found in Chap. 3 of [38].

In [47, 48], the mutation of an individual (a sector configuration) was done by first picking at random one ATC sector and one of its neighbors. The volume of airspace made up of the two chosen ATC sectors was then repartitioned. This partial reconfiguration of the sectors was also random, with the constraint that the result should not contain more than three ATC sectors. The new ATC sectors then replaced the two initial sectors in the mutated individual.

The crossover operator removed some ATC sectors from each of the two parents and tried to form a new partition from each amputated partition, using ATC sectors from the other parent. This did not usually result in a complete partition of the airspace. A full partition was obtained by randomly choosing control sectors compatible with the incomplete partition.

The fitness criterion depended on the following factors, in decreasing order of priority: excessive overloads, the number of working positions (i.e., the number of ATC sectors in the configuration), excessive underloads, and small overloads or underloads. For any ATC sector, the workload was assessed by considering the difference between the flow of incoming traffic and a threshold value, called the sector capacity. The capacity values were the ones that were actually used in operations at the time. Once computed, the raw fitness criterion was modified using clusterized sharing and sigma truncation (see [54], or [38] p. 59), so as to leave a chance even for the least fit individuals to reproduce, thus allowing a better exploration of the search space. For the sharing operator, a difficulty arises in defining a distance criterion between partitions of the set of airspace sectors. A pseudo-distance between two partitions, similar to the Hamming distance, was specifically designed for this sharing operator. The only difference from the Hamming distance was that the sequence of symbols (ATC sectors) that were compared—counting the differences between the two partitions—need not have the same length.

An elitist strategy was applied in order to preserve the best individuals of the old population when building a new one. The new population was made of the fittest elements of the previous population, of the mutated individuals, and of the offspring resulting from the crossover operator. Both the mutation and the crossover operator were applied to individuals randomly chosen from a pool of parents, with probabilities P_c (crossover) and P_m (mutation). The population was then completed according to the stochastic remainder without replacement mechanism (see [38]), so as to attain the same fixed size as the previous population.

This approach using genetic algorithms was compared, using real instances, with two tree-search methods. Other authors have used constraint programming on a similar problem. We shall now briefly present these exact approaches that exhaustively explore the search space of possible airspace partitions.

16.3.3.3 Tree-Search Methods, Constraint Programming

Two tree-search strategies were presented in [47, 48]. One is a depth-first search, illustrated in Fig. 16.20, using our toy example with five airspace sectors. The other is a best-first search inspired by an A^* algorithm that develops first the nodes that have the best estimate of the total cost for the path from the root to a leaf of the tree.

In his Ph.D. thesis [6], Barnier successfully applied constraint programming methods to a similar problem of airspace partitioning (although not with the same capacity values). The partitioning problem was formalized as a constraint satisfaction

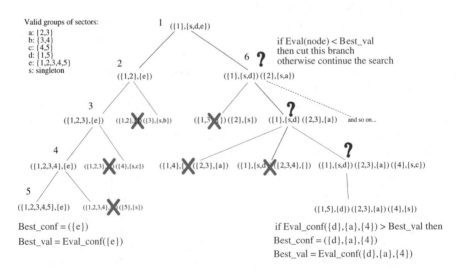

Fig. 16.20 Search for an optimal partition by a depth-first tree-search algorithm

problem. The solution of this problem also relied on a tree-search method (back-tracking) that iteratively reduced the domain of each variable.

All these tree-search methods were tested on real instances, using the airspace sectors of the five French air traffic control centers. The results showed that, on these real instances of relatively small size, when taking into account some operational constraints such as a list of restrictions concerning the valid groups of sectors, the global optimum could be reached in a very short time (a few seconds at most, with a 1.8 GHz Pentium IV).

In [47, 48], the depth-first and best-first strategies were compared with the genetic algorithm presented in Sect. 16.3.3.2. With 220 elements in the population, evolving over 300 generations, and with a crossover probability of 0.6 and a mutation probability of 0.2, the genetic algorithm found the global optimum in nearly all cases. The computation times were, however, much longer (several minutes).

16.3.3.4 A Neural Network for Workload Prediction

In [6, 47, 48], the chosen variables (input traffic flow) and the ATC sector capacities, which were the values actually used in operations at the time, did not provide a reliable estimate of the air traffic controllers' workload. Further studies [44, 52, 53] by Gianazza and Guittet were aimed at selecting more relevant indicators, from among the multitude of ATC complexity metrics proposed in the literature, to better explain the controller workload.

In these studies the dependent variables that were chosen to represent the actual workload were related to the status of the ATC sector. Considering past sector openings, the following observations can be used to assess the workload in any given sector:

- when the sector is "collapsed" (merged) with other sectors to form a larger sector, we can assume that this is due to a low workload;
- when the sector is "opened" (i.e., actually operated on a controllers' working position), we can assume a normal workload;
- when the sector is "split" into several smaller sectors, this reflects an excessive workload in the initial sector.

The basic assumption is that this observed sector status ("collapsed," "opened," or "split") is statistically related to the actual workload perceived by the controllers.

A neural network was used to compute a triple (p_1, p_2, p_3) representing the probabilities for a sector to be in the above states. The network inputs were the ATC complexity indicators computed from aircraft trajectories, and metrics of the sector geometry (the sector volume). The neural network was first trained on a set of examples, based on recorded traffic and historical data on sector openings from the five French air traffic control centers.

Training a neural network consists in adjusting the weights assigned to the network connections so as to minimize the error in the output when compared with the desired output in the examples. This requires the use of an optimization method operating in the space of the weights. The first methods that were designed to train multilayer perceptrons relied on the gradient of the error to search iteratively for the optimal weight vector. In these methods, starting from an initial point in the space of the weights, every step consists in computing a new iterate from the current one, following a descent direction based on the error gradient. Subject to several conditions on the objective function, these descent method converge to a local minimum. Such methods require the computation of the error gradient, which can be done efficiently using backpropagation of the error in the network [14]. More recently, several metaheuristics have also been proposed, either to optimize the network topology or to tune the weights: genetic algorithms [68], particle swarm optimization [57], ant colonies [15], differential evolution [84], etc.

The results presented in [44, 52, 53] on the prediction of ATC controllers' workload were obtained using a quasi-Newton method (specifically, BFGS) to train the network. Some preliminary results using particle swarm optimization and differential evolution showed fairly similar results.

In [45, 46], the depth-first tree-search algorithm that computed optimal airspace partitions (see Sect. 16.3.3.3) was combined with the neural network for workload prediction in order to provide realistic predictions of ATC sector openings. This prediction of the workload and airspace partitioning is illustrated in Fig. 16.21.

An initial evaluation of this research approach was done by comparing the number of working positions computed by these algorithms with the number of positions that were actually open on the same day. In Fig. 16.22, the two dotted lines representing these quantities are quite close. The continuous line above the dotted lines shows the total traffic in the ATC center, and is given here only as an indication of the evolution of traffic during the day.

16.3.3.5   Conclusions About the Prediction of Sector Openings

We have seen that the difficulty of the problem of partitioning an airspace into ATC sectors assigned to controllers' working positions, which is in essence highly

Fig. 16.21 Prediction of workload and airspace partitioning

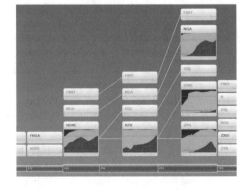

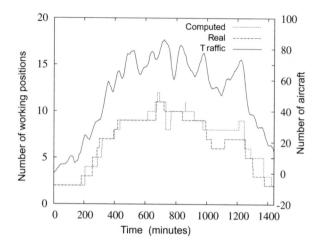

Fig. 16.22 Computed versus actual number of controllers' working positions

combinatorial, is reduced when operational constraints are taken into account, such as by restricting the number of ways to group airspace sectors to an existing list of valid ATC sectors. We have also seen that a realistic prediction of ATC sector configurations requires a reliable workload prediction model.

Metaheuristics can be useful for both of these problems (airspace partitioning and workload prediction). For large instances that cannot be addressed by exact tree-search methods, metaheuristics are often the only option: they rely on a random walk in the search space, guided by a heuristic that introduces a bias toward good solutions. Metaheuristics can also be used to tune the weights of a neural network for predicting the air traffic controllers' workload.

In conclusion, it must be noticed that in this specific example based on real instances of airspace sectors and ATC sectors from the French airspace, metaheuristics are not the fastest and most efficient methods. For such instances of relatively small size, optimal partitions can be obtained in a short time using exact tree-search methods.

However, exact methods can become impracticable for larger instances with more airspace sectors or more ATC sectors. In such cases, using a metaheuristic can be a good alternative for finding optimal or near-optimal partitions of ATC sectors.

16.4 Departure Slot Optimization

In order to prevent saturation of controlled airspace in Europe, departure slots are sometimes imposed on aircraft. A departure slot is a 15 min time window during which an aircraft must takeoff. The Network Manager Operations Centre (NMOC), formerly called the CFMU,[5] tries to optimize the delays that aircraft, are subjected to. This optimization problem has been studied by several research teams around the world, using different models and algorithms. In the United States, delays are mainly due to congestion at the arrival airport: instead of making an aircraft stack before landing, it is better to delay its departure. This generates two types of problems. In Europe, which aircraft should be delayed, and for how long, in order to respect control sector capacities? In the United States, which aircraft should be delayed, and for how long, in order to prevent them from stacking at their destination?

The first approaches to dealing with these problems mainly used integer linear programming [71, 76]. Similar approaches were used by Bertsimas and Patterson [10] at the end of the 1990s and by Bertsimas et al. [9] in 2008.

The first article introducing the use of evolutionary algorithms to optimize takeoff slots was written by Delahaye and Odoni [23]. At first, Delahaye used a simple toy problem in which the route and takeoff time were optimized. Later, Oussedik et al. [77–79] adapted this approach to real traffic data. Cheng et al. [17] solved a small example with a genetic algorithm. In 2007, Tandale and menon [87] used a genetic algorithm on the FACET[6] simulator developed by NASA[7] in order to solve problems in which sector capacities were respected. They compared their algorithm with an exhaustive method using an example dealing with two airports, and generalized their approach to a problem involving 10 airports.

In 2000, Barnier and Brisset [7] gave a more accurate definition of a sector capacity and used a constraint programming approach in order to optimize slots. Once again, comparing methods is challenging because research teams do not share data. In his Ph.D. thesis, Allignol [2] resolved conflicts by modifying takeoff slots. An initial calculation was done to detect all potential conflicting trajectories. This calculation generated constraints on the takeoff times for aircraft pairs: the difference between the takeoff times for aircraft pairs should not belong to some time interval.

Two approaches are being used to solve the problem. The first one is based on constraint programming, and the second on an evolutionary algorithm. In the constraint programming approach, the problem is to find an instantiation for every delay that resolves every conflict, and to minimize the total delay. In the evolutionary algorithm, approach, separation constraints are taken into account in the fitness function. Numerical results on real French data [29] show that constraint programming generally gives better results and is faster than evolutionary algorithms, but the latter penalize fewer aircraft with a larger mean delay. Su et al. [86] adapted a cooperative coevolution approach to Chinese data. Unfortunately, it is impossible to compare results on different data sets.

---

[5]Central Flow Management Unit.

[6]Future ATM Concepts Evaluation Tool.

[7]National Aeronautics and Space Administration.

16.5   Airport Traffic Optimization

Many optimization problems can be formulated in the field of airport traffic management: indeed, airports have to be highly reactive to many kinds of events, that may be more or less usual (delays to passengers and flights, meteorological phenomena, equipment failures, surface congestion, terrorism risks, etc.), which makes their traffic difficult to predict. For these reasons, the various stakeholders often need to adapt their planning and operations in real time. All the decisions that are taken in this way can induce various positive or negative effects in the global situation of the airport, and result in very variable operating costs.

In this domain, the problems of gate assignment, scheduling of aircraft on the runway, strategic surface routing, and, more generally, the development of decision support tools that can help operations planning are major concerns for all airport services (Fig. 16.23).

16.5.1   Gate Assignment

Assigning gates (and stands) to aircraft appears to be the first important step in the planning process at an airport. It involves many operational aspects, such as constraints related to each gate and all the connections between flights.

Hu and Paolo [59] modeled the problem with a global minimization criterion, defined by a balance between three measures: the waiting times of aircraft on the aprons, the walking distances, of passengers, and the baggage transport distances.

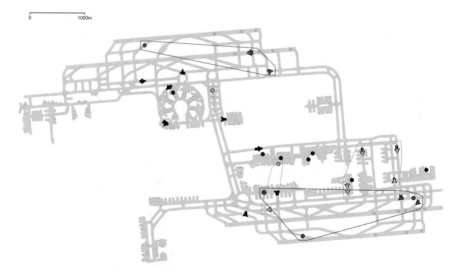

Fig. 16.23 Simulation of traffic at Roissy-Charles-De-Gaulle airport

The variables must assign not only a parking spot to each aircraft but also the order in which each aircraft will access the gate. For these reasons, these authors compared different possible encodings for solving the problem with a genetic algorithm. They showed that a binary encoding, which seemed at first to be more complex than other possibilities, associated with a specific uniform crossover, made the genetic algorithm more efficient.

16.5.2   Scheduling the Aircraft on the Runway

Runways are often seen as the main bottlenecks in an airport, because some important separation times (over one minute) are needed between movements, in order to keep a following aircraft free from the wake vortex turbulence of the previous aircraft. These separation times depend on the type of movement (takeoff or landing) and on the categories of the two aircraft (the heavier an aircraft is, the stronger its vortex is, but the less it is penalized by the vortex of the previous aircraft). Thus, the separation time after an aircraft A depends not only on A but also on the following aircraft B, as illustrated in Fig. 16.24. This makes the problem less symmetrical than many other classical scheduling problems.

In [62], Hu and Di Paolo focused on the optimization of an arrival sequence, and compared the efficiency of two different encodings for a genetic algorithm:

- The first one, rather intuitive, was integer based and consisted of a rank assigned to each arrival in the sequence.
- The second one was based on a binary matrix, specifying all of the Boolean priority relationships between each pair of arrivals.

The second encoding, associated with a uniform crossover (which makes each child inherit a specific part of the priority relationships of its parents) gave the best results, especially by avoiding premature convergence toward local optima. This kind of

Fig. 16.24 Scheduling of aircraft on the runway

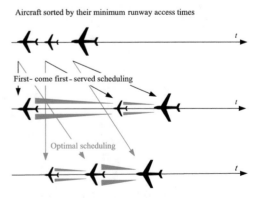

encoding maintains some promising subsequences across several generations, while still favoring a good exploration of all the possible sequences.

Hu and Di Paolo confirmed the efficiency of this kind of encoding for the genetic algorithm by extending it to a problem of arrivals that have to be distributed over several runways [63]: here, the arrivals have to be scheduled on each runway, but they also have to be assigned to one of the available runways. In [61], these authors improved their results further with a new *ripple spreading* genetic algorithm: in this model, each chromosome encodes an epicenter point in a two-dimensional artificial space, and a method to project each aircraft into this space (depending on its wake vortex category and its soonest landing time). The ripple spreading procedure is a simple algorithm that assigns a runway to each aircraft and defines the sequence on each runway from the set of points in the artificial space (by sorting each point by increasing distance from the epicenter). Thus, each chromosome is reduced to five numbers $(x, y, \delta_1, \delta_2, \delta_3)$, where (x, y) are the coordinates of the epicenter and $(\delta_1, \delta_2, \delta_3)$ are the coefficients defining the projection in the artificial space. A big advantage of this method is that the size of the chromosomes does not depend anymore on the number of aircraft, but only on the number of parameters used to characterize them.

Particle swarm methods can also be used to optimize the departure flow of aircraft that have to be scheduled on a runway and can use different routes to access that runway [40, 66]: in this model, each departure route is seen as a first-in-first-out queue (aircraft using the same route cannot change their order). The problem is to find the gate departure times and the takeoff times that minimize the time spent in offloading all of the traffic (while maintaining separation constraints between taxiing aircraft). Using an evolution function based on an oscillating equation of second order (inherited from control theory) [66], or by controlling the evolution with a simulated annealing method [40], the authors of those publications improved the convergence of the particle swarm, while avoiding local optima.

In Europe, departure scheduling appears to be more complex, because some of the departures are also constrained by a takeoff slot assigned by the European Network Manager Operations Centre (because these flights fly through overloaded airspace). For these constrained departures, a specific takeoff time is specified, and the corresponding flight can only takeoff five minutes before or ten minutes after the given time. In his Ph.D. thesis [18], Deau provided a global formulation for the aircraft scheduling problem on a mixed runway (on which both landing and takeoff may be scheduled), where some of the departures are constrained by a specific takeoff slot: the variables are the takeoff and landing times, and the minimization criterion is a balance between the deviations from the constrained slots (for the constrained departures) and the delays (of the other flights). By taking advantage of some particular properties of the problem (symmetries, equivalences, between aircraft, and detection of suboptimal scheduling as illustrated in Fig. 16.25), this author produced a branch and bound algorithm that found and proved an optimal solution in a few seconds, for a large sample of problems involving more than 50 aircraft.

Applying the same scheduling algorithm to shifting periods in a whole day of traffic at Roissy-CDG airport [19], Deau et al. found a global schedule for all movements, on all runways, compliant with all the constrained takeoff slots, that generated

Fig. 16.25 Detection of a
suboptimal scheduling

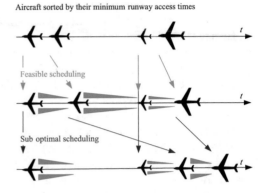

a global delay that appeared to be half of that measured from a complete simulation
of the same traffic. These results show that the runways are not the only source of
delay at an airport such as Roissy-CDG, and that the traffic also needs to be optimized
during taxiing.

16.5.3 Surface Routing Optimization

Airport studies often ignore the problem of taxiing aircraft, although this step causes
serious issues to airport controllers and can generate important delays (especially
in the stand area, where aircraft have to maneuver or be maneuvered at low speed,
without the possibility of overtaking other aircraft).

The first detailed study concerning airport surface routing optimization was provided in [80]: the authors of that study modeled the taxiways of the airport as an
oriented graph connecting the stands to the runways (and conversely). Classical path
enumeration algorithms were used to compute a set of alternative routes for each
aircraft. The routing problem was then formulated as the choice of the routes associated with some optional holding points, in such a way that a minimum distance
is ensured between each aircraft pair in each time step, while minimizing a global
criterion based on the total delay (due to route lengthening and waiting times). To
solve this very combinatorial problem, the authors compared two strategies:

- The first strategy consists in simplifying the problem by attaching priority levels
  to the aircraft: a total order allows the aircraft to be sorted and to be considered one
  after the other. Each aircraft is assigned a trajectory (a route and some optional
  holding points) in the given order. Thus, the nth aircraft has to avoid the $n-1$
  previous ones, once their trajectories have been fixed. The problem is thus split
  into a succession of best-path searches with avoidance of obstacles, which can be
  performed very quickly by a simple A^* algorithm.

Fig. 16.26 Encoding in genetic algorithm for the aircraft routing problem

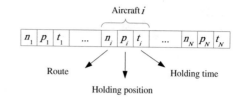

Aircraft i

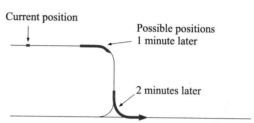

Route

Holding position

Holding time

Fig. 16.27 Trajectory prediction with speed uncertainties

Current position

Possible positions 1 minute later

2 minutes later

- The second strategy is based on a genetic algorithm that deals with the whole problem, without presuming any priority levels of aircraft: each chromosome describes a route and a holding position (associated with a holding time) for each aircraft (see Fig. 16.26). More efficient mutation and crossover operators can be defined with this kind of encoding, taking advantage of some partial fitnesses (one per aircraft) that allow the parts of the chromosomes that are the least promising to be changed more often.

Measured by simulation of some actual traffic at Roissy-CDG airport, the genetic algorithm appeared more efficient, as it reduced the mean aircraft delay by one minute (from 4 min), compared with the strategy based on priority levels.

In his Ph.D. thesis [55], Gotteland developed and refined this study:

- Aircraft trajectories <u>where</u> we predicted with a given rate of uncertainty in their speeds (see Fig. 16.27), and conflicts were detected between all the possible positions of each aircraft.
- The criterion to be minimized measured the deviations from the takeoff slots assigned to the constrained departures.
- Conflicts caused by arrivals crossing the departure runway after landing were also considered.

With this formulation, the problem is a mix between the aircraft routing problem, the management of the arrivals that have to cross the departure runway, and the scheduling of departures on the runway. A new genetic algorithm was introduced, in the form of a hybridization of the two previous routing strategies (priority levels and genetic algorithm):

- Each chromosome described a route and a priority level (or rank) for each aircraft (see Fig. 16.28).

Fig. 16.28 Encoding in hybrid genetic algorithm for the routing problem

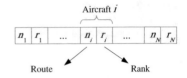

- To evaluate such a chromosome, the aircraft were considered one after the other (by increasing rank), and were assigned their specified route. For each aircraft, a branch and bound algorithm (which appeared more efficient than the previous A* algorithm once the choice of the route had been made) was run to find the corresponding best trajectory, avoiding the ones already computed.

The efficiency of this hybrid genetic algorithm was compared with the two previous strategies, using the same simulator with an actual sample of traffic at Roissy-CDG airport. The delays due to surface conflicts were decreased by more than one minute from 5 min during heavy periods, and the assigned takeoff slots were all respected (in the 15 min tolerance time window) and better scheduled (more than 80 % happened at less than one minute from the specified time).

Dealing with the aircraft routing problem at Madrid-Barajas airport, García et al. [41], combined a deterministic flow management algorithm with a genetic algorithm to assign a route and a beginning time to each movement (a landing time for arrivals and an off-block time for departures).

In [82], for simpler, fictional airport (with fewer taxiways and fewer movements), Roling and Visser succeeded in modeling and globally solving the airport surface traffic planning problem, using mixed integer linear programming (where the variables described the times at which each aircraft traveled on each portion of taxiway). They obtained a route assignment process associated with some specific aircraft holding positions that globally minimized the taxi times.

16.5.4 Global Airport Traffic Planning

In the more global framework of traffic planning at busy airports, several different concepts or systems are often studied:

- Arrival management (AMAN) includes all the predictions that can be made about the arrival flow, taking into account the constraints of the approach sectors (which are sometimes shared by different airports), in order to evaluate aircraft landing times with the best possible accuracy.
- Departure management (DMAN) starts with the prediction of the takeoff sequences, taking into account the departure times targeted by the airlines, potential constraints on takeoff slots, and the separation times needed on the runways. By considering the taxi-out times of aircraft and the takeoff sequences, it is also possible to delay some off-block times for departures, in order to make them hold

at the gate (with engines off) rather than in a queue for the runway (with engines on).

- Surface management (SMAN) deals with the routing of aircraft at the airport (taking into account all the AMAN and DMAN information): the goal is to assign strategic routes that are compliant with the predicted landing or takeoff times of aircraft, while keeping the ground traffic situation as fluent as possible.

Deau et al. [19], pointed out the obvious dependency problems that arise in these predictive systems: the delay of an arrival can affect the time of its subsequent departure, and decisions made while handling taxiing aircraft can quickly result in situations where the takeoff sequences must be updated (as the off-block and landing times must also be, when the runway is shared by both types of movements). Moreover, the uncertainties that exist in the speed of aircraft during taxiing (which can easily reach 50 % of their average speed on each taxiway portion) make the ground traffic situation hard to predict (the possible positions of an aircraft 5 min later extend over one kilometer). Thus, the predictions of the different systems cannot share the same magnitude: it is over 30 min for the AMAN–DMAN system, but under 10 min for the SMAN. Deau et al. proposed an iterative process that would allow coordination between the different systems, in which some optimal takeoff and landing sequences are computed, taking into account the current positions of the aircraft (with a runway time window TW_R of 30 min). These sequences are then used to resolve the ground conflicts more efficiently (with a surface time window TW_S of 5 min), as illustrated in Fig. 16.29.

By carrying out fast simulations of some actual traffic at Roissy-CDG, Deau et al. measured how the mean delay of aircraft could be decreased, first by the optimization of runway sequences, and then by the use of a hybrid genetic algorithm (rather than a sequential method using fixed priority levels) to solve ground conflicts (see Fig. 16.30).

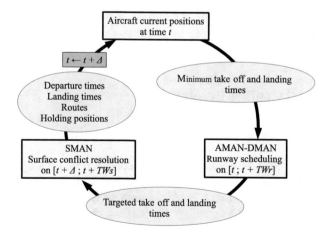

Fig. 16.29 Coordination of AMAN–DMAN and SMAN systems

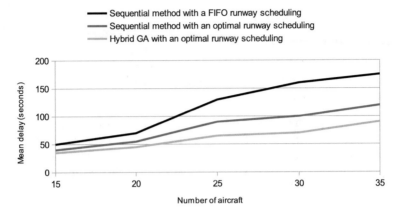

Fig. 16.30 Mean delay of aircraft in Roissy-CDG simulations (FIFO, first-in first-out; GA, genetic algorithm)

Still on the topic of integrating different predictive systems, the management of the capacity of several neighboring airports has also been studied: Hu et al. [60], considered a set of airports made up of one main airport surrounded by other satellite airports, between which arrivals could be exchanged. The capacity of each airport varied, owing to meteorological conditions, its configuration (the runways used and the distribution of arrivals and departures on each runway), and its traffic composition (aircraft types). The problem was modeled as follows:

- The variables described the airports' successive configurations on one hand and the assignment of airports to arriving aircraft on the other hand.
- The minimization criterion was formulated as a balance between the size of the various aircraft queues (for arrival and for departure, at each airport) and the number of airport changes (compared with the initial airport assignments).

Hu et al. showed that a genetic algorithm could find some efficient solutions to the problem for a one-day traffic sample, using successive resolutions of the situations (for shifting periods of the day).

16.6 Aircraft Conflict Resolution

An air traffic controller is charged with the task of separating aircraft in order to prevent conflicts.[8]

Alliot et al. [5] first introduced, in 1993, a conflict resolution method using a genetic algorithm. The model was very simple: time was discretized in 16 steps of 40 seconds each. Each aircraft could, in each of the 16 time steps, either go straight, turn right, or turn left, with a 30° heading change. Each maneuver was

---

[8]Two aircraft are conflicting if the horizontal distance between them is less than 5 nautical miles and the vertical distance is less than 1000 feet.

encoded with two bits (00 and 01 = go straight; 10 = turn right; 11 = turn left). Each trajectory was encoded with 32 bits. For a two-aircraft problem, 64 bits were necessary. Results obtained with the genetic algorithm were compared with an $A*$ algorithm and a simulated annealing method. The genetic algorithm showed good efficiency on simple examples.

In his Ph.D. thesis, Durand [27] modeled the problem differently: the maneuvers were not encoded as bit strings but as reals and quantitative values: each aircraft could execute at most one maneuver starting at time t_0 and ending at time t_1. This could be a heading change of 10, 20, or 30° to the right or to the left of the initial heading. An n-aircraft conflict was thus encoded by $3n$ variables. Durand and Alliot defined a crossover operator adapted to partially separable problems [32]: from two parents, two children are built using the "best" characteristics of their parents. Figures 16.31 and 16.32 detail the principle of this operator for a seven-aircraft conflict. The objective of the operator is to copy from each parent the part that resolves the largest number of conflicts.

Thanks to this operator, an evolutionary algorithm was able to resolve large conflicts involving up to 30 aircraft in a very short time (less than a minute). Durand and Alliot tested the method on a fast time simulator using real traffic data and showed that they could resolve every conflict, even with important uncertainty margins on the trajectory predictions [4, 30, 33, 36].

Granger et al. [56] adapted the previous results to direct routes by modeling existing routes. Akker et al. [1] used a free-route approach. Malaek et al. [70] used a model close to Durand's approach and took the impact of wind into account. A genetic algorithm was used to coordinate continuous aircraft maneuvers.

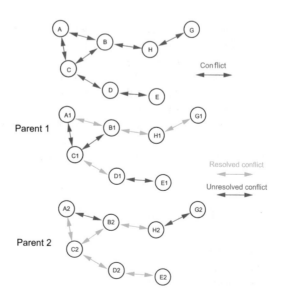

Fig. 16.31 Aircraft cluster; structure of the two parents

Fig. 16.32 Adapted
crossover operator

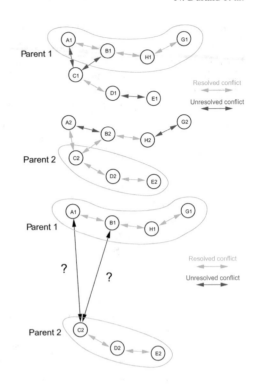

16.6.1 Ant Colony Optimization Approach

Other metaheuristics have been tested on the conflict resolution problem. Durand
and Alliot [32] introduced an ant colony optimization algorithm to resolve complex
conflicts. Here, in every generation, each aircraft is represented by an ant. Ants
which have been able to reach their destination without creating any conflict with
other ants deposit pheromones according to the shortness of the path found. The other
ants do not deposit pheromones. For difficult problems, the separation constraints
between aircraft can be relaxed: ants deposit pheromones even if they do not respect
constraints. The amount of pheromone is inversely proportional to the number of
conflicts generated. This idea was used by Meng and Qi [73] with a naive formulation.

16.6.2 Free-Flight Approaches

Evolutionary algorithms have been used in distributed approaches. In the United
States, Mondoloni et al. [75] and Vivona et al. [89] introduced free-flight models for
optimizing coordinated trajectories.

Free-flight models were also used in the reactive approach introduced by Durand et al. [34, 37]. This approach uses a neural network for each aircraft in order to avoid intrusive aircraft. The parameters of the neural network parameters are optimized by an evolutionary algorithm for a set of conflicts representing different configurations.

Figure 16.33 shows the data used as an input for the neural network, and its structure. Figure 16.34 gives examples of the conflicts used to optimize the weights of the neural network. The fitness function used in the evolutionary algorithm takes into account the fact that conflicts are resolved and that trajectories generate little

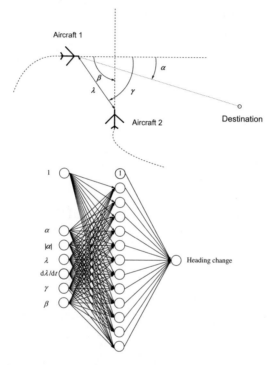

Fig. 16.33 Inputs for aircraft 1; structure of the neural network

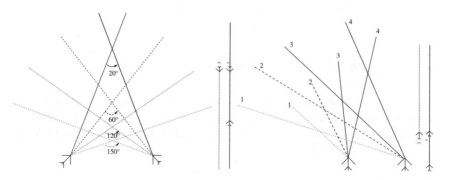

Fig. 16.34 Learning examples

Fig. 16.35 Solution comparison: *top* local method; *bottom*, neural network

delay. Figure 16.35 compares maneuvers obtained with the neural network (bottom) and a classical optimization tool (top).

16.6.3 A Framework for Comparing Different Approaches

It is very challenging to compare results obtained by different teams when reading articles on conflict resolution methods. Research teams generally use different data, they do not offer free access to the data they use, and they are often experts in one optimization method only.

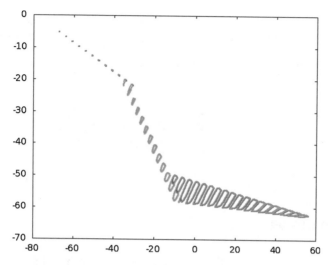

Fig. 16.36 Trajectory prediction with uncertainty

Recent studies have tried to answer this issue by offering benchmarks that can be downloaded to test different algorithms. Vanaret et al. [88] compared three metaheuristics on the conflict resolution problem: a differential evolution method, an evolutionary algorithm, and a particle swarm optimization approach. These authors showed that, most of the time, differential evolution was as efficient as the evolutionary algorithm and sometimes even better, and always better than particle swarm optimization in many examples.

In [3], Allignol et al. proposed a benchmark that can easily be used by anyone by accessing it on the link http://clusters.recherche.enac.fr. It does not require any knowledge of air traffic control. The benchmark contains 120 different scenarios of conflicts involving n aircraft (n varying from 5 to 20) and three levels of uncertainties ε_{low}, ε_{medium} and ε_{high}. Uncertainties in speeds and also in headings and turning points are considered. Future positions of aircraft are represented by convex hulls, the sizes of which evolve with time (see Fig. 16.36). In each scenario, aircraft can choose from among $m = 151$ different trajectories.

In the benchmark, one file contains a description of the trajectory of each aircraft maneuver and another file describes the four dimensional conflict matrix. For each aircraft–maneuver pair (i, k) and aircraft–maneuver pair (j, l), the matrix returns 1 if there is a conflict, and 0 otherwise. The file also gives the cost of each maneuver. The model is completely separated from the problem to be solved. The problem can thus be solved with constraint programming methods as well as evolutionary algorithms. Results (Table 16.2) show that the approach is often more efficient when the problems are not too large, and it has the great advantage that it can prove the optimality of the solution, or prove that no solution can be found.

Table 16.2 Mean cost of the best solutions for different conflict sizes and different levels of uncertainties. Conflicts for which optimality was not proven are shaded. The cells with only one number for both CP (constraint programming) and EA (evolutionary algorithm) correspond to instances for which both algorithms reached the optimum

	n			
	5	10	15	20
	CP EA	CP EA	CP EA	CP EA
ε_{low}	5.3	29.8	86.3 86.8	185.8 176.9
ε_{med}	4.2	46.6	104.0 104.0	267.6 282.8
ε_{high}	5.1	45.7	170.4 156.3	299.0 305.0

16.7 Conclusion

In this chapter, we have presented many applications of metaheuristics to air traffic management problems. We have focused on the different possible models that have been explored, and detailed the solution methods that were used.

When it was possible, we have tried to compare the different methods used. More particularly, some problems could be solved with exact methods as well as with metaheuristics, and in those cases we have given some elements of a comparison. The complexity of the problems, their connection with external problems, their huge size, and the uncertainties that are involved, make these problems very challenging and exciting to deal with, but they also limit the possibility of a rigorous scientific approach in which one compares many different methods on series of freely accessible benchmarks. As a consequence, it is not easy to find exhaustive comparisons of methods on problems, that are reproducible by other research teams with publicly available data. However, a few benchmarks have been put online recently.

For some problems, we have shown that it was possible to use exact optimization methods, especially on highly constrained problems such as allocation of sectors to teams of controllers. On other problems, such as the creation of a route network, geometrical methods can give good solutions, even if they do not optimize the solution.

In many cases, however, metaheuristics are the most efficient existing methods, sometimes the only applicable methods to deal with difficult combinatorial problems for which the criteria to be optimized require one to run a simulation. Metaheuristics are useful tools and sometimes are necessary to tackle air traffic management problems. They allow us to model problems in a realistic way instead of using a simplified mathematical model that is often unable to handle realistic constraints.

References

1. van den Akker, J., Van Kemenade, C., Kok, J.: Evolutionary 3D-Air Traffic Flow Management. Citeseer (1998)
2. Allignol, C.: Planification de trajectoires pour l'optimisation du trafic aérien. Ph.D. thesis, INPT (2011)
3. Allignol, C., Barnier, N., Durand, N., Alliot, J.M.: A new framework for solving en-route conflicts. In: *10th USA/Europe Air Traffic Management Research and Development Seminar* (2013)
4. Alliot, J., Bosc, J., Durand, N., Maugis, L.: CATS: A complete air traffic simulator. In: *16th DASC* (1997)
5. Alliot, J.M., Gruber, H., Schoenauer, M.: Genetic algorithms for solving ATC conflicts. In: *Proceedings of the Ninth Conference on Artificial Intelligence Application*. IEEE (1992)
6. Barnier, N.: Application de la programmation par contraintes à des problèmes de gestion du trafic aérien. Ph.D. thesis, INPT (2002)
7. Barnier, N., Brisset, P.: Slot allocation in air traffic flow management. In: *PACLP'2000* (2000)
8. Barnier, N., Brisset, P.: Graph coloring for air traffic flow management. In: *CPAIOR'02: Fourth International Workshop on Integration of AI and OR Techniques in Constraint Programming for Combinatorial Optimisation Problems,* Le Croisic, France, pp. 133–147 (2002)
9. Bertsimas, D., Lulli, G., Odoni, A.: The air traffic flow management problem: An integer optimization approach. In: *IPCO*, pp. 34–46 (2008)
10. Bertsimas, D., Patterson, S.S.: The air traffic flow management problem with enroute capacities. Operations Research **46**, 406–420 (1998)
11. Bichot, C.E.: Élaboration d'une nouvelle métaheuristique pour le partitionnement de graphe : la méthode de fusion-fission. Application au découpage de l'espace aérien. Ph.D. thesis, INPT (2007)
12. Bichot, C.E., Alliot, J.M., Durand, N., Brisset, P.: Optimisation par fusion et fission. Application au problème du découpage aérien européen. Journal Européen des Systèmes Automatisés **38**(9–10), 1141–1173 (2004)
13. Bichot, C.E., Durand, N.: A tool to design functional airspace blocks. In: *7th USA/Europe Air Traffic Management Research and Development Seminar* (2007)
14. Bishop, C.M.: *Neural Networks for Pattern Recognition*. Oxford University Press (1996)
15. Blum, C., Socha, K.: Training feed-forward neural networks with ant colony optimization: An application to pattern classification. In: *Fifth International Conference on Hybrid Intelligent Systems* (2005)
16. Cai, K., Zhang, J., Zhou, C., Cao, X., Tang, K.: Using computational intelligence for large scale air route networks design. Applied Soft Computing **12**(9), 2790–2800 (2012)
17. Cheng, V., Crawford, L., Menon, P.: Air traffic control using genetic search techniques. In: *1999 IEEE International Conference on Control Applications*, Hawaii, August 22–27 (1999)
18. Deau, R.: Optimisation des séquences de pistes et du trafic au sol sur les grands aéroports. Ph.D. thesis, INPT (2010)
19. Deau, R., Gotteland, J.B., Durand, N.: Airport surface management and runways scheduling. In: *8th USA/Europe Air Traffic Management Research and Development Seminar* (2009)
20. Delahaye, D.: Optimisation de la sectorisation de l'espace aérien par algorithmes génétiques. Ph.D. thesis, ENSAE (1995)
21. Delahaye, D., Alliot, J.M., Schoenauer, M., Farges, J.L.: Genetic algorithms for partitioning airspace. In: *Proceedings of the Tenth Conference on Artificial Intelligence Application*. IEEE (1994)
22. Delahaye, D., Alliot, J.M., Schoenauer, M., Farges, J.L.: Genetic algorithms for automatic regroupment of air traffic control sectors. In: *Evolutionary Programming 95* (1995)
23. Delahaye, D., Odoni, A.: Airspace congestion smoothing by stochastic optimization. In: *Evolutionary Programming 97* (1997)

24. Delahaye, D., Puechmorel, S.: 3D airspace sectoring by evolutionary computation: Real-world applications. In: *Proceedings of the 8th Annual Conference on Genetic and Evolutionary Computation, GECCO '06*, pp. 1637–1644. ACM, New York (2006). doi:10.1145/1143997. 1144267. URL http://doi.acm.org/10.1145/1143997.1144267

25. Delahaye, D., Puechmorel, S.: 3D airspace design by evolutionary computation. In: *Digital Avionics Systems Conference, 2008. DASC 2008*, pp. 3.B.6-1–3.B.6-13. IEEE (2008). doi:10. 1109/DASC.2008.4702803

26. Delahaye, D., Schoenauer, M., Alliot, J.M.: Airspace sectoring by evolutionary computation. In: *IEEE International Congress on Evolutionary Computation* (1998)

27. Durand, N.: Optimisation de trajectoires pour la résolution de conflits en route. Ph.D. thesis, ENSEEIHT, Institut National Polytechnique de Toulouse (1996)

28. Durand, N.: Algorithmes Génétiques et autres méthodes d'optimisationappliqués 'a la gestion de trafic aérien. Institut National Polytechnique de Toulouse (2004). Thse d'habilitation

29. Durand, N., Allignol, C., Barnier, N.: A ground holding model for aircraft deconfliction. In: *29th DASC* (2010)

30. Durand, N., Alliot, J.M.: Optimal resolution of en route conflicts. In: *Séminaire Europe/USA*, Saclay, June 1997 (1997)

31. Durand, N., Alliot, J.M.: Genetic crossover operator for partially separable functions. In: *Proceedings of the Third Annual Genetic Programming Conference* (1998)

32. Durand, N., Alliot, J.M.: Ant colony optimization for air traffic conflict resolution. In: *8th USA/Europe Air Traffic Management Research and Development Seminar* (2009)

33. Durand, N., Alliot, J.M., Chansou, O.: An optimizing conflict solver for ATC. Air Traffic Control (ATC) Quarterly (1995)

34. Durand, N., Alliot, J.M., Medioni, F.: Neural nets trained by genetic algorithms for collision avoidance. In: Applied Artificial Intelligence, Vol. 13, Number 3 (2000)

35. Durand, N., Alliot, J.M., Noailles, J.: Algorithmes genetiques: un croisement pour les problemes partiellement separables. In: *Proceedings of the Journees Evolution Artificielle Francophones*. EAF (1994)

36. Durand, N., Alliot, J.M., Noailles, J.: Automatic aircraft conflict resolution using genetic algorithms. In: *Proceedings of the Symposium on Applied Computing, Philadelphia*. ACM (1996)

37. Durand, N., Alliot, J.M., Noailles, J.: Collision avoidance using neural networks learned by genetic algorithms. In: *Ninth International Conference on Industrial & Engineering (IEA-AEI 96)*, Nagoya, Japan (1996)

38. Eiben, A., Smith, J.: *Introduction to Evolutionary Computing*. Springer (2003)

39. Fortune, S.: Voronoi diagrams and Delaunay triangulations. In: *Computing in Euclidean Geometry* (1995)

40. Fu, A., Lei, X., Xiao, X.: The aircraft departure scheduling based on particle swarm optimization combined with seamulated annealing algorithm. In: *2008 IEEE World Congress on Computational Intelligence*, Hong Kong, June 1–6 (2008)

41. García, J., Berlanga, A., Molina, J.M., Casar, J.R.: Optimization of airport ground operations integrating genetic and dynamic flow management algorithms. AI Communications **18**(2), 143–164 (2005)

42. Gianazza, D.: Optimisation des flux de trafic aérien. Ph.D. thesis, INPT (2004)

43. Gianazza, D.: Algorithme évolutionnaire et a* pour la séparation en 3d des flux de trafic aérien. Journal Européen des Systèmes Automatisés, **38**(9), 10/2004, Special Issue "Métaheuristiques pour l'optimisation difficile" (2005)

44. Gianazza, D.: Smoothed traffic complexity metrics for airspace configuration schedules. In: *Proceedings of the 3rd International Conference on Research in Air Transportation*. ICRAT (2008)

45. Gianazza, D.: Forecasting workload and airspace configuration with neural networks and tree search methods. Submitted to Artificial Intelligence Journal (2010)

46. Gianazza, D., Allignol, C., Saporito, N.: An efficient airspace configuration forecast. In: *Proceedings of the 8th USA/Europe Air Traffic Management R & D Seminar* (2009)

47. Gianazza, D., Alliot, J.M.: Optimal combinations of air traffic control sectors using classical and stochastic methods. In: *2002 International Conference on Artificial Intelligence, IC-AI'02*, Las Vegas (2002)
48. Gianazza, D., Alliot, J.M.: Optimization of air traffic control sector configurations using tree search methods and genetic algorithms. In: *Digital Avionics Systems Conference 2002* (2002)
49. Gianazza, D., Durand, N.: Separating air traffic flows by allocating 3D-trajectories. In: *23rd DASC* (2004)
50. Gianazza, D., Durand, N.: Assessment of the 3D-separation of air traffic flows. In: *6th USA/Europe Seminar on Air Traffic Management Research and Development* (2005)
51. Gianazza, D., Durand, N., Archambault, N.: Allocating 3D trajectories to air traffic flows using a* and genetic algorithms. In: *CIMCA04* (2004)
52. Gianazza, D., Guittet, K.: Evaluation of air traffic complexity metrics using neural networks and sector status. In: *Proceedings of the 2nd International Conference on Research in Air Transportation*. ICRAT (2006)
53. Gianazza, D., Guittet, K.: Selection and evaluation of air traffic complexity metrics. In: *Proceedings of the 25th Digital Avionics Systems Conference, DASC* (2006)
54. Goldberg, D.: *Genetic Algorithms*. Addison-Wesley (1989)
55. Gotteland, J.B.: Optimisation du trafic au sol sur les grands aéroports. Ph.D. thesis, INPT (2004)
56. Granger, G., Durand, N., Alliot, J.M.: Optimal resolution of en route conflicts. In: *ATM 2001* (2001)
57. Gudise, V., Venayagamoorthy, G.: Comparison of particle swarm optimization and backpropagation as training algorithms for neural networks. In: *Proceedings of the 2003 IEEE Swarm Intelligence Symposium* (2003)
58. Holland, J.: *Adaptation in Natural and Artificial Systems*. University of Michigan Press (1975)
59. Hu, X., Di Paolo, E.: An efficient genetic algorithm with uniform crossover for the multi-objective airport gate assignment problem. In: *IEEE Congress on Evolutionary Computation, 2007, CEC 2007*, pp. 55–62 (2007). doi:10.1109/CEC.2007.4424454
60. Hu, X.B., Chen, W.H., Di Paolo, E.: Multiairport capacity management: Genetic algorithm with receding horizon. IEEE Transactions on Intelligent Transportation Systems **8**(2), 254–263 (2007)
61. Hu, X.B., Di Paolo, E.: A ripple-spreading genetic algorithm for the aircraft sequencing problem. Evolutionary Computation **19**(1), 77–106 (2011)
62. Hu, X.B., Di Paolo, E.: Binary-representation-based genetic algorithm for aircraft arrival sequencing and scheduling. IEEE Transactions on Intelligent Transportation Systems **9**(2), 301–310 (2008). doi:10.1109/TITS.2008.922884
63. Hu, X.B., Di Paolo, E.: An efficient genetic algorithm with uniform crossover for air traffic control. Comput. Computers and Operations Research **36**(1), 245–259 (2009). doi:10.1016/j.cor.2007.09.005. URL http://dx.doi.org/10.1016/j.cor.2007.09.005
64. Kicinger, R., Yousefi, A.: Heuristic method for 3D airspace partitioning: Genetic algorithm and agent-based approach. AIAA Paper **7058** (2009)
65. Kirkpatrick, S., Gelatt, C., Vecchi, M.: Optimization by simulated annealing. Science **220**(4598), 671–680 (1983)
66. Lei, X., Fu, A., Shi, Z.: The aircraft departure scheduling based on second-order oscillating particle swarm optimization algorithm. In: *2008 IEEE World Congress on Computational Intelligence*, Hong Kong, June 1–6 (2008)
67. Letrouit, V.: Optimisation du réseau des routes aériennes en Europe. Ph.D. thesis, Institut National Polytechnique de Grenoble (1998)
68. Leung, F., Lam, H., Ling, S., Tam, P.: Tuning of the structure and parameters of a neural network using an improved genetic algorithm. IEEE Transactions on Neural Networks **14**(1), 79–88 (2003)
69. Liang, J., Qin, A., Suganthan, P., Baskar, S.: Comprehensive learning particle swarm optimizer for global optimization of multimodal functions. IEEE Transactions on Evolutionary Computation **10**(3), 281–295 (2006)

70. Malaek, S.M., Alaeddini, A., Gerren, D.S.: Optimal maneuvers for aircraft conflict resolution based on efficient genetic webs. IEEE Transactions on Aerospace and Electronic Systems **47**(4), 2457–2472 (2011)

71. Maugis, L.: Mathematical programming for the air traffic flow management problem with en-route capacities. In: *XVI World Conference of the International Federation of Operational Research Societies* (1996)

72. Mehadhebi, K.: A methodology for the design of a route network. In: *Proceedings of the Third Air Traffic Management R & D Seminar, ATM-2000, Naples.* Eurocontrol & FAA (2000)

73. Meng, G., Qi, F.: Flight conflict resolution for civil aviation based on ant colony optimization. In: *Fifth International Symposium on Computational Intelligence and Design* (2012)

74. Michalewicz, Z.: *Genetic algorithms + Data Structures = Evolution Programs.* Springer (1992)

75. Mondoloni, S., Conway, S., D, P.: An airborne conflict resolution approach using a genetic algorithm (2001)

76. Odoni, A.: The flow management problem in air traffic control. In: *Flow Control of Congested Network*, pp. 269–288 (1987)

77. Oussedik, S.: Application de l'évolution artificielle aux problèmes de congestion du trafic aérien. Ph.D. thesis, Ecole Polytechnique (2000)

78. Oussedik, S., Delahaye, D.: Reduction of air traffic congestion by genetic algorithms. In: *Parallel Problem Soving from Nature (PPSN98)* (1998)

79. Oussedik, S., Delahaye, D., Schoenauer, M.: Dynamic air traffic planning by genetic algorithms. In: *Proceedings of the 1999 Congress on Evolutionary Computation, CEC 99*, vol. 2 (1999). doi:10.1109/CEC.1999.782547

80. Pesic, B., Durand, N., Alliot, J.M.: Aircraft ground traffic optimisation using a genetic algorithm. In: *GECCO 2001* (2001)

81. Riviere, T.: Redesign of the European route network for sector-less. In: *23rd DASC* (2004)

82. Roling, P., Visser, H.G.: Optimal airport surface traffic planning using mixed-integer linera programming. International Journal of Aerospace Engineering (2008)

83. Siddiquee, M.: Mathematical aids in air route network design. In: *IEEE Conference on Decision and Control, 12th Symposium on Adaptive Processes* (1973)

84. Slowik, A., Bialko, M.: Training of artificial neural networks using differential evolution algorithm. In: *Conference on Human System Interactions* (2008)

85. Sridhar, B., Grabbe, S., Sheth, K., Bilimoria, K.: Initial study of tube networks for flexible airspace utilization. In: *Proceedings of the AIAA Guidance, Navigation, and Control Conference and Exhibition*, Keystone, CO (2006)

86. Su, J., Zhang, X., Guan, X.: 4D-trajectory conflict resolution using cooperative coevolution. In: *Proceedings of the 2012 International Conference on Information Technology and Software Engineering* (2012)

87. Tandale, M., Menon, P.: Genetic algorithm based ground delay program computations for sector density control. In: *AIAA Guidance Navigation and Control Conference*, Hilton Head, SC, 20–23 August (2007)

88. Vanaret, C., Gianazza, D., Durand, N., Gotteland, J.B.: Benchmarking conflict resolution algorithms. In: *5th International Conference on Research in Air Transportation (ICRAT 2012)*, May 22–25, 2012, University of California, Berkeley (2012)

89. Vivona, R.A., Karr, D.A., Roscoe, D.A.: Pattern-based genetic algorithm for airborne conflict resolution. In: *AIAA Guidance Navigation Control Conference Exibition* (2006)

90. Xue, M.: Airspace sector redesign based on Voronoi diagrams. Journal of Aerospace Computing, Information and Communication **6**(12), 624–634 (2009)

91. Xue, M., Kopardekar, P.: High-capacity tube network design using the Hough transform. Journal of Guidance, Control, and Dynamics **32**(3), 788–795 (2009)

92. Zelinski, S.: A comparison of algorithm generated sectorization. In: *8th USA/Europe Air Traffic Management Research and Development Seminar* (2009)

Index

© Springer International Publishing Switzerland 2016
P. Siarry (ed.), *Metaheuristics*, DOI 10.1007/978-3-319-45403-0

Printed in the United States
By Bookmasters